Physical
Geology

Physical Geology

Fifth Edition

Charles C. Plummer

California State University, Sacramento

David McGeary

California State University, Sacramento

WCB **Wm. C. Brown Publishers**

Book Team

Editor *Jeffrey L. Hahn*
Developmental Editor *Lynne M. Meyers*
Production Editor *Daniel Rapp*
Designer *David C. Lansdon*
Art Editor *Janice M. Roerig*
Photo Editor *Mary Roussel*
Visuals Processor *Jodi Wagner*

 Wm. C. Brown Publishers

President *G. Franklin Lewis*
Vice President, Publisher *George Wm. Bergquist*
Vice President, Publisher *Thomas E. Doran*
Vice President, Operations and Production *Beverly Kolz*
National Sales Manager *Virginia S. Moffat*
Advertising Manager *Ann M. Knepper*
Marketing Manager *John W. Calhoun*
Editor in Chief *Edward G. Jaffe*
Managing Editor, Production *Colleen A. Yonda*
Production Editorial Manager *Julie A. Kennedy*
Production Editorial Manager *Ann Fuerste*
Publishing Services Manager *Karen J. Slaght*
Manager of Visuals and Design *Faye M. Schilling*

Cover image © Peter L. Kresan Photography

The credits section for this book begins on page 537, and is considered an extension of the copyright page.

Library of Congress Catalog Card Number: 90–80176

ISBN 0–697–09826–5 (Cloth)
ISBN 0–697–09827–3 (Paper)

Printed in the United States of America by Wm. C. Brown Publishers, 2460 Kerper Boulevard, Dubuque, IA 52001

10 9 8 7 6 5 4 3 2 1

TABLE OF

Contents

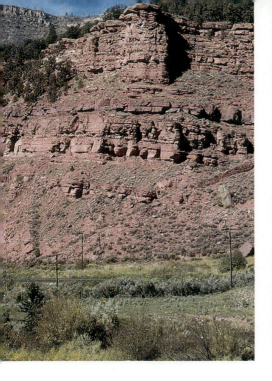

✳Chapter 12
Glaciers and Glaciation 255

✳Chapter 13
Deserts and Wind Action 283

Chapter 14
Waves, Beaches, and Coasts 303

This fifth edition of P[...]
straightforward, eas[...]
geology for both ge[...]
nonmajors. The organizatio[...]
traditional and matches the[...]
manuals. Each chapter has[...]
self-contained as possible [...]
reorganize the chapter seq[...]

The most obvious chang[...]
the illustrations. All chapter[...]
numerous line drawings an[...]
been replaced. The improv[...]
especially obvious in the ch[...]
weathering, mass wasting, [...]
the earth's interior, mounta[...]

New boxes have been a[...]
including radon, asbestos, [...]
shields, transgressions, de[...]
xenoliths, and unit cells of [...]

The boxes within the tex[...]
(1) topics of special human[...]
concern, such as the dange[...]
or the amount of fresh wate[...]
and (2) topics slightly more [...]
the text, such as the electro[...]
minerals. The boxed materi[...]
interesting, should be consi[...]
the text.

We have updated inform[...]
mentioning recent events s[...]
oil spill, the storm surge fro[...]
the 1989 earthquake near S[...]

Physical Geology is acco[...]
instructor's manual, a stude[...]
laboratory manual. The Inst[...]
by the authors of the text, g[...]
objectives for the twenty-on[...]
as well as numerous sugge[...]
demonstrations, discussion[...]
exam questions. The manua[...]
outlines and a lab schedule [...]
suppliers of films, slides, ro[...]
information for the course a[...]

The Student Study Guide[...]
foundation for the beginning[...]
Written by Esther Tuttle, a s[...]
Sherwood D. Tuttle, profess[...]
University of Iowa, the guide[...]
fundamentals of geology, th[...]
science, and the techniques[...]
in the field of geology.

The Laboratory Manual, [...]
and Robert Rutford, has bee[...]
to be used with *Physical Ge*[...]
good selection of experimer[...]
laboratory.

LIST OF
Boxes

We have tried to write a book that will be useful to both students and instructors. We would be grateful for any comments by users, especially regarding mistakes within the text or sources of good geological photographs.

Charles C. Plummer/David McGeary
Geology Department
California State University, Sacramento
Sacramento, CA 95819

Acknowledgments

Susan Slaymaker wrote the boxed material on planetary geology. We are grateful for her assistance.

The successful completion of this edition of *Physical Geology* is largely due to the efforts of our reviewers, who gave us invaluable advice throughout the revision of the manuscript. We extend our special thanks and appreciation to those who reviewed all or part of the manuscript, including Donald K. Marchand, Jr., *Old Dominion University;* Albert M. Kudo, *University of New Mexico;* Frank M. Hanna, *California State University–Northridge;* Charles R. Singler, *Youngstown State University;* David N. Lumsden, *Memphis State University.* Reviewers of earlier editions include: Albert M. Kudo, *University of New Mexico;* Peter L. Kresan, *University of Arizona;* M. G. Frey, *University of New Orleans;* Robert E. Behling, *West Virginia University;* Ronald H. Konig, *University of Arkansas;* Robert J. Elias, *University of Manitoba;* Jacqueline Patterson, *California State University, Fullerton;* Kenneth J. Van Dellen, *Macomb Community College;* R. Scott Babcock, *Western Washington State University;* Victor R. Baker, *The University of Texas;* Joan Baldwin, *El Camino College;* Alexander R. Ball, *Los Angeles Valley College;* Peter E. Borella, *Riverside City College;* David P. Bucke, *University of Vermont;* Greg S. Conrad, *Sam Houston State University;* Paul A. Dike, *Glassboro State College;* Peter Fischer, *California State University;* Ronald C. Flemal, *Northern Illinois University;* Richard A. Flory, *California State University at Chico;* Jon S. Galehouse, *San Francisco State University;* Stephen L. Harris, *California State University at Sacramento;* Barry Haskell, *Los Angeles Pierce College;* Miles O. Hayes, *University of South Carolina;* Roy L. Ingram, *University of North Carolina;* Howard Level, *Ventura College;* Lawrence W. Knight, *William Rainey Harper College;* C. Daniel Miller, *U.S. Geological Survey;* Gary D. Rosenberg, *Indiana University–Purdue University at Indianapolis;* Charles R. Singler, *Youngstown State University;* Richard Smosna, *University of West Virginia;* Norman W. Ten Brink, *Grand Valley State College;* Sherwood D. Tuttle, *University of Iowa;* W. R. Van Schmus, *University of Kansas;* Stephen H. Watts, *Sir Sandford Fleming College;* William J. Wayne, *University of Nebraska.*

We also wish to thank Jeffrey Hahn and Lynne Meyers of Wm. C. Brown Publishers for their patience and assistance in the publication of this fifth edition.

Physical Geology

G eology uses the scientific method to explain natural aspects of the earth—for example, how mountains form and valleys develop, or why oil resources are concentrated in some rocks and not in others. This chapter briefly explains how and why the earth's surface, and its interior, are constantly changing. It relates this constant change to the major geological topics of the modern theory of plate tectonics and geologic time. These concepts form a framework for the rest of the book. Understanding them will aid you in studying the chapters that follow.

1

Introduction to Physical Geology

Mt. Robson, highest peak in the Canadian Rocky Mountains.
SuperStock.

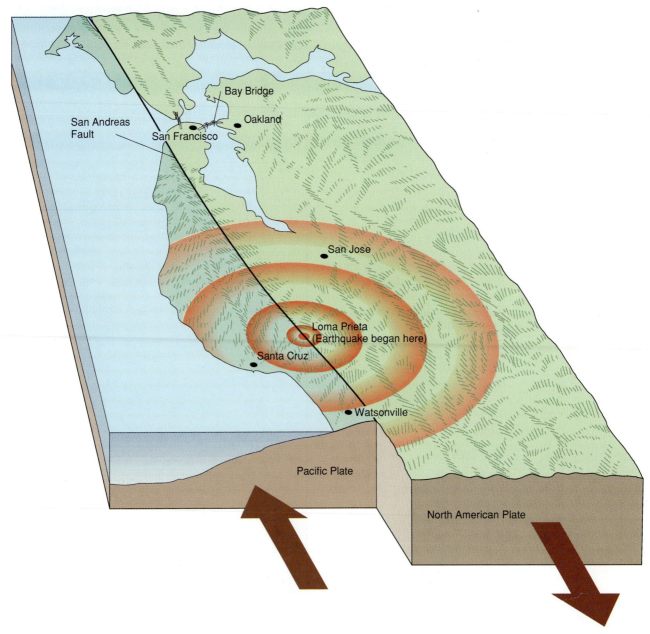

Figure 1.1
California's 1989 earthquake was due to motion between the North American plate and the Pacific plate.

The Earth: A Giant Machine

The earth does not stand still.

According to the theory of plate tectonics, the earth's rigid outer shell is broken into a series of *plates*. Adjoining plates may slide past, move away from, or collide with one another. Plates generally move from 1 to 10 centimeters a year. But the motion may not be smooth and continuous. Plates may be "locked" against one another for many years and move suddenly.

One such movement took place on October 17, 1989 (see figure 1.1). In San Francisco, the third game of baseball's World Series was about to begin. This was the first ever World Series between northern California's two Major League teams, the San Francisco Giants and the Oakland "A"s. The excitement grew as the capacity crowd settled into San Francisco's Candlestick Park. Shortly after 5 P.M., minutes before the game was to start, the stadium began moving. It shook violently. Fifteen seconds later it stopped. As the crowd realized they had just experienced an earthquake, they began cheering. The earth and California had put on a pre-game spectacle worthy of the Golden State's reputation. Many fans had portable radios or television sets and as the reports came in the magnitude of the disaster became apparent. A segment of the Bay Bridge, connecting Oakland and San Francisco, collapsed. In Oakland, the upper level of a double-decked freeway pancaked onto the lower lanes over a length of more than a kilometer crushing cars and their passengers. In San Francisco's Marina District, buildings collapsed or crashed into each other (figure 1.2) when the soil upon

Figure 1.2
Buildings in San Francisco's Marina District destroyed by the October 17, 1989 Loma Prieta earthquake.
Photo by Diane Stagner.

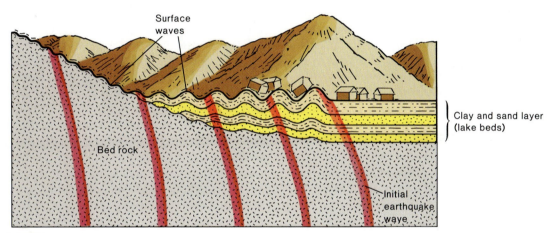

Surface waves

Clay and sand layer (lake beds)

Bed rock

Initial earthquake wave

Figure 1.3
Earthquake waves are amplified while traveling through layers of sand and clay.

which they were built was shaken into mud. Natural gas lines were severed and fires raged. Further south, closer to Loma Prieta, the hill along the San Andreas fault where the quake originated, towns and smaller cities suffered severe damage as entire blocks of buildings were destroyed or severely damaged. The Loma Prieta earthquake's death toll was 67 and its cost in the billions of dollars.

While the loss in lives and property is regrettable, it could have been far worse. (Compare, for example the 1985 Mexico City earthquake that killed 8,000 people, injured 30,000 and resulted in $5 billion in damage.) The Loma Prieta quake's toll was relatively low partially due to luck and partially to good planning. Many lives were saved because of the World Series. Normally, all bridges and highways in the Bay area are clogged with rush-hour traffic in the afternoon, but people left work early to go watch the baseball game on television. Freeway traffic was the lightest in anyone's memory when the quake hit. Be-

cause the winter rains had not begun, the soil was dry in most places and few landslides were triggered by the quake.

Planning for a major earthquake has been an ongoing process for coastal California for decades. The quake was expected (as is a much larger one) because of what geologists have learned about moving plates. None of the skyscrapers in downtown San Francisco were damaged because they were built to meet high quake-resistant standards—the result of engineers studying the effects of countless earthquakes worldwide.

The most severe damage was to older structures and to those built on bay fill (soil or unconsolidated sand and clay dumped into the water to add more land for a city to build on). If the bay fill has a high water content it liquifies when shaken hard (as in the Marina District). Unconsolidated *sediment,* such as bay fill, amplifies an earthquakes's waves (figure 1.3), much as if they were

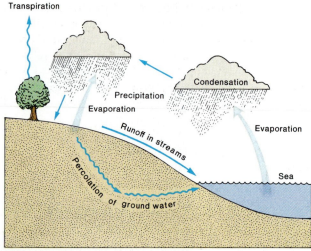

Figure 1.5
The hydrologic cycle. Water vapor evaporates from the sea and land, condenses to form clouds, and falls as precipitation. Water falling on land runs off over the surface as streams or percolates into the ground to become ground water. It returns to the atmosphere again by evaporation and transpiration.

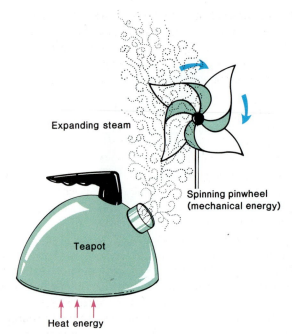

Figure 1.4
A pinwheel held over steam is an example of a heat engine.

traveling through Jell-O. (Try shaking a bowl of Jell-O and notice how the Jell-O's surface wobbles; if you shake a similar-sized block of wood, you won't notice any distortion of its surface.) The collapsed section of freeway in Oakland was built on bay fill. San Francisco's bay fill is man-made. Mexico City is built on the naturally deposited sand and clay layers of an ancient lake bed. For millions of years, loose material washed in from the surrounding mountains and accumulated on the lake floor as layers of sediment—clay and sand in this case. The great amount of damage in the Mexico City quake was largely due to the amplification of the earthquake's waves by the sediment.

The awesome energy released by an earthquake is a product of the earth's machinery, machinery driven by forces within the earth. Earthquakes are only one consequence of the ongoing changing of the earth. Ocean basins open and close. Mountain ranges rise and are worn down to plains through slow, but dramatic, processes. Studying how the earth's machinery works can be as exciting as watching a great theatrical performance. Understanding the changes that take place in and on the earth, and the reasons for those changes, is the challenging objective of **geology,** the scientific study of the earth.

The earth can be visualized as a giant machine driven by two engines, one internal and the other external. Both are **heat engines,** devices that convert heat energy into mechanical energy. A simple heat engine is shown in figure 1.4. An automobile is powered by a heat engine. When gasoline is ignited in the cylinders, the resulting hot gases expand, driving pistons to the far end of cylinders. In this

way, the heat energy of the expanding gas has been converted to the mechanical energy of the moving pistons, then transferred to the wheels, where the energy is put to work moving the car.

The *internal* heat engine of the earth is powered by heat flowing from the hot interior of the earth toward the cooler exterior. Moving plates and earthquakes are products of this heat engine.

The earth's *external* heat engine is essentially solar powered. Heat from the sun provides the energy for the motion of the oceans and atmosphere, producing clouds, rain, and snowfall, as well as daily and seasonal weather. Water, especially in the oceans, is evaporated by solar heating. The water vapor becomes part of the air. When the air cools, it loses its ability to hold and transport the water vapor, and clouds form. If the droplets become large enough, they are attracted to the earth's surface by gravity and fall as rain or snow. The movement of water and water vapor from the sea to the atmosphere to the land and back to the sea and atmosphere is known as the **hydrologic cycle** (see figure 1.5).

Over long periods of time, moisture at the earth's surface helps rock disintegrate. Water washing down hillsides and flowing in streams loosens and carries away the rock particles. In this way mountains originally raised by the earth's internal forces are worn away by processes driven by the external heat engine.

The force of gravity causes streams to flow and pulls material downward. We can think of **gravity** as mutual attraction between bodies. The greater the masses of the bodies, the greater the force. Because the mass of the earth is vastly greater than any object on its surface, material is strongly attracted toward the earth's center.

The earthquake in Mexico and its effects show how the two heat engines were at work. The internal forces were not only responsible for the sudden release of energy during the earthquake but also for the building of the mountains surrounding Mexico City. During and after the time when mountains were forming, external forces were at work breaking down the mountains' rocks into small particles. These particles, or **sediment,** in this case clay and sand, were carried by running water into the ancient lake to settle into the layers of clay and sand that would eventually become the site of Mexico City.

Events such as earthquakes, volcanic eruptions, and landslides are, of course, of great interest to geologists. But geologists are equally fascinated by more subtle geologic events, such as the way solid rock will break down when exposed to the atmosphere or how clay will slowly accumulate into thick layers of sediment in a lake or ocean. Like all scientists, geologists are curious. They want to understand how the earth works. However, geology is pursued not only because of curiosity. Geology has many practical applications. An earthquake in 1976 in China is estimated to have killed nearly 600,000 people; most of the deaths were caused by loose debris falling off buildings or by structures collapsing. In the Bay area, there would have been far greater destruction if engineers had ignored what geologists had learned about earthquakes or what they knew about the geological setting of the city. Geologists expect that detailed studies of the Loma Prieta earthquake will further increase scientific understanding of earthquakes; that knowledge should be applied to improve the design of buildings built on different kinds of material.

Before we return to the topic of how the earth works, it is worth pointing out some other ways in which you, as an inhabitant of this planet, can benefit from understanding geology.

Understanding Our Surroundings

Understanding geology leads to a greater appreciation of scenery. If, for instance, you were traveling through the Canadian Rockies, you might see the scene in the chapter opening photograph and wonder how the landscape came to be.

You might wonder: (1) why there are layers in the rock exposed in the cliffs; (2) why the peaks are so jagged; (3) why there are glaciers; (4) why this mountain belt extends northward and southward for thousands of kilometers; (5) why there are mountain ranges here and not in the central part of the continent. After completing a course in physical geology, you should be able to answer these questions as well as understand how other kinds of landscapes formed.

People who have completed a physical geology course find that their understanding of geological processes makes travel much more interesting. Figure 1.6 is a map showing the landforms of the contiguous United States and southernmost Canada. Major features discussed or depicted in this book are highlighted. In your future air or surface travel, you may want to take along your book with the map to enhance your appreciation of the scenery.

Supplying Things We Need

The earth's heat engines, at work for billions of years, have concentrated in different parts of the earth the materials that people want and need for survival, comfort, and pleasure. By learning how the earth works and how different kinds of substances are distributed and why, we can intelligently search for metals, sources of energy, gems, and sand and gravel for construction purposes.

The economic systems of Western civilization depend currently on abundant and cheap energy sources. Nearly all *our* engines—powered mainly by gasoline, coal, or nuclear power—depend on concentrations of energy derived from the earth and *its* engines. The United States economy in particular is geared to petroleum as a cheap source of energy. In a few decades Americans have used up much of the country's known petroleum reserves, which took nature hundreds of millions of years to store in the earth. Americans are now heavily dependent on imported oil. To find more of this diminishing resource will require more money and increasingly sophisticated knowledge of geology. Although many people are not aware of it, we face similar problems with diminishing sources of other materials, notably metals such as iron, aluminum, copper, and tin, each of which has been concentrated in a particular environment by the action of the earth's engines.

Protecting the Environment

Our demands for more energy and metals have, in the past, led us to extract them with little regard for effects on the environment, on the balance of nature within the earth, and therefore on us, earth's residents. Mining of coal, for example, can release acids into water supplies. Understanding geology can lessen or prevent damage to the environment—just as it can find the resources in the first place.

The environment is further threatened because these are nonrenewable resources. Petroleum and metal deposits do not grow back after being harvested. As demands for these resources increase, so does the pressure to disregard the ecological damage caused by the extraction of the remaining deposits.

Problems involving petroleum illustrate this. Oil companies employ geologists to discover new oil fields, while the public and government depend on other geologists to assess the potential environmental impact of petroleum's removal from the ground, the transportation of petroleum (see Box 1.1), and disposal of any toxic wastes from petroleum products.

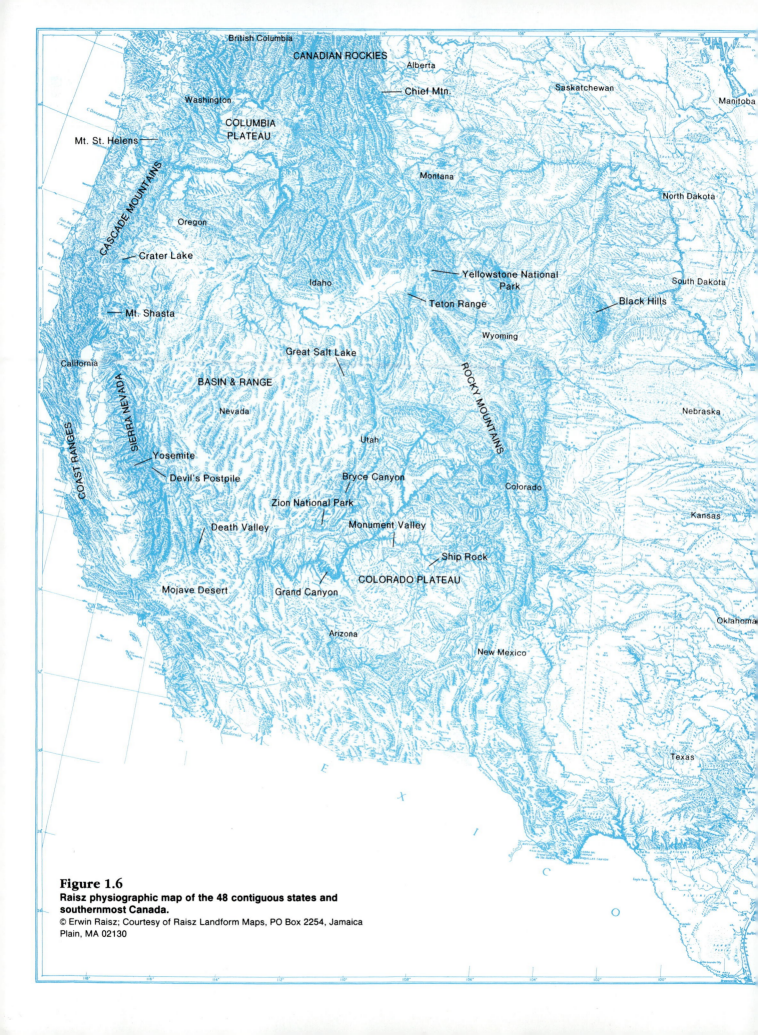

Figure 1.6
Raisz physiographic map of the 48 contiguous states and southernmost Canada.
© Erwin Raisz; Courtesy of Raisz Landform Maps, PO Box 2254, Jamaica Plain, MA 02130

CANADIAN SHIELD

Laurentian upland of low hills and many lakes

Ontario

Quebec

Newfoundland

New Brunswick

Maine

Minnesota

Wisconsin

Michigan

ADIRONDACK MOUNTAINS

Vermont

New Hampshire

Massachusetts

Rhode Island

Connecticut

New York

Iowa

Illinois

Indiana

Ohio

Pennsylvania

New Jersey

Maryland

Delaware

West Virginia

Missouri

Kentucky

Virginia

APPALACHIAN MOUNTAINS

North Carolina

Tennessee

Arkansas

MISSISSIPPI RIVER

South Carolina

OUACHITA MOUNTAINS

Mississippi

Alabama

Georgia

Louisiana

Florida

Mississippi Delta

Miami

LANDFORMS OF THE UNITED STATES

by ERWIN RAISZ   Sixth revised edition 1957

Scale   0   50   100   150   200   Miles
         0   100   200   Kilometers

BOX 1.1

Delivering Alaskan Oil— the Environment vs. the Economy

When the tanker *Exxon Valdez* ran aground in 1989, over 240,000 barrels of crude oil were spilled into the waters of Alaska's Prince William Sound. It was the worst ever oil spill in U.S. waters. The spill, with its devastating effects on wildlife and the fishing industry, dramatically highlighted the conflicts between maintaining the energy demands of the American economy and conservation of resources and the environment. The *Exxon Valdez* was only one of a fleet of tankers that carry oil from the southern end of the Alaskan pipeline to the refineries on the west coast of the United States. Oil spills, such as this one, were predicted by the 1972 environmental impact statement for the Alaska pipeline prepared by the U.S. Geological Survey.

In the late 1970s the United States was importing almost half its petroleum, at a loss of billions of dollars per year to the national economy. This drain on the country's economy and the increasing cost of energy are major causes of inflation, lower industrial productivity, unemployment, and the erosion of standards of living.

Over a decade earlier, geologists discovered oil beneath the shores of the Arctic Ocean on Alaska's North Slope. It is now the United States' largest oil field. Thanks to the Alaska pipeline, completed in 1977, Alaska now supplies about 20% of the United States' domestic oil.

Despite its important role in the American economy, some people consider the Alaska pipeline and the use of tankers as unacceptable threats to the area's ecology. The 1989 oil spill demonstrated the hazards of the marine portion of the oil transportation system. Oil spilled from a ruptured pipe would have a devastating effect on the fragile Arctic plant and animal life. Crude oil would stay on the surface of permanently frozen ground rather than be washed away by running water as in a temperate climate.

Geologists with the U.S. Geological Survey conducted the official environmental impact investigation of the proposed pipeline route. They recommended against its construction, partially because of the hazards to oil tankers and partially because of the geologic hazards of the pipeline route.

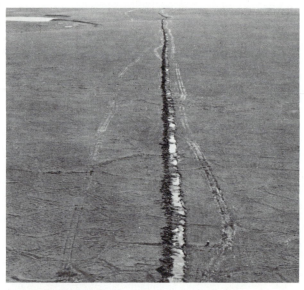

Box 1.1 Figure 1
Road made on Alaska's North Slope. Bulldozing of vegetation caused thawing, which led to ponds developing.
Photo by O. J. Ferrians, Jr., U.S. Geological Survey.

(They favored a much longer pipeline going through Canada to Chicago.) Their report was overruled. The Congress and the President of the United States exempted the pipeline from laws that require a favorable environmental impact statement before a major project can begin. The 1.52 million barrels of oil a day that now flow from the Arctic oil fields mean that over $10 billion a year that would have been lost through the purchase of foreign oil instead remains in the American economy.

The 1,250-kilometer-long pipeline crosses regions of ice-saturated, frozen ground and major earthquake-prone mountain ranges that geologists regard as serious hazards to the structure. Building anything on frozen ground creates problems. For example, the road in box figure 1 was built during the exploration of Alaska's North Slope oil fields. When the road was being built, the protective vegetation was scraped off. During the summer thaw the road became a quagmire. As thawing continued, the ponds shown in the photo grew so big that the road will never be passable.

Building the pipeline over such terrain also presented enormous engineering problems. If the pipeline were placed on the ground, the hot oil flowing through it could melt the frozen ground. On a slope, mud could easily slide and rupture the pipeline. Careful (and costly) engineering minimized these hazards. Much of the pipeline is elevated above the ground. Radiators conduct heat out of the structure. In some places refrigeration equipment in the ground protects against melting.

Box 1.1 Figure 2
The Alaska pipeline.
Photo by © Steve McCutcheon/Alaska Pictorial Service.

Records indicate that a strong earthquake can be expected every few years in the earthquake belts crossed by the pipeline. An earthquake would surely rupture a pipeline—at least, a conventional pipe as in the original design. However, when the Alaska pipeline was built, in several places sections were specially jointed to allow the pipe to shift as much as 6 meters without rupturing.

The original estimated cost of the pipeline was $900 million, but the final cost was $7.7 billion, making it the costliest privately financed construction project in history. The redesigning and construction that minimized the potential for an environmental disaster were among the many reasons for the increased cost. There have been some minor breaks in the pipeline. For instance, in January 1981, 5,000 barrels of oil were lost when a valve ruptured. However, the pipeline company maintains that there is virtually no chance of a major oil spill from the pipe. Environmentalists are not so optimistic. It remains to be seen whether the best technology money can buy can prevent a disaster. But even if problems with the pipeline remain minor, the risks involving marine transportation remain.

Avoiding Geologic Hazards

Geology can have a direct application in ensuring people's safety and well-being. For example, if you were building a house in an earthquake-prone area, you would want to know how to minimize danger to yourself and your home. You would want to build the house on a type of ground not likely to be shaken apart by an earthquake. You would want the house designed and built to absorb the kind of vibrations given off by earthquakes. Hundreds or perhaps thousands of lives were saved because of what geologists had learned from years of studying Mount St. Helens and her sister volcanoes. As explained in chapter 3, Mount St. Helens erupted harmlessly for several months in 1980 before the devastating May 18 eruption in which the top and one side of the mountain were blasted away. When the eruptions began, U.S. Geological Survey scientists told state and federal officials what to expect. Because of this advice, the potentially most dangerous areas were closed to public access—despite the outcry from many residents who could not conceive of the eruptions becoming more violent. When the big eruption came on May 18 most of the 63 people killed were those who had ignored the warnings and gone into closed areas.

By contrast, Nevado del Ruiz, a glacier-covered volcano in Colombia erupted in December of 1985, and some 23,000 people died. No one was killed by the eruption itself but by indirect effects. The hot rocks blasted out of the volcano caused part of the ice and snow capping the peak to melt. The water mixed with loose rock on the flank of the mountain and flowed down stream channels as a *mudflow*. At the base of the volcano, the mudflow overwhelmed the town of Armero, killing most of its inhabitants (figure 1.7). Colombian geologists had previously predicted such a mudflow could occur and published maps showing the location and extent of expected mudflows. The actual mudflow that wiped out the town matched that shown on the geologists' map almost exactly. Unfortunately, government officials had ignored the map and the geologists' report; otherwise the tragedy could have been averted.

Geologic hazards, other than earthquakes and volcanoes, that geologists investigate include floods, wave erosion at coastlines, collapsing ground surfaces where the bed rock is limestone, and landslides. (In the United States and Canada, far more property and lives have been lost due to landslides and floods than to earthquakes and volcanoes.)

A

Figure 1.7

(A) Nevado del Ruiz before the November 1985 eruption that melted snow and ice causing the disastrous mudflow. Mud was channeled into a gully such as shown on the photo. (B) Armero, Colombia after the 1985 mudflow. The buildings are the only portion of the town that survived the mudflow.
Photos by U.S. Geological Survey.

B

Figure 1.7
continued

An Overview of Physical Geology

Physical geology is the large division of geology concerned with earth materials, changes in the surface and interior of the earth, and the dynamic forces that cause those changes. We will look at how the earth's heat engines work and then show how some of the major topics of physical geology are related to the *surficial* (on the earth's surface) and *internal* processes powered by the heat engines.

Internal Processes: How the Earth's Internal Heat Engine Works

The earth's internal heat engine works because hot material deep within the earth moves slowly upward toward the cool surface. **Convection**—the upward movement of low-density material and downward movement of high-density material—probably accounts for most of the motion in the earth's interior. You can see the effects of convection by heating a pot of water. As the water at the bottom heats, it rises (figure 1.8). This is because most substances, when heated, become less dense (a given volume of the material will weigh less). As the hot water rises, colder and denser water sinks along the sides of the pot, creating a circular pattern known as a **convection current.** We can see convection currents moving in liquids. Surprisingly, convection can also take place in solids, if the solid is capable of very slow flow (comparable, for instance, to that of toothpaste). If a solid behaves in this manner, we say that it behaves **plastically,** meaning it is capable of being molded.

The Earth's Interior

Geologists believe that convection currents occur in the interior of the earth in the zone known as the **mantle,** the largest, by volume, of the earth's three major concentric zones (see figure 1.9). The other two zones are the **crust**

Figure 1.8
Convection currents in a pot of water.

and the **core.** The mantle is solid (except in a few spots) and probably composed of rock not very different from some kinds of rock found at the earth's surface. The crust of the earth is analogous to the skin on an apple. The thickness of the crust is insignificant compared to the whole earth. We have direct access to only the crust, and not much of the crust at that. We are like microbes crawling on an apple, without the ability to penetrate its skin. We are concerned more with the crust than with the inaccessible mantle and core. The crust varies in thickness. Two major types of crust are *oceanic crust* and *continental crust.* The crust under the oceans is much thinner. It is made of rock that is somewhat denser than the rock that underlies the continents.

The lower parts of the crust and the entire mantle are inaccessible to direct observation. No mine or oil well has penetrated the crust, so our concept of the earth's interior is based on indirect evidence (see the chapter titled "The Earth's Interior").

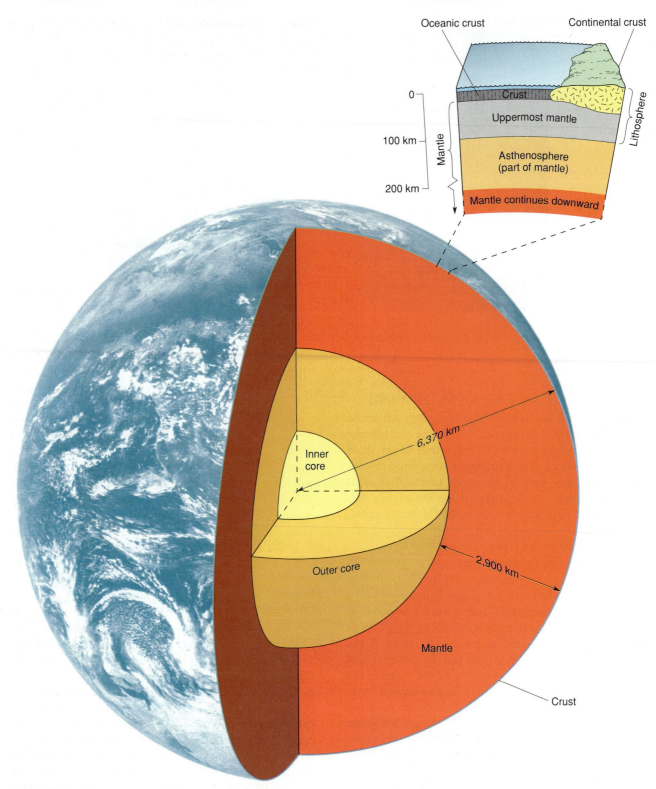

Figure 1.9
Cross section through the earth. Expanded section shows relationship between the two types of crust, the lithosphere and the asthenosphere, and the mantle. The crust ranges from five to fifty kilometers thick.

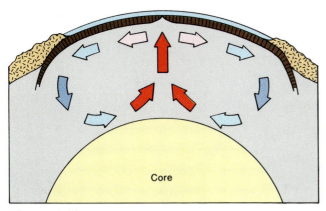

Figure 1.10
Mantle convection.

The crust and the uppermost part of the mantle are relatively rigid. Collectively they make up the **lithosphere** (*lith* means "rock" in Greek). The upper mantle underlying the lithosphere behaves plastically and is called the **asthenosphere.** Convection is believed to take place within the asthenosphere as well as within the lower mantle (figure 1.10). The lithosphere seems to be moving, probably as a result of the underlying mantle convection. The effect of the internal processes on the crust is of great significance to geology. The **tectonic forces,** which are forces generated from within the earth, cause deformation of rock as well as vertical and horizontal movement of portions of the earth's crust. Mountain ranges are evidence of tectonic forces strong enough to outdo gravitational forces. (Mount Everest, the world's highest peak, is made of rock that formed beneath an ancient sea.) Mountain ranges are built over extended periods, as portions of the earth's crust are compressed and raised.

Most tectonic forces are mechanical forces. Some of the energy from these forces is put to work deforming rock, bending and breaking it, and raising mountain ranges. The mechanical energy may be stored (an earthquake is a sudden release of stored mechanical energy) or converted to heat energy (rock may melt, resulting in volcanic eruptions). The way the machinery of the earth works is called **plate tectonics.**

The Theory of Plate Tectonics

From time to time a theory emerges within a science that revolutionizes that field. The plate tectonic theory, currently accepted by most geologists, is a unifying theory that accounts for many seemingly unrelated geological phenomena. The theory is perhaps as important to geology as the theory of relativity is to physics, the atomic theory to chemistry, or the theory of evolution to biology.

Plate tectonics was seriously proposed as a hypothesis in the early 1960s, though the idea was based on earlier work. We will present a brief overview here. In the chapter titled "Plate Tectonics" we will discuss it more thoroughly and show how material covered in intervening chapters is incorporated into the composite theory of plate tectonics.

As we proceed, we will note in appropriate chapters how the theory explains particular phenomena, notably in connection with earthquakes and the origin of different types of rocks.

According to the theory, the lithosphere is broken into *plates* (figure 1.11). The plates, which are much like segments of cracked shell on a boiled egg, move relative to one another, sliding on the underlying asthenosphere. Much of what we observe in the rock record can be explained by what takes place along *plate boundaries,* where two plates are pulling away from each other, sliding past each other, or moving toward each other.

According to plate tectonics, **spreading centers,** or **diverging boundaries,** exist where plates are moving apart (figure 1.12). Most spreading centers coincide with the crests of submarine mountain ranges, called **mid-oceanic ridges** (figure 1.12).

A mid-oceanic ridge is higher than deep ocean floor (figure 1.13) partly because the upward flow of hot mantle material pushes the lithosphere upward and partly because the rocks, being hotter there, are less dense. Tensional cracks develop along the ridge crest (figure 1.14). These cracks tap localized **magma** (molten rock) chambers in the underlying hot mantle, and the magma squeezes into fissures (cracks through the lithosphere). Some magma erupts along the ridge crest, and the rest solidifies in the fissure. Continued pulling apart of the ridge crest develops new cracks, and the process of filling and cracking continues indefinitely. Thus, new oceanic crust is continuously created at a spreading center. Not all of the mantle material melts; a solid residue remains under the newly created crust. New crust and residual mantle make up the lithosphere that moves away from the ridge crest, traveling like the top of a conveyor belt. The rate of motion is generally 1 to 10 centimeters per year—slow in human terms but quite fast by geologic standards.

As the lithosphere moves away from the spreading center, the material slowly cools. As it cools it contracts and becomes denser. The contraction of the lithosphere and its slow sinking (because of the increased density) cause the floors beneath oceans to deepen away from ridge crests.

The top of a plate may be composed exclusively of oceanic crust or include a continent or part of a continent. For example, if you live on the North American plate, you are riding westward relative to Europe because the plate's spreading center is along the mid-oceanic ridge in the North Atlantic Ocean. The western half of the North Atlantic sea floor and North America are moving together in a westerly direction relative to the mid-Atlantic ridge plate boundary.

A second type of boundary, a **transform boundary,** occurs where two plates slide past each other. The San Andreas fault in California is an example of this type of boundary, and the earthquakes along the fault (such as the Loma Prieta quake) are a result of plate motion.

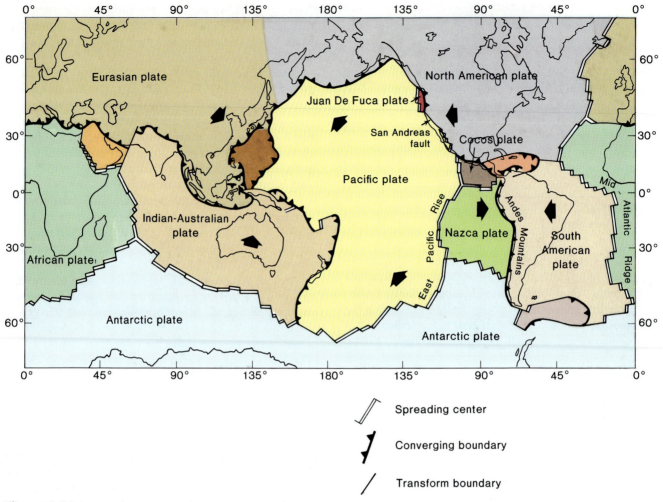

Spreading center

Converging boundary

Transform boundary

Figure 1.11
**Plates of the world. Double lines are spreading centers. Lines
with barbs are converging boundaries. Single lines are
transform boundaries.**
After W. Hamilton, U.S. Geological Survey.

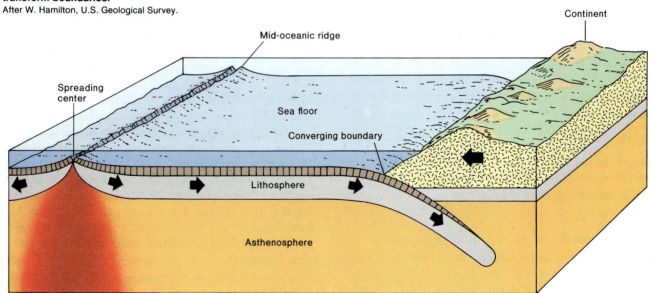

Figure 1.12
**Plate motion away from a spreading center toward a
converging boundary.**

Figure 1.13
The sea floors of the world.
The whole-world map from "World Ocean Floor Panorama" by Bruce C. Heezen and Marie Tharp. Copyright 1977 © Marie Tharp.

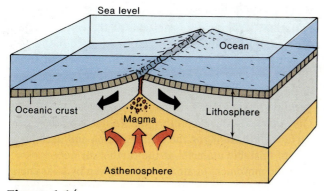

Figure 1.14
A spreading center at a mid-oceanic ridge. Hot asthenosphere wells upward beneath the ridge crest. Magma forms and squirts into fissures. Solid material that does not melt remains as mantle in lower part of lithosphere. As lithosphere moves away from spreading center, it cools, becomes denser, and sinks to a lower level.

The third type of boundary, one resulting in a wide range of geologic activities, is a **converging boundary,** where plates move toward each other (figure 1.15). If one plate is capped by oceanic crust and the other by continental crust, the less dense, more buoyant continental plate will override the denser, oceanic plate. The oceanic plate sinks along what is known as a **subduction zone,** a zone where an oceanic plate descends into the mantle beneath an overriding plate. The entire oceanic plate becomes increasingly hotter as it descends deeper in the earth. Where

the two plates grind past each other, temperatures increase even more because of the friction between the two plates (mechanical energy converts to heat energy). In the region where the top of the subducting plate comes in contact with the asthenosphere, melting takes place and magma is created. Magma is less dense than the overlying solid rock. Therefore, the magma created along the subduction zone works its way upward and either erupts on the earth's surface to solidify as *extrusive* **igneous rock** or solidifies within the crust to become *intrusive* igneous rock. Rock near the subduction zone that does not melt may, due to the high pressure and high temperature, change in the solid state to a new rock—**metamorphic rock.**

In addition to containing igneous and metamorphic rocks, major mountain belts show the effects of squeezing caused by plate convergence (for instance, the "folded sedimentary rocks" shown on figure 1.15). In the process, rock that may have been below sea level is pushed upward to become part of a mountain range.

Surficial Processes: The Earth's External Heat Engine and the Hydrologic Cycle

When tectonic forces shove a portion of the earth's crust above sea level, rocks are exposed to the atmosphere. The earth's external heat engine then comes into play. The earth's external heat engine, driven by solar power and

Introduction to Physical Geology   **17**

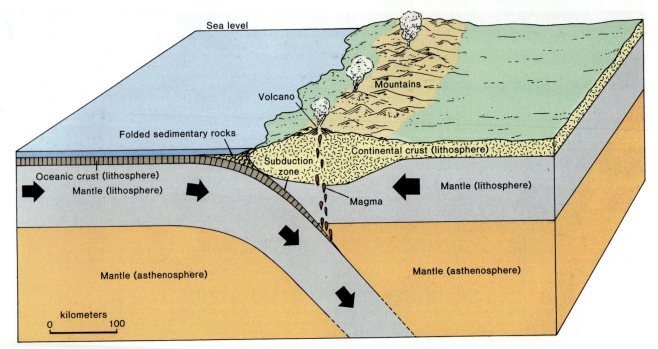

Figure 1.15
A converging boundary.

gravity, is exemplified by the hydrologic cycle, described earlier (see figure 1.5). When rain or snow falls on the land surface, more than half the water returns, rather rapidly, to the atmosphere by evaporation or by transpiration from plants. The remainder flows over the land surface as *runoff* in streams, or it trickles or percolates down into the ground to become *ground water.*

There is much more to the earth's external heat engine than is suggested by the hydrologic cycle. Our weather patterns are largely a product of this heat engine. Hot air rises, near the equator, for instance, and sinks in cold regions nearer the poles. Ocean waves and currents are largely caused by wind, generated by solar heating. Glaciers, produced by abundant snowfall at high elevations where it is cold, are pulled downhill by gravity.

The interaction of surface waters and the atmosphere with the lithosphere is important to physical geology. Streams running toward oceans remove and transport some of the land over which they run. Landslides powered by gravity move material at high elevations to lower levels. Waves crashing along a shoreline cut back the coast. Glaciers moved downward by gravity grind away at underlying rock. In each case, rock originally raised to high elevations by the earth's internal processes is worn down by surficial processes.

Rocks formed at high temperature and under high pressure deep within the earth and pushed upward by tectonic forces are unstable in their new environment (figure 1.16). Air and water tend to cause the once deep-seated rocks to break down and form new materials. The new materials, stable under conditions at the earth's surface, are said to be in **equilibrium,** that is, adjusted to the physical and chemical conditions of their environment so that they do not change or alter with time. For example, most of the minerals in igneous rock that formed at a high temperature tend to break down chemically to clay minerals. Clay minerals are in equilibrium, are stable, at the earth's surface. **Weathering** is the disintegration and decomposition of unstable material at the earth's surface.

The movement of air and water driven by solar energy and gravity results in erosion of the weathered material. **Erosion** is the loosening and removing of material. Erosion requires a transporting agent such as running water. Eventually the weathered material is deposited as loose sediment when the transporting agent loses its carrying power. For example, when a river slows down as it meets the sea, the sand it carries is deposited near the mouth as part of a layer of sediment.

In time a layer of sediment deposited on the sea floor becomes buried under another layer. This process may continue until sedimentary layers of great thickness accumulate. The pressure from overlying layers compresses the sediment, helping to consolidate the loose material. With the cementation of the loose particles, the sediment becomes **lithified** (cemented or otherwise consolidated) into a **sedimentary rock.**

BOX 1.2

Plate Tectonics and the Scientific Method

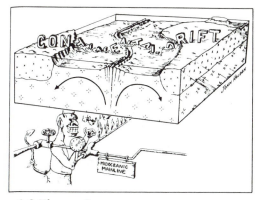

Box 1.2 Figure 1
Continental drift, sea-floor spreading, and the internal heat engine according to John Holden.
By permission of J. C. Holden and *Sea Frontiers.*

Our description of plate tectonics implies little doubt about the existence of the process. The theory of plate tectonics has only recently been accepted by a majority of geologists (this does not mean it is "true"). Plate tectonic theory, like all knowledge gained by science, has evolved through the processes of the **scientific method.** We will illustrate the scientific method by showing how plate tectonics has evolved from a vague idea into a plausible theory.

The basis for the scientific method is the *belief* that the universe is orderly and that by *objectively* analyzing phenomena, we can discover its workings. The technique is best illustrated as a series of steps, although a scientist does not ordinarily go through a formal checklist.

1. A question is raised or a problem is presented, usually after the gathering of facts, which scientists call **data.**
2. After the available data on the subject have been analyzed, tentative explanations or solutions, called **hypotheses,** are proposed.
3. One predicts what would occur in given situations if a hypothesis were correct.
4. Predictions are tested. Incorrect hypotheses are discarded.
5. A hypothesis that passes the testing becomes a **theory,** which is regarded as having an excellent chance of being true. In science, however, nothing is considered proven absolutely. All theories remain open to scrutiny, further testing, and refinement.

Like any human endeavor, the scientific method is not infallible. Objectivity is needed throughout. Someone can easily become attached to the hypothesis he or she has created and so tend subconsciously to find only supporting evidence. As in a court of law, every effort is made to have disinterested observers examine the logic of both procedures and results. Courts sometimes make wrong decisions; science, likewise, is not immune to error.

How the concept of plate tectonics evolved into a theory is outlined below.

Step 1: A question asked or problem raised
Actually, a number of questions were being asked about seemingly unrelated geological phenomena. What caused the submarine ridge that extends through most of the oceans of the world? Why are rocks in mountain belts intensely deformed? What sets off earthquakes? What causes rock to melt underground and erupt as volcanoes?

Step 2: Hypotheses proposed Most of the questions being asked were treated as separate problems, although some appeared interrelated. Hypotheses were proposed for each of the problems or for sets of problems. One hypothesis, **continental drift,** proposed by several workers, was best explained by Alfred Wegener, a German scientist, in a book published in the early 1900s.

Wegener postulated that the continents all were once a single supercontinent called Pangaea. The hypothesis explained why the coastlines of Africa and South America look like separated parts of a jigsaw puzzle. Some 200 million years ago this supercontinent broke up, and the various continents slowly drifted into their present positions. The hypothesis suggested that the rock within mountain belts becomes deformed as the leading edge of a continental crust moves against and over the stationary oceanic crust. Earthquakes were presumably caused by continuing movement of the continents.

It was not until the 1960s, after new data on the nature of the sea floor became available, that the idea of continental drift was incorporated into the concept of plate tectonics. What was added in the plate tectonic hypothesis was the idea that oceanic crust, as well as continental crust, was shifting.

Box 1.2 *continued*

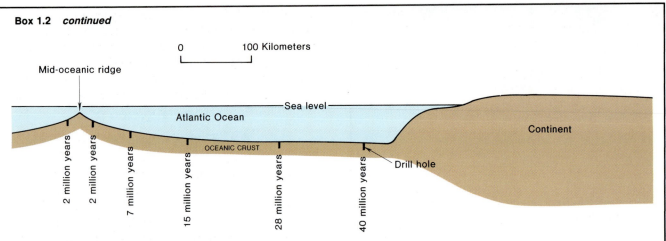

Box 1.2 Figure 2
Ages of rocks from holes drilled into the oceanic crust.
(Vertical scale of diagram is exaggerated.)

Step 3: Prediction An obvious prediction, if plate tectonics is correct, is that since Europe and North America are moving away from each other, the distance between them will be greater in ten years. To test this prediction, satellites are being used to try to make such measurements; preliminary results indicate that the continents are indeed moving. Another prediction was that the rocks of the oceanic crust will be progressively older the farther they are from the spreading center, the crest of the mid-oceanic ridge.

Step 4: Predictions are tested Experiments were conducted in which holes were drilled in the deep sea floor from a specially designed ship. Rocks and sediment were collected from these holes, and the ages of these materials were determined. As the hypothesis predicted, the youngest sea floor (generally less than a million years old) is near the mid-oceanic ridges, whereas the oldest sea floor (up to about 200 million years old) is farthest from the ridges (box figure 2).

This test was only one of a series. Various other tests, described in some detail later in this book, tended to confirm the hypothesis of plate tectonics. Some tests did not work out exactly as predicted. Because of this, and more detailed study of data, the original concept was modified. The basic premise, however, is generally regarded as valid.

Step 5: The hypothesis becomes a theory Most geologists in the world consider the results of the testing positive enough to imply that the concept is probably true. It can now be called the plate tectonic theory. This, of course, does not mean that it is "proved." It would be unscientific not to be open to other explanations that might account for the observed phenomena. Furthermore, treating the theory as dogma would require ignoring aspects of geology that cannot easily be reconciled with plate tectonics. We should never become so arrogant as to think nature behaves as we want it to.

Like other new theories, the plate tectonic theory raises questions that call for further scientific investigation.

In this book we describe the earth's materials in chapters 1 through 7. We consider the factor of time in chapter 8. We discuss in detail the work of the various agents of erosion and deposition, such as landsliding and related downslope movements, water, glaciers, wind, and ocean waves (chapters 9 through 14). We examine how these agents sculpture landscapes and deposit sediment in distinctive styles. The jagged peak in the chapter opening photograph, for example, indicates erosion by glaciers.

Chapters 15 through 20 deal with the earth's interior, tectonic forces generated within the earth, and the effect these forces have on rocks of the crust.

Geologic Time

We have mentioned the great amount of time required for geological processes. As humans, we think in units of time related to personal experience—seconds, hours, years, a human lifetime. It stretches our imagination to contemplate ancient history that involves 1,000 or 2,000 years.

To be sure, some geological processes occur quickly, such as a great landslide or a volcanic eruption. These events occur when stored energy (like the energy stored in a car battery), sometimes stored for centuries, is suddenly released. Most geological processes, however, are slow but relentless, reflecting the pace at which the heat

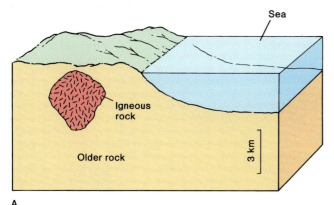

A

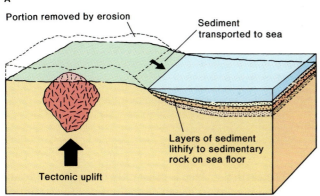

B

Figure 1.16
Uplift, erosion, and deposition. (*A*) Magma has solidified underground to become igneous rock. (*B*) Land is uplifted. Upper portion is weathered and eroded. Sediment is transported to the sea to become sedimentary rock.

engines work. It is unlikely that a hill will visibly change in shape or height during your lifetime (unless through natural catastrophe or human activity). However, in a geologic time frame, the hill probably is eroding away quite rapidly. "Rapidly" to a geologist may mean that within a few million years the hill will be reduced nearly to a plain. Similarly, in the geologically "recent" past of several million years ago, a sea may have existed where the hill is now. Some processes are regarded by geologists as "fast" if they are begun and completed within a million years.

The rate of plate motion is relatively fast. If new magma erupts and solidifies along a mid-oceanic ridge, we can easily calculate how long it will take that igneous rock to move 1,000 kilometers away from the spreading center. At the rate of 1 centimeter per year, it will take 100 million years for the presently forming part of the crust to travel the 1,000 kilometers.

Uniformitarianism

Two hundred years ago you would have been declared a heretic in Christian countries if you believed the earth existed before 4004 B.C. Some astute observations and reasoning by an eighteenth-century Scotsman named James Hutton profoundly changed people's thinking. He noted (as had others before his time) that layers of rock several thousand meters thick possessed textural and compositional features like those of loose sediment at the bottom of lakes or along shorelines. Therefore, he reasoned, the rocks must have formed by consolidation of similar sediments. Fossil sea shells in the rocks seemed to support this conclusion. But how could layers several thousand meters thick have accumulated? The standard answer at the time was that all the sediment had been deposited during the great Biblical flood of Noah's time.

Hutton's revolutionary hypothesis was that the layered rock could be explained by the same processes he observed taking place around him. He could see sediment settling slowly through water to collect in layers. The rock must have formed after a slow process of settling, too. The implication was that in very large periods of time huge amounts of sediment settled to the sea floor and later consolidated into rock. Hutton's hypothesis, which has become known as the **principle of uniformitarianism,** says that the geological processes operating now are the same processes that operated in the geologic past. This has been concisely paraphrased as "The present is the key to the past." Applying the principle, we can determine the past history of the earth by observing processes that are taking place in the present.

The principle of uniformitarianism should not be interpreted as excluding sudden or catastrophic events. For instance, a violent volcanic explosion may alter the earth's weather patterns for years. Similarly, one hurricane on the Gulf Coast of North America moves more sediment in a few hours than is moved in hundreds or even thousands of years by day-to-day processes. Nor can we apply the principle too rigidly. Processes that are slow today may have been more rapid in the past. For instance, today it is extremely rare for a meteorite large enough to form a crater to hit the earth. Yet, based on our studies of the moon, it is very likely that 3.8 to 4.0 billion years ago the earth was bombarded by meteorites. The moon was bombarded by meteorites at that time, and the earth certainly could not have escaped similar bombardment. The earth's meteorite craters were erased long ago by weathering and erosion, but those on the moon are preserved for eternity because the moon lacks an atmosphere and therefore an external heat engine.

What the principle of uniformitarianism *does* tell us is that we cannot violate the physical and chemical laws of the universe—we cannot attribute the earth's features and processes to magic and supernatural powers.

Although we will discuss geologic time in detail in the chapter titled "Time and Geology," table 1.1 shows some reference points to keep in mind. The earth is estimated to be about 4.5 billion years old (4,500 million years). Fossils in rocks indicate that complex forms of animal life

TABLE 1.1

Some Important Ages in the Development of Life on Earth

Millions of Years Before Present	Animal Life on Earth	Eras	Periods
3	First humans	**Cenozoic**	Quaternary
			Tertiary
65	First important mammals; extinction of dinosaurs		
		Mesozoic	Cretaceous
			Jurassic
			Triassic
245	First dinosaurs		
		Paleozoic	Permian
			Pennsylvanian
300	First reptiles		Mississippian
400	Fish become abundant		Devonian
			Silurian
			Ordovician
			Cambrian
570	First abundant fossils		
3,500	Earliest single-celled fossils	**Precambrian**	
4,500	Origin of the earth		

have existed in abundance on the earth for about the past 600 million years. Reptiles—most notably, dinosaurs—became abundant about 230 million years ago. Dinosaurs became extinct about 65 million years ago. Humans have been here only about the last 3 million years. The eras and periods shown in table 1.1 comprise a kind of calendar for geologists into which geologic events are placed (as explained in the chapter on geologic time).

Not only are the immense spans of geologic time difficult to comprehend, but very slow processes are impossible to duplicate. A geologist who wants to study a certain process cannot repeat in a few hours a chemical reaction that takes a million years to occur in nature. As Mark Twain wrote in *Life on the Mississippi,* "Nothing hurries geology."

The earth's surface is constantly being altered by air and water in motion. The hydrologic cycle consists of rain falling on the land surface, flowing toward a sea, evaporating to the atmosphere, and eventually precipitating once again, as rain (or snow).

Although the earth is changing constantly, the rates of change are generally extremely slow by human standards. To a geologist, a million years is not a very long time. By the principle of uniformitarianism we can determine what happened in the past by understanding present-day processes.

Summary

Geology is the scientific study of the earth. Geological investigations indicate that the earth is changing because of internal and surficial processes. Internal processes are probably driven mostly by convection currents within the earth's mantle. Surficial processes are caused by solar energy and gravity. Internal forces cause the crust of the earth to move. Plate tectonic theory visualizes the lithosphere (the crust and uppermost mantle) as broken into plates that move relative to each other. The plates are moving *away* from spreading centers usually located at the crests of mid-oceanic ridges where new crust is being created. Plates are moving *toward* boundaries with other plates. Continuous convergence may result in one plate being overridden and forced back into the mantle while the other plate is deformed into mountains.

Terms to Remember

asthenosphere
continental drift
convection
convection current
converging boundary
core
crust
equilibrium
erosion
geology
gravity
heat engine
hydrologic cycle
hypothesis
igneous rock
lithified
lithosphere
magma

mantle
metamorphic rock
mid-oceanic ridge
physical geology
plastically
plate tectonics
principle of
   uniformitarianism
scientific method
sediment
sedimentary rock
spreading center (diverging
   boundary)
subduction zone
tectonic forces
theory
transform boundary
weathering

Questions for Review

1. Describe three ways in which rocks form.
2. What is the distinction between weathering and erosion?
3. What is meant by tectonic activity?
4. What would the surface of the earth be like if there were no tectonic activity?
5. Explain why cavemen never saw a dinosaur.
6. What are the three types of plate boundaries described in this chapter and how do they differ from one another?
7. What is meant by *equilibrium*? What happens when rocks are forced out of equilibrium?
8. What are the relationships among the mantle, the crust, the asthenosphere, and the lithosphere?
9. What role does gravity play in each of the heat engines?

Questions for Thought

1. According to plate tectonic theory, where are crustal rocks created? Why doesn't the earth keep getting larger if rock is continually created?
2. What evidence is there that geologic processes operated in the past at about the same rate and in about the same way as they do now?
3. What percentage of geologic time is accounted for by the last century?
4. What would the earth be like without solar heating?
5. What are some of the technical difficulties you would expect to encounter if you tried to drill a hole to the center of the earth?

Supplementary Readings

Allègre, C. 1988. *The behaviour of the earth.* Boston: Harvard University Press.

Drake, E. T., and W. M. Jordan, eds. 1985. *Geologists and ideas: A History of North American Geology.* Boulder, Colo.: Geological Society of America.

Gardner, M. 1981. *Science: Good, bad and bogus.* Buffalo, N.Y.: Prometheus Books.

McPhee, J. 1981. *Basin and range.* New York: Farrar, Straus & Giroux.

Pirsig, R. M. 1974. *Zen and the art of motorcycle maintenance.* New York: Bantam Books (paperback). This book contains an exceptionally good exposition of the scientific method as well as considerable insight into the philosophy of science.

Rhodes, F. H. T., and R. O. Stone, eds. 1981. *Language of the earth.* New York: Pergamon Press.

Sinderman, C. J. 1985. *The joy of science.* New York: Plenum Publishing Corp.

Sullivan, W. 1974. *Continents in motion.* New York: McGraw-Hill.

U.S. Geological Survey. 1990. The Loma Prieta, California earthquake: An anticipated event. *Science* 247: 286–93.

Wegener, A. 1966. *The origin of the continents and oceans.* New York: Dover Publications (paperback reprint of the original translation from German).

Wyllie, P. J. 1976. *The way the earth works.* New York: John Wiley & Sons.

Knowing the names and important characteristics of the most common elements and minerals is essential to understanding how they combine to form the rocks of the earth. Familiarity with this basic information will help you throughout your study of geology.

Knowing some basic principles of chemistry is also necessary for understanding material covered in later chapters, such as the various types of rocks, weathering, and the composition of the earth's interior and crust. You need to know what a mineral is; how each mineral is composed of certain chemical elements in a remarkably orderly arrangement; and how the arrangement and characteristics of atoms control the physical properties of minerals. You will learn how to readily determine physical properties and use them to identify common minerals. (Appendix A is a further guide to identifying minerals.)

The chapters that follow are about various types of rocks. To help you keep rocks and related topics in perspective, we close this chapter by introducing you to a conceptual device called the rock cycle.

2

Atoms, Elements, and Minerals

Quartz Crystals.
© Doug Sherman.

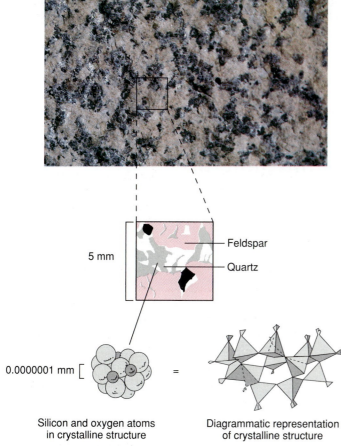

5 mm

Feldspar

Quartz

0.0000001 mm

=

Silicon and oxygen atoms
in crystalline structure

Diagrammatic representation
of crystalline structure

Figure 2.1
**Specimen of granite showing the relationship among rock,
minerals, and crystal structures of atoms. The diagrammatic
representation as tetrahedrons is explained later in the
chapter.**

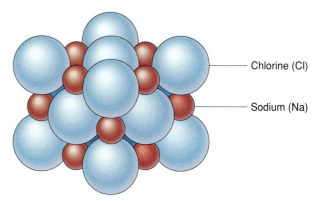

Chlorine (Cl)

Sodium (Na)

Figure 2.2
Model of the crystal structure of halite (or table salt).

Rock is naturally formed, consolidated material composed of grains of one or more minerals (this definition has a few exceptions).

Figure 2.1 shows a specimen of the common rock *granite*. It forms from magma solidifying within the earth's crust. Granite is made up mostly of the minerals *feldspar* and *quartz*. What, then, is a mineral? A *mineral* is composed of atoms arranged in a very orderly, three-dimensional structure (which is to say, it is *crystalline*). Quartz is made up exclusively of oxygen and silicon atoms. More precisely, there are twice as many oxygen atoms in the structure as silicon atoms (therefore, the chemical formula for quartz is SiO_2).

Rock salt is another rock, quite different from granite. It forms when salt water is evaporated. Unlike granite, it is made of grains of only one mineral. Rock salt is a consolidated aggregate of grains of the mineral *halite,* familiar to everyone as table salt. The formula for halite is NaCl. This means that it is composed of equal numbers of sodium (Na) and chlorine (Cl) atoms. These atoms are arranged in a simple, orderly crystalline pattern. Each sodium atom is surrounded by six chlorine atoms and each chlorine atom is surrounded by six sodium atoms. Billions

of each type of atom are necessary to form a salt crystal the size of a pinhead. Halite crystals tend to be cubic because of the particular orderly way in which chlorine and sodium atoms are packed together (figure 2.2). This and other physical and chemical properties of halite are caused by (1) the pattern of repeating atoms, (2) the way the atoms are bonded to neighboring atoms, and (3) the characteristics of the elements chlorine and sodium.

Atoms and Elements

Halite (or table salt) crystals can be separated chemically into sodium and chlorine, both of which are elements. An **element** is a substance that cannot be broken down into other substances by ordinary chemical methods.

An **atom** is the smallest possible particle of an element that retains the properties of that element. All atoms of the element chlorine are essentially identical to all other chlorine atoms; the same is true for sodium. Water (H_2O) can be chemically broken down to the elements oxygen and hydrogen (two hydrogen atoms for every oxygen atom).

Atoms are far too small to see, even with the most powerful microscope. Our pictures of atoms are really models. A *model* in science is an image—graphic, mathematical, or verbal—that is consistent with the known data.

Models of atoms constructed by chemists and physicists show three types of subatomic particles—protons, neutrons, and electrons. A **proton** is a subatomic particle that contributes mass and a single positive electrical charge to an atom. A **neutron** is a subatomic particle that contributes mass to an atom but is electrically neutral. An **electron** is a single negative electric charge that contributes virtually no mass to an atom. Electrons are regarded as moving very rapidly within specific energy levels, which are depicted as shells (figure 2.3*A*).

Protons and neutrons form the **nucleus** of an atom. Although the nucleus occupies an extremely tiny fraction of the volume of the entire atom, practically all the mass of the atom is concentrated in the nucleus. The **atomic mass number** is the total number of neutrons and protons in an atom. The atomic mass number of the oxygen atom in

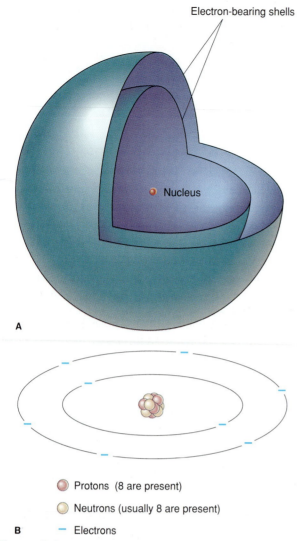

Electron-bearing shells

Nucleus

A

Protons (8 are present)

Neutrons (usually 8 are present)

B — Electrons

Figure 2.3
(**A**) Model of an oxygen atom. The nucleus, composed of neutrons and protons is actually much smaller than indicated relative to the volume of the atom. The hollow spheres represent the two electron-bearing shells. (**B**) Schematic representation of the oxygen atom. The two circles containing electrons represent the electron-bearing shells.

common isotope of oxygen has 8 neutrons, but oxygen isotopes with 10 neutrons are sometimes detected. In geology, isotopes are important in radioactive dating of rocks (chapter 8).

An element's atomic weight is closely related to the mass number. **Atomic weight** is the weight of an *average* atom of an element, given in atomic mass units. Since sodium has only one isotope, its atomic mass number and its atomic weight are the same—23. On the other hand, chlorine has two common isotopes, with mass numbers of 35 and 37. The atomic weight of chlorine, which takes into account the abundance of each isotope, is 35.5 (the lighter isotope is more common than the heavier one).

Electrons, though having almost no mass, are visualized as occupying shells around the nucleus. Virtually the entire volume of an atom is taken up by the space in which these tiny, electrically negative charges move. The number of electrons in an atom is related to the number of protons in the nucleus. Each electron is opposite in charge but equal in strength to a proton.

If the number of electrons in orbit around a nucleus is equal to the number of protons in the nucleus, the positive and negative charges balance each other and the atom is electrically neutral. Helium has 2 electrons, exactly balancing its 2 protons. Most elements, however, are not able to maintain an electrical balance between protons and electrons within a single atom.

Chemical Activity

Many geological processes can be explained as chemical reactions. Some rocks form as a result of chemical reactions between substances. Chemical reactions also play a part in weathering (the chapter titled "Weathering and Soil"). Understanding a few basic concepts of chemistry will clarify why and under what conditions chemical reactions occur.

Atoms that are not electrically neutral tend to react (or combine) with other atoms to neutralize the electrical imbalance. Each atom not only seeks electrical neutrality but wants each of its shells to be full of electrons. The innermost shell is full when it possesses 2 electrons; outer shells generally each require 8 electrons to be complete.

Helium, for example, is a *stable* element because 2 protons are balanced by 2 electrons, and the 2 electrons exactly fill one shell. Neon is also stable; its 10 protons are balanced by 2 electrons in the inner shell and 8 electrons in the next shell (see figure 2.4). Normally, neither of these elements reacts with other elements.

Ions

Chlorine and sodium are more typical elements in that if an electron shell is complete, the atom is electrically out of balance. A sodium atom (figure 2.5) has a complete inner shell with 2 electrons and a second shell, also filled, with 8 electrons. One more electron would neutralize all 11 protons in the nucleus, but an eleventh electron alone in a shell is extremely unstable, so the sodium atom normally does without it. In each sodium atom, then, the 11

figure 2.3*B* is 16 (8 protons plus 8 neutrons). Heavier elements have more neutrons and protons than do lighter ones. For example, the heavy element gold has an atomic mass number of 197, whereas helium has 4.

The number of protons controls the "character" of an element more than does the number of other subatomic particles. The **atomic number** of an element is the number of protons in each atom. We can refine our earlier definition of an element by adding that each atom of an element has the *same number of protons*. Gold has an atomic number of 79, or 79 protons per atom; oxygen always has 8 protons; hydrogen always has 1 proton; chlorine has 17; and sodium has 11. (Other atomic numbers are listed in Appendices C and D.)

The number of neutrons (and therefore the mass of an element) can vary within limits. **Isotopes** of an element are atoms containing different numbers of neutrons but the *same number of protons*. For example, the most

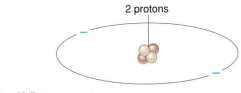

A Helium

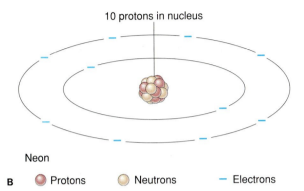

10 protons in nucleus

Neon

B 🔴 Protons   🟡 Neutrons   — Electrons

Figure 2.4
(**A**) Helium atom and (**B**) neon atom.

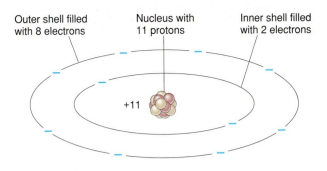

Outer shell filled with 8 electrons

Nucleus with 11 protons

Inner shell filled with 2 electrons

+11

A Sodium (Na⁺)

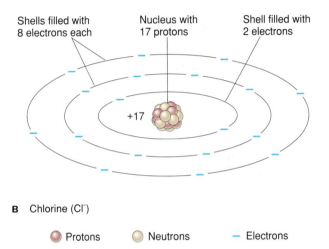

Shells filled with 8 electrons each

Nucleus with 17 protons

Shell filled with 2 electrons

+17

B Chlorine (Cl⁻)

🔴 Protons   🟡 Neutrons   — Electrons

Figure 2.5
(**A**) Sodium (Na⁺) ion. Ten electrons fill two shells. The nucleus contains eleven protons. (**B**) Chlorine (Cl⁻) ion. Electron shown in color completes the outer shell of the chlorine atom, making it an ion.

protons ($11+$) and 10 electrons ($10-$) add up to a single excess positive charge ($+1$). Such an atom is an **ion**, an electrically charged atom or group of atoms. The sodium ion can be abbreviated as Na^+.

Chlorine, with an atomic number of 17, has a complete inner shell with 2 electrons and a complete second shell of 8 electrons around this. A neutral chlorine atom would have only 7 electrons in the third shell, but since this shell requires 8 electrons an extra electron is captured and incorporated in it. The chlorine ion then contains 18 electrons and 17 protons, and so has a single excess negative charge (Cl^-).

Positive and negative ions are attracted to each other. In a crystal structure the mutual attraction is one way atoms are held in place or bonded to one another. **Bonding** is the attachment of an atom to one or more adjacent atoms (see Box 2.1).

Chemical Composition of the Earth's Crust

Estimates of the chemical composition of the earth's crust are based on many chemical analyses of the rocks exposed on the earth's surface. (Models for the composition of the interior of the earth—the core and the mantle—are based on more indirect evidence.) Table 2.1 lists the generally accepted estimates of the abundance of elements in the earth's crust. At first glance, the chemical composition of the crust (and, therefore, the average rock) may be quite surprising.

Most people think of oxygen in terms of the air we breathe. Yet most rocks are composed largely of oxygen, which is the most abundant element in the earth's crust. Unlike the oxygen gas in air, oxygen in minerals is strongly bonded to other elements. By weight, oxygen accounts for about half the crust, but it takes up 93% of the volume of

TABLE 2.1

Crustal Abundance of Elements

Element	Symbol	Percentage by Weight	Percentage by Volume	Percentage of Atoms
Oxygen	O	46.6	93.8	60.5
Silicon	Si	27.7	0.9	20.5
Aluminum	Al	8.1	0.8	6.2
Iron	Fe	5.0	0.5	1.9
Calcium	Ca	3.6	1.0	1.9
Sodium	Na	2.8	1.2	2.5
Potassium	K	2.6	1.5	1.8
Magnesium	Mg	2.1	0.3	1.4
All other elements		1.5	—	3.3

an average rock. This is because the oxygen electron shells take up a large amount of space relative to their weight. It is not an exaggeration to regard the crust as a mass of oxygen with other elements occupying positions in crystal structures between oxygen atoms.

Silica is a term for *oxygen plus silicon*. Because silicon is the second most abundant element in the crust, most

BOX 2.1

Bonding

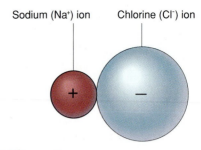

Box 2.1 Figure 1
Ionic bonding between sodium (Na⁺) and chlorine (Cl⁻).

Ions may be regarded as tiny spheres that behave much like magnets. Positively charged ions attract negatively charged ions so that their electrical charges can be neutralized. In salt water, equal numbers of sodium ions (Na⁺) and chlorine ions (Cl⁻) move about freely. The electrical neutrality of the water is maintained because positive sodium ions exactly balance negative chlorine ions. If the water evaporates, the sodium and chlorine are electrically attracted to each other and crystallize into halite. The crystal is the most orderly way for chlorine and sodium ions to pack themselves together and neutralize their collective charges.

A chlorine ion and a sodium ion are fixed in place by their electrical attraction to each other. This is called **ionic bonding** because it is brought about by an attraction between positively and negatively charged ions (box figure 1).

Ionic bonding is the most common type of bonding in minerals. However, in most minerals the bonds between ions are not purely ionic. Atoms are also commonly bonded together by **covalent bonding,** or bonding in which adjacent atoms *share* electrons. Diamond is composed exclusively of covalently bonded carbon atoms (box figure 2). Carbon has an atomic number of 6, which means that the innermost shell is full with 2 electrons. Four more electrons are required to maintain electrical neutrality. In a diamond, each carbon atom has 4 electrons in the outer shell to maintain neutrality, while the need for 8 electrons in that shell is satisfied by electrons that are shared with adjacent carbon atoms. Neighboring carbon atoms are so close together that each of the outer-shell electrons

Box 2.1 Figure 2
Carbon atoms covalently bonded, as in diamond.

spends half its time orbiting one atom and half orbiting an adjacent atom. Electrical neutrality is maintained, and each atom, in a sense, has 8 electrons in the outer shell (even though they are not all there at the same time). Covalent bonds in the diamond are extremely strong, and diamond is the hardest natural substance on earth. However, covalent bonds are not necessarily stronger than ionic bonds.

A third type of bonding, *metallic bonding,* is not as important to geology. In metals, such as iron or gold, the atoms are closely packed together and the electrons move freely throughout the crystal. The ease with which electrons move accounts for the high electrical conductivity of metals.

Finally, residual forces (called van der Waal's forces) remaining after atoms have bonded together may result in very weak bonds, such as those that hold adjacent sheets of graphite or mica together.

---

minerals contain silica. The common mineral quartz (SiO_2) is pure silica that has crystallized. Quartz is one of many minerals that are **silicates,** substances that contain silica (as indicated by their chemical formulas). Most silicate minerals also contain one or more other elements.

Note that the third most abundant element is aluminum, which is more common in rocks than iron. Knowing this, one might assume that aluminum would be less expensive than iron, but of course this is not the case. Common rocks are not mined for aluminum because it is so strongly bonded to oxygen and other elements. The amount of energy required to break these bonds and separate the aluminum makes the process too costly for com-

mercial production. Aluminum is mined from the uncommon deposits where aluminum-bearing rocks have been weathered, producing compounds in which the crystalline bonds are not so strong.

Collectively, the eight elements listed in table 2.1 account for more than 98% of the weight of the crust. All the other elements total only about 1.5%. Absent from the top eight elements are such vital elements as hydrogen (tenth by weight) and carbon (seventeenth by weight).

The element copper is only twenty-seventh in abundance, but our industrialized society is highly dependent on this metal. Most of the wiring in electronic equipment is copper, as are many of the telephone and power cables

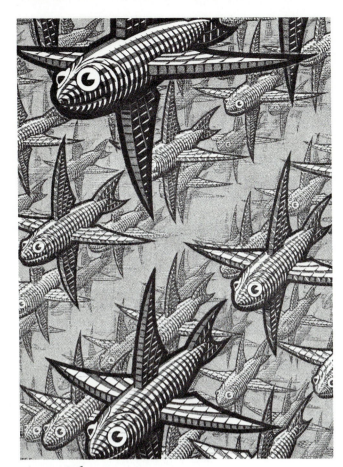

Figure 2.6
Depth, **print by M. C. Escher.**
© M. C. Escher Heirs/Cordon Art—Baarn—Holland.

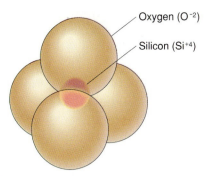

A Arrangement of atoms in silica tetrahedron

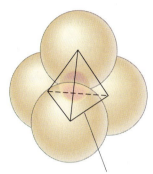

B Diagrammatic representation of a silica tetrahedron

Figure 2.7
(A) The silica tetrahedron. (B) The silica tetrahedron showing the corners of the tetrahedron coinciding with the centers of oxygen ions.

that crisscross the continent. However, the earth's crust is not homogeneous, and geological processes have created concentrations of elements such as copper in a few places. Exploration geologists are employed by mining companies to discover where ore deposits of copper and other metals occur as well as why.

Crystals and Crystallinity

Most solids are crystalline. A **crystalline** substance is one in which the atoms are arranged in a regularly repeating, orderly pattern. The print by M. C. Escher (figure 2.6) vividly expresses the principle of crystallinity. You can visualize what crystallinity is in nature by mentally substituting identical clusters of atoms for each fish and imagining the clusters packed together.

As you can tell from figure 2.2, halite (or table salt) is crystalline. What controls the type (or "architecture") of a crystal is the relative size of adjacent atoms. Thus, if all the sodium atoms in figure 2.2 were about the same size as the chlorine atoms, the particular crystal structure would be different (bearing in mind that neighboring atoms must "touch" one another). Liquids are not crystalline because the atoms are free to move about; nor is glass, because the atoms in glass are as randomly arranged as those in a liquid, only "frozen" into place. The structure of glass is comparable to what would happen if

fish like those in the Escher drawing (figure 2.6) were swimming freely and randomly distributed when the water suddenly froze.

Of particular importance are the crystal structures derived from the two most common elements in the earth's crust—oxygen and silicon.

The Silica Tetrahedron

The two most abundant elements, silicon and oxygen, combine to form the basic building block for most common minerals. In each "building block," four oxygen atoms are packed together around a single, much smaller, silicon atom, as shown in figure 2.7A. The four-sided, pyramidal, geometric shape called a *tetrahedron* is used to represent the four oxygen atoms surrounding a silicon atom. Each *corner* of the tetrahedron represents the *center* of an oxygen atom (figure 2.7B). This basic building block of a crystal is called a **silica tetrahedron.**

The atoms of the tetrahedron are strongly bonded together. Within a silica tetrahedron the negative charges exceed the positive charges (see figures 2.7A and 2.8A). A single silica tetrahedron has a formula of SiO_4^{-4} because silicon has an ionic charge of $+4$ and the four oxygen ions have 8 negative charges (-2 for each oxygen atom).

For the silica tetrahedron to be stable within a crystal structure, it must either (1) be balanced by enough positively charged ions or (2) share oxygen atoms with adjacent tetrahedrons (as shown in figure 2.8C & D) and therefore reduce the need for extra, positively charged ions.

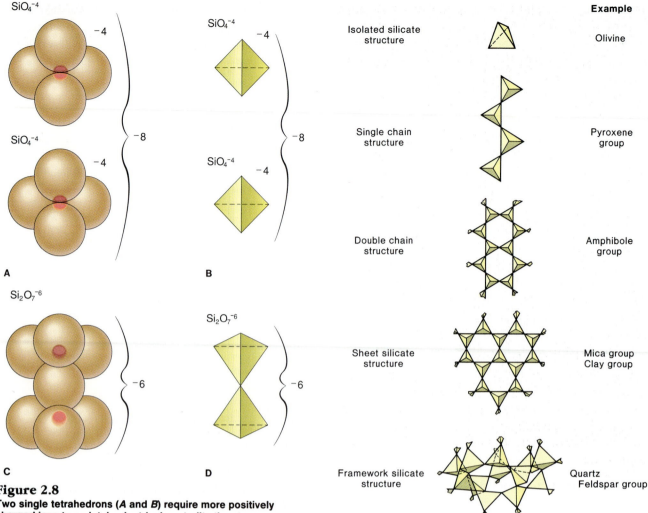

Figure 2.8
Two single tetrahedrons (*A* and *B*) require more positively charged ions to maintain electrical neutrality than two tetrahedrons sharing an oxygen atom (*C* and *D*). *B* and *D* are the schematic representations of *A* and *C* respectively.

Figure 2.9
Common silicate structures.

The structures of silicate minerals range from an *isolated silicate structure,* which depends entirely on positively charged ions to hold the tetrahedrons together, to *framework silicates* (quartz, for example), in which all oxygen atoms are shared by adjacent tetrahedrons. The various types of silicate structures are shown diagrammatically on figure 2.9 and are discussed next.

Isolated silicate structure   Silicate minerals that are structured so that none of the oxygen atoms are shared by tetrahedrons have an **isolated silicate structure.** The individual silica tetrahedrons are bonded together by positively charged ions (figure 2.10). The common mineral **olivine,** for example, contains two ions of either magnesium (Mg^{+2}) or iron (Fe^{+2}) for each silica tetrahedron. The formula for olivine is $(Fe, Mg)_2SiO_4$.

Chain silicates   If two oxygen atoms of each tetrahedron are shared with adjacent tetrahedrons, the result may be a chain of tetrahedrons, a **chain silicate structure.** Each chain, which extends indefinitely, has a net excess of negative charges. The ratio of silicon to oxygen (as figure 2.11 shows) is 1 : 3; therefore, each mineral in this group (the

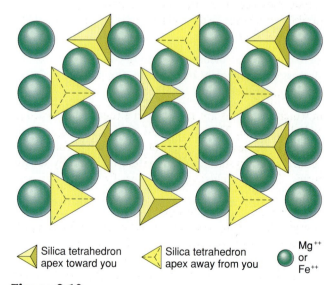

Figure 2.10
Diagram of the crystal structure of olivine, as seen from one side of the crystal.

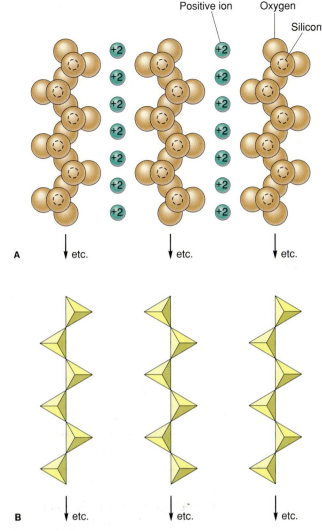

Figure 2.11
Single-chain silicate structure. (*A*) Model of a single-chain silicate mineral. (*B*) The same chain silicate shown diagrammatically as linked tetrahedrons; positive ions between the chains are not shown.

Sheet silicates

Sheet silicates When each tetrahedron shares three oxygen ions, the result is a **sheet silicate structure,** characteristic of the *mica* group and the *clay* group of minerals (figure 2.9). The positive ions that hold the sheets together are "sandwiched" between the silicate sheets.

Framework silicates When all four oxygen ions are shared by adjacent tetrahedrons, a **framework silicate structure** is formed. Quartz is a framework silicate mineral. A feldspar is a framework silicate as well. However, the structure is slightly complicated because aluminum substitutes for some of the silicon atoms in some of the tetrahedrons. The same kind of substitution also takes place in pyroxenes, amphiboles, and micas, which helps account for the wide variety of silicate minerals.

Minerals

Although we have been discussing minerals, we have not fully defined what a mineral is. To be a **mineral** in the geologic sense of the term, a substance must satisfy five conditions:

1. It must be a *crystalline solid*.
2. It must *occur naturally*.
3. It must be *inorganic*.
4. It must have a *definite chemical composition*.
5. It must possess *characteristic physical properties*.

There are other meanings for the word *mineral* that conflict with the geologist's definition. The "minerals" listed on cereal boxes, for instance, have nothing to do with what geologists or chemists mean when they talk about minerals. Nor, for that matter, does the geologist's definition agree with the miner's definition of a mineral. To a miner, a "mineral" is anything of commercial value that is extracted from the ground.

In this book, the term *mineral* is used strictly in the geologic sense.

Crystalline Solids

We have already mentioned that crystallinity, the first criterion for a mineral, is an orderly arrangement of atoms. Nature is not always accommodating to definitions, however, and some substances exist that are not crystalline but otherwise meet the criteria for a mineral. These not-quite-minerals are called *mineraloids*. An example of a mineraloid is *opal*.

Natural and Inorganic Substances

The second and third criteria need little explanation. Man-made crystalline compounds are not regarded as minerals. Substances that are part of plants or animals—that is, organic—are also excluded. Maple sugar that has crystallized from the sap of a maple tree is not a mineral, although it is crystalline and occurs naturally.

pyroxene group) incorporates SiO_3^{-2} in its formula, and it must be electrically balanced by the positive ions that hold the parallel chains together.

One pyroxene mineral, for example, has a formula of $MgSiO_3$. This pyroxene may form in a cooling magma when earlier formed crystals of olivine, Mg_2SiO_4, react with silica (SiO_2) in the remaining melt. To accommodate the additional silica, olivine's isolated silicate structure is rearranged into the single-chain silicate structure of pyroxene. In this case, Mg^{+2} ions occupy the positive ion positions between chains as shown in figure 2.11*A*.

The *amphibole* group is characterized by *two* parallel chains (double-chain silicate structure) in which every other tetrahedron along a chain shares an oxygen ion with the adjacent chain (see figure 2.9). In even a small amphibole crystal, millions of parallel double chains are bonded together by positively charged ions.

Minerals with chain silicate structure tend to be shaped like columns, needles, or even fibers (see Box 2.2). The long structure of the external form corresponds to the linear dimension of the chain structure.

BOX 2.2

Asbestos

Asbestos is a generic name for fibrous aggregates of minerals (box figure 1). Because it does not ignite or melt in fire, asbestos has a number of valuable industrial applications. Woven into cloth, it is used to make suits for firefighters. It is also used as a fireproof insulation for homes and other buildings and has commonly been used in plaster for ceilings. Five of the six commercial varieties of asbestos are amphiboles, known commercially as "brown" and "blue" asbestos. The fifth variety is *chrysotile,* which belongs to the *serpentine* family of minerals and is more commonly known as "white asbestos." White asbestos is, by far, the most commonly used in North America (about 95% of that used in the United States).

Public fear of asbestos in the United States has resulted in its being virtually outlawed by the federal government. Hundreds of millions of dollars are being spent to remove or seal off asbestos from schools. If proposed further regulations are put into effect, billions could be spent eliminating asbestos from all public buildings. Churches, school districts, and towns could go bankrupt.

The bad reputation of asbestos comes from the high death rate among asbestos workers exposed, without protective attire, to extremely high levels of asbestos dust. Some of these workers, whose bodies were covered with fibers, were called "snowmen." In Manville, New Jersey, children would catch the "snow" (white asbestos dust released from the asbestos factory there) in their mouths. The high death rates among asbestos workers are attributed to *asbestosis* and lung cancer. Asbestosis is similar to silicosis contracted by miners; essentially, the lungs become clogged with asbestos dust after prolonged heavy exposure. The incidence of cancer has been especially high among asbestos workers who were also smokers. It's not clear that heavy exposure to white asbestos caused cancer among nonsmoking asbestos workers. However, the amphibole asbestos (brown and blue asbestos) has been linked to cancer for heavy exposure (even if for a short term).

What are the hazards of asbestos to an individual in a building where walls or ceilings contain asbestos? Recent studies from a wide range of scientific disciplines indicate that the risks are minimal to nonexistent, at least for exposure to white asbestos.

Box 2.2 Figure 1
Chrysolite asbestos.
Photo by C. C. Plummer.

The largest asbestos mines in the world are at Thetford Mines, Quebec. A study of longtime Thetford Mines residents, whose houses border the waste piles from the asbestos mines, indicated that their incidence of cancer was no higher than that of Canadians overall. Nor have studies in the United States been able to link nonoccupational exposure to asbestos and cancer. One estimate of the risk of death from cancer due to exposure to asbestos dust is one per 100,000 lifetimes. (Compare this to the risk of death from lightning of 4 per 100,000 lifetimes or automobile travel—1600 deaths per 100,000 lifetimes).

In light of the best information available, the current United States government's acceptable level of 0.25 asbestos fibers per cubic centimeter in the air of public buildings is not reasonable. Malcolm Ross, U.S. Geological Survey expert on asbestos, has suggested that 2 fibers per 1 cubic centimeter of air would be a safe standard for asbestos levels in a building. The regulations do not, however, distinguish between white and amphibole asbestos. Banning and removing amphibole asbestos also makes sense. (The amphibole varieties of asbestos are not mined in North America anyway.) The huge sums being spent by Americans to remove white asbestos from buildings could be put to better use. Moreover, the health effects of asbestos substitutes, such as fiberglass, are not known.

Further Reading: Skinner, H. C. W., M. Ross and C. Frondel. 1988. *Asbestos and other fibrous materials—mineralogy, crystal chemistry, and health effects.* New York: Oxford University Press.

Definite Chemical Composition

The fourth criterion, definite chemical composition, is to be expected because of the inherent orderliness of crystalline substances. Essentially, this means that chemical analysis of any sample of a given mineral will always produce the same ratios of elements (in quartz, for example, two atoms of oxygen for every atom of silicon). In other words, the composition of any mineral can be expressed as a chemical formula. Quartz has a composition of SiO_2; halite, $NaCl$. Some leeway is allowed. For instance, we gave the formula for the mineral olivine as $(Mg, Fe)_2SiO_4$. Because magnesium and iron ions are about the same size, they can substitute freely for each other without distorting the crystal structure and significantly altering the properties of the mineral. Some chemical formulas appear much more complex than the crystal structures they represent because several such substitutions can occur.

Chemical analysis can aid the identification of minerals. But, for a variety of reasons, including the cost and difficulty of chemical analysis, physical properties (which reflect chemical composition) are more generally used in identifying minerals.

Physical Properties

If two substances have identical chemical compositions and the same arrangement of atoms, it follows that they should possess the same physical properties. So the final condition is really a consequence of conditions 1 and 4, crystallinity and definite chemical composition.

Characteristic physical properties are the bases by which minerals are usually identified. Later in this chapter we will describe those properties that allow us to identify common minerals without using special equipment. The physical properties generally useful in identification are color, streak, luster, hardness, external crystal form, fracture, cleavage, and specific gravity. Occasionally helpful are the properties of taste, smell, magnetism, and striations.

To identify a sample of an unknown mineral, begin by systematically checking its physical properties. By comparing the properties you find in an unknown mineral with a mineral identification table (such as Appendix A in this book), you should be able to identify the mineral. With a bit of experience, you may get to know the few diagnostic tests for each common mineral and no longer need to refer to an identification table.

The Important Minerals

It is useful to be able to associate the names of important minerals with the physical properties that identify them. Of course, what constitutes an "important" mineral depends on your perspective. To a miner or prospector, an important mineral is one that is commercially valuable (and is, by implication, relatively uncommon). A "rock hound" is interested in collecting any mineral that is pretty or unusual. A gemologist specializes in those varieties of minerals that are of gem quality (diamonds, emeralds, etc.). A mineralogist is a scientist who studies the chemistry and crystallographic structure of minerals. In this

TABLE 2.2

Minerals of the Earth's Crust

Name	Chemical Composition	Type of Silicate Structure or Chemical Group
The most common rock-forming minerals. *(These make up more than 90% of the earth's crust.)*		
Feldspar group		
Plagioclase	Ca and Na Al silicate	Framework silicate
Orthoclase	K Al silicate	Framework silicate
Pyroxene group (augite most common)	Fe, Mg silicate (some with Al, Na, Ca)	Single-chain silicate
Amphibole group (hornblende most common)	Complex Fe, Mg, Al silicate hydroxide	Double-chain silicate
Quartz	Silica	Framework silicate
Mica group		
Muscovite	K Al silicate hydroxide	Sheet silicate
Biotite	K Fe, Mg Al silicate hydroxide	Sheet silicate
Other common rock-forming minerals.		
Silicates		
Olivine (especially common in the mantle)	Mg, Fe silicate	Isolated silicate
Garnet group	Complex silicates	Isolated silicate
Clay minerals group (especially common at the earth's surface)	Complex Al silicate hydroxides	Sheet silicate
Nonsilicates		
Calcite	$CaCO_3$	Carbonate
Dolomite	$CaMg(CO_3)_2$	Carbonate
Gypsum	$CaSO_4 \cdot 2H_2O$	Sulfate
Much less common minerals of commercial value.		
Halite	NaCl	Chloride
Diamond	C	Native element
Gold	Au (gold)	Native element
Hematite	Iron oxide (Fe_2O_3)	Oxide
Magnetite	Iron oxide (Fe_3O_4)	Oxide
Chalcopyrite	Cu, Fe sulfide	Sulfide
Sphalerite	Zn sulfide	Sulfide
Galena	Pb sulfide	Sulfide

book, the minerals we regard as important are those that help us understand the nature of the earth. We are particularly interested in the rock-forming minerals because they make up most of the rocks of the earth's crust.

Of the several thousand identifiable minerals on earth, most are rare and not important to geology (many occur at only a single site on the globe). Only a few hundred are classified as rock-forming minerals. Even most of these are relatively uncommon in comparison with the few minerals that make up the vast bulk of the earth's crust. The five mineral groups listed in the upper third of table 2.2 account for well over 90% of the earth's crust. These are the minerals whose names recur most often in this book. All are silicates.

BOX 2.3

Clay Minerals That Swell

Clay minerals are very common at the earth's surface; they are a major component of soil. There are a great number of different clay minerals. What they all have in common is that they are sheet silicates. They differ in which ions hold sheets together and in the number of sheets "sandwiched" together. Surprisingly, some clay minerals are edible; some are used in the manufacturing of pills. *Kaolinite,* a clay mineral, is the main ingredient in Kaopectate, a remedy for upset stomachs. Popular fast-food chains use clay minerals as a thickener for shakes (you can tell which ones, because the chains do not call them "milk shakes"— they do not use milk).

Montmorillonite is one of the more interesting clay minerals. It is better known as *expansive clay* or *swelling clay.* If water is added to the montmorillonite, the water molecules are adsorbed into the spaces between silicate layers (box figure 1). This results in a large increase in volume, sometimes up to several hundred percent. The pressure generated can be up to 50,000 kilograms per square meter. This is sufficient to lift a good-sized building.

If a building is erected on expansive clay that subsequently gets wet, a portion of the building will be shoved upward. In all likelihood the building will break. Some people regard expansive soils as having resulted in more damage than earthquakes and landslides combined.

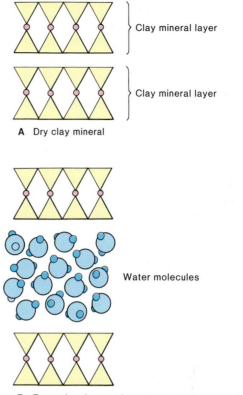

A Dry clay mineral

Water molecules

B Expansion due to adsorption of water

Box 2.3 Figure 1
Expansive clays.

On the other hand, swelling clays can be put to use. Montmorillonite, mixed with water, can be pumped into fractured rock or concrete. When the water is adsorbed, swelling clay expands to fill and seal the crack. The technique is particularly useful where dams have been built against fractured bed rock. Sealing the cracks with expansive clays ensures that water will stay in the reservoir behind the dam.

Quartz may be the only familiar name among the most common minerals, unless you have already had some exposure to geology. However, like people, each mineral has its own character or physical properties. As you become more familiar with them, they will become more than just strange names.

As shown in table 2.2, minerals with similar crystal structures and compositions are grouped under a common name. For instance, the **feldspar group,** the most common minerals in the crust, all have similar crystal structures of oxygen, silicon, and aluminum atoms. Minerals within the group are named according to whether potassium (**orthoclase** or **potassium feldspar**) or sodium and calcium (**plagioclase feldspar**) are incorporated into this basic crystal structure. A geologist doing field work cannot always distinguish the two varieties.

The **pyroxene group** and the **amphibole group,** which are single- and double-chain silicates, respectively, each contain a number of minerals. However, only one mineral from each group is important for our purposes. **Augite** is the most common pyroxene, and **hornblende** is the most common amphibole.

The **mica group** is characterized by minerals with a sheet silicate structure. The two most common micas are biotite and muscovite. **Biotite** is a dark-colored, iron/magnesium-bearing mica. **Muscovite** mica lacks iron and magnesium and is transparent or white.

The **clay mineral group** is also comprised of sheet silicates. Clays are abundant on the earth's surface and in sedimentary rocks but make up only a minor percentage of the crust as a whole.

Nonsilicate minerals include *native elements,* which are minerals composed of only one element. Gold is a native element, as are diamond and graphite, both of which are composed solely of carbon. Other nonsilicates are classified according to the predominant negatively charged ions in their crystal structures. For instance, halite is a chloride because the negatively charged ions in the crystal are Cl^-. If the mineral contains CO_3^{-2} ions, it is a *carbonate.* *Sulfides* have S^{-2} ions, *sulfates* SO_4^{-2}, and *oxides* O^{-2} (but without Si, S, or C bonded to the oxygen atoms).

Nonsilicate minerals are also more abundant on the earth's surface than in the crust as a whole. **Calcite** (calcium carbonate, or $CaCO_3$) is the most common nonsilicate mineral and is usually found at or near the earth's surface. Limestone and marble are rocks composed mainly of calcite.

Ore minerals, or economic minerals, are minerals of commercial value; most are not silicates. Among the ore minerals are iron oxides (the minerals magnetite and hematite) mined for iron and a copper-iron sulfide (the mineral chalcopyrite) that is the main source of copper. Lead and zinc come from galena (lead sulfide) and sphalerite (a zinc sulfide).

The Physical Properties of Minerals

The best approach to understanding physical properties of minerals is to obtain a sample of each of the most common rock-forming minerals named in table 2.2. The properties described can then be identified in these samples.

To identify an unknown mineral, you should first determine its physical properties, then match the properties with the appropriate mineral, using a mineral identification key or chart such as the ones included in Appendix A of this book.

Color

The first thing most people notice about a mineral is its color. Because color is so obvious, beginning students tend to rely too heavily on it as a key to mineral identification. Unfortunately, color is also apt to be the most ambiguous of physical properties. If you look at a number of quartz crystals, for instance, you may find specimens that are white, pink, black, yellow, or purple. Color is extremely variable in quartz and many other minerals because even minute chemical impurities can strongly influence it. Obviously, it is poor procedure to attempt to identify quartz strictly on the basis of color.

In some minerals, however, color is a useful property. Muscovite mica is white or colorless. Most of the **ferro-magnesian minerals** (iron/magnesium-bearing), such as augite, hornblende, olivine, and biotite, are either green or black.

Streak

A pulverized mineral gives a color, called a **streak,** that usually is more reliable than the color of the specimen itself. Scraping the edge of a mineral sample across an unglazed porcelain plate leaves a streak that may be diagnostic of the mineral. For instance, hematite always leaves a reddish brown streak though the sample may be brown or red or silver.

Unfortunately, few of the silicate minerals—the most common minerals—leave an identifying streak because most are harder than the porcelain streak plate.

Luster

The quality and intensity of *light* that is reflected from the surface of a mineral is termed **luster.** (A photograph cannot show this quality.) The luster of a mineral is described by comparing it to familiar substances.

Luster is either *metallic* or *nonmetallic.* A **metallic luster** gives a substance the appearance of being made of metal. Metallic luster may be very shiny, like a chrome car part, or less shiny, like the surface of a broken piece of iron.

Nonmetallic luster is more common. The most important type is **glassy** (also called **vitreous**) luster, which gives a substance a glazed appearance, like glass or porcelain. Most silicate minerals have this characteristic. The feldspars, quartz, the micas, and the pyroxenes and amphiboles all have a glassy luster.

Less common is an **earthy luster.** This resembles the surface of unglazed pottery and is characteristic of the various clay minerals. Some uncommon lusters include *resinous* luster (appearance of resin), *silky* luster, and *pearly* luster.

Hardness

The property of "scratchability," or **hardness,** can be tested fairly reliably. For a true test of hardness, the harder mineral or substance must be able to make a groove or scratch on a smooth, fresh surface of the softer mineral. For example, quartz can always scratch calcite or feldspar. Substances can be compared to **Mohs' hardness scale** (table 2.3), on which ten minerals are designated as standards of hardness. The softest mineral, talc (used for talcum powder because of its softness), is designated as 1. Diamond, the hardest natural substance on earth, is 10 on the scale.

Rather than carry samples of the ten standard minerals, a geologist doing field work usually relies on common objects to test for hardness (table 2.3). A fingernail usually has a hardness of about 2½. If you can scratch the smooth surface of a mineral with your fingernail, the hardness of the mineral must be less than 2½ (figure 2.12). A copper coin, such as a penny, has a hardness between 3 and 4; however, the brown oxidized surface of most pennies is much softer, so using a penny for hardness tests requires caution. A knife blade generally has a hardness slightly greater than 5, but it depends on the particular

T A B L E 2 . 3

Mohs' Hardness Scale

1. Talc	6. Orthoclase feldspar
2. Gypsum	**File**
Fingernail	7. Quartz
3. Calcite	8. Topaz
Copper coin	9. Corundum
4. Fluorite	10. Diamond
5. Apatite	
Knife blade	
Glass	

A

B

C

Figure 2.13
Crystals. (A) Quartz. (B) Feldspar. (C) Garnet.
Photos by C. C. Plummer.

Figure 2.12
**Fingernail (hardness of 2½) easily scratches gypsum
(hardness of 2).**
Photo by C. C. Plummer.

steel alloy used for the blade. A geologist uses a knife blade to distinguish between softer minerals, such as calcite, and harder minerals similar in appearance, such as quartz. Ordinary window glass, usually slightly harder than a knife blade (although some glass, such as that containing lead, is much softer), can be used in the same way as a knife blade for hardness tests. A file (one made of tempered steel for filing metal, not a fingernail file) can be used for a hardness of between 6 and 7.

External Crystal Form

The **crystal form** of a mineral is a set of faces that have a definite geometric relationship to one another. If minerals were always able to develop their characteristic crystal forms, mineral identification would be a much simpler task. In rocks, however, most minerals grow while competing for space with other minerals. In fact, the orderly sets of faces that make up a crystal form can develop only under rather specialized conditions. Specifically, most minerals are able to develop their characteristic crystal faces only if they are surrounded by a fluid that can be easily displaced as the crystal grows. On the other hand, a few minerals, notably garnet, are able to overpower and displace surrounding solid material during growth, so that they almost always develop their characteristic crystal

faces. Some well-developed crystals of common minerals are shown in figure 2.13. (The term **crystal** is used by some to mean a substance exhibiting crystal faces; others use the term interchangeably with "crystalline substance.")

Crystals of minerals have played an important role in the development of chemistry and physics. Steno, a Danish naturalist of the seventeenth century, first noted that the angle between two adjacent faces of quartz is always exactly the same, no matter what part of the world the quartz sample comes from or the color or size of the quartz crystal. As shown in figure 2.14, the angle between any two adjacent sides of the six-sided "pillar" (which is called a prism by mineralogists) is always exactly 120°, while between a face of the "pillar" and one of the "pyramid" faces (actually part of a rhombohedron) the angle is always exactly 141°45'.

The discovery of such regularity in nature usually has profound implications. When minerals other than quartz were studied, they too were found to have sets of angles for adjacent faces that never varied from sample to sample. This observation became formalized as the *law of constancy of interfacial angles,* or *Steno's law.* Later the discovery of X-ray beams and their behavior in crystals confirmed Steno's theory about the structure of crystals.

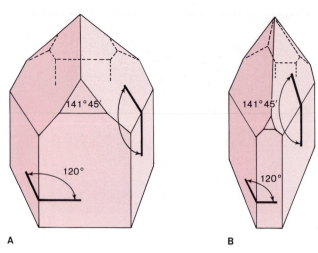

A   B

Figure 2.14
Quartz crystals showing how two interfacial angles remain the same in perfectly proportioned crystals (*A*) and misshapen crystals (*B*).

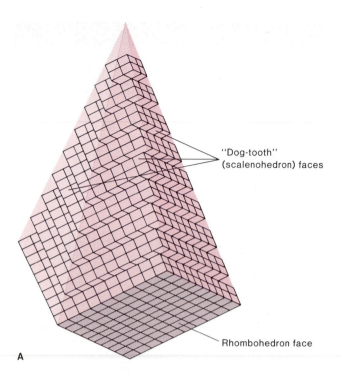

"Dog-tooth" (scalenohedron) faces

Rhombohedron face

A

B

Figure 2.16
Geometric forms built by stacking (*A*) rhombohedrons and (*B*) bricks. (*A*) From a diagram published in 1801 by Haüy, a French mathematician. It shows how rhombohedrons are stacked to result in the "dog-tooth" (scalenohedron) form (the lightly shaded planes) for the mineral calcite. The base represents a rhombohedral face.

Figure 2.15
Quartz City on Mercury by science-fiction artist Frank R. Paul. Crystals have been a recurring theme in science fiction. According to the story, this city on Mercury was built entirely of quartz crystals by insect men.
Amazing Stories, September 1941; reprinted by permission of Ultimate Publishing Company.

Steno suspected that each type of mineral was composed of many tiny, identical building blocks, with the geometric shape of the crystal being a function of how these building blocks are put together. If you are stacking rectangular bricks, the stack can take only a few shapes. Likewise, stacking rhombohedrons in a three-dimensional pattern limits the number of shapes (figure 2.16).

Steno's law was really a precursor of atomic theory, developed centuries later. Our present concept of crystals is that atoms are clustered into geometric forms—cubes, bricks, hexagons, and so on—and that a crystal is essentially an orderly three-dimensional stacking of these tiny geometric forms (see Box 2.4). Halite, for example, may be regarded as a series of cubes stacked in three dimensions (figure 2.17). Because of the cubic "building block," crystal faces on halite are at 90° angles to each other.

BOX 2.4

Unit Cells and Crystal Systems

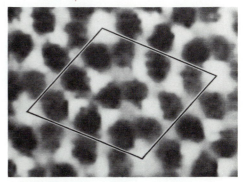

Box 2.4 Figure 1
The surface atoms of silicon metal are shown magnified about ten million times by a Scanning Tunneling Microscope. One side of a unit cell is outlined.
Courtesy IBM Corporation.

The cubes in figure 2.17 and rhombohedrons of figure 2.16 are known to crystallographers as *unit cells,* small volumes of identical shape, size, and orientation that collectively form the crystal. Each unit cell has a chemical composition identical to its neighboring unit cells.

Because a crystal is a three-dimensionally repeating pattern of atoms, each unit cell must share its faces and corners with adjoining unit cells (as shown in figure 2.16). Therefore, unit cells must be six-sided polyhedrons (such as cubes or rhombohedrons). The faces of a unit cell may be square, rectangular, or rhomboidal (see box figure 1). These restrictions result in only six basic shapes of unit cells being possible. The shape of a mineral's unit cell determines its *crystal system.* Halite, with its cubic unit cell, belongs in one system, whereas calcite, with its rhombohedral unit cells belongs in another. For minerals, the most

common of the six crystal systems has a unit cell with two opposite sides being rhombs while the remaining sides are rectangles.

Unit cells cannot, of course, be seen. Their shape must be inferred by the symmetry of patterns that result from X rays beamed through the crystal (explained later in this chapter). Each of the crystal systems will have a particular symmetry to distinguish it from the other systems. When a perfect crystal (such as those in figure 2.13) is found, a trained person can infer its crystal system by visually determining its symmetry.

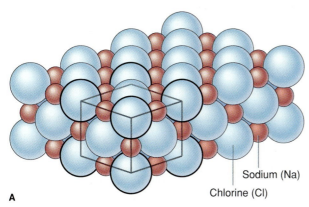

Sodium (Na)

Chlorine (Cl)

Figure 2.17
(*A*) Relationship of cube to halite structure. (*B*) Halite crystal represented as cubes stacked together.

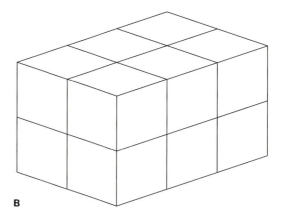

B

Cleavage

The internal order of a crystal may be expressed externally by crystal faces, or it may be indicated by the mineral's tendency to split apart along certain preferred directions. **Cleavage** is the ability of a mineral to break along preferred planes.

A mineral tends to break along certain planes because the bonding between atoms is weaker there. In quartz, the bonds are equally strong in all directions; therefore quartz has no cleavage. The micas, however, are easily split apart into sheets (figure 2.18). If we could look

at the arrangement of atoms in the crystalline structure of micas, we would see that the individual silica tetrahedrons are strongly bonded to one another *within* each of the silicate sheets. The bonding *between* adjacent sheets, however, is very weak; therefore it is easy to pull the mineral apart parallel to the plane of the sheets.

Cleavage is one of the most useful diagnostic tools because it is identical for a given mineral from one sample to another. Cleavage is especially important for identifying minerals in rocks when they are grains.

A

Because of weak bonds, mica splits easily between "sandwiches"

Positive ions, sandwiched between two sheet silicate layers

Sheet silicate layer

B

Figure 2.18
(*A*) Mica pulled apart along cleavage planes. (*B*) Relationship of mica to cleavage. Mica crystal structure is simplified in this diagram.
Photo by C. C. Plummer.

The wide variety of combinations of cleavage and *quality* of cleavage also increases the diagnostic value of this property. Mica has a single direction of cleavage, and its quality is perfect. Other minerals are characterized by one, two, or more cleavage directions; the quality can range from perfect to poor (poor cleavage is very hard for anyone but a well-trained mineralogist to detect). In determining how many cleavage directions a mineral has, be sure that you are counting directions using a single grain or crystal of that mineral (a rock is composed of many grains or crystals).

Three of the most common mineral groups—the feldspars, the amphiboles, and the pyroxenes—have two directions of cleavage (figures 2.19*B* and *C*). In feldspars, the two directions are at angles of about 90° to each other, and both directions are of very good quality. In pyroxenes, the two directions are also at about right angles, but the

quality is only fair. In amphiboles (figure 2.20), the quality of the cleavage is very good and the two directions are at an angle of 56° (or 124° for the obtuse angle).

Halite is an example of a mineral with three excellent cleavage directions, all at 90° to each other. This is called *cubic* cleavage (figures 2.19*D* and 2.21). Halite's cleavage tells us that the bonds are weak in the planes parallel to the cube faces shown in figure 2.17.

Calcite also has three cleavage directions, each excellent. But the angles between them are clearly not right angles. Calcite's cleavage is known as *rhombohedral* cleavage (figure 2.19*E*).

Some minerals have more than three directions of cleavage (figures 2.19*F* and *G*). Diamond has very good cleavage in four directions (ironically, the hardest natural substance on earth can be easily shattered into small cleavage fragments). Sphalerite, the principal ore of zinc, has six directions.

Recognizing cleavage and determining angular relationships between cleavage directions takes some practice. Students new to mineral identification tend to ignore cleavage because it is not as immediately apparent to the eye as color. But determining cleavage is frequently the key to identifying a mineral, so the small amount of practice needed to develop this skill is worthwhile.

Fracture

Fracture is the way a substance breaks where not controlled by cleavage. Minerals such as quartz, olivine, and garnet, which have no cleavage, usually have an irregular fracture. This is the most common type of fracture for minerals.

Some minerals show **conchoidal fracture,** curved fracture surfaces (figure 2.22). These look rather like the inside of a clam shell (hence the name). This type of fracture is sometimes seen in quartz but is more common in noncrystalline substances, such as glass.

Specific Gravity

The density or "heaviness" of a mineral is generally given in comparison to the weight of water. **Specific gravity** is defined as the ratio of the mass of a substance to the mass of an equal volume of water.

Liquid water has a specific gravity of 1. (Ice, being lighter, has a specific gravity of about 0.9.) Most of the common silicate minerals weigh about two and a half times as much as equal volumes of water: quartz has a specific gravity of 2.65; the feldspars range from 2.56 to 2.76. Special scales are needed to determine specific gravity precisely. However, a person can easily distinguish by hand very heavy minerals such as galena (a lead sulfide with a specific gravity of 7.5) from the much lighter silicate minerals.

Gold, with a specific gravity of 19.3, is much heavier than galena. Because of its high specific gravity, gold can be collected by "panning." While the lighter clay and silt particles in the pan are sloshed out with the water, the gold dust lags behind in the bottom of the pan.

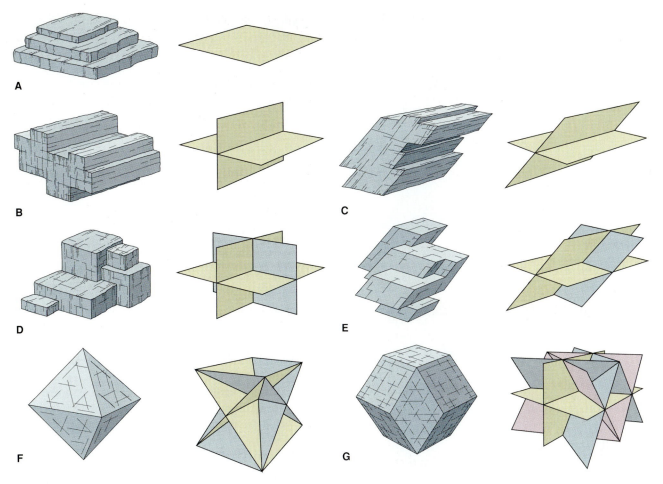

Figure 2.19

Possible types of mineral cleavage. (*A*) One direction of cleavage. (*B*) Two directions of cleavage that intersect at 90° angles. Feldspar is an example. (*C*) Two directions of cleavage that do not intersect at 90° angles. Amphibole is an example. (*D*) Three directions of cleavage that intersect at 90° angles. Halite is an example. (*E*) Three directions of cleavage that do not intersect at 90° angles. Calcite is an example. (*F*) Four directions of cleavage. Diamond is an example. (*G*) Six directions of cleavage. Sphalerite is an example.

Reprinted by permission from R. D. Dallmeyer, *Physical Geology Laboratory Manual*, Dubuque, Iowa: Kendall-Hunt Publishing Company, 1978.

Figure 2.20
Amphibole cleavage as seen under a microscope.

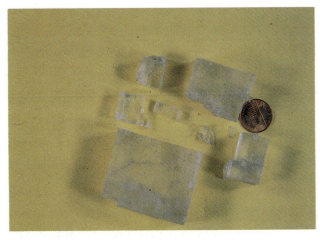

Figure 2.21
Cleavage fragments of halite.
Photo by C. C. Plummer.

Atoms, Elements, and Minerals　**41**

Figure 2.22
Conchoidal fracture in glass.
Photo by C. C. Plummer.

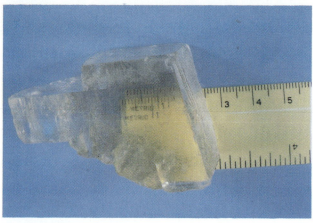

Figure 2.24
Double refraction in calcite. Two images of the letters are seen through the transparent calcite crystal.
Photo by C. C. Plummer.

Figure 2.23
Plagioclase striations.
Photo by C. C. Plummer.

Other Properties

Properties that are useful in only a few instances include taste and smell. Halite obviously tastes salty; few other minerals have any taste at all. An "earthy" smell is characteristic of some clay minerals when they are moistened.

Plagioclase feldspar commonly exhibits **striations**—straight, parallel lines on the *flat* surfaces of one of the two cleavage directions (figure 2.23). The lines appear to be etched by a delicate scriber. In plagioclase, they are caused by a systematic change within the pattern of crystalline structure.

The mineral **magnetite** (an iron oxide) owes its name to its characteristic physical property of being attracted to a magnet. Where large bodies of magnetite are found in the earth's crust, compass needles point toward the magnetite body rather than to magnetic north. Airplanes navigating by compass have become lost because of the influence of large magnetite bodies. Some other minerals are weakly magnetic, but the magnetism cannot be detected except by specialized instruments.

A clear crystal of calcite exhibits an unusual property. If you place a calcite crystal over an image on paper, you will see two images (figure 2.24). This phenomenon is known as *double refraction* and is caused by light splitting into two components when it enters some crystalline materials.

Very specialized equipment is needed to detect some properties. Perhaps most important are the characteristic effects of minerals on X rays, which we can explain only briefly here. X rays penetrate between the atoms in a crystal until they are reflected or "bounced off" planes of atoms within the crystalline pattern. The X rays leave the crystal at precise and measurable angles controlled by the planes of atoms that make up the internal crystalline structure (figure 2.25). The pattern of X rays leaving the crystal can be recorded on photographic film or by various recording instruments. Each mineral has its own pattern of reflected X rays, which serves as an identifying "fingerprint."

Simple Chemical Tests

One chemical reaction is routinely used for identifying minerals. The mineral calcite, as well as some other carbonate minerals (those containing CO_3^{-2}), reacts with a weak acid to produce carbon dioxide gas. In this test, a drop of dilute hydrochloric acid applied to the sample of calcite bubbles vigorously, indicating that CO_2 gas is being formed. Normally this is the only chemical test that geologists do during field research.

BOX 2.5

On Time with Quartz

If you have bought a watch recently, chances are it says "quartz" on it. It may have been quite inexpensive, yet it keeps time much more accurately than even the most expensive spring-driven watch of a decade or more ago. What is a quartz watch, and why is it so much more accurate than the older kinds of watches?

Each quartz watch contains a thin wafer of quartz sliced along a particular crystallographic orientation. The quartz acts as a "pacemaker" to keep the hands or the electronic display precisely on time. The property of quartz that makes this work is called *piezoelectricity,* which means that pressure applied to a crystal creates an electric current (box figure 1A). Quartz is used to make gauges that measure high pressures (for instance, the water pressure on a deep-ocean submersible). The greater the pressure, the stronger the electric current. With tension (outward pulling), the electric current flows in the opposite direction (box figure 1B). Electrons move one way in an electric circuit if the quartz crystal is under pressure and the other way if the quartz is under tension.

If you reverse the process and apply an electric current to a quartz crystal, the crystal changes shape slightly. More precisely, it expands and compresses (box figure 1C); and it will do so extremely rapidly and with remarkable regularity. The expansions and compressions are tiny vibrations that occur at the rate of about 100,000 vibrations per second. Thus for each minute that your watch runs, the quartz wafer in it shrinks and expands 6 million times (the electricity for this is supplied by the watch's battery). These vibrating crystals are so accurate that they are off by no more

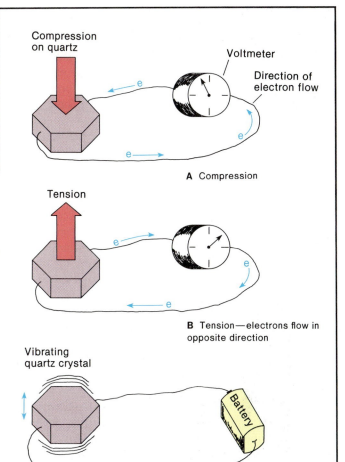

A Compression

B Tension—electrons flow in opposite direction

C Electric current applied to quartz crystal causes it to vibrate rapidly

Box 2.5 Figure 1
Piezoelectricity.

than one vibration out of 10 billion. Precision-manufactured quartz clocks used in observatories lose or gain no more than one second every ten years. Your watch is probably not that accurate because of imperfections in the mechanical parts or the electronic circuitry. But even if your watch is a few seconds off each month, it is a vast improvement over the windup watches, even the finest of which are now obsolete.

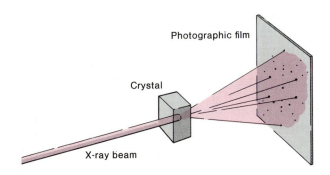

Figure 2.25
An X-ray beam passes through a crystal and is deflected by the atoms into a pattern of beams. The dots exposed on the film reveal the orderly pattern for the particular mineral.

The Rock Cycle

With few exceptions, rocks are made of minerals. You will see how some minerals break down chemically and form new minerals when a rock finds itself in a new physical setting. For instance, feldspars that may have formed at high temperatures deep within the earth can become clay minerals at the earth's surface.

As mentioned in chapter 1, the earth changes because of its internal and external heat engines. If the earth's internal engine had died (and tectonic forces had therefore stopped operating), the external engine plus gravity would long ago have leveled the continents, and the resulting sediment would have been deposited on the sea floor. Everything would be at rest. Nothing would be changing. That is to say, everything would be in equilibrium (and geology would be a dull subject). But this is not the case. The internal and external forces continue to interact, forcing substances out of equilibrium. Therefore, the earth has a highly varied and ever-changing surface. Minerals and rocks change as well.

A useful aid in visualizing these relationships is the **rock cycle** shown in figure 2.26. The three major rock types—igneous, metamorphic, and sedimentary—are shown. As you see, each may form at the expense of another if it is forced out of equilibrium with its physical or climatic environment by either internal or surficial forces.

As described in chapter 1, *magma* is molten rock. *Igneous rocks* form when magma solidifies. If the magma is brought to the surface by a volcanic eruption, it may solidify into an *extrusive* igneous rock. Magma may also solidify very slowly beneath the surface. The resulting *intrusive* igneous rock may be exposed later after uplift and erosion remove the overlying rock (as shown in figure 1.15). The igneous rock, being out of equilibrium, may then undergo weathering, and the debris produced is transported and ultimately deposited (usually on a sea floor) as *sediment*. If the unconsolidated sediment becomes *lithified* (cemented or otherwise consolidated into a rock), it becomes a *sedimentary rock*. As the rock is buried by additional layers of sediment and sedimentary rock, heat and pressure increase. Tectonic forces may also increase the temperature and pressure. If the temperature and pressure become high enough, usually at depths greater than several kilometers below the surface, the original sedimentary rock is no longer in equilibrium and recrystallizes. The new rock that forms is called a *metamorphic rock*. If the temperature gets very high, the rock melts and becomes magma again, completing the cycle.

The cycle can be repeated, as implied by the arrows in figure 2.26. However, there is no reason to expect all rocks to go through each step in the cycle. For instance, sedimentary rocks might be uplifted and exposed to weathering, creating new sediment.

The rock cycle diagram will reappear on the opening pages of chapters 3 through 7. The highlighted portion of the diagram will indicate where the material covered in each chapter fits into the rock cycle.

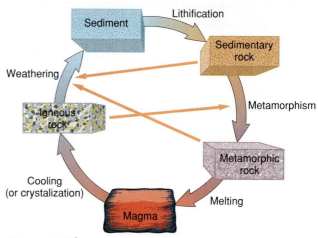

Figure 2.26
The rock cycle.

Summary

Atoms are composed of *protons* ($+$), *neutrons,* and *electrons* ($-$). A given element always has the same number of protons. An atom in which the positive and negative electric charges do not balance is an *ion*.

Ions or atoms bond together in very orderly, three-dimensional structures called *crystals*.

A crystalline substance is considered a mineral (in geologic terms) if it is naturally occurring and inorganic and has a definite chemical composition and characteristic physical properties.

The three most abundant elements in the earth's crust are oxygen, silicon, and aluminum. Most minerals are silicates, with the silica tetrahedron the basic building block.

Feldspars are the most common minerals in the earth's crust. The next most abundant minerals are quartz, the pyroxenes, the amphiboles, and the micas. All are silicates.

Minerals are usually identified by their physical properties. Cleavage is generally the most useful physical property for identification purposes. Other important physical properties are external crystal form, fracture, hardness, luster, color, streak, and specific gravity.

The interaction between the internal and external forces of the earth is illustrated by the rock cycle, a conceptual device relating igneous, sedimentary, and metamorphic rocks to each other, to surficial processes such as weathering and erosion, and to internal processes such as tectonic forces. Changes take place when one or more processes force earth's material out of equilibrium.

Terms to Remember

Questions for Review

1. Answer the following about quartz.
   What elements does it contain?
   What is its chemical formula?
   What type of silicate structure does it have?
   How many protons does each of its elements have? (See Appendix C.)
   Why is it regarded as a mineral?
2. Explain the difference between ionic and covalent bonding.
3. How do the various feldspars differ from one another chemically?
4. Distinguish the following terms:
   silica   silicate
   silicon   silica tetrahedron
5. What is the distinction between cleavage and external crystal form?
6. How would you distinguish the following on the basis of physical properties? (You might refer to Appendix A.)
   feldspar/quartz   amphibole/pyroxene
   muscovite/feldspar   pyroxene/feldspar
   calcite/feldspar

7. Using triangles to represent tetrahedrons, start with a single triangle (to represent isolated silicate structure) and, by drawing more triangles, build on the triangle to show a single-chain silicate structure. By adding more triangles, convert that to a double-chain structure. Turn your double-chain structure into a sheet silicate structure.
8. What major factor controls chemical activity between atoms?
9. What are the three most common elements (by number and approximate percentage) in the earth's crust?
10. What are the next five most common elements?

Questions for Thought

1. Why are there more nonsilicate minerals on the surface of the earth than within the crust?
2. How does oxygen in the atmosphere differ from oxygen in rocks and minerals?
3. Using Appendix A, can you make any generalization relating physical properties to chemical composition? (For instance, carbonate minerals tend to be softer than most silicate minerals.)
4. What happens to the atoms in water when it freezes? Is ice a mineral?
5. How would you expect the appearance of a rock high in iron and magnesium to differ from a rock with very little iron and magnesium?

Supplementary Readings

Blackburn, W. H., and W. H. Dennen. 1988. *Principles of mineralogy.* Dubuque, Iowa: Wm. C. Brown Publishers.

Bloss, F. D. 1971. *Crystallography and crystal chemistry: An introduction.* New York: Holt, Rinehart & Winston.

Chesterman, C. W. 1978. *The Audubon Society field guide to North American rocks and minerals.* New York: Alfred A. Knopf.

Gribble, C. D. 1988. *Rutley's elements of mineralogy.* 27th ed. London: Unwin Hyman.

Klein, C., and C. S. Hurlbut. 1985. *Manual of mineralogy (after J. D. Dana).* 20th ed. New York: John Wiley & Sons.

Mason, B., L. G. Berry, and R. V. Dietrich. 1983. *Mineralogy: Concepts, descriptions, determinations.* 2d. ed. New York: W. H. Freeman.

Prinz, M., G. Harlow, and J. Peters. 1978. *Simon & Schuster's guide to rocks and minerals.* New York: Simon & Schuster.

Zim, H. S., and P. R. Shaffer. 1967. *Rocks and minerals.* New York: Golden Press.

Volcanic eruptions, while awesome natural spectacles, also provide important information on the workings of the earth's interior. Volcanic eruptions vary in nature and in degree of explosive violence. A strong correlation exists between the chemical composition of the magma (or lava) of a volcanic event and the violence of an eruption. The size and shape of volcanoes and lava flows and their pattern of distribution on the earth's surface also correspond to the chemistry of the lavas.

Understanding volcanism provides a background for theories relating to mountain building, the development and evolution of continental and oceanic crust, and how the crust is deformed. Our observations of volcanic activity fit nicely into plate tectonic theory. The kinds of eruptions that take place along spreading centers are radically different from those associated with converging plate boundaries. In the next chapter we complete the story of igneous rocks and what they tell us about the earth's interior by extending our discussion to rocks that form from magma solidifying underground.

The names of and a classification scheme for the common extrusive rocks are introduced here. With practice with real rocks, you should be able to identify the common extrusive rocks using the information in this chapter and the identification table in Appendix B.

It is worth bearing in mind how the rocks and processes discussed in this chapter fit into the rock cycle (see diagram). As you progress to later chapters, referring to the rock cycle will help you maintain a perspective relative to geology's "big picture."

3

Volcanism and Extrusive Rocks

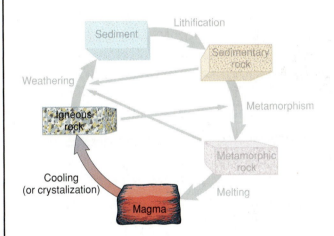

Lava flow in Hawaii.
© Doug Sherman.

BOX 3.1

Mount St. Helens, 1980—North America's Eruption of the Century

Box 3.1 Figure 1
Mount St. Helens before the 1980 eruption, as seen from the north.
Photo by Jim Nieland, U.S. Forest Service.

It has been a decade since Mount St. Helens blew up. Before 1980 not many people living in Washington or Oregon thought that the volcanic cones of the Cascade Range might be hazardous. Peaks such as Mount Rainier, Mount Hood, and Mount Baker (and about ten others) were considered merely parts of the scenery. These picturesque snow-capped cones towering over evergreen forests seemed unlikely to change. Geologists thought otherwise. They knew it was exceedingly unlikely that volcanoes that had shown intermittent activity over hundreds of thousands of years were dead. In fact, California's Lassen Peak at the southern extreme of the Cascade Range had erupted from 1914 to 1917.

Mount St. Helens (box figure 1), in southern Washington, had last erupted in 1857. Geologists were not surprised when it erupted in 1980. Its inactivity had lasted little more than a century—a trivial amount of time in geology. Between March and May of 1980 Mount St. Helens was transformed from a serene, snow-covered cone to a darkened, gouged-out stump of a peak; from a thing of beauty to a killer.

The 1980 eruptions were preceded by thousands of small earthquakes. On March 27 ash and steam eruptions began and continued for the next six weeks. Water that had seeped into the volcano became superheated steam. As in an exploding boiler, the water vapor expanded suddenly and blasted through the overlying rock, carrying pulverized rock skyward. The steam explosions and the pattern of earthquakes indicated that magma was working its way upward beneath the volcano.

Box 3.1 Figure 2
A steam and ash eruption. Mount St. Helens from the south. Mt. Rainier, another Cascade volcano is to the left.
Photo by Lyn Topinka, U.S. Geological Survey, David A. Johnston Cascades Volcano Observatory, Vancouver, Washington.

During this early stage of steam and ash eruptions, St. Helens provided a fine show (box figure 2). "Volcano fever" spread, particularly in Portland, Oregon, the largest city with a view of the eruptions. Ice-cream dishes and cocktails resembling volcanoes and T-shirts and commemorative trinkets sold briskly to residents and tourists alike.

A few years earlier a team from the U.S. Geological Survey completed a study of the potential hazards of renewed eruption, based on the peak's geologic record for the past 38,000 years. On the advice of the survey geologists, U.S. Forest Service and state

The May 18, 1980 eruption of Mount St. Helens was a spectacular reminder of the energy in the earth's interior. If plate tectonic theory is correct, parts of North America are overriding a portion of the Pacific ocean floor. Some of the mechanical energy is transformed to heat energy along the subduction zone (as described in chapter 1). At depth, previously solid rock melts at high temperatures. At least some of the **magma** (molten rock or liquid that is mostly silica) works its way upward to the earth's surface to erupt. Magma does not always reach the earth's surface before solidifying, but when it does it is called **lava.**

Lava may erupt quietly or, as in the case of Mount St. Helens, violently. Volcanic activity, or **volcanism,** is not restricted to the eruption of lava. Rock fragments may erupt instead of lava (as happened during the early 1980 eruptions of Mount St. Helens). Gases also play an important role in volcanism. Gas explosions at Mount St. Helens were particularly violent; in other cases, the escaping gases merely force magma out of the ground and produce lava flows.

Material blown out of a volcano is called **pyroclastic debris,** or **tephra**, both terms meaning rock fragments produced by volcanic explosions. Tephra, such as the volcanic officials blocked off the potentially most hazardous areas on and near the mountain. During the steam and ash stage of the eruptions, geologists were concerned about the hazards of large mudflows. Numerous small mudflows had already been observed. The mudflows were slurries of loose ash and water (from rainfall or melting glacier ice or snow) that flowed down the volcano. The fear was that a large mudflow might continue along a stream valley beyond the mountain and overwhelm a populated area. For this reason, water levels were lowered in reservoirs on the flanks of the volcano so that mudflows might be impounded.

The steam-driven eruptions continued. But geologists could not determine exactly how deep beneath the surface the magma was or when (or even if) it would break through to the surface. Another indication that the magma was moving upward was that the peak was swelling—like a balloon being inflated. The northern flank of the mountain bulged outward at a rate of 1.5 meters per day. Bulging continued until the surface of the northern slope was displaced outward over a hundred meters from its original position (see box figure 3). The bulge was gravitationally unstable and the Geological Survey warned of another hazard—a truly large landslide.

Although geologists knew that the magma was responsible for the bulge, they considered the potential landslide far more dangerous than a magmatic eruption. They knew that the magma might blast out of the north side of the volcano, but based on the past behavior of St. Helens, they thought that the first eruptions of magma would probably be through the summit, with the energy released relatively harmlessly skyward.

The huge blast (box figure 4) that marked the arrival of the magma at the surface on May 18 destroyed the summit and north flank of St. Helens almost instantly. Seconds after the eruption began, an area extending outward 10 kilometers was stripped of all vegetation and soil. Downslope forests were leveled; from the air the leafless, scorched trunks looked like thousands of jackstraws strewn on the ground.

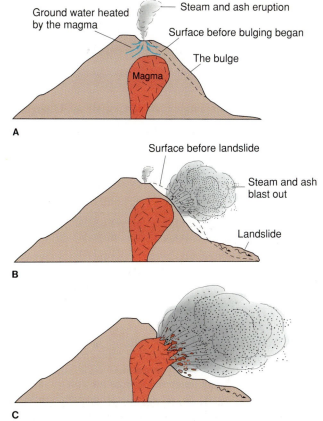

Box 3.1 Figure 3
Sequence of events at Mount St. Helens, May 18, 1980. (A) Just before the May 18, 1980 eruption. (B) The landslide relieves the pressure on the underlying magma. (C) Magma blasts outward.

Although the sequence of events was exceedingly rapid, it is now clear what happened (box figure 3). A fairly strong earthquake shook the bulging north slope loose. The resulting landslide stripped away the rock that sealed in the magma. With the protective lid removed, gases that had been dissolved in the magma were suddenly released. On a grand scale, it was comparable to shaking a bottle of warm beer or pop

ash that littered much of the Pacific Northwest in 1980, and rock formed by solidification of lava are collectively regarded as **extrusive rock,** surface rock resulting from volcanic activity.

The most obvious landform created by volcanism is a **volcano,** a hill or mountain formed by the extrusion of lava or ejection of rock fragments from a vent. However, volcanoes are not the only landforms created by volcanism. Very fluid lavas may flow out of the earth and flood an area, solidifying into a nearly horizontal layer of extrusive rock. Successive layers of lava flows may accumulate into a lava plateau.

Volcanism

Volcanic activity is important to geology for several reasons. Landforms are created and portions of the earth's surface built up. Less commonly, landforms are destroyed by violent eruptions. Volcanoes are important to the science of geology because they provide clues about the nature of the earth's inaccessible interior and help us understand how the earth's internal processes work. By studying the magma, gases, and rocks from eruptions, we can infer the chemical conditions as well as the temperatures and pressures within the earth's crust or underlying mantle.

Box 3.1 *continued*

Box 3.1 Figure 4
May 18, 1980. The side of the volcano has been blasted away as magmatic explosions continue.
© 1980 Keith Ronnhom.

and removing the top. The magma exploded into a froth while blasting out of the north flank of the volcano. The mass of gases and volcanic rock fragments roared down the side of the volcano and continued beyond its base, destroying trees and all organic matter close to the volcano and knocking over forests farther downslope.

For the next thirty hours, exploding gases continued to propel frothing magma and volcanic ash vertically into the atmosphere (box figure 5). The dense, mushroom-shaped cloud of ash was blown northeastward by the winds. Fallout of ash went on for days, causing serious damage as far away as Montana. In some towns, there was nightlike darkness at midday. Crops were destroyed. Breathing became difficult. Automobile engines were destroyed by fine volcanic particles that were sucked into cylinders and wore away working parts.

Damage was estimated to be in the hundreds of millions of dollars, and 63 people were killed or presumed dead. The death toll might have been much worse. For comparison, 30,000 people were killed during an eruption of Mt. Pelée (described later in this chapter), and 23,000 lives were lost in the 1985 volcanically triggered mudflow in Colombia.

Countless lives were saved, no doubt, by the U.S. Geological Survey advice to public officials. Just as important, the officials heeded the advice (perhaps it helped that Washington's governor was a scientist) and resisted pressure from the public, notably the people that wanted access to the closed lands. Only two days before the catastrophic eruption, a group whose homes were in the area north of the peak staged a protest demanding access. They were escorted in and out of the area to pick up their belongings. We assume they knew little about geology. They owe their lives to scientists whose knowledge kept them out of an area that would be utterly destroyed.

Since the eruption on May 18, 1980, there have been occasional but less intense eruptions. A volcanic dome has grown as magma periodically welled upward into the floor of the new, large crater.

A decade has gone by since the big eruption and the mountain is quiet. We do not know when eruptions will resume—perhaps in a few years, perhaps in a century. Nor do we know which of the other Cascade volcanoes will erupt next. Seattle, with Mount Rainier nearby, and Portland, with Mount Hood almost in its suburbs, could be endangered by eruptions from these mountains. Vancouver might suffer if either Mount Garibaldi in British Columbia or Mount Hood in Washington erupt (see figure 3.5).

Effects on Humans

Volcanism is also significant in human affairs. Its effects can be catastrophic or, surprisingly, beneficial.

Hawaii One region where the overall effects of volcanism have been favorable to the human inhabitants is Hawaii. Occasionally a field or village is overrun by outpourings of lava. During the 1980s, Kilauea volcano was especially active. Over 800 million cubic meters of lava erupted—enough to cover a four-lane highway extending from the east coast to the west coast of North America with a 10-meter-thick layer of lava. Nearly forty houses were destroyed during this period, but no one was killed or injured. Nevertheless, the weathered volcanic ash and lava produce excellent fertile soil. Moreover, Hawaii's periodically erupting volcanoes (which are relatively safe to watch) are great spectacles that attract both tourists and scientists, benefiting the islands' economy (figure 3.1).

Were it not for volcanic activity, Hawaii would not exist. The islands are the crests of a series of volcanoes that have been built up from the bottom of the Pacific Ocean over millions of years (the vertical distance from the summit of Mauna Loa volcano to the ocean floor greatly exceeds the height of Mount Everest). When lava flows into the sea and solidifies, more land is added to the islands. Hawaii is, quite literally, growing.

Box 3.1 Figure 5
Vertical eruptions blast volcanic ash high into the atmosphere.
Photo by Austin Post, U.S. Geological Survey, David A. Johnston Cascades Volcano Observatory, Vancouver, Washington.

Geothermal energy In some other areas of geologically recent volcanic activity, underground heat generated by volcanism is harnessed for human needs. In Italy, Mexico, New Zealand, Argentina, and California, geothermal installations produce electric power. Steam or superheated water trapped in layers of hot volcanic rock is tapped by drilling and then piped out of the ground to power turbines that generate electricity. Naturally heated geothermal fluids can also be tapped for space or domestic water heating or industrial use, as in paper manufacturing.

Volcanic catastrophes While the eruption of Mount St. Helens in 1980 was indeed awesome, its effects were not nearly as disastrous as a number of historical eruptions elsewhere in the world. For instance, the Roman city of Pompeii and at least four other towns near Naples in Italy were destroyed in A.D. 79 when Mount Vesuvius erupted (figure 3.2). Before that time vineyards on the flanks of the apparently "dead" volcano extended to the summit. Pompeii was buried under 5 to 8 meters of hot ash from the surprise eruption of Vesuvius. Seventeen

Figure 3.1
Volcanic eruptions in Hawaii, 1969. A lava fountain is supplying the lava cascading over the cliff.
Photo by D. A. Swanson, U.S. Geological Survey.

Figure 3.2
Pompeii with Mt. Vesuvius in the background.
Photo by R. W. Decker.

centuries later the town was rediscovered. Excavation revealed molds of people suffocated by the ashfall, many with facial expressions of terror. This eruption was not the end of Vesuvius's activity. The volcano was active almost continually from 1631 to 1944, with major twentieth-century eruptions in 1906, 1929, and 1944.

The island of Krakatoa in the western Pacific, composed of three apparently inactive volcanoes, erupted in 1883 with the force of several hydrogen bombs. This Indonesian island, which formerly rose 800 meters above sea level, was blown apart. Only one-third of the island remained after the eruption. An estimated 13 cubic kilometers of rock collapsed into the subsurface magma chamber that had been emptied by the eruption, leaving an underwater depression 300 meters deep where the major part of the island had been. The explosion was heard 5,000 kilometers away. On nearby Java, tens of thousands of people died as a result of the giant sea waves (tsunamis) generated by the explosion.

A similar series of explosions in prehistoric time (about 6,600 years ago) was at least partially responsible for creating Crater Lake in Oregon. Volcanic debris covering more than a million square kilometers in Oregon and neighboring states has been traced to those eruptions. The original volcano, named Mount Mazama by geologists, is estimated to have been about 2,000 meters higher than the present rim of Crater Lake. Collapse of the volcano, as well as explosions, accounts for the depression the present-day lake occupies (figures 3.3 and 3.4).

The southern Cascade Mountains, where Crater Lake is located, have been built up by eruptions over the past 30 to 40 million years (figure 3.5). Only the youngest peaks (those built within the past 2 million years), such as Mount St. Helens, Mount Rainier, Mount Shasta, and Mount Hood, still stand out as cones. As we know from Mount St. Helens, any of these could again become active.

Eruptive Violence and Physical Characteristics of Lava

What determines the degree of violence associated with volcanic activity? Why can we state confidently that active volcanism in Hawaii poses only slight danger to humans when violent explosions, such as at Mount St. Helens, occur in the Cascade Mountains? Whether eruptions are very explosive or relatively "quiet" is largely determined by two factors: (1) the amount of gas in the lava or magma and (2) the ease or difficulty with which the gas escapes to the atmosphere. The **viscosity**, or resistance to flow, of a lava determines how easily the gas escapes. The more viscous the lava and the greater the volume of gas trying to escape, the more violent the eruption will be. Later we will show how these factors not only determine the degree of violence of an eruption but also influence the shape and height of a volcano.

Scientific Investigation of Volcanism

Volcanoes and lava flows, unlike many other geologic phenomena, can be observed directly, and samples can be collected without great difficulty (at least for the quiet, Hawaiian type of eruption). We can measure the temperature of lava flows, collect samples of gases being given off, observe the lava solidifying into rock, and take rock samples into the laboratory for analysis and study. By comparing rocks observed solidifying from lava with similar ones from other areas of the world (and even with samples from the moon) where volcanism is no longer active, we can determine the nature of volcanic activity that took place in the geologic past.

Gases From active volcanoes we have learned that most of the gas released during eruptions is water vapor, which condenses as steam. Other gases, such as sulfur dioxide, hydrogen sulfide (which smells like rotten eggs), carbon dioxide, and hydrochloric acid, are given off in lesser amounts with the steam.

Figure 3.3
Crater Lake, Oregon.
Photo by D. A. Rahm, courtesy Rahm Memorial Collection, Western Washington University.

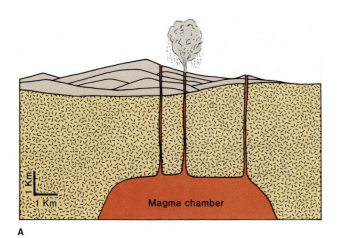

A

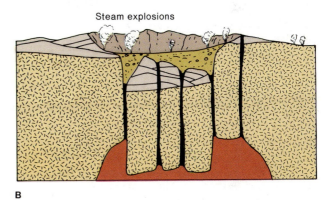

B

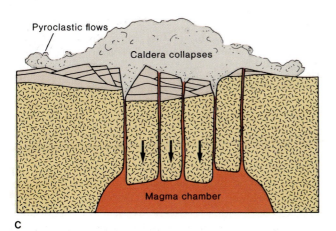

C

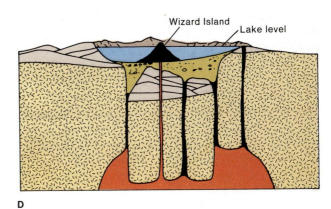

D

Figure 3.4
The development of Crater Lake. (*A*) Cluster of overlapping volcanoes form. (*B*) Collapse into the partially emptied magma chamber is accompanied by violent eruptions. (*C*) Volcanic activity ceases, but steam explosions take place in the caldera. (*D*) Water fills the caldera to become Crater Lake, and minor renewed volcanism builds a cinder cone (Wizard Island).
After C. Bacon U.S. Geological Survey.

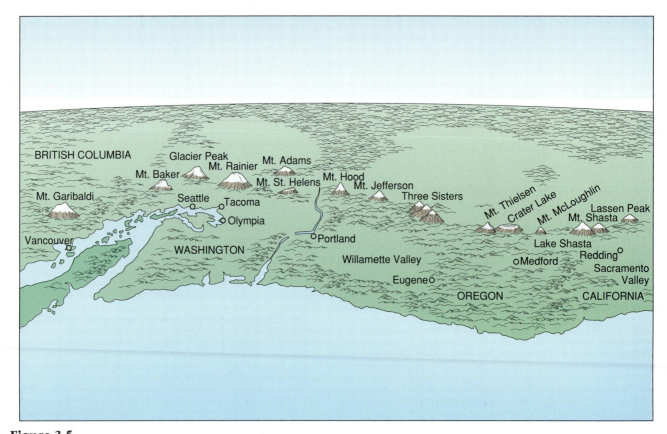

Figure 3.5
The Cascade volcanoes.
Map by Gary Tands, from *Fire and Ice, the Cascade Volcanoes* by
S. L. Harris, courtesy of The Mountaineers.

Chemistry of Volcanic Rocks

Chemical analyses show that silica (SiO_2) is the most abundant component of virtually all volcanic rocks. The amount of silica, however, can vary from about 45% to about 75% of the total weight of volcanic rocks (figure 3.6). The variations between these extremes account for striking differences in the appearance and mineral content of the rocks as well as in the behavior of the parent lava.

Mafic rocks   Rocks with a silica content close to 50% (by weight) are considered *silica-poor,* even though silica is, by far, the most abundant constituent. Chemical analyses show that the remainder is composed mostly of the oxides of aluminum (Al_2O_3), calcium (CaO), magnesium (MgO), and iron (FeO and Fe_2O_3). (These oxides generally combine to form the silicate minerals described in the chapter titled "Atoms, Elements, and Minerals"—they are not usually found as isolated oxides in a rock.) Rocks in this group are called **mafic**—silica-poor igneous rocks with a relatively high content of magnesium, iron, and calcium. (The term *mafic* comes from magnesium and ferric.) The most common mafic extrusive rock is *basalt,* which is dark in color.

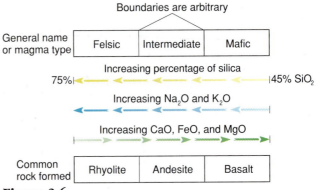

Figure 3.6
Relationship between volcanic rock names and the composition of the lava from which they formed.

Felsic rocks   At the other extreme, the *silica-rich* (65% or more of silica) rocks tend to have only very small amounts of the oxides of calcium, magnesium, and iron. The remaining 25% to 35% of these rocks is mostly aluminum oxide (Al_2O_3) and oxides of sodium (Na_2O) and potassium (K_2O). These are called **felsic rocks**—silica-rich igneous rocks with a relatively high content of potassium and sodium (the name comes from the generally high amount of *feldspar*, which crystallizes from the potassium, sodium, aluminum, and silicon oxides). *Rhyolite,* the most abundant volcanic rock with a felsic composition, is light in color because of the low iron and magnesium content.

Intermediate rocks Rocks with a chemical content between that of felsic and mafic are classified as **intermediate rocks.** *Andesite,* usually medium to dark gray in color, is the most common intermediate volcanic rock.

Relationship of minerals to chemistry The chemical composition of the magma determines which minerals and how much of each will crystallize to form an igneous rock. Therefore, mafic, intermediate, and felsic types of rock, being of different chemical compositions, are made up of quite different combinations of minerals. How mineral content is used to identify extrusive rocks is described later in this chapter.

Viscosity of Lava

The degree of violence with which a volcano erupts is, as mentioned earlier, closely related to the viscosity of the lava. The two most important factors that influence viscosity are (1) the temperature of the lava relative to the cooler temperature at which it solidifies and (2) the silica content of the lava. If the lava being extruded is considerably hotter than its solidification temperature, the lava is less viscous (more fluid) than when its temperature is near its solidification point. Temperatures at which lavas solidify range from about 700°C for felsic rocks to 1500°C for mafic rocks.

Mafic lavas, being low in silica, tend to flow easily. Conversely, felsic lavas, which are high in silica, are very viscous and flow sluggishly. Felsic lavas are more viscous because even before they have cooled enough to allow crystallization of minerals, silica tetrahedrons have begun to form small frameworks in the lava. Although too few atoms are involved for the structures to be considered crystals, the total effect of these silicate structures is to make the liquid lava more viscous, much the way that flour or cornstarch thickens gravy.

Types of Volcanoes

Volcanic material that is ejected from and deposited around a central vent produces the conical shape typical of volcanoes. The **vent** is the opening through which an eruption takes place. The **crater** of a volcano is a basinlike depression over a vent at the summit of the cone (figure 3.7). Material is not always ejected from the central vent. In a **flank eruption,** lava pours from a vent on the side of a volcano.

Figure 3.7
Crater on Cotopaxi volcano in Ecuador.
Photo by C. C. Plummer.

Figure 3.8
The top of Mauna Loa, a shield volcano in Hawaii, and its summit caldera. The smaller depressions are pit craters. In the distance is Mauna Kea, another shield volcano which last erupted about 3,000 years ago.
Photo by D. W. Peterson, U.S. Geological Survey.

A **caldera** (figure 3.8) is a volcanic depression much larger than the original crater. (The most famous caldera in the United States is misnamed "Crater Lake.") A caldera can be caused when a volcano's summit is blown off by exploding gases, as occurred at Mount St. Helens in May 1980, or, as in the case of Crater Lake, when the crater floor collapses into a vacated magma chamber beneath the volcano (figure 3.4).

The three major types of volcanoes (shield, cinder cone, and composite) discussed below are markedly distinct from one another in size, shape, and composition. Although volcanic domes are not cones, they can be considered volcanoes and are also discussed in this section.

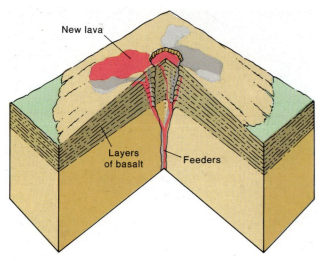

Figure 3.9
Cutaway view of a shield volcano.

Figure 3.10
Pahoehoe from a 1972 eruption in Hawaii.
Photo by D. W. Peterson, U.S. Geological Survey.

Shield Volcanoes

Shield volcanoes are broad, gently sloping cones constructed of solidified lava flows. During eruptions, the lava spreads widely and thinly due to its low viscosity. Because the lava flows from a central vent, without building up much near the vent, the slopes are usually between 2° and 10° from the horizontal, producing a volcano in the shape of a flattened dome or "shield" (figure 3.9).

The islands of Hawaii are essentially a series of shield volcanoes built upward from the ocean floor by intermittent eruptions over millions of years (figure 3.8). Although spectacular to observe, the eruptions are relatively nonviolent because the lavas are fairly fluid (less viscous). By implication, then, the shield volcanoes of the Hawaiian Islands are composed of a series of layers of basalt (low in silica and relatively high in iron, magnesium, and calcium oxides).

There are two types of basalt flows, each of which solidifies with a characteristic surface. Both types have Hawaiian names. **Pahoehoe** (pronounced *pah-hoy-hoy*) is a lava flow characterized by a ropy or billowy surface (figure 3.10). The surface is formed by the quick cooling and solidification from the surface downward of a lava flow or pool of lava that was fully liquid. By contrast, flowing basalt that is cool enough to have partially solidified moves as a slow, pasty mass. Its largely solidified front is shoved forward as a pile of rubble. A flow such as this is called **aa** (pronounced *ah-ah*) and solidifies with a spiny, rubbly surface (figure 3.11). The surface of a solidifying lava flow sometimes develops a minor feature called a **spatter cone**, a small, steep-sided cone built from lava sputtering out of a vent (figure 3.12). When a small local pocket of gas is trapped in a cooling lava flow, the gas seeks to escape and belches the lava up out of a vent through the already hard surface of the flow. Falling lava plasters itself onto the developing cone and solidifies. The sides of a spatter cone can be very steep, but the height rarely exceeds 10 meters.

Figure 3.11
An aa flow in Hawaii, 1983.
Photo by J. D. Griggs, U.S. Geological Survey.

Figure 3.12
A spatter cone (approximately 1 meter high) erupting in Hawaii.
Photo by J. B. Judd, U.S. Geological Survey.

B

Figure 3.13
**Cerro Negro, a cinder cone in Nicaragua. (A) View from the air;
(B) night-time eruption of pyroclastics at the summit.**
Photo A by Mark Hurd Aerial Surveys Corp. courtesy California Division of
Mines and Geology; Photo B by R. W. Decker.

Cinder Cones

A **cinder cone** is a volcano constructed of loose rock fragments ejected from a central vent (figure 3.13). In contrast to the gentle slopes of shield volcanoes, cinder cones
commonly have slopes of about 30°. Most of the ejected
material lands near the vent during an eruption, building
up the cone to a peak. The steepness of slopes of accumulating loose material is limited by gravity to about 33°.
Cinder cones tend to be very much smaller than shield
volcanoes. Few of them exceed a height of 500 meters.

The fragments of volcanic rock that make up the
cinder cones are *pyroclastic* (from the Greek *pyro,* "fire,"
and *clastic,* "broken"). *Pyroclasts,* or *tephra,* the fragments formed by volcanic explosion, can be almost any
size. *Dust* and *ash* are the finest particles; *cinders* range
from about 4 to 32 millimeters; *bombs* and *blocks* are large
pyroclasts. When solid rock has been blasted apart by a
volcanic explosion, the pyroclastic fragments are *angular,*
with no rounded edges or corners. If lava is ejected into
the air, a molten blob becomes streamlined during flight,
solidifies, and falls to the ground as a **bomb,** a spindle- or
lens-shaped pyroclast (figure 3.14).

Figure 3.14
Volcanic bombs.
Photo by J. P. Lockwood, U.S. Geological Survey.

Local concentrations of gas in a magma cause eruptions that build cinder cones. Cinder cones are not necessarily related to the silica content of the extrusive rock
or lava. Some cinder cones are found in Hawaii, perched
on the flanks of the much larger shield volcanoes. Although there are numerous cinder cones in parts of the

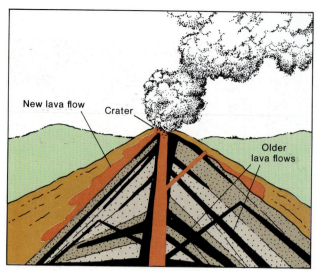

Figure 3.15
Cutaway view of a composite volcano. Stippled layers are pyroclastics.

world where the crust is geologically active, such as in western North America, most of the cinder cones themselves are apparently extinct.

The life span of an active cinder cone tends to be short. The local concentration of gas is depleted rather quickly during the eruptive periods. Moreover, as landforms, cinder cones are temporary features in terms of geologic time. They are made of unconsolidated material and are eroded away relatively easily.

Composite Volcanoes

A **composite volcano** (also called **stratovolcano**) is one constructed of alternating layers of pyroclastics and rock solidified from lava flows (figure 3.15). The slopes are intermediate in steepness compared with cinder cones and shield volcanoes. Pyroclastic layers build up steep slopes as debris collects near the vent, just as in cinder cones. However, subsequent lava flows partially flatten the profile of the cone as the downward flow builds up the height of the flanks more than the summit area (figure 3.16). The solidified lava acts as a protective cover over the loose pyroclastic layers, making composite volcanoes less vulnerable to erosion than cinder cones.

Composite volcanoes are built up over long spans of time. Eruption is intermittent, with hundreds or thousands of years of inactivity separating a few years of violent activity. During the quiet intervals between eruptions, composite volcanoes may be eroded by running water, landslides, or glaciers. These surficial processes tend to alter the surface, shape, and form of the cone. But because of their long lives and relative resistance to erosion, composite cones can become very large. Aconcagua, a composite volcano in the Andes, is 6,960 meters above sea level and the highest peak in the western hemisphere.

The extrusive material that builds composite cones is predominantly of intermediate composition, although there may be local minor felsic and mafic eruptions. Therefore,

Figure 3.16
Mount Shasta, a composite volcano in California. Shastina on Mt. Shasta's flanks is a subsidiary cone, largely made of pyroclastics. Note the lava flow that originated on Shasta and extends beyond the volcano's base.
Photo by B. Amundson.

andesite is the rock most associated with composite volcanoes. If the temperature of an andesite lava is considerably above the temperature at which it would be completely solidified, the relatively nonviscous fluid flows easily from the crater down the slopes. On the other hand, if enough gas pressure exists, an explosion may litter the slopes with pyroclastic andesite, particularly if the lava has fully or partially solidified and clogged the volcano's vent.

The composition as well as eruptive history of individual volcanoes can vary considerably. For instance, Mount Rainier is composed of 90% lava flows and only 10% pyroclastic layers. On the other hand, Mount St. Helens was built mostly from pyroclastic eruptions—reflecting a more violent history. As would be expected, the composition of the rocks formed during the 1980 eruptions of Mount St. Helens is somewhat higher in silica than average for Cascade volcanoes.

Distribution of composite volcanoes Nearly all the larger and better known volcanoes of the world are composite volcanoes. They tend to align along two major belts on the earth (figure 3.17). The **circum-Pacific belt,** or "Ring of Fire," is the larger. The Cascade Range volcanoes described earlier make up a small segment of the circum-Pacific belt.

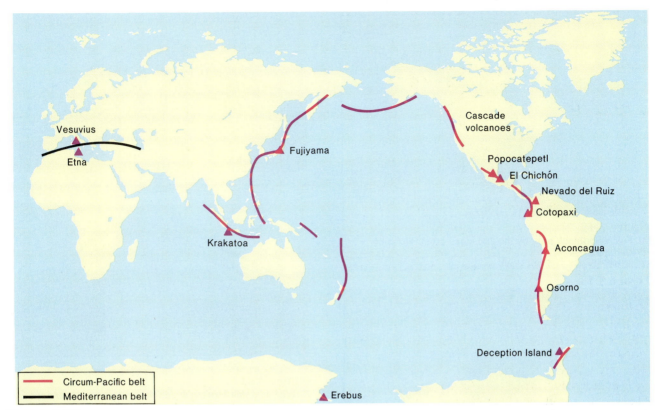

Figure 3.17
Map of the world showing the major volcanic belts.

Legend:
— Circum-Pacific belt
— Mediterranean belt

Labels on map: Vesuvius, Etna, Fujiyama, Krakatoa, Cascade volcanoes, Popocatepetl, El Chichón, Nevado del Ruiz, Cotopaxi, Aconcagua, Osorno, Deception Island, Erebus

Several composite volcanoes in Mexico rise higher than 5,000 meters, including Orizaba (third highest peak in North America) and Popocatepetl. Cortez sent his men to climb Popocatepetl during his conquest of Mexico in 1521. They were lowered into the smoking crater and returned with sulfur needed to make gunpowder.

Mexico also possesses North America's most recently active killer volcano. In 1982, two years after the Mount St. Helens eruption, an apparently insignificant, jungle-covered, 1,000-meter-high volcano called *El Chichón* (meaning "the breast") erupted with a series of violent explosions. Towns near the previously inactive volcano were buried by the heavy fall of ash or blasted by searing, gas-charged ash flows. The number of dead could only be estimated and was placed in the thousands. Most of the volcano's explosive force was directed upward (rather than outward as at St. Helens), propelling exceptional amounts of fine particles and gas into the upper atmosphere. The large amounts of sulfurous gases that reached the uppermost atmosphere resulted in less of the sun's radiation reaching the earth's surface, and climatologists thought that world weather patterns were significantly altered by this solar filtering effect.

The circum-Pacific belt includes many volcanoes in Central America, western South America (including Nevado del Ruiz in Colombia), and Antarctica. Deception Island in the Antarctic, a caldera similar to Crater Lake, began erupting again in 1967, forcing the evacuation of three scientific research stations that were set up on the island. Mount Erebus, also in Antarctica, is the southernmost active volcano in the world (figure 3.18).

Figure 3.18
Mount Erebus, Antarctica, the southernmost active volcano in the world.
Photo by Philip R. Kyle.

The western portion of the Pacific belt includes volcanoes in New Zealand, Indonesia, the Philippines, and Japan. The beautifully symmetrical Fujiyama, in Japan, is probably the most frequently painted volcano in the world (figure 3.19). The northernmost part of the circum-Pacific belt includes active volcanoes on Alaska's Aleutian Islands.

The second major volcanic belt is the **Mediterranean belt,** which includes Mount Vesuvius. An exceptionally violent eruption of Mount Thera, an island in the Mediterranean, may have destroyed an important site of early Greek civilization. (Some archeologists consider Thera the original "lost continent" of Atlantis.)

Volcanism and Extrusive Rocks   **59**

Figure 3.19
Mt. Fuji, woodblock print by the Japanese artist Hiroshige (1797–1858).

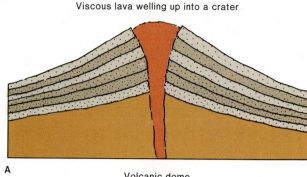

Viscous lava welling up into a crater

A

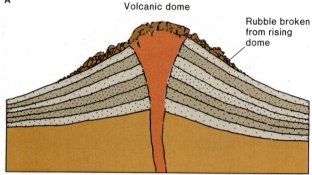

Volcanic dome

Rubble broken from rising dome

B

Figure 3.21
A volcanic dome forming within the crater of a cinder cone.

Figure 3.20
The volcanic dome in Mount St. Helens from the north, May, 1982.
Photo by Lyn Topinka, U.S. Geological Survey, David A. Johnston Cascades Volcano Observatory, Vancouver, Washington.

Volcanic Domes

Volcanic domes are steep-sided, dome- or spine-shaped masses of volcanic rock formed from viscous lava that solidifies in or immediately above a volcanic vent. A volcanic dome grew within the caldera of Mount St. Helens after the climactic eruption of May 1980 (figure 3.20). This was expected because of the exceptionally high viscosity of the lava from the St. Helens eruptions. In 1983 alone, the dome increased its elevation by 200 meters. Most of the viscous lavas that form volcanic domes are very high in silica (felsic). They solidify as *rhyolite* or, less commonly, *andesite* if minerals crystallize, or as *obsidian* (volcanic glass) if no minerals crystallize.

Because the thick, pasty lava that squeezes from a vent is too viscous to flow, it builds up a steep-sided dome or spine (figure 3.21). Some volcanic domes act like champagne corks, keeping gases from escaping. If the plug is removed or broken, the gas escapes suddenly and violently. Some of the most destructive volcanic explosions known have been associated with volcanic domes. However, volcanic domes are much less common than the three major types of volcanic cones.

Sometimes magma solidifies beneath a vent and clogs the throat of a volcano formed earlier, and the mass is shoved upward like a piston by the gas trapped beneath. The spine that grew around the time of the 1902 eruption of Mount Pelée on the Caribbean island of Martinique is a famous example. A virtually solid plug of lava was pushed up out of the crater as rapidly as 20 meters per day by gases trying to escape. As the rock mass rose, large chunks of hot rock broke off, carving the plug into a spine that attained a height of over 300 meters before collapsing into a pile of rubble (figure 3.22).

Before the spine developed on Pelée, gas explosions blew out clouds of red-hot ash and dust called **nuées ardentes** (French for "glowing clouds"). The climax came suddenly on the morning of May 8, when great exploding clouds descended like an avalanche down the mountainside, engulfing the port town of St. Pierre (figure 3.23) in minutes and incinerating everything in its path. Thirty thousand people were burned to death or suffocated (of the four survivors, one was a condemned prisoner in a poorly ventilated dungeon). During its sweep through the city, the avalanche reached temperatures of 700°C.

Figure 3.22
The spine of Mount Pelée.
Historical Pictures Service, Chicago.

Figure 3.23
The ruins of St. Pierre in 1902. Mount Pelée is in the clouds.
Photo by Underwood & Underwood, courtesy Library of Congress.

Such a very hot flow of pyroclastics is called a **glowing avalanche.** Being heavier than air, a glowing avalanche, composed of fiery debris mixed with hot gases, descends swiftly downward and outward. After it comes to a halt, the particles may weld together to form a hard rock called a **welded tuff.**

There is a plateau in eastern California built of successive deposits of welded tuff. Nearby are Mono Craters, a number of cinder cones whose craters are filled by volcanic domes. Some may have erupted as little as two centuries ago. Could a Pelée type of eruption occur here? The possibility is strong. In 1982, the U.S. Geological Survey issued a volcanic warning after scientists determined that a body of magma was moving upward in the earth's crust and seemed to be within a few kilometers of the surface. They could not tell how rapidly the magma body was rising nor precisely when an eruption would occur. The warning stated only that the eruption would be within the next fifty years.

Lava Floods

Not all extrusive rocks are associated with volcanoes. Large regions of the continents are now covered by extrusive rock that did not originate in a volcano. This lava was apparently so fluid, or nonviscous, that it flowed almost as easily as water, building up no cone at all around the vents. Such lava is, of course, mafic (low in silica). It was extruded at a temperature well above the melting point of basalt, the rock that the lava formed as it solidified.

Plateau basalts were produced during the geologic past by vast outpourings of lava. The Columbia plateau area of Washington, Idaho, and Oregon (figures 3.24 and 3.37), for example, is constructed of layer upon layer of basalt, in places as thick as 3,000 meters. Each individual flood of lava added a layer generally between 15 and 100 meters thick and hundreds of square kilometers in extent.

Basalt layers give the landscape a striking appearance in most places where they are exposed. Instead of stacked-up slabs or tablets of solid, unbroken rock, the individual layers appear to be formed of parallel, vertical columns, mostly six-sided. This characteristic of basalt is called **columnar structure** or **columnar jointing** (figure 3.25). The columns can be explained by the way in which basalt contracts as it cools *after* solidifying. Basalt solidifies completely at temperatures below about 1200°C. The hot layer of rock then continues to cool to temperatures normal for the earth's surface. Like most solids, basalt contracts as it cools. The layer of basalt is easily able to accommodate the shrinkage in the narrow vertical dimension; but the cooling rock cannot "pull in" its edges, which may be hundreds of kilometers away. The tension fractures the rock into an orderly hexagonal pattern.

Submarine Eruptions

Submarine eruptions, notably those occurring along mid-oceanic ridges, almost always consist of mafic lavas that create basalt. In fact, basaltic rock, thought to have been formed from lava erupting along mid-oceanic ridges, or solidifying underground beneath the ridges, makes up virtually the entire crust underlying the oceans. In a few places—Iceland, for example—volcanic islands rise above the otherwise submerged system (see Box 3.2).

Figure 3.24
Basalt layers in the Columbia plateau, Washington.
Photo by P. Weis, U.S. Geological Survey.

A

B

Figure 3.25
Columnar jointing at Devil's Postpile, California. (*A*) The
columns as seen from above. (Scratches were caused by
glacial erosion, as explained in chapter 12.) (*B*) Side view.
Photos by C. C. Plummer.

Figure 3.26
Pillow basalt in Iceland.
Photo by R. W. Decker.

Pillow Basalts

Figure 3.26 shows **pillow structure**—rocks, generally basalt, occurring as pillow-shaped rounded masses closely
fitted together. From observations of submarine eruptions
by divers, we know how the pillow structure is produced.
Elongate blobs of lava break out of a thin skin of solid
basalt over the top of a flow that is submerged in water.
Each blob is squeezed out like toothpaste, and its surface
is chilled to rock almost instantly. A new blob forms as
more lava inside breaks out. Each new pillow settles down
on the pile, with little space left in between. When pillow
basalts are found exposed in mountain ranges in various
parts of the world, they are usually indications of submarine eruption in the geologic past followed, much later,
by uplift.

BOX 3.2

An Icelandic Community Battles a Volcano—and Wins

A

The inhabitants of Iceland are in daily contact with volcanism. Volcanic heat mitigates the harsh cold of these sub-Arctic islands through the geothermal water that provides warmth for homes.

In 1963 submarine eruptions resumed in this region after a period of no volcanic activity. Within a few months, the volcano Surtsey had built itself up above sea level, giving Iceland a new island.

Ten years later, in 1973, eruptions began on one of the tiny Vestmann Islands south of Iceland's main island. Heimaey, one of Iceland's most important and prosperous fishing ports, was nearly destroyed by the eruptions. First ash and then lava flows erupted from a new fissure on the edge of the town. The ashfall quickly buried nearby parts of the town and accumulated on the streets and houses. Flaming spatter ignited roofs, and houses burned. Within a week a cinder cone 150 meters high marked the edge of the town. Some residents were evacuated, but the townspeople who remained, along with rescue workers, worked hard to save the rest of the town. Ash was shoveled off roofs before a heavy accumulation could cause them to collapse. Windows were sealed off with sheet metal to prevent drifting ash from breaking them. Despite the constant volcanic activity, water and electrical services were maintained. The greatest fear of the residents was that the harbor would be blocked by a 40-meter-high tongue of rubbly *aa* lava that began moving slowly away from the main flow and toward the harbor, engulfing twelve houses in one day. The town's livelihood depended on the survival of the harbor.

To halt the lava's advance, fire hoses and large pumps were put to work to cool down the lava with sea water. This unprecedented step apparently worked; the lava front hardened enough so that new lava was deflected elsewhere. The harbor was saved, but the town and its surroundings bore little resemblance to their former selves.

Five months after they began, the eruptions stopped. Ash cleared from the streets and buildings was used to make an extension of the airport runway. Nearly all the evacuated residents returned to resume their lives on the volcano-built island.

B

Box 3.2 Figure 1
(*A*) Lava fountaining behind the town on Heimaey. The glow behind the town in the left part of the photo is the lava flow advancing to the harbor. (*B*) Spraying the advancing flow with water.
Photo *A* © Solarfilma. Photo *B* U.S. Geological Survey.

The town benefited by having a unique heating system engineered and built for it. By 1982, nearly every home in the town was being heated by the lava flows. Water is pumped to the lava flow where it percolates through the pyroclastic layers overlying the flows. The water turns to steam and rises. The steam is collected and goes to a central station. From there, hot water condensed from the steam is circulated through individual homes. The cooled down water is recycled through the lava flows.

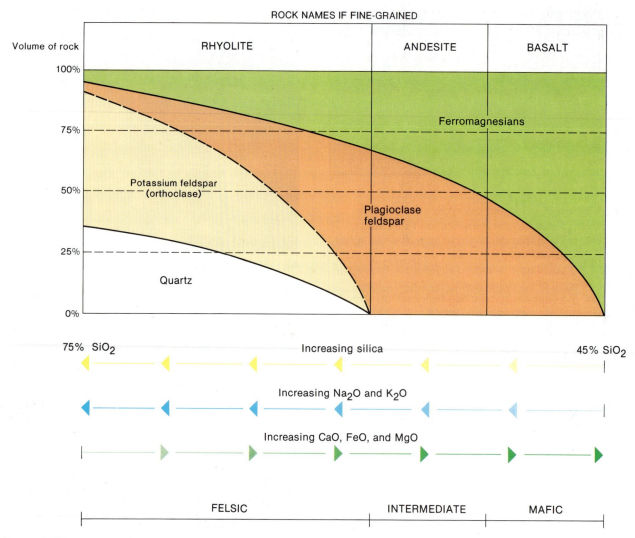

Figure 3.27
Classification of the common extrusive rocks. Rocks with names based entirely on texture (obsidian, breccia, pumice, tuff) are not included. Coarse-grained equivalent rocks are added to this diagram in the next chapter.

Identification of Extrusive Rocks

Extrusive rocks are named and identified on the basis of their composition and texture.

Composition

The amount of silica in a lava strongly determines not only the viscosity of lava and the violence of eruptions but also which particular rock is formed. Fortunately, we do not have to make a chemical analysis of a volcanic rock to determine whether it is a rhyolite, an andesite, or a basalt. We can indirectly determine the approximate chemical composition of most igneous rocks by identifying the minerals present and the relative abundance of each mineral in the rock. Because extrusive igneous rocks are generally fine-grained, a microscope is usually needed for precise identification of the component minerals. In most cases, however, we can guess the probable mineral content by noting how dark or light in color an extrusive rock is. Most felsic rocks are light-colored because they contain more feldspar and quartz (which have a high percentage of

silica) and fewer of the dark minerals (which contain iron and magnesium). Mafic rocks, on the other hand, tend to be dark because of the abundance of ferromagnesian minerals.

Rhyolite, a felsic rock, is usually cream-colored, tan, or pink; it is made up mostly of feldspar but always includes some quartz. Note that the rhyolite portion of figure 3.27 is larger than the areas shown for andesite and basalt. Geologists commonly subdivide this portion of the classification system. For example, *dacite,* the rock associated with the 1980 St. Helens eruptions, contains more ferromagnesian minerals and plagioclase but less potassium feldspar and quartz than the average rhyolite. In our classification system, dacite corresponds to the right portion of the area in figure 3.27 assigned to rhyolite.

A **basalt** has a relatively low amount (about 50% by weight) of silica. Much of that silica is bonded to iron and magnesium to form ferromagnesian minerals, such as *olivine* or *augite,* which are dark green or black. The remaining silica plus aluminum is bonded predominantly with calcium to form calcium-rich *plagioclase feldspar*

Figure 3.28
Obsidian.
Photo by C. C. Plummer.

(which tends to be darker gray than the white or pink potassium or sodium feldspars associated with felsic rocks). Basalt does not contain quartz because no silica is left over after the other minerals have formed. Because of the preponderance of dark minerals in basalt, this rock is usually dark gray to black.

Andesite, which crystallizes from an intermediate lava, can be recognized by its moderately gray or green color. It is this color because a little over half the rock is composed of light- to medium-gray plagioclase feldspar, while the rest of its components are ferromagnesian minerals. Rarely is there sufficient silica in the lava for quartz to form in an andesite.

The chemical and mineralogical relationships of the common extrusive igneous rocks are shown in figure 3.27. This classification chart can be used along with the rock identification table in Appendix B to identify extrusive rocks.

Textures

Grain size   Some extrusive rocks (such as obsidian and pumice) are classified solely on the basis of their textures, but most are classified by composition *and* texture. *Grain* size is a rock's most important textural characteristic. For the most part, extrusive rocks are fine-grained or else made of glass.

A **fine-grained rock** is one in which most of the mineral grains are smaller than 1 millimeter. In some, the individual minerals are distinguishable only with a microscope. **Obsidian** (figure 3.28), which is volcanic glass, is one of the few rocks that is not composed of minerals. A fine-grained or glassy texture distinguishes extrusive rocks from most intrusive rocks. Intrusive igneous rocks solidify underground and generally are coarse-grained.

Two critical factors determine grain size during the solidification of igneous rocks: rate of cooling and viscosity. If lava cools rapidly, the atoms have time to move only a short distance; they bond with nearby atoms, forming only small crystals. With extremely rapid or almost instantaneous cooling, individual atoms in the lava are "frozen" in place, forming glass rather than crystals.

A

B

Figure 3.29
(*A*) Porphyritic andesite. (*B*) Photomicrograph of the same rock. The photograph was taken through a microscope that uses polarized light. The black and white striped grains are plagioclase, green grains are ferromagnesian minerals.
Photos by C. C. Plummer.

Grain size is controlled to a lesser extent by the viscosity of the lava. The atoms in a very viscous lava cannot move as freely as those in a very fluid lava. Hence, a rock formed from viscous lava is more likely to be obsidian or of finer grains than one formed from more fluid lava. Most obsidian, when chemically analyzed, has a very high silica content and is the chemical equivalent of rhyolite.

Porphyritic textures   Volcanic extrusive rock that does not have a uniformly fine-grained texture throughout is described as porphyritic. A **porphyritic rock** is one in which large crystals are enclosed in a *matrix* (or *groundmass*) of much finer-grained minerals or obsidian. The large crystals are termed **phenocrysts.** A porphyritic rock looks rather like raisin bread; the matrix or groundmass is the bread, the phenocrysts are the raisins. In the porphyritic andesite shown in figure 3.29, phenocrysts of feldspar and ferromagnesian minerals are enclosed in a matrix of crystals too fine-grained to distinguish with the naked eye but visible under a microscope.

Two stages of solidification are represented in porphyritic texture. Slow cooling takes place while the magma is underground. Minerals that form at higher temperatures crystallize and grow to form phenocrysts in the still partly fluid magma. If the entire mass is then erupted, the remaining liquid portion cools rapidly at the earth's surface and forms the fine-grained matrix.

Extraterrestrial Volcanic Activity

Astrogeology Box 3.1 Figure 1
The smooth area at the bottom of the photograph is part of a lunar maria. Note the rilles and wrinkle ridges.
NASA.

Volcanic activity has been one of the most common geologic processes operating on the Moon and on several other bodies in the solar system. Approximately one-third of the Moon's surface consists of nearly circular, dark-colored, smooth, relatively flat lava plains. The lava plains, found mostly on the near side of the Moon, are called *maria* (singular, *mare;* literally, ''seas'') (box figure 1). This kind of terrain represents a significant period in the Moon's early history when large lava flows flooded parts of the Moon's surface, filling in low places including some earlier formed meteorite impact craters.

Many of the largest craters (called basins) are partially or completely filled by extensive maria lava flows. In some instances, as in Oceanus Procellarum, the central depressions of the basins have been completely filled and the lava has covered the surrounding areas as well, completely hiding the ringlike structures that normally surround basins of this size.

Rilles are elongate trenches or cracklike valleys found mainly in the smoother portions of the lunar maria. They range in length from a few kilometers to hundreds of kilometers and are often quite deep. Some are arc-shaped or crooked and may be collapsed lava tubes, channels eroded by ash flows, or fractures along which gas has escaped.

Wrinkle ridges are wrinkles on maria surfaces, frequently found near the maria edges. Some are hundreds of kilometers long and several kilometers wide. The ridges may be compressional features produced as the lava cooled and subsided or may be the surface expression of underlying intrusions.

There are a few shield *volcanoes* on the Moon, notably a group of small volcanoes named the Marius Hills. These volcanoes are probably not active at the present time.

Mercury also has extensive areas covered by lava flows, corresponding to the lunar maria. Mercury's maria, like those on the Moon, apparently formed during a widespread outpouring of lava that filled the ringed basins, covered the floors of many smaller craters, and spread to form a thin and discontinuous layer over larger areas outside the basins. Wrinkle ridges are very common in the maria on Mercury.

The surface of Venus is covered with thick layers of clouds, and its surface features must be observed by the use of indirect methods such as radar. One of the most intriguing features discovered is Maxwell Montes, a high mountainous area over 11 kilometers high, which is higher than Mount Everest. A caldera-like feature suggests that Maxwell Montes may be volcanic. Other large shield volcanoes, some in chains along a great fault, have been identified, and molten lava lakes may exist. Several of the volcanoes may be active, emitting large amounts of sulfur gases and being responsible for the almost continuous lightning that has been observed by spacecraft.

Nearly half the planet Mars may be covered with volcanic material. There are areas of extensive lava flows similar to the lunar maria and a number of volcanoes, some with associated lava flows.

Most of the northern half of Mars seems to have been flooded by a series of thin lava flows during an early period of extensive volcanism. These lava flows partially or completely filled low areas such as ringed basins and covered over large numbers of craters formed earlier. Wrinkle ridges are common.

The largest Martian volcano, Olympus Mons (box figure 2), is three times the height of Mount Everest and wider than the state of California. Its caldera is more than 65 kilometers across. Olympus Mons and at least eighteen other large Martian volcanoes are shield volcanoes, probably composed of basalt.

There are a number of smaller Martian volcanoes with steeper slopes than the shield volcanoes. These steeper slopes imply that the lava that formed these volcanoes was more viscous than that of the shield volcanoes.

Astrogeology Box 3.1 Figure 2
The rampart crater Arandas on Mars. Material around it appears to have flowed along the surface.
NASA.

Martian volcanoes are widely spread over the planet, but the largest shield volcanoes appear to be concentrated in one area, within an uplifted region known as the Tharsis Ridge.

Martian volcanoes vary widely in age. This can be determined by comparing crater densities on their surfaces and by noting differences in the amount of erosion they have undergone. The youngest volcanoes are Olympus Mons and the other shield volcanoes of the Tharsis Ridge area.

Jupiter's moon Io is the only object in the solar system besides the Earth that is known to have currently active volcanoes (box figure 3). Ten have been observed by spacecraft and at least seven of those have erupted for more than four months. Material very rich in sulfur compounds is thrown at least 500 kilometers into space at speeds of up to 3,200 kilometers per hour. This material often forms umbrella-shaped clouds as it spreads out and falls back to the surface. Lakes of molten sulfur and huge multicolored (black, yellow, red, orange, and brown) lava flows of sulfur or a sulfur-silicate mixture are common. More than 100 calderas larger than 25 kilometers across have been observed, including one that vents sulfur gases. Clouds of sulfur and sulfur dioxide often form and precipitate reddish sulfur dioxide "snow." The energy source for Io's volcanoes may be the gravitational pulls of Jupiter and two of its other larger satellites, causing Io to heat up much as a piece of wire will do if it is flexed continuously. Below Io's solid crust a layer of molten sulfur may exist. Neptune's moon Triton may have active volcanoes that erupt nitrogen.

Astrogeology Box 3.1 Figure 3
A volcanic eruption on Jupiter's moon Io.
NASA.

Figure 3.30
Photomicrograph of a tuff. Fragments of different rocks and minerals are angular and variously colored.

Fragmental textures   When loose pyroclastic material (ash, bombs, etc.) is cemented or otherwise consolidated, the new rock is called *tuff* or *volcanic breccia,* depending on the size of the fragments. A **tuff** (figure 3.30) is a rock composed of fine-grained pyroclastic particles (ash and dust). A **volcanic breccia** is a rock formed of larger pieces of volcanic rock (cinders, blocks, bombs).

How welded tuff is formed from a glowing avalanche was described earlier. A volcanic breccia is produced when pyroclastic debris on the flank of a volcano is overrun or engulfed by a lava flow that then solidifies.

Textures due to trapped gas   A magma deep underground is under high pressure, generally high enough to keep all its gases in a dissolved state. On eruption, the pressure is suddenly released and the gases come out of solution. This is analogous to what happens when a bottle of beer or soda is opened. Because the drink was bottled and capped under pressure, the gas (carbon dioxide) is in solution. Uncapping the drink relieves the pressure, and the carbon dioxide separates from the liquid as gas bubbles. If you froze the newly opened drink very quickly, you would have a piece of ice with small, bubble-shaped holes. Similarly, when a lava solidifies while gas is bubbling through it, holes are trapped in the rock, creating a distinctive *vesicular* texture. A **vesicle** is a cavity caused by gas in a lava. A vesicular rock has the appearance of Swiss cheese (whose texture is also caused by trapped carbon dioxide gas). *Vesicular basalt* is quite common (figure 3.31). *Scoria,* a highly vesicular basalt, actually contains more gas space than rock.

In more viscous lavas, where the gas cannot escape as easily, the lava is churned into a froth (like the head of a glass of beer). When cooled quickly, it forms **pumice** (figure 3.32), a frothy glass with so much void space that it floats in water. Powdered pumice is used as an abrasive because it can scratch metal or glass.

Figure 3.31
Vesicular basalt.

Figure 3.32
Pumice.

The Source of Lava

Geologists agree that magma, which becomes lava if it reaches the earth's surface, forms locally within the outer 100 kilometers or so of the earth. Pockets of magma form, apparently, from the melting of rock within the crust or the uppermost part of the mantle. Why do rocks within the crust or upper mantle melt? What controls the composition of the magmas and, by inference, the rocks that form from them? Plate tectonics explains much of these and other questions.

Any theory that deals with the origin and composition of magma must explain the data summarized in this chapter. The important aspects of volcanism that must be

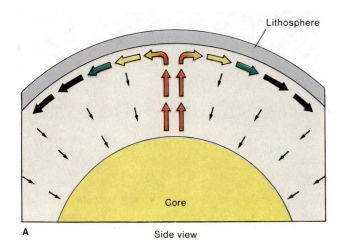

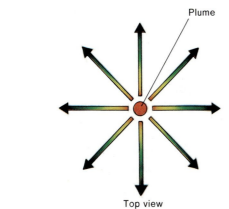

Figure 3.33
Mantle plume, showing schematic pattern of heat flow. (*A*) Side view. (*B*) Top view.

accounted for include the following: (1) basalt, which makes up virtually the entire oceanic crust, is the most abundant extrusive rock, but felsic rocks are the most common intrusives; (2) geologically young andesite volcanoes are remarkably aligned along the circum-Pacific and Mediterranean belts; (3) most contemporary submarine volcanism occurs along the mid-oceanic ridges.

Plate Tectonics and the Origin of Basalt

Geologists agree that most basaltic magma is produced by partial melting of the asthenosphere. The *asthenosphere,* as described in chapter 1, is the plastic zone of the mantle beneath the rigid *lithosphere* (the upper mantle and crust that make up a plate). **Partial melting** means that only the chemical components of the asthenosphere with the lowest melting temperatures liquify, while minerals with higher melting temperatures remain crystalline.

The probable reason the asthenosphere is plastic or "soft" is that temperatures there are only slightly lower than the temperatures required for partial melting of mantle rock that is deeply buried. Partial melting may take

place either because extra heat raises the temperature above the melting points for some components of the asthenosphere or because pressure is reduced. In general, the greater the pressure on a substance, the higher the temperature required to melt that substance.

Hotter material from the lower mantle wells upward, either by convection currents or by more localized hot **mantle plumes,** narrow columns of hot mantle rock that rise from deep within the earth and spread outward beneath the lithosphere (figure 3.33). Spreading centers may overlie upwelling convection currents or a series of hot mantle plumes or perhaps both. Pressure decreases as the plastic, mantle rock travels upward. With the decreasing pressure, the melting temperatures for the rock's components are lowered. When the material reaches the upper part of the asthenosphere, reduced pressure has lowered the melting temperature for the rock to the point where partial melting takes place. Basaltic magma is created.

As shown in figure 3.34, the asthenosphere is much closer to the earth's surface beneath a spreading center. Newly formed magma separates from the solid residue and collects in a magma chamber beneath the crust. Some of the magma works its way into the cracks or series of cracks that develop where the two plates of the diverging boundary are moving away from each other.

Part of the basaltic magma erupts along the surface of the spreading center (usually a mid-oceanic ridge crest), and part of the magma solidifies in the underlying fissures. As shown in figure 3.34, the process is relatively continuous as spreading keeps pulling the lithosphere apart and creating new fissures. Magma in the chamber is replenished by continued upwelling and partial melting of material coming from the depths of the mantle.

In 1974 geologists in a remarkable program called project FAMOUS (an acronym for French American Mid-Ocean Underseas Study) made deep dives to the mid-Atlantic ridge in small research submarines. A submarine actually entered the fracture that splits the ridge, and scientists observed pillow structure on the floor and walls of the rift (figure 3.35).

Great stacks of basalt layers on land, such as the plateau basalts of eastern Washington and Oregon and part of Idaho, are believed to have formed because a spreading center developed and a continent began splitting. Fractures extended through the continental lithosphere. Basaltic magma from the asthenosphere rose through the fissures to flow over the surface of the continent. For some reason, after a few million years the process stopped. According to plate tectonics, the split between the present continents of North America and Europe began in a similar manner. However, unlike in the Pacific Northwest, spreading continued for hundreds of millions of years, creating the present Atlantic Ocean basin. The spreading continues today, widening the Atlantic slightly each year.

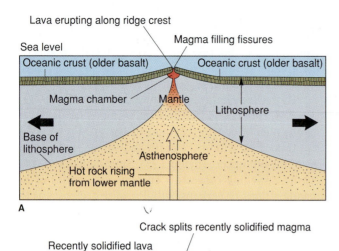

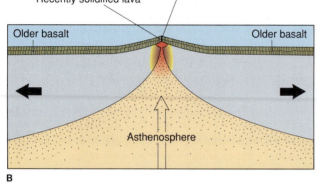

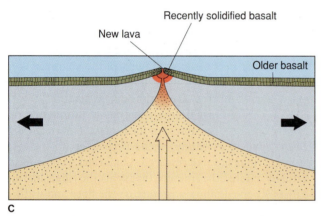

Figure 3.34
Basaltic crust is continuously being created along a spreading center. (*A*) A fissure develops due to sea-floor spreading. Magma fills the fissure and erupts on the sea floor. (*B*) Magma solidifies and a new fissure develops through the recently formed igneous rock. (*C*) Magma again fills the fissure and erupts on the sea floor.

Not all basalt erupts along spreading centers. Hawaiian eruptions occur within the Pacific plate far from a mid-oceanic ridge. Most geologists attribute the basaltic magma to a mantle plume or "hot spot" beneath the plate (figure 3.36). As the Pacific plate overrides the plume (currently beneath the island of Hawaii), magma is created in the asthenosphere and eventually erupts, adding new land to the Hawaiian islands.

Figure 3.35
Pillow basalt on a mid-oceanic ridge. Photo taken from a submersible.
Courtesy of Woods Hole Oceanographic Institution.

Plate Tectonics and the Origin of Andesite

According to plate tectonic theory, the remarkable alignment of composite volcanoes along the circum-Pacific belt is due to subduction at converging plate boundaries. For example, the source of magma for Mount St. Helens and the other Cascade volcanoes is the Juan de Fuca plate, a small plate whose spreading center is in the Pacific a little west of the Washington and Oregon coastline (figure 3.37). The Juan de Fuca plate moves eastward and is subducted beneath the North American continent, which is moving westward. The oceanic lithosphere, capped by basalt with a veneer of overlying marine sediment, sinks diagonally beneath the continental lithosphere (figure 3.38). As it descends, the upper portion of the oceanic plate comes in contact with increasingly hotter material. Once it has slid beneath the continental lithosphere, it continues under the asthenosphere. Here temperatures are high enough to cause partial melting of the upper portion of the descending plate. The depth at which this happens is fairly uniform, as implied by the alignment of active volcanoes parallel to the Pacific coastline. The magma, being less dense than the surrounding rocks, rises toward the earth's surface, eventually erupting and building composite cones.

Although rhyolite and basalt eruptions sometimes take place when composite cones are built, the vast majority of eruptions involve andesitic lava or tephra. Geologists are not certain why andesite predominates. Perhaps the melt is a combination of basalt and silica-rich sedimentary material (subducted on top of the basalt of the oceanic crust). Alternatively, only part of the basaltic crust may melt, leaving behind a more ferromagnesian-rich residue. Or perhaps a mafic magma is created because of partial melting of the wedge of asthenosphere overlying the descending crust. As the mafic magma ascends, it absorbs felsic material as it passes through the lower part of the continental crust.

To understand more thoroughly how magma is generated, we need to know about *intrusive rocks,* igneous rocks that solidify underground. In the next chapter we describe intrusive rocks and expand on the topic of how magma forms.

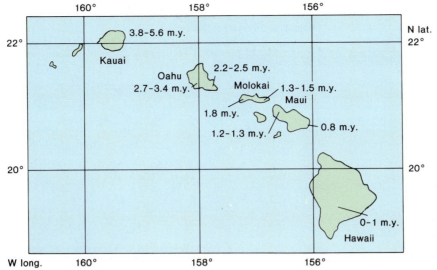

Figure 3.36
How plate motion over a "hot spot" has created the Hawaiian
Islands. Map showing the sequence of ages of volcanic rocks
in the Hawaiian island group. Kauai, the northernmost Hawaiian
island, formed 3.8–5.6 million years ago. Plate motion over the
"hot spot" has continued creating the various volcanic islands.
Present eruptions take place along the edge of the island of
Hawaii.

From I. McDougall, 1964, *Geological Society of American Bulletin*.

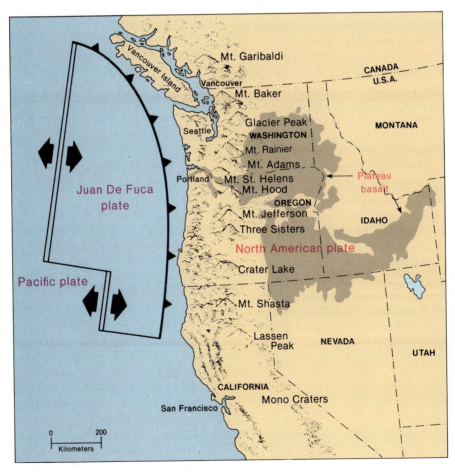

Figure 3.37
The Juan de Fuca plate, the Cascade volcanoes, and the
Columbia Plateau Basalts.

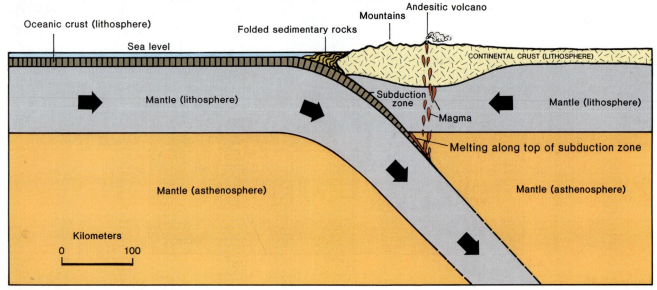

Figure 3.38
Magma forming at a converging plate boundary.

Summary

Lava is molten rock that reaches the earth's surface, having been formed as *magma* from rock within the earth's crust or from the uppermost part of the mantle.

Lava contains 45% to 75% *silica* (SiO_2). The more silica, the more viscous the lava. Its viscosity is also determined by the temperature of the lava. Viscous lavas are associated with more violent eruptions than are fluid lavas. *Volcanic domes* form from the extrusion of very viscous lavas.

A *mafic* lava, relatively low in silica, crystallizes into *basalt,* the most abundant extrusive igneous rock. Basalt, which is dark in color, is composed of minerals that are relatively high in iron, magnesium, and calcium.

Rhyolite, a light-colored rock, forms from *felsic* lavas that are high in silica but contain little iron, magnesium, or calcium. Because potassium and sodium are important elements in rhyolite, its constituent minerals are mostly potassium- and sodium-rich feldspars and quartz.

A lava with a composition between mafic and felsic crystallizes to *andesite,* a moderately dark rock. Andesite contains about equal amounts of ferromagnesian minerals and sodium- and calcium-rich feldspars.

Extrusive rocks are characteristically fine-grained. *Porphyritic* rock contains some larger crystals in an otherwise fine-grained rock. Rocks that solidified too rapidly for crystals to develop are called *obsidian.* Gas trapped in rock forms *vesicles.*

Pyroclasts, or *tephra,* are the result of volcanic explosions. *Tuff* is volcanic ash that has reconsolidated into a new rock. If large pyroclastic fragments have reconsolidated, the rock is termed a *volcanic breccia.*

A *cinder cone* is composed of loose pyroclastic material that forms steep slopes as it falls back around the crater. Cinder cones are not as large as the other two major types of cones.

A *shield volcano* is built up by successive eruptions of mafic lava. Its slopes are gentle but its volume generally large.

Composite cones are made of alternating layers of pyroclastic material and solidified lava flows. They are not as steep as cinder cones but steeper than shield volcanoes. Young composite volcanoes, predominantly composed of andesite, are aligned along the circum-Pacific belt and, less extensively, in the Mediterranean belt. These belts are located at converging plate boundaries, which accords with the theory of plate tectonics.

Plateau basalts and basalts formed at mid-oceanic ridges appear to be produced by magmas that originate in the upper mantle and follow fissures upward through the earth's crust. The *mid-oceanic ridges* are regarded in plate tectonic theory as present-day spreading centers. The newly erupted basalt becomes part of the oceanic crust. Plateau basalts may be due to discontinued spreading centers on continental crust.

Terms to Remember

aa	nuée ardente
andesite	obsidian
basalt	pahoehoe
bomb	partial melting
caldera	phenocryst
cinder cone	pillow structure (pillow
circum-Pacific belt	basalts)
columnar structure	plateau basalts
(columnar jointing)	porphyritic rock
composite volcano	pumice
(stratovolcano)	pyroclastic debris (tephra)
crater	rhyolite
extrusive rock	shield volcano
felsic rock	spatter cone
fine-grained rock	tuff
flank eruption	vent
glowing avalanche	vesicle
intermediate rock	viscosity
lava	volcanic breccia
mafic rock	volcanic dome
magma	volcanism
mantle plume	volcano
Mediterranean belt	welded tuff

Questions for Review

1. Name the minerals, and the approximate percentage of each, that you would expect to be present in each of the following rocks: andesite, rhyolite, basalt.
2. What clues would you look for if you were asked to predict whether a violent and disastrous eruption might occur in a volcanic area that is not presently active?
3. What property (or characteristic) of obsidian makes it an exception to the usual geologic definition of *rock?*
4. What determines the viscosity of a lava?
5. What determines whether a series of volcanic eruptions builds a shield volcano, a composite volcano, or a cinder cone? Describe each type of volcanic cone.
6. What might be the origin of basalt found on the sea-floor a thousand or more kilometers from an active mid-oceanic ridge?
7. Explain how a vesicular porphyritic andesite might have formed.
8. What gases are liberated by volcanic action?
9. Why are extrusive igneous rocks fine-grained?
10. Why don't flood basalts build volcanic cones?

Questions for Thought

1. Why does only a small portion of the mantle rock melt, and why does the melting take place below fissures that extend upward through the crust?
2. Why are continental igneous rocks richer in silica than oceanic igneous rock?
3. If a hot mantle plume were under the interior of a continent, what might happen?
4. Why are most spreading centers part of ocean floors rather than on continents?

Supplementary Readings

Bullard, F. M. 1984. *Volcanoes of the earth.* 2d ed. Austin: University of Texas Press.

Condie, K. C. 1989. *Plate tectonics and crustal evolution.* 3d ed. Oxford, England: Pergamon Press.

Decker, R. W., and B. Decker. 1980. *Volcano watching.* Hawaii Volcanoes National Park: Hawaii Natural History Association.

Decker, R. W., and B. Decker. 1989. *Volcanoes.* 2d ed. New York: W. H. Freeman.

Green, J., and N. M. Short. 1971. *Volcanic landforms and surface features: A photographic atlas and glossary.* New York: Springer-Verlag.

Harris, S. L. 1989. *Fire mountains of the west: The Cascade and Mono Lake volcanoes.* Missoula, Montana: Mountain Press.

Lipman, P. W., and D. R. Mullineaux, eds. 1981. *1980 eruptions of Mount St. Helens, Washington.* U.S. Geological Survey Professional Paper 1250.

Macdonald, G. A. 1983. *Volcanoes.* 2d ed. Englewood Cliffs, N.J.: Prentice-Hall.

Macdonald, G. A., A. T. Abbott, and F. L. Peterson. 1983. *Volcanoes in the sea: The geology of Hawaii.* 2d ed. Honolulu: University of Hawaii Press.

McPhee, J. 1989. *The control of nature.* New York: Farrar, Straus & Giroux.

Video

Eruption at sea. VHS. Ka'io Productions; P.O. Box 476, Volcano, Hawaii 96785.

Eruptive phenomena of Kilauea's East Rift Zone. VHS. Ka'io Productions; P.O. Box 476, Volcano, Hawaii 96785.

Rivers of fire. VHS. Hawaii Natural History Association; P.O. Box 74, Hawaii Volcanoes National Park, Hawaii 96718.

I n this chapter we continue to discuss igneous activity, now including the rock-forming processes of magmas that do not reach the earth's surface but solidify underground. Although intrusive and extrusive rocks are similar in many ways, intrusive rocks cannot be thought of simply as underground equivalents of volcanic rocks. There are important differences between extrusive and intrusive rocks, both in distribution and in properties. As later chapters show (chapters 17 through 20), intrusive rocks are far more important to the continental crust and mountain belts, whereas extrusive rocks are more important to the oceanic crust.

This chapter begins with an explanation of the structural relationships between bodies of intrusive rock and other rocks in the earth's crust. Intrusive rocks are then described and classified (the classification system is an elaboration of the one in chapter 3). Following that is a summary of theories of the origin of extrusive as well as intrusive rocks from which you can see how various hypotheses, some old, some new, relate to the theory of plate tectonics.

CHAPTER

4

Intrusive Activity and the Origin of Igneous Rocks

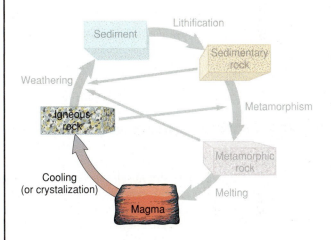

Granite Peaks, Sequoia National Park, California.
David Muench Photography.

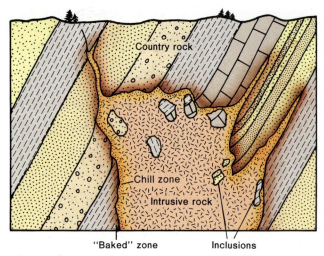

Figure 4.1
Igneous rock apparently intruded preexisting rock (country rock) as a liquid.

Figure 4.2
Granite (light-colored rock) has intruded dark-colored country rock. A geologist is standing on snow at the base of the cliff. Northern Victoria Land, Antarctica.
Photo by C. C. Plummer.

Intrusive rocks are rocks that appear to have crystallized from magma emplaced in surrounding rock. Bodies of intrusive rocks are exposed to us only after erosion and, usually, uplift.

Some intrusive rocks formed from magma that was solidifying underground while volcanic activity was taking place above. It would be a misconception, however, to regard all intrusive events as subsurface counterparts of volcanism. If this were true, the intrusive equivalent of basalt (the most abundant extrusive rock) ought to be the predominant intrusive rock. Quite the opposite is true. **Gabbro,** a mafic, coarse-grained, igneous rock composed predominantly of ferromagnesian minerals and calcium-rich plagioclase feldspar, is the intrusive equivalent of basalt. But gabbro is not very abundant. **Granite,** a felsic, coarse-grained, igneous rock composed mostly of potassium- and sodium-rich feldspars and quartz, is the most abundant intrusive rock. Coarse-grained granite is, in turn, chemically and mineralogically similar to rhyolite, a not-very-abundant extrusive rock. This striking contrast between the most abundant intrusive and extrusive igneous rocks has long puzzled geologists. Any comprehensive theory of the origin of magmas and igneous rocks must take this paradox into account.

What evidence suggests that large bodies of granite (and other intrusive rocks) might have formed from the underground solidification of magma? (1) Mineralogically and chemically, intrusive rocks are essentially identical to volcanic rocks (but are of different textures). (2) Experiments have confirmed that most of the minerals in these rocks can form only at high temperatures. Other experiments indicate that some of the minerals could have formed only under high pressures, implying they were deeply buried. (3) In many places preexisting solid rock, *country rock,* appears to have been forcibly broken by an intruding liquid, with the magma flowing into the fractures that developed (figures 4.1 and 4.2). **Country rock,** incidentally, is an accepted term for any older rock into which an igneous body intruded. (4) The country rock adjacent to the intrusive rock is seen to have been "baked" (*metamorphosed* is the correct term) along the contact with the intrusive rock. (5) The rock types of the country rock match **inclusions,** fragments of rock that are distinct from the body of igneous rocks in which they are enclosed. (6) In the intrusive rock adjacent to contacts with country rock are **chill zones,** finer-grained rocks that indicate magma solidified more quickly here because of the rapid loss of heat to cooler rock.

Although laboratory experiments have greatly increased our understanding of how igneous rocks form, geologists have not been able to create in the laboratory an artificial rock identical to granite. Only very fine-grained rocks containing the minerals of granite have been made from artificial magmas, or "melts." The temperature and pressure at which granite apparently forms can be duplicated in the laboratory—but not the time element. According to calculations, a large body of magma requires over a million years to solidify completely. This very gradual cooling is thought to cause the coarse-grained texture of most intrusive rocks.

Chemical processes involving silicates are known to be exceedingly slow. But another problem in trying to apply experimental procedures to real rocks is determining the role of water vapor and other gases in the crystallization of rocks such as granite. Gases are not retained in rock crystallized underground from a magma, but large amounts of gas (especially water vapor) are released during volcanic eruptions. No one has seen an intrusive rock forming; hence we can only speculate about the role these gases might have played before they escaped. One example shows why gases are important. Laboratory studies have shown that granite can melt at temperatures as low as 650°C if water (dissolved in the melt) is present and under high pressure. Without the water, the melting temperature is several hundred degrees higher. Not knowing how much water was present during crystallization makes accurate determination of temperatures difficult and speculative.

Intrusive Bodies

Intrusions, or **intrusive structures,** are bodies of intrusive rock. Their size and shape, as well as their relationship to surrounding rocks, are important aspects of the architecture, or *structure,* of the earth's crust. The various intrusions are named and classified on the basis of the following considerations: (1) Is the body large or small? (2) Does it have a particular geometric shape? (3) Did the rock form at a considerable depth or was it a shallow intrusion? (4) What is the geometric relationship of the intrusion to the country rock?

Shallow Intrusive Structures

Some igneous bodies apparently solidified near the surface of the earth (at depths of probably less than 2 kilometers). These bodies probably clogged the subsurface "plumbing systems" of volcanoes or lava flows. Shallow intrusive structures tend to be relatively small compared with those that formed at considerable depth. Because the country rock near the earth's surface generally is cool, intruded magma tends to chill and solidify relatively rapidly. Many shallow intrusive rocks are as fine-grained as their volcanic counterparts, although some are slightly coarser-grained. If a rock is fine-grained, even if it solidified underground, it has the same name as the corresponding extrusive rock. For example, a fine-grained rock of intermediate composition is an andesite whether it formed from a lava flow or from rapid cooling within the

A

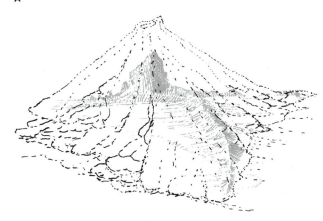

B

Figure 4.3
(A) Ship Rock in New Mexico. **(B)** Relationship to the former volcano.
Photo by D. A. Rahm, courtesy Rahm Memorial Collection, Western Washington University.

throat of a volcano. A **volcanic neck** is an intrusive structure apparently formed from magma that solidified within the throat of a volcano.

One of the best examples is Ship Rock in New Mexico (figure 4.3). Here is how geologists interpret the history of this feature: A volcanic cone formed. As eruptions ceased, the throat beneath the vent became clogged with magma that solidified into a more or less cylindrical body. In time the volcano (probably a cinder cone) eroded away; the more resistant plug remained and eroded more slowly into its present shape. Weathering and erosion are continuing (falling rock has been a serious hazard to rock climbers), and eventually Ship Rock will erode to the level of the desert floor.

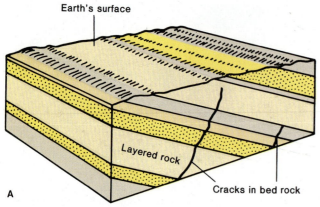

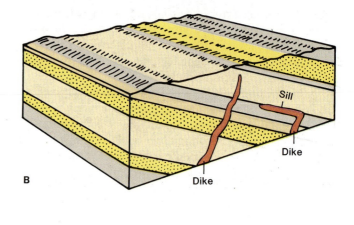

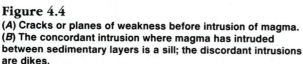

Figure 4.4
(A) Cracks or planes of weakness before intrusion of magma.
(B) The concordant intrusion where magma has intruded between sedimentary layers is a sill; the discordant intrusions are dikes.

Dikes and sills Another, and far more common, intrusive structure can also be seen at Ship Rock. The low, wall-like ridge extending outward from Ship Rock is an eroded dike. A **dike** is a tabular (shaped like a table top), discordant, intrusive structure (figure 4.4). **Discordant** means that the body is not parallel to any layering in the country rock. (Think of a dike as cutting across layers of country rock.) Dikes may form at shallow depths and be fine-grained, such as those at Ship Rock, or form at greater depths and be coarse-grained. Dikes need not appear as walls protruding from the ground (figure 4.5). The ones at Ship Rock do so only because they are more resistant to weathering and erosion than the country rock.

A **sill** is also a tabular intrusive structure, but it is **concordant.** That is, sills, unlike dikes, are parallel to any planes or layering in the country rock (figures 4.4 and 4.6). Typically, the country rock bounding a sill is layered sedimentary rocks. As magma squeezes into a crack between two layers, it solidifies into a sill.

A rarely found close relative of a sill is a blister-shaped structure called a **laccolith.** It also is a concordant intrusive body, but the central portion is thicker and domed upward like a mushroom.

Intrusives That Crystallize at Depth

Igneous rocks that formed at great depth—more than several kilometers—are called **plutonic rocks** (after Pluto, the Greek god of the underworld). Characteristically, these rocks are coarse-grained, reflecting the slow cooling and solidification of magma. A **pluton** is an igneous body that crystallized at considerable depth. Most plutons are irregular in shape, unlike dikes and sills. The larger masses of magma had to break through country rock or shoulder it aside as they moved upward and so developed an irregular shape. Discordant plutons that have been exposed at the earth's surface by erosion are arbitrarily distinguished by size. A **stock** is a small discordant pluton with an outcrop area (i.e., the area over which it is exposed to the atmosphere) of less than 100 square kilometers. If the outcrop area is greater than 100 square kilometers, the body is called a **batholith** (figure 4.7), a large discordant pluton.

Most batholiths crop out over areas vastly greater than the minimum 100 square kilometers. Although batholiths can be mafic, intermediate, or felsic, most are felsic; that is, they are composed of granite. Many major mountain ranges are carved largely from granite, examples being found in the Appalachians in the eastern part of North America and in the ranges that parallel the West Coast. Over half the rock exposed in the Sierra Nevada of California is plutonic (figure 4.8). Geologists used to think that batholiths probably extended as continuous bodies from surface exposures to the base of the crust. Indirect evidence now suggests that few batholiths extend more than 20 kilometers below the surface.

Figure 4.5
**Dikes (light-colored rocks) in northern Victoria Land,
Antarctica.**
Photo by C. C. Plummer.

Figure 4.6
**Sill (dark rock) in sedimentary layered rock, Grand Canyon,
Arizona.**

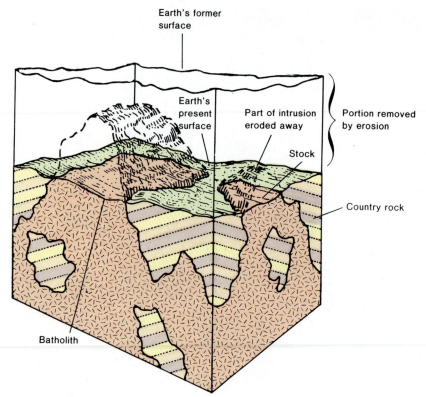

Earth's former surface

Earth's present surface

Part of intrusion eroded away

Portion removed by erosion

Stock

Country rock

Batholith

Figure 4.7
A batholith and a stock. The cross-sectional view indicates that the stock and batholith are part of the same intrusion.

Figure 4.8
Part of the Sierra Nevada batholith. All light-colored rock shown here (including that under the distant snow-covered mountains) is granite.
Photo by C. C. Plummer.

Identification of Intrusive Igneous Rocks

A coarse-grained texture is the most significant difference between plutonic rocks and finer-grained extrusive and shallow intrusive rocks. (For our purposes, **coarse-grained rocks** are defined as those in which most of the grains are larger than 1 millimeter.) The crystalline grains of plutonic rocks are commonly interlocked in a mosaic pattern (figure 4.9).

As with extrusive rocks, the names of plutonic rocks are based on texture and mineralogical composition. Because of their larger mineral grains, plutonic rocks are easier to identify than extrusive rocks. The physical properties of each mineral in a plutonic rock can be determined more easily. And, of course, knowing the minerals makes rock identification a simpler task. In figure 4.10, the classification chart presented earlier for extrusive rocks has been expanded to include intrusive rocks. Use this chart, along with table 4.1, to identify common igneous rocks. You may also find it helpful to turn to Appendix B, which includes a key for identifying common igneous rocks. (Photos of typical plutonic rocks are shown in figure 4.11.)

A

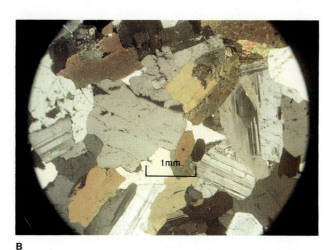

B

Figure 4.9

(A) Coarse-grained texture characteristic of plutonic rock. Plagioclase is white and quartz is grey; potassium feldspar has been stained yellow. Small gradations on ruler are millimeters. **(B)** A similar rock seen through a polarizing microscope. Note the interlocking crystal grains of individual minerals.

Photos by C. C. Plummer.

Coarse-grained	GRANITE		DIORITE	GABBRO	ULTRA-MAFIC
Fine-grained	RHYOLITE		ANDESITE	BASALT	

Ferromagnesians

Potassium feldspar (orthoclase)

Plagioclase feldspar

Quartz

100%

75%

50%

25%

0%

75% SiO$_2$ Increasing silica   45% SiO$_2$

Increasing K$_2$O and Na$_2$O

Increasing CaO, FeO and MgO

FELSIC	INTERMEDIATE	MAFIC	ULTRA-MAFIC

Figure 4.10

Classification chart for the most common igneous rocks. Rock names based on special textures are not shown.

Identification of Igneous Rocks

Coarse-Grained	Granite	Diorite	Gabbro	Ultramafic Rocks
Fine-Grained	Rhyolite	Andesite	Basalt	—
Mineral Content	Quartz, feldspars (white, light gray, or pink). Minor ferromagnesian minerals.	Feldspars (white or gray) and about 30–50% ferromagnesian minerals. No quartz.	Predominance of ferromagnesian minerals. Rest of rock is plagioclase feldspar (medium to dark gray).	Entirely ferromagnesian minerals (usually olivine and pyroxene).
Color of Rock (most commonly)	Light-colored	Medium-gray or medium-green	Dark gray to black	Very dark green to black

ROCKS NAMED SOLELY ON THE BASIS OF TEXTURE:
Obsidian—volcanic glass.
Pumice—frothy volcanic glass.
Volcanic breccia—coarsely fragmental volcanic rock.
Tuff—volcanic rock composed of fine fragments (a microscope is usually necessary to distinguish tuff from the fine-grained rocks described above).

Pegmatite—very coarse-grained rock generally having the same mineral content as granite.
The following adjectives may be applied to igneous rocks to give a more complete description:
Porphyritic—some of the grains are very much larger than the majority of the grains.
Vesicular—a rock containing holes created by gas trapped in the cooling lava.

A

Figure 4.11
Plutonic rocks. (A) Granite. (B) Diorite. (C) Gabbro.
Photos by C. C. Plummer.

B

C

BOX 4.1

Pegmatite—A Rock Made of Giant Crystals

Box 4.1 Figure 1
Pegmatite in northern Victoria Land, Antarctica. The knife is 8 centimeters long. The black crystals are tourmaline. Quartz and feldspar are light colored.
Photo by C. C. Plummer.

Pegmatites are extremely coarse-grained igneous rocks. In some pegmatites, crystals are as large as 10 meters across. Strictly speaking, a pegmatite can be of diorite, gabbro, or granite. However, the vast majority of pegmatites are felsic, with very large crystals of potassium feldspar, sodium-rich plagioclase feldspars, and quartz, plus very minor amounts, if any, of ferromagnesian minerals. Hence the term *pegmatite* generally refers to a rock of granitic composition (if otherwise, a term like *gabbroic pegmatite* is used). Pegmatites are interesting as geological phenomena and important as minable resources.

The extremely coarse texture of pegmatites is attributed to both slow cooling and the low viscosity of the fluid from which they form. Since lava solidifying to rhyolite is very viscous, we would expect that magma solidifying to granite, being chemically similar, would also be viscous.

Pegmatites, however, probably crystallize from a fluid composed largely of water under high pressure. Water molecules and ions derived from the parent, granitic magma make up this residual magma. Geologists believe the following sequence of events accounts for most pegmatites.

As a granite pluton cools, increasingly more of the magma solidifies into the minerals of a granite. By the time the pluton is well over 90% solid, the residual magma contains a very high amount of silica and ions of elements that will crystallize into potassium and sodium feldspars. Also present are elements that could not be accommodated into the crystal structures of the common minerals that formed during the normal solidification phase of the pluton. Fluids, notably water, that were in the original magma are left over as well, sometimes in considerable quantities. If no fracture above the pluton permits the fluids to escape as gases, they are sealed in, as in a pressure cooker. The watery residual magma has a low viscosity. Moreover, the water retards nucleation by preventing silica tetrahedrons in the melt from combining into incipient crystalline structures. When a crystal does start to form during slow cooling of the watery solution, the appropriate atoms within the melt are able to move freely in the fluid and become part of that growing crystal. The crystal adds more and more atoms and grows very large.

Pegmatite bodies are generally quite small. Many are podlike structures, located either within the upper portion of a granite pluton or within the overlying country rock near the contact with granite, the fluid body evidently having squeezed into the country rock before solidifying. Pegmatite dikes are fairly common, especially within granite plutons, where they apparently filled cracks that developed in the already solid granite. Some pegmatites form small dikes along contacts between granite and country rock, filling cracks that developed as the cooling granite pluton contracted.

Most pegmatites contain only quartz, feldspar, and perhaps mica. Minerals of considerable commercial value are found in a few pegmatites. Large crystals of muscovite mica are mined from pegmatites. These crystals are called "books" because the cleavage flakes (tens of centimeters across) look like pages. Because muscovite is an excellent insulator, the cleavage sheets are used in electrical devices, such as toasters, to separate uninsulated electrical wires. Even the large feldspar crystals in pegmatites are mined for various industrial uses, notably the manufacture of ceramics.

Many rare elements are mined from pegmatites. These elements were not absorbed by the minerals of the main pluton and so were concentrated in the residual pegmatitic magma, where they crystallized as constituents of unusual minerals. Minerals containing the element lithium are mined from pegmatites. Lithium becomes part of a sheet silicate structure to form a pink or purple variety of mica (called lepidolite). Uranium ores, similarly concentrated in the residual melt of magmas, are also extracted from pegmatites.

Some pegmatites are mined for gemstones. Emerald and aquamarine, varieties of the mineral beryl, occur in pegmatites that crystallized from a solution containing the element beryllium. A large number of the world's very rare minerals are found only in pegmatites, many of these in only one known pegmatite body. These rare minerals are mainly of interest to collectors and museums.

Hydrothermal veins (described in the chapter titled "Metamorphism, Metamorphic Rocks, and Hydro- thermal Rocks") are closely related to pegmatites. Veins of quartz are common in country rock near granite. Many of these are believed to be caused by water that escapes from the magma. Silica dissolved in the very hot water cakes on the walls of cracks as the water cools while traveling surfaceward. Sometimes valuable metals such as gold, silver, lead, zinc, and copper are deposited with the quartz in veins.

Meteorites

Small solid particles of ice, stone, and/or metal orbiting the Sun are called **meteoroids.** When these particles encounter the Earth and pass through the Earth's atmosphere, they are heated to incandescence by friction and partially or totally vaporized. We sometimes call them "shooting stars" or "falling stars," but the correct term is *meteor.* About a million meteors that can be seen without a telescope enter the Earth's atmosphere each hour. Most of these glowing particles are about the size of a pinhead.

Occasionally, larger objects (such as asteroids, fragments of asteroids, and comets) collide with the Earth. About 150 of these objects per year are able to pass all the way through the atmosphere without being consumed and thus strike the Earth's surface. These objects are called *meteorites* (box figure 1). The largest fragment of a meteorite found (in South Africa) weighs 50 tons, but much larger meteorites have hit the Earth's surface in the past.

Meteorites, like igneous rocks, are usually classified on the basis of their textures and compositions. The three basic types are stones, stony-irons, and irons.

Stones are by far the most abundant type of meteorite, but because they are very similar in appearance to ordinary Earth rocks, they are not often collected unless they are actually seen to fall. Irons, on the other hand, are often found long after they fall and so appear more often than stones in museum collections.

Iron meteorites are principally iron-nickel alloy, with small amounts of other minerals.

Stony-iron meteorites are made up of silicate minerals and nickel-iron alloy in approximately equal amounts.

Stony meteorites are made up mostly of plagioclase and iron-magnesium silicates such as olivine and pyroxene; but they may be as much as 20% iron-nickel alloy. About 90% of stony meteorites contain round silicate grains called *chondrules* and are referred to as *chondrites.* The remaining 10% are called *achondrites.*

Chondrules, composed largely of olivine and pyroxene, vary in appearance from distinct spheres in a very fine-grained groundmass of silicates and iron-nickel alloy to indistinct bodies with fuzzy outlines imbedded in a coarse-grained groundmass. Chondrites with coarse-grained textures and indistinct chondrules are thought to have been reheated and partially recrystallized after forming. The terrestrial rock most similar in composition to chondrites is probably the ultramafic rock peridotite; but peridotite does not have either a chondritic texture or the nickel-iron content of stony meteorites.

One kind of chondrite is composed mostly of serpentine or pyroxene and contains up to 5% organic materials, including carbon, hydrocarbon compounds, and amino acids. These meteorites are called *carbonaceous chondrites.* All available evidence indicates that the "organic" compounds were in fact produced by inorganic processes. Carbonaceous chondrites are of particular interest to scientists because they are believed to have the same composition as the original material from which the solar system was formed.

Achondrites are generally similar to terrestrial rocks in composition and texture. In composition they are probably most similar to basalts. Some have textures like ordinary igneous rocks, and others are breccias with fragments of different compositions and textures.

Ultramafic Rocks

As you may have noted in the chart (figure 4.10), ultramafic rocks do not have a fine-grained counterpart. This curious fact seems to indicate that ultramafic magma doesn't reach the surface. An **ultramafic rock** is composed entirely or almost entirely of ferromagnesian minerals. No feldspars are present and, of course, no quartz. Most ultramafics are composed of coarse-grained pyroxene and/or olivine. Chemically, these rocks contain less than 45% silica.

Experiments show that very high temperatures (almost 2000°C) are needed to melt ultramafic rocks. Such high temperatures could not be sustained by a magma traveling upward through the considerably cooler earth's crust. For this reason, geologists believe that most ultramafic rocks form far below the surface, probably in the mantle. This view is supported by experiments that have found that some of the minerals in ultramafic rocks could have formed only at very high pressure. Where we find large bodies of ultramafic rocks, the usual interpretation is that a piece of the mantle has traveled upward as solid rock.

Astrogeology Box 4.1 Figure 1
A large meteorite from Meteor Crater Arizona.
© Tom Fridmann/Industrial Visuals.

Many meteorites have coarse-grained textures, and such textures are formed only by slow cooling inside a large body. Coarse-grained textures and the fact that the iron-nickel alloys in iron meteorites are very similar in composition suggest that some meteorites represent fragments of larger bodies that were partially melted and differentiated to form an iron-rich core and a silicate-rich outer zone. Meteorites have been dated using radiometric methods, and all meteorites seem to be about the same age, about 4.5 billion years old.

Varieties of Granite

Granite and rhyolite occupy a larger area in the classification chart than do the other rocks. This reflects granite's greater variation in composition. For instance, a granite whose composition corresponds to the right side of the field (that is, nearer to diorite/andesite) contains much more plagioclase than potassium feldspar and somewhat more ferromagnesian minerals than does a rock whose composition plots in the left side of the granite field. Geologists have arbitrarily subdivided the field of granite and named each of the varieties; a rock in the right portion of the field, for example, is called granodiorite.

Any classification system is, of course, a human device, and for this reason, classification systems differ somewhat among groups of geologists. We define the boundary between granite and **diorite** (a coarse-grained rock composed of somewhat more plagioclase feldspar than ferromagnesian minerals) by the presence or absence of quartz; but we could just as easily have placed the boundary slightly to the left, so that a rock with 10% or less quartz would be diorite.

Abundance and Distribution of Plutonic Rocks

Compared with the plutons that occupy large portions of the earth's crust, the shallow intrusive bodies are minor in extent and significance.

We mentioned earlier the paradox that basalt is the most abundant extrusive rock whereas granite is the most common intrusive rock. We also pointed out that most basalt is associated with the oceanic crust (although plateau basalts cannot be ignored) and that granite is associated with mountain ranges.

Granite is not only the most abundant igneous rock in mountain ranges but the predominant igneous rock throughout the continental crust. In most of the non-mountainous parts of North America—the plains of the central interior, for instance—no igneous rock is exposed at the surface. But underneath a veneer of sedimentary rock only a few kilometers thick lies the much older "basement" rock, much of which was orginally plutonic. Some very old granite plutons are not covered by sedimentary rocks, as, for example, in eastern Canada. Plutons have also been found in places like Grand Canyon where local erosion has cut through the sedimentary veneer and exposed the continental basement. These plutons have intruded metamorphic rocks that are even older. Some parts of the basement underlying the continent are thought to be mountain ranges that long ago eroded to plains (as explained in the chapter titled "Mountain Belts and the Continental Crust").

It is apparent that *granite* is the igneous rock of the continents, *basalt* of the oceans, and *andesite* (usually on or near continental margins) the building material of young volcanic mountains.

How Magma Forms

A rock melts and becomes a magma when the temperature is higher than the melting points of minerals in the rock. A rock becomes entirely molten if the temperature is higher than a melting point controlled by all the minerals in that rock. If the temperature is not high enough to melt some of the minerals, then only part of the rock melts. The temperature at which a given mineral melts can vary. Pressure, amount of gas (particularly water) present, and the kind of neighboring minerals can all influence the melting point of a mineral. The factors that control the melting of rocks are discussed later in this chapter.

Sources of Heat for Melting

Geothermal gradient   A miner descending a mine shaft notices a rise in temperature. This is due to the **geothermal gradient,** the rate at which temperature increases with increasing depth beneath the surface. Data show the geothermal gradient, on the average, to be about 2.5°C for each 100 meters (25°C/km) of depth in the upper part of the crust. The geothermal gradient is not the same

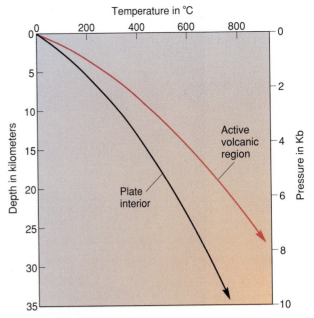

Figure 4.12
Geothermal gradients at two parts of the earth's crust.

everywhere. Figure 4.12 shows geothermal gradients for two regions. The curve for the volcanic region indicates a higher geothermal gradient than that for the continental interior. Temperatures high enough to melt rock would be expected at a relatively shallow depth beneath the volcanic region. You would have to go deeper in the continental interior to reach the same temperature; however, the rock there would not melt because of the increased pressure at that depth.

Hot mantle plumes   "Hot spots" in the crust (where the geothermal gradient is locally very high) may be caused by hot mantle plumes, which are narrow upwellings of hot material within the mantle. Hot mantle plumes help account for some igneous activity, such as the mid-oceanic eruptions that built up the Hawaiian islands. Volcanism in the middle of continents may also be attributable to mantle plumes. Yellowstone National Park has had considerable volcanic activity in the recent geologic past, and some geologists have suggested that a hot plume may exist in the mantle below the park.

Friction   Rock grinding past rock may also generate heat. In regions where active mountain building is taking place, the friction caused by the moving and shifting of large rock masses generates heat that may combine with heat from other sources and melt rocks.

Heat transfer   Heat from a body of magma may be conducted into overlying rocks. As the magma cools, the overlying country rock becomes heated. Especially significant is the extra heat generated when magma crystallizes. As atoms move from the random disorder of the liquid to the orderly structure of a crystal, heat is given off. (Approximately 100 calories of heat are released when 1 gram of a mineral crystallizes.)

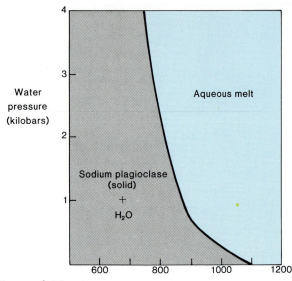

Figure 4.13
Melting temperature of a mineral relative to water pressure.
After Tuttle and Brown, Geologic Society of America, 1958.

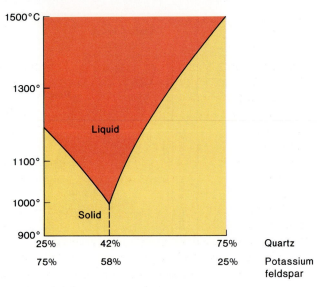

Figure 4.14
Melting temperatures for mixtures of quartz and potassium feldspar at atmospheric pressure.
Modified from Schairer and Bowen, 1956, V 254, p. 16, *American Journal of Science.*

Radioactivity Some elements, such as uranium, produce heat through radioactive decay. Normally these elements are in rocks in very small amounts.

Radioactive elements are present in greater amounts in felsic rocks than in mafic rocks. Heat from radioactivity is more significant in continental rocks, where granite predominates. But even in granitic regions, radioactivity only raises the geothermal gradient slightly.

Factors That Control Melting Temperatures

Pressure The melting point of a mineral generally *increases* with increasing pressure. Pressure increases with depth in the earth's crust, just as temperature does. So a rock that melts at a given temperature at the surface of the earth requires a higher temperature to melt deep underground.

Water under pressure If enough gas, especially water vapor, is present and under high pressure, the effect of pressure is dramatically changed. Water vapor sealed in under high pressure by overlying rocks helps break down crystal structures. High water pressure can significantly lower the melting points of minerals. (Figure 4.13 shows the relationship between water pressure and the melting temperature of a plagioclase.)

Experiments have shown that, under moderately high pressure, water mixed with granite lowers the melting point of granite from about 900°C (when dry) to as low as 625°C when saturated with water under pressure equivalent to that of 10,000 atmospheres or *bars.* (Pressure at depth is usually expressed in *kilobars;* one kilobar is equal to 1,000 bars.)

Effect of mixed minerals Two metals—as in solder—can be mixed in a ratio that lowers their melting temperature far below that of the melting points of the pure metals. Minerals behave similarly. Experiments have shown that in some cases mixed fragments of two minerals melt at a lower temperature than either mineral alone. Figure 4.14 shows the melting temperatures for quartz and potassium feldspar mixed in various proportions. If the mixture is 42% quartz (58% potassium feldspar), melting takes place at just above 1000°C. On the other hand, if the mixture is 75% quartz and 25% potassium feldspar (corresponding to the right edge of the diagram), 1500°C is needed to turn that mixture into a liquid. Pure quartz requires even higher temperatures to melt.

How Magmas of Different Compositions Evolve

Differentiation and Bowen's Reaction Theory

Differentiation is the process by which different ingredients separate from an originally homogenous mixture. An example is the separation of whole milk into cream and skimmed milk. In the early part of the twentieth century, N. L. Bowen conducted a series of laboratory experiments demonstrating that differentiation is a plausible way for felsic and mafic rocks to form from a single parent magma.

Bowen's reaction series, shown in figure 4.15, is the sequence in which minerals crystallize from a cooling magma, as demonstrated by Bowen's laboratory experiments. In simplest terms, Bowen's reaction series shows that those minerals with the highest melting temperatures crystallize from the cooling magma before those with lower melting points. However, the concept is a bit more complicated than that.

Crystallization begins along two branches, the *discontinuous branch* and the *continuous branch.* In the discontinuous branch, one mineral changes to another at discrete temperatures during cooling and solidification of the magma. Changes in the continuous branch occur gradationally through a range in temperatures and affect only

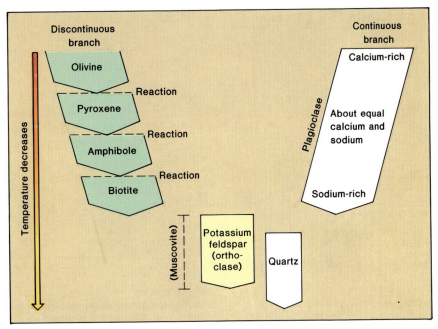

Figure 4.15
Bowen's reaction series.

the one mineral, plagioclase. Although crystallization takes place simultaneously along both branches, we must explain each separately.

Discontinuous branch All the minerals in the discontinuous branch are ferromagnesian. In this branch, as the completely liquid magma slowly cools, it reaches the temperature at which *olivine* begins to crystallize from the magma. Olivine is a mineral with an exceptionally high proportion (2:1) of iron and magnesium to silicon—its formula is $(Fe, Mg)_2SiO_4$. The liquid left after olivine has crystallized is relatively depleted in iron and magnesium and relatively enriched in silicon (because only one part of Si is used for two parts of Fe and Mg).

As the melt cools further, the melting temperature for the next mineral of the series is reached, and *pyroxene* begins to crystallize. The previously formed olivine now *reacts* with the remaining melt, and the original crystal structure of olivine rearranges into that of pyroxene (from isolated silica tetrahedrons to single chains of tetrahedrons). The crystal structure of pyroxene accommodates a higher amount of silicon relative to the iron and magnesium—a ratio of 1 to 1. Its formula is $(Mg, Fe) SiO_3$. As the temperature decreases, more pyroxene crystallizes directly from the melt until the point at which *amphibole* becomes stable.

Now pyroxene reacts with the melt (if the magma is not yet completely solidified). Its crystal structure rearranges into amphibole's double chains of silica tetrahedrons. More of the silicon leaves the melt (along with aluminum and minor amounts of sodium and calcium) and is incorporated into the newly developing amphibole crystals.

If still more melt is left after amphibole has formed, on further cooling amphibole reacts with the melt to produce *biotite* (which is a sheet silicate). Biotite is the last

of the ferromagnesian minerals to crystallize. Any magma remaining after biotite has finished crystallizing contains neither iron nor magnesium.

Continuous branch Plagioclase feldspar is the only mineral in the continuous branch. As the completely liquid magma cools, the plagioclase with the highest melting point begins to crystallize. This plagioclase has a very high amount of calcium. Silicon and aluminum, which are part of all feldspars, combine with calcium to form the variety of plagioclase associated with mafic igneous rocks. Upon further cooling, plagioclase develops at the expense of the magma. As the temperature slowly drops, the total amount of plagioclase increases, and more and more sodium is incorporated into each growing plagioclase crystal. Plagioclase continues to crystallize during cooling until all the calcium and sodium in the magma are used up.

Any magma left after the crystallization is completed along the two branches is richer in silicon than the original magma and also contains potassium and aluminum. The potassium and aluminum combine with silicon to form *potassium feldspar*. If the water pressure is high, *muscovite* may also form at this stage. Any leftover melt is pure silica, which crystallizes as *quartz*.

Bowen used this experimentally determined reaction series to support the hypothesis that all magmas (mafic, intermediate, and felsic) derive from a single parent magma (mafic) by differentiation. The early-developing minerals are separated from the remaining magma. Normally a newly erupted cooling basalt lava progresses only a short distance down the reaction series before all the ingredients are used up. Olivine develops, but only part of it reacts with the melt to form pyroxene before all the magma is used up. Simultaneously, calcium-rich plagioclase grows and becomes increasingly sodic; but its growth ceases when the minor amount of sodium is used up and

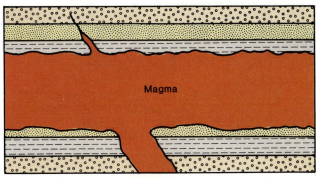

A

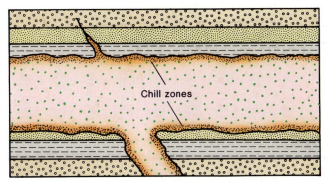

Chill zones

B

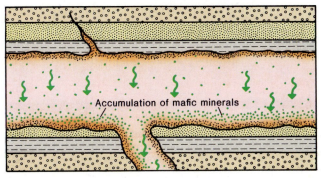

Accumulation of mafic minerals

C

Figure 4.16
Cross section showing the process of differentiation by crystal settling in a sill and dike. (*A*) Recently intruded magma is completely liquid. (*B*) Upon slow cooling, minerals such as olivine crystallize first. (*C*) The heavier crystals that formed early sink, leaving the remaining magma depleted in mafic constituents.

the liquid is consumed. The rock becomes a completely solid aggregate of calcium-rich plagioclase, pyroxene, and olivine—in other words, what one expects to find in a basalt or gabbro.

Crystal settling   Only if the original basaltic magma cools slowly, and the earliest-formed minerals physically separate from the magma, can the minerals on the lower part of the reaction series crystallize. **Crystal settling** is the downward movement of minerals that are denser (heavier) than the magma from which they crystallized. What is pictured happening is that as the olivine crystallizes from the magma, the crystals settle to the bottom of the magma chamber (figure 4.16). Calcium-rich plagioclase also separates as it forms. The remaining magma is,

therefore, depleted of calcium, iron, and magnesium. Because these minerals were economical in using the relatively abundant silica, the remaining magma becomes richer in silica as well as in sodium and potassium. If enough mafic ingredients are removed in this manner, the remaining residue of magma eventually solidifies into a granite.

Undoubtedly this method of differentiation does take place in nature, though probably not to the extent that Bowen envisioned. The lowermost portions of some large sills are composed entirely of calcium-rich plagioclase and olivine, whereas upper levels are considerably less mafic. Even in large sills, however, differentiation has rarely progressed far enough to produce any granite within the sill.

Even assuming that the mafic minerals settle ever deeper in large magma bodies, there is a problem in trying to explain the origin of granite by Bowen's theory. Calculations show that to produce a given volume of granite, about ten times as much mafic rock first has to form and settle out. If this is true, we would expect to find far more mafic plutonic rock than granite in the earth's crust. But, as we know, granite is far more abundant than gabbro. Unless the mafic rocks simply settled completely out of the crust, it seems unlikely that granite batholiths could have been produced by differentiation.

This is not to say that Bowen's work is discredited. Quite the opposite. His work has led to other theories on the behavior of magmas. Moreover, differentiation does occur and can explain relatively minor compositional variations within intrusive bodies, even if it does not satisfactorily explain the origin of large granite bodies.

Assimilation

A very hot magma may melt some of the country rock and **assimilate** the newly molten material into the magma (figure 4.17). This is like putting a few ice cubes into a cup of hot coffee. The ice melts and the coffee cools as it becomes diluted. Similarly, if a hot basaltic magma, perhaps generated from the mantle, melts portions of the continental crust, the magma simultaneously becomes richer in silica and cooler. Possibly the intermediate magmas that are associated with circum-Pacific andesite volcanoes may derive from assimilation of some crustal rocks by the basalt.

Mixing of Magmas

The idea that some of our igneous rocks may be "cocktails" of different magmas is currently receiving more attention by geologists than it did in the past. The concept is quite simple. If two magmas meet and merge within the crust, the combined magma will be compositionally intermediate (figure 4.18). If you had approximately equal amounts of a granitic magma mixing with a basaltic magma, the resulting magma should consolidate underground as diorite or on the surface as andesite.

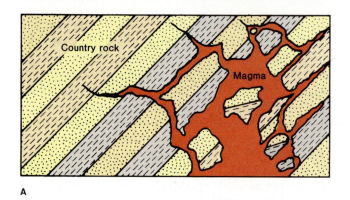

A

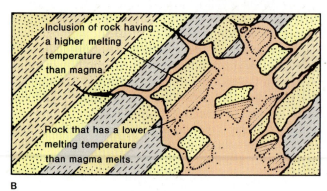

B

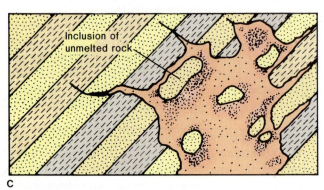

C

Figure 4.17
Assimilation. Magma formed is intermediate in composition between the original magma and the absorbed country rock. (A) Ascending magma breaks off blocks of country rock (the process is called *stoping*). (B) Inclusions of country rock with melting temperatures lower than the magma melt. (C) The molten country rock blends with the original magma, leaving unmelted portions as inclusions.

Partial Melting

A granitic magma could be created by partial melting of rock—visualize a progression *upward* through Bowen's reaction series (going from cool to hot). As might be expected, the first portion of a rock to melt as temperatures rise forms a liquid with the chemical composition of quartz and potassium feldspar. The silicon plus potassium and aluminum "sweated out" of the solid rock could accumulate into a pocket of felsic magma. If higher temperatures prevailed, more mafic magmas would be created. Small pockets of magma could consolidate and form a large mass.

The lower part of the continental crust is a very plausible source for granite batholiths. Partial melting of these rocks could be caused by combinations of heat sources and

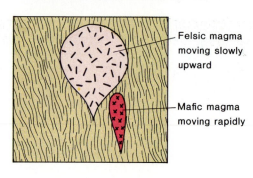

A

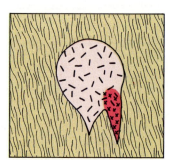

B

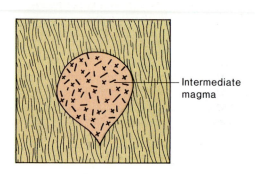

C

Figure 4.18
Mixing of magmas. (A) Two bodies of magma moving surfaceward. (B) The mafic magma catches up with the felsic magma. (C) The two magmas combine and become an intermediate magma.

other factors that control melting. The magma created, being less dense than the surrounding rock, would well upward in large viscous blobs. The magma would make room for itself as it rises, either by shouldering aside the overlying country rock or by stoping. **Stoping** means fracturing country rock above the magma chamber. The blocks of rock would settle to the bottom of the magma chamber (if they are denser than the magma) or perhaps be partially assimilated by the upward moving magma. Upon cooling, the magma eventually solidifies into granite plutons. Rarely will a granitic magma reach the surface. When it does, it will erupt violently and is likely to result in huge deposits of rhyolitic tuffs, such as those found at Yellowstone National Park.

Many geologists regard basalt as the product of partial melting within the mantle, at temperatures hotter than those in the crust. Since basaltic magma is much more fluid than rhyolitic magma, it travels rapidly and easily through fissures to the surface rather than solidifying in the crust. (In the few places where basaltic magma does solidify as a pluton within the crust, gabbro forms.) The solid residue left behind in the mantle when the basaltic magma is removed is ultramafic rock.

Extraterrestrial Minerals and Igneous Rocks

Astrogeology Box 4.2 Figure 1
Astronaut on the lunar surface. Photographed in a mountainous region during the Apollo 17 flight, 1972.
NASA.

For geologists, one of the most exciting aspects of sending spacecraft to the Moon was the ability to collect rocks and bring them to the Earth for study (box figure 1).

Over 100 minerals, many especially rich in iron and titanium compounds, have been found in lunar rocks. The four most common minerals on the Moon are pyroxenes such as augite, Ca-rich plagioclase, Mg-rich olivine, and ilmenite, an iron-titanium oxide. Several common Earth minerals are not found on the Moon. These include the amphiboles, the micas, the clay minerals, hematite, and limonite. Minerals such as these are hydrated and/or formed by weathering processes and would not be expected on the Moon, with its lack of an atmosphere and the absence of water. Also found on the Moon have been a few new minerals such as pyroxferrite (a mineral related to pyroxene but richer in iron) and armalcolite (an iron-magnesium-titanium oxide whose name is derived from the first letters in the names of the astronauts of the first manned landing mission).

Moon rocks are of two types, crystalline rocks and breccias (see Astrogeology Box 7.1 for information about lunar breccias). Most of the crystalline rocks appear to be simple igneous rocks that have crystallized and solidified from magma at or near the surface of the Moon. A few of the crystalline rocks are apparently made of material that melted during meteorite impacts and then recrystallized as it cooled.

The rock type most common in the maria seems to be very similar to basalt. Mare basalts are composed primarily of calcium-rich plagioclase feldspar, pyroxene, and ilmenite. Many of the rocks, especially the finer-grained ones, also contain olivine as a major or minor component. The rocks have a wide range of textures, including porphyritic, but most have rather even textures with crystals of about the same size. Vesicles are common. Lunar basalts appear to have greater quantities than Earth basalts of elements that do not evaporate easily, such as titanium, chromium, and zirconium, and smaller amounts of easily evaporated elements, such as potassium and sodium. In addition, lunar basalts contain no water. These chemical differences probably indicate that the lunar materials were heated to temperatures higher than the highest temperature reached by Earth materials. The easily vaporized material was driven off, leaving behind a higher percentage of materials able to withstand high temperatures.

Crystalline rocks from the heavily cratered regions of the Moon known as the lunar highlands are of two principal types: (1) a basalt often called *KREEP* because it is enriched in *potassium* (K), the *rare earth elements* (REE), and *phosphorus* (P); and (2) *anorthosite,* a gabbroic rock composed almost entirely of calcium-rich plagioclase feldspar.

Lunar maria basalts range in age from about 3.9 to 3.3 billion years, indicating that most lava flows occurred during that span. Anorthosite rocks and KREEP basalts have been dated at more than 4 billion years old, confirming an earlier age for the highlands than for the maria.

Current theories about the origin of the three principal crystalline lunar rock types suggest that a surface layer of anorthosite was produced during the initial differentiation of the Moon into crust, mantle, and core; that the KREEP was formed by partial melting of the lower part of the anorthosite crust and subsequent eruption of material onto the surface; and that the maria basalt was formed by partial melting of the upper mantle and subsequent eruption.

We know much less about rocks from other parts of the solar system than we know about Moon rocks. Soviet spacecraft that landed on the surface of Venus analyzed and photographed angular to rounded rocks abundantly scattered about the landing sites (box figure 2). These rocks are slablike, and some appear to be layered. They are chemically similar in composition to terrestrial igneous rocks such as granite and basalt. Photographs show that the surface of Mars is covered by a fine-grained, orange-pink dust. Limonite and clay minerals are known to be present. Several different kinds of rocks have been observed (box figure 3). They vary greatly in size, texture, and color and include breccia and basalt. Rocks on Mercury are probably similar to lunar anorthosites.

Astrogeology Box 4.2 Figure 2
Venus surface with rocks.
TASS from Sovfoto.

Astrogeology Box 4.2 Figure 3
Martian landscape showing rocks and dunes. Boulder at left is 3 meters long. Viking lander antenna obscures middle of photo, taken in 1976.
NASA.

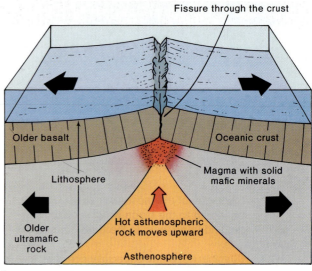

A

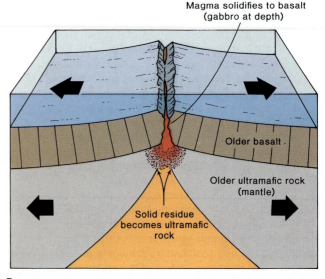

B

Figure 4.19
Schematic representation of how basaltic oceanic crust and the underlying ultramafic mantle rock form at a spreading center. The process is more continuous than the two-step diagram implies. (*A*) Partial melting of asthenosphere takes place beneath a spreading center. (*B*) The magma squeezes into the fissure system. Solid mafic minerals are left behind as ultramafic rock.

Explaining Igneous Activity by Plate Tectonics

The Origin of Basalt and Ultramafic Rocks

One of the appealing aspects of the theory of plate tectonics is that it accounts reasonably well for the variety of igneous rocks and their distribution patterns. As explained in the chapter titled "Volcanism and Extrusive Rocks," most basalt erupts along spreading centers, notably along the mid-oceanic ridge system. The basaltic magma originates from partial melting of the underlying mantle (figure 4.19). Some of the basaltic magma erupts along the submarine ridge to form pillow basalt, while some fills the fissures as dikes. The dikes are pulled apart

Harnessing Magmatic Energy

Geothermal energy, at the few sites in the world where it is being used for generating electricity, indirectly uses the heat from a buried body of magma. As explained in the chapter titled "Ground Water," water becomes heated in hot rocks. The heat source is usually presumed to be an underlying magma chamber. The rocks containing the hot water are penetrated by drilling. Steam exiting the hole is used to generate electricity.

Why not drill into and tap magma itself for energy? The amount of energy stored in a body of magma is enormous. The U.S. Geological Survey estimates that magma chambers in the United States within 10 kilometers of the earth's surface contain about 5,000 times as much energy as the country consumes each year. Our energy problems could largely be solved if significant amounts of this energy were harnessed.

There are some formidable technical difficulties in drilling into a magma chamber and converting the heat into useful energy. Despite these difficulties, the United States is seriously considering developing magmatic energy in the future. The U.S. government is sponsoring a feasibility study to identify more precisely the problems of tapping magmatic energy. Experimental drilling has been carried out in Hawaii through the basalt crust of a lava lake that formed in 1960.

As drill bits approach a magma chamber, they must penetrate increasingly hotter rock. The drill bit must be made of special alloys to prevent it from becoming too soft to cut rock. The rock immediately adjacent to a basaltic magma chamber is around 1000°C, even though that rock is solid. Drilling into the magma would require a special technique. One device currently being experimented with is a jet-augmented drill. As the drill enters the magma chamber, it simultaneously cools and solidifies the magma in front of the drill bit. Thus the drill bit creates a column of rock that extends downward into the magma chamber and simultaneously bores a hole down the center of this column. Once the hollow column is deep enough within the magma chamber, a boiler is placed in the hole. The boiler is protected from the magma by the jacket of the column of rock. Water would be pumped down the hole and turned to water vapor in the boiler by heat from the magma. Steam emerging from the hole would be used to generate electricity.

In principle, the idea is fairly simple, but there are serious technical problems. For one thing, high pressures would have to be maintained on the drill bit during drilling and while the boiler system was being installed; otherwise, gases within the magma might blast the magma out of the drill hole and create a man-made volcano. (The closest thing to a man-made volcano occurred in Iceland in 1977 when a small amount of magma broke into a geothermal steam well and erupted briefly at the well head, showering the area with a few tons of tephra.)

---

by spreading plates; and more magma fills the new fracture and erupts on the sea floor. Solidified basalt is carried away from the ridge as newer basalt is added. Sediments slowly cover the older basalt as it moves farther from the ridge.

The basalt magma that builds the oceanic crust is removed from the underlying mantle, depleting the mantle beneath the ridge of much of its calcium, aluminum, and silicon oxides. The unmelted residue (olivine and pyroxene) becomes depleted mantle, but it is still a variety of ultramafic rock. The rigid ultramafic rock, the overlying basalt, and any sediment that may have deposited on the basalt compose the lithosphere of an oceanic plate, which moves away from a spreading center over the asthenosphere.

The Origin of Granite and Andesite

Intermediate and felsic magmas are thought to be products of convergence between two plates (figure 4.20). A cool slab of oceanic lithosphere (mostly basalt and ultramafic rock that probably formed millions of years earlier, plus younger overlying sedimentary rock) descends beneath the continental crust. The descending slab gets hotter with increasing depth. Extra heat is created by friction between the descending plate and the overriding continent. Water trapped in the descending oceanic crust probably helps a magma form by lowering the melting temperatures for the rocks. Melting (more likely, partial melting) of rock takes place.

Exactly how intermediate magmas are created in one place and felsic magmas in another is uncertain. A number of hypotheses have been proposed. There is general agreement that the composite volcanoes aligned along the circum-Pacific belt are tapping into magma being generated at a fairly uniform depth beneath the earth's surface. But why is the intermediate magma created at such a level? And how does subduction relate to the formation of a granitic magma? Several hypotheses dealing with these questions are currently being debated and are briefly described here.

Partial melting of basalt As temperature increases with the descent of the oceanic slab, the basalt begins to melt. The initial melt is less mafic than basalt. The resulting magma is of felsic or, more likely, intermediate composition. Some geologists think that this is how the

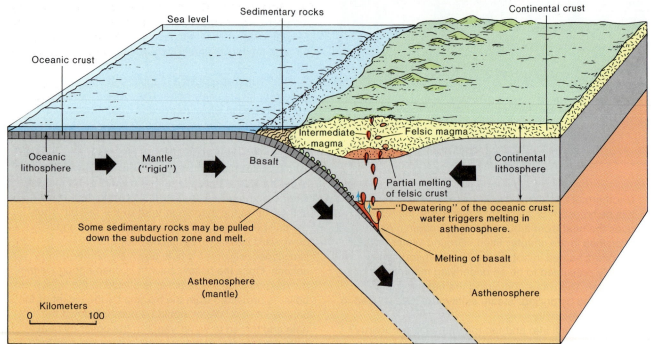

Figure 4.20
Some processes that may contribute to magma generation at a converging boundary.

andesite of the composite volcanoes forms. However, because this process, at best, could create only a minor amount of granitic melt, it is not regarded as the source of magma for the great granite batholiths.

Melting of sedimentary rock The sedimentary rock that rests on the basalt of the descending oceanic plate is much richer in silica than the basalt itself. The sedimentary rock therefore melts at a lower temperature than the basalt and forms a felsic magma. (This alone could probably not account for the huge volume of granitic batholiths in the continental crust.)

Partial melting of mantle Figure 4.20 shows part of the asthenosphere wedged between the descending oceanic crust and the overlying continental lithosphere.

In figure 4.20, note the zone where "dewatering" of the oceanic crust takes place. Here, water that had collected in the oceanic crust when it was beneath the ocean is driven out because of the heating of the descending plate. The water rises into the overlying asthenosphere and lowers the melting temperature of this part of the mantle sufficiently so that partial melting creates a magma; probably a basaltic or an andesitic magma. The magma rises (leaving behind a solid residue of ultramafic rock). Other processes, such as differentiation or assimilation of crustal rocks, may modify the mafic magma on its surfaceward journey.

Assimilation of crustal rocks Rocks of the overlying continental crust are more felsic than those of the descending oceanic plate. A mafic magma is generated, as described above, either in the asthenosphere or from oceanic crust beneath the continental lithosphere. While moving upward into the overriding continental crust, the more mafic magma can absorb some of the more silica-rich rocks, becoming intermediate in composition.

Partial melting of the lower crust Continental crust is thicker under the mountains on the edge of the continent. Because this "root" zone of mountains is at a lower level, the rock is hotter. A high enough temperature causes partial melting, creating a very felsic magma. (This process was regarded as a likely source for granitic magmas even before plate tectonics became a popular theory.)

Heating by transient magma or by underplating Intermediate or mafic magma formed beneath the continent would travel relatively quickly to the earth's surface. Perhaps a large number of fissures lead it through the lower continental crust (like a radiator heating system in which hot-water pipes disperse heat to surroundings), and the already hot rocks of the lower continental crust are triggered into melting into granitic magma (figure 4.21A). The granitic magma, being more viscous, wells up much more slowly than the rapidly moving andesite or basalt magma.

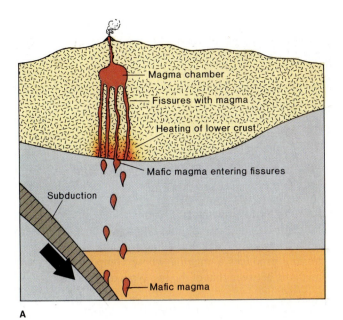

A

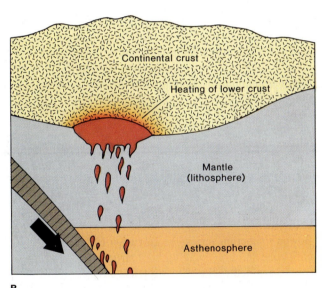

B

Figure 4.21
Two ways in which mafic magma could add heat to the lower crust and result in partial melting to form a granitic magma. (*A*) Mafic magma from the asthenosphere rises through closely spaced fissures in the lower crust (widths are highly exaggerated in diagram). (*B*) Magmatic underplating of the continental crust.

Alternatively, mafic magmas may pool up beneath the continental crust in a process known as *magmatic underplating* (figure 4.21*B*). The mafic magma acts like a hot plate beneath the continental crust. Partial melting of the overlying crust results in a granitic magma.

Combinations of processes None of these hypotheses alone seems to account for the granitic and andesitic igneous phenomena. Most hypotheses currently being considered combine two or more of these to account for what happens.

What occurs might happen as follows. Perhaps the subducted basaltic crust melts as it slides under the hot asthenosphere that underlies the continental lithosphere. Or, perhaps more likely, the overlying asthenosphere is partially melted when water is given off from the descending oceanic slab. Basaltic magma is created. As the basaltic magma ascends, differentiation takes place and some mafic minerals are removed from the magma, resulting in an intermediate magma. The intermediate magma works its way relatively quickly through the continental lithosphere to erupt and help build an andesitic composite volcano. As an alternative to differentiation, the basaltic magma becomes contaminated by felsic material when it reaches the lower part of the continental crust. Contamination may be caused by the assimilation of felsic rocks by the basaltic magma or by the mixing of a felsic magma (generated from the continental crust) with the basaltic magma that had formed at a lower level. In either case, the resulting magma would be intermediate in composition.

To explain the great volumes of granitic plutonic rocks, most geologists think that partial melting of the lower crust must take place. Since the felsic rocks of the continental crust have relatively low melting temperatures, partial melting of the lower continental crust is likely. Currently, geologists think that magmatic underplating plays an important role as the extra heat source needed to generate granitic magmas in the lower continental crust. Initially, some of the mafic magma coming from the asthenosphere works its way through fissures systems in the lower crust (figure 4.21A). As the lower crust gets hotter, the rock becomes more plastic and melting begins. Fissures are sealed. The denser, mafic magma then pools under the lighter, partially molten, lowest crust (figure 4.21B). Heat from the cooling and crystallizing mafic magma is conducted upward to create larger volumes of felsic magma by partially melting more of the continental crust. The felsic magma, in turn, separates from its solid residue and works its way upward to a higher level of the crust where it slowly solidifies to a pluton.

We should emphasize that the picture we have presented is not an observation but a plausible interpretation of available data. There are a number of variations of this picture based on different interpretations of the data. Plate tectonics has not provided final answers to the problems raised by analyzing igneous phenomena. But it has provided a broad framework in which to work toward solutions to the problems.

Summary

Intrusive rocks are igneous rocks that formed underground. Some intrusive rocks have solidified near the surface as a direct result of volcanic activity. Volcanic *necks* solidified within volcanoes. Fine-grained *dikes* and *sills* may also have formed in cracks during local extrusive activity. A sill is *concordant*—parallel to the planes within the country rock. A dike—which is *discordant*—cuts across these planes. Both are tabular bodies. Coarser grains in either a dike or a sill indicate that it probably formed at considerable depth.

Most intrusive rock is *plutonic*—that is, coarse-grained rock that solidified slowly at considerable depth. Most plutonic rock is in *batholiths*—large plutons with no particular shape. A smaller, irregular body is called a *stock*.

Plutonic rocks are, like extrusive rocks, named on the basis of their mineral content, which in turn reflects the chemical composition of the magmas from which they formed. *Granite, diorite,* and *gabbro* are the coarse-grained equivalents of *rhyolite, andesite,* and *basalt,* respectively. *Ultramafic* rocks, made entirely of ferromagnesian minerals, have no fine-grained equivalent, and most of them probably originate in the mantle.

The pattern of worldwide distribution of intrusive rock contrasts sharply with the distribution of extrusive rocks. Basalt strongly predominates in the oceanic crust. Granite strongly predominates in the continental crust. Younger granite batholiths occur mostly within younger mountain belts. Andesite is largely restricted to narrow zones along convergent plate boundaries.

Each mineral within a rock has a different melting temperature. Minerals within a basalt or a gabbro have higher melting points than those within a granite or a rhyolite. If the temperature rises, not all of a rock melts at once; a magma forms from only the components of the rock that melt at that particular temperature. Water dissolved in a magma and kept there by high pressure reduces the melting temperature. High pressure without water present has the opposite effect. If very hot rock without water is solid under high pressure, reducing the pressure may allow the rock to melt partially or totally.

The heat for melting of rocks comes from several sources. The *geothermal gradient* is the increase in temperature with increase in depth. *Friction* from rocks sliding past other rocks can generate heat, as can *radioactive decay* of certain elements. Hot *mantle plumes,* from the lower mantle, may bring in more heat along the mid-oceanic ridges and perhaps elsewhere.

No single process can satisfactorily account for all igneous rocks. Several hypotheses adequately explain some igneous rocks. In the process of *differentiation,* based on *Bowen's reaction series,* a residual magma more felsic than the original mafic magma is created when the early-forming minerals separate out of the magma. In *assimilation,* a hot, original magma is contaminated by picking up and absorbing rock of a different composition. *Magma mixing* produces a magma whose composition is intermediate, between that of the two types of magma that were mixed.

Partial melting produces a felsic magma from more mafic rocks because the temperature attained is not enough to melt the mafic minerals.

The theory of *plate tectonics* incorporates various parts of previous theories. Basalt is generated where hot mantle rock partially melts. The fluid magma rises easily through fissures, if present. The ferromagnesian portion that stays solid remains in the mantle as ultramafic rock. Granite and andesite are associated with subduction. Differentiation, assimilation, partial melting, and mixing of magmas may each play a part in creating the appropriate rocks.

Terms to Remember

assimilation	granite
batholith	inclusion
Bowen's reaction series	intrusion (intrusive structure)
chill zone	
coarse-grained rocks	intrusive rock
concordant	laccolith
country rock	pegmatite
crystal settling	pluton
differentiation	plutonic rock
dike	sill
diorite	stock
discordant	stoping
gabbro	ultramafic rock
geothermal gradient	volcanic neck

Questions for Review

1. Why do mafic magmas tend to reach the surface much more often than felsic magmas?
2. If basalt were subjected to differential melting, what would you expect the magma to be composed of? What would be retained as solid material?
3. What rock would probably form if magma that was feeding composite volcanoes solidified at considerable depth?
4. Why is a higher temperature required to form magma at the oceanic ridges than in the continental crust?
5. What is the difference between feldspar found in gabbro and feldspar found in granite?
6. What is the difference between a dike and a sill?
7. How are most ultramafic rocks believed to have formed?
8. Describe the differences between the continuous and the discontinuous branches of Bowen's reaction series.

Questions for Thought

1. In parts of major mountain belts there are sequences of rocks that geologists interpret as slices of ancient oceanic lithosphere. Assuming that such a sequence formed at a spreading center and was moved toward a converging boundary by plate motion, what rock types would you expect to make up this sequence, going from the top downward?
2. What would happen, according to Bowen's reaction series, under the following circumstances: olivine crystals form and only the surface of each crystal reacts with the melt to form a coating of pyroxene that prevents the interior of olivine from reacting with the melt?

Supplementary Readings

Barker, D. S. 1983. *Igneous rocks*. Englewood Cliffs, N.J.: Prentice-Hall.

Best, M. G. 1982. *Igneous and metamorphic petrology*. New York: W. H. Freeman.

Cox, K. G., J. D. Bell, and R. J. Pankhurst. 1979. *The interpretation of igneous rocks*. London: Unwin Hyman.

Ernst, W. G. 1969. *Earth materials*. Englewood Cliffs, N.J.: Prentice-Hall.

Hess, P. C. 1989. *Origins of igneous rocks*. Boston: Harvard University Press.

Hyndman, D. W. 1985. *Petrology of igneous and metamorphic rocks*. 2d ed. New York: McGraw-Hill.

MacKenzie, W. S., C. H. Donaldson, and C. Guilford. 1982. *Atlas of igneous rocks and their textures*. New York: Halsted Press.

Wilson, M. 1988. *Igneous petrogenesis, a global tectonic approach*. London: Unwin Hyman.

Zim, H. S., and P. R. Shaffer. 1967. *Rocks and minerals*. New York: Golden Press.

In this chapter, you will study several visible signs of weathering in the world around you, such as the cliffs and slopes of the Grand Canyon and the rounded edges of boulders. As you study these features, keep in mind that weathering processes made the planet suitable for human use. From the weathering of rock eventually came the development of soil, on which the world's food supply depends.

How and why does rock weather? You learned in the chapters on volcanic and intrusive rocks that the minerals making up igneous rocks crystallize at relatively high temperatures and sometimes at high pressures as magma and lava cool. Although these minerals are stable when they form, most of them are not stable during prolonged exposure to the much lower temperatures and pressures at the earth's surface. In this chapter you see how minerals and rocks change when they are subjected to the physical and chemical conditions existing at the earth's surface. Rocks undergo mechanical weathering (physical disintegration) and chemical weathering (decomposition) as they are attacked by air and water. Your knowledge of the chemical composition and atomic structure of minerals will help you understand the reactions that occur during chemical weathering.

Weathering processes create sediments (primarily mud and sand) and soil. Sedimentary rocks, which form from sediments, are discussed in a later chapter. In a general sense, weathering prepares rocks for erosion and is a fundamental part of the rock cycle, transforming rocks into the raw material that eventually becomes sedimentary rocks.

CHAPTER

5

Weathering and Soil

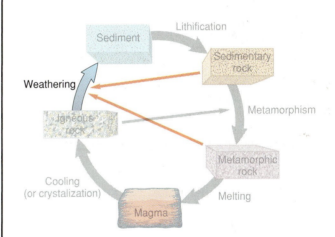

Rock arches formed by weathering of sandstone, Arches National Park, Utah.
David Muench Photography.

Weathering, Erosion, and Transportation

Rocks exposed at the earth's surface are constantly being altered by water, air, changing temperature, and other environmental factors. The term **weathering** refers to the group of destructive processes that change the physical and chemical character of rock at or near the earth's surface. The tightly bound crystals of an igneous rock can be loosened and altered to new minerals by weathering.

It is important to distinguish between weathering and *erosion,* and between erosion and *transportation.* Weathering breaks down rocks that are either stationary or moving. **Erosion** is the picking up or *physical removal* of rock particles by an agent such as streams or glaciers. Weathering helps break down a solid rock into loose particles that are easily eroded. Most eroded rock particles are at least partially weathered, but rock can be eroded before it has weathered at all. A stream can erode weathered or unweathered rock fragments.

After a rock fragment is picked up (eroded), it is transported. **Transportation** is the movement of eroded particles by agents such as rivers, waves, glaciers, or wind. Weathering processes continue during transportation. A boulder being transported by a stream can be physically worn down and chemically altered as it is carried along by the water.

How Weathering Alters Rocks

Rocks undergo both mechanical weathering and chemical weathering. **Mechanical weathering** (or physical disintegration) includes several processes that break rock into smaller pieces. The change in the rock is physical; there is little or no chemical change. For example, water freezing and expanding in cracks can cause rocks to disintegrate physically. **Chemical weathering** is the decomposition of rock from exposure to water and atmospheric gases (principally carbon dioxide and water vapor). As rock is decomposed by these agents, new chemical compounds form.

Mechanical weathering breaks up rock but does not change the composition. A large mass of granite may be broken into smaller pieces by frost action, but its original crystals of quartz, feldspar, and ferromagnesian minerals are unchanged. On the other hand, if the granite is being chemically weathered, some of the original minerals are chemically changed into different minerals. Feldspar, for example, will change into a clay mineral (with a crystal structure similar to mica). In nature, mechanical and chemical weathering usually occur together, and the effects are interrelated.

Weathering is a relatively long, slow process. Typically, cracks in rock are enlarged gradually by frost action or plant growth (as roots pry into rock crevices), and as a result, more surfaces are exposed to attack by chemical agents. Chemical weathering initially works along contacts between mineral grains. Tightly bound crystals are loosened as weathering products form at their contacts. Mechanical and chemical weathering then proceed together, until a once tough rock slowly crumbles into individual grains.

Solid minerals are not the only products of chemical weathering. Some minerals—calcite, for example—dissolve when chemically weathered. We can expect limestone and marble, rocks consisting mainly of calcite, to weather chemically in quite a different way than granite.

Effects of Weathering

The results of chemical weathering are easy to find. Look along the edges or corners of old stone structures for evidence. The inscriptions on statues and tombstones that have stood for several decades may no longer be sharp (figure 5.1). Building blocks of limestone or marble exposed to rain and atmospheric gases may show solution effects of chemical weathering in a surprisingly short time. Granite buildings may also show the effects of weathering, although the effects may not become apparent for centuries. Mineral grains in granite may be loosened, cracks enlarged, and the surface discolored and dulled by the products of weathering. Surface discoloration is also common on rock *outcrops,* where rock is exposed to view, with no plant or soil cover. That is why field geologists carry rock hammers—to break rocks to examine unweathered surfaces.

We tend to think of weathering as destructive because it mars statues and building fronts. As rock is destroyed, however, valuable products can be created. Soil is produced by rock weathering, so most plants depend on weathering for the soil they need in order to grow. In a sense, then, all agriculture depends upon weathering. Weathering products dissolved in the sea make sea water salty and serve as nutrients for many marine organisms. Some metallic ores, such as those of copper and aluminum, are concentrated into economic deposits by chemical weathering.

Many weathered rocks display interesting shapes. A **spheroidally weathered boulder** is a rock that has been rounded by weathering from an initial blocky shape. It is rounded because chemical weathering acts more rapidly or intensely on the corners and edges of a rock than on the smooth rock faces (figures 5.2 and 5.3).

Differential weathering is the term for varying rates of weathering in an area where some rocks are more resistant to weathering than others. Resistant rocks weather slowly, and may protrude above softer rocks that weather rapidly. Figures 5.4, 5.5, and 5.6 show some striking landforms produced by erosion of rocks that weather at different rates.

Figure 5.1
(*A*) Chemical weathering in a humid climate has pitted and discolored this marble tombstone since it was erected in 1882.

(*B*) This marble statue has lost most of the fine detail on the face by chemical weathering.

Figure 5.2
Spheroidal weathering of granite. (*A*) The edges and corners of an angular rock are attacked by weathering from more than one side and retreat faster than flat rock faces.

(*B*) In this way, angular rocks become rounded, often shedding partially weathered minerals in onionlike layers.

Figure 5.4
Pedestal rocks in Monument Park, Colorado. The caps are formed of a resistant rock that weathers more slowly than the softer rock of the pedestals.
Photo by W. H. Jackson, U.S. Geological Survey.

Figure 5.3
Water penetrating along cracks at right angles to one another in this igneous rock produces spheroidal weathering of once-angular blocks underground. Salt River Canyon, Arizona.

Figure 5.5
Differential weathering effects at Bryce Canyon National Park, Utah. Resistant rock layers protrude from the spires and cliffs. Weak rock layers weather rapidly, forming grooves.

Figure 5.6
Sedimentary rocks in the Grand Canyon, Arizona. In the foreground layers of sandstone resist weathering and form steep cliffs. Less resistant layers of shale weather to form gentler slopes between cliffs.

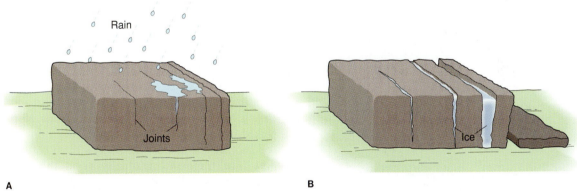

A   B

Figure 5.7
Frost wedging occurs when (*A*) water fills joints (cracks) in a rock and then freezes. (*B*) Expanding ice wedges the rock apart.

Mechanical Weathering

Of the many processes that cause rocks to disintegrate, the most effective are frost action, abrasion, and pressure release.

Frost Action

Frost action—the mechanical effect of freezing water on rocks—commonly occurs as frost wedging or frost heaving. In **frost wedging** the expansion of freezing water pries rock apart. Most rock contains a system of cracks (or *joints*—

see the chapter on "Geologic Structures"). Water that has trickled into a crack in a rock freezes and expands when the temperature drops below 32°F (0°C). The expanding ice wedges the rock apart, extending the crack or even breaking the rock into pieces (figures 5.7 and 5.8). Frost wedging is most active in regions with many days of freezing and thawing. Partial thawing adds new water to the ice in the crack; refreezing adds new ice to the old ice. The pressure of the expanding ice wedge is the most effective agent of mechanical weathering.

Figure 5.8
Frost wedging. This granite has broken as ice expanded in its cracks, some of which are sheet joints (see Pressure Release section).
Photo by Frank M. Hanna.

Figure 5.9
When these granite boulders are carried by floodwaters, they are reduced in size by abrasion as they grind against one another and against the rocky stream bed.

In **frost heaving** a layer of loose rock or soil is lifted by the expansion of freezing water. An ice layer can form just below the surface of the ground. Rain and melting snow add new water to the soil, and when this water freezes, the ice layer thickens. As the thickening ice layer expands, the ground bulges upward. Frost heaving can break up road surfaces and leave lawns spongy and misshapen after the ice thaws in spring.

Abrasion

Another process that can mechanically weather rock is **abrasion,** the grinding away of rock by friction and impact during transportation. As loose fragments of rock are picked up and moved by a stream, they tumble against one another and against the rocky stream bottom (figure 5.9). The fragments gradually grind themselves into smaller and smaller pieces and also grind away the rock of the stream bottom. Glaciers, waves, and even wind are other agents that carry and abrade rock fragments.

Pressure Release

The reduction of pressure on a body of rock can cause it to crack as it expands; **pressure release** is a significant type of mechanical weathering. A large mass of rock, such as a batholith, may originally form under great pressure from the weight of several kilometers of rock above it. This batholith is gradually exposed by uplift of the region and erosion of the overlying rock (figure 5.10). The removal of the great weight of rock above the batholith, usually termed *unloading,* allows the granite to expand upward.

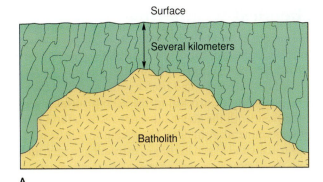

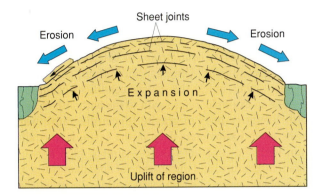

Figure 5.10
Exfoliation caused by pressure release. A granite batholith is exposed by regional uplift and the erosion of the overlying rock. Unloading reduces pressure on the granite and causes outward expansion. Sheet joints are closely spaced at the surface where expansion is greatest. Exfoliation of rock layers produces rounded exfoliation domes.

Figure 5.11
Exfoliation layers in a granite outcrop near the top of the Sierra Nevada, California. The granite formed several kilometers below the surface, and expanded outward when it was exposed by uplift and erosion. Note that the sheet joints are closer together near the top of the outcrop, where the pressure release is the greatest.

Cracks called **sheet joints** develop parallel to the outer surface of the rock as the outer part of the rock expands more than the inner part (figures 5.10 and 5.11). On slopes, gravity may cause the rock between such joints to break loose in concentric slabs from the underlying granite mass. This process of spalling off of rock layers is called **exfoliation;** it is somewhat similar to peeling layers from an onion. **Exfoliation domes** (figure 5.12) are large, rounded landforms developed in massive rock, such as granite, by exfoliation.

Several other processes mechanically weather rock but in most environments are less effective than frost action, abrasion, and pressure release. *Plant growth,* particularly roots growing in cracks (figure 5.13A), can break up rocks, as can *burrowing animals.* Such activities help to speed up chemical weathering by enlarging passageways for water and air. The *pressure of salt crystals* formed as water evaporates inside small spaces in rock also helps to disintegrate desert rocks (figure 5.13B). *Extreme changes in temperature,* as in a forest fire, can cause a rock to expand or contract until it cracks. Whatever processes of mechanical weathering are at work, as rocks disintegrate into smaller fragments the total surface area increases (figure 5.14), allowing more extensive chemical weathering by water and air.

Figure 5.12
Exfoliation dome, Yosemite National Park, California. Onion-like layers of rock are peeling off the dome.

Weathering and Soil   **105**

A

B

Figure 5.13
(*A*) Birch trees are slowly prying this rock apart as they grow within the rock's joints, Spearfish Canyon, Black Hills, South Dakota. (*B*) This rock is being broken by the growth of salt crystals, which precipitate as water evaporates within the cracks in the rock, Death Valley, California.
Photo *B* by Frank M. Hanna.

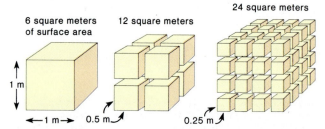

Figure 5.14
Mechanical weathering can increase the surface area of a rock, increasing the rate of chemical weathering. As a cube breaks up into smaller pieces, its volume remains the same but its surface area increases.

Chemical Weathering

The processes of chemical weathering, or *rock decomposition,* transform rocks and minerals exposed to water and air into new chemical products. A mineral that crystallized deep underground from a water-deficient magma may eventually be exposed at the earth's surface, where it can react with the abundant water there to form a new, different mineral. A mineral containing very little oxygen may react with oxygen in the air, extracting oxygen atoms from the atmosphere and incorporating them into its own crystal structure, thus forming a different mineral. These new minerals are weathering products. They have adjusted to physical and chemical conditions at (or near) the earth's surface. Minerals change gradually at the surface until they come into *equilibrium,* or balance, with the surrounding conditions. Once in equilibrium, they do not change further, unless the surface conditions change.

For an example, refer again to the batholith in figure 5.10. The large body of granite (upper diagram), which forms several kilometers deep in the earth, is in equilibrium with the conditions there—namely, high pressure, high temperature, and the absence of abundant water and atmospheric gases. The minerals within the granite are *stable* under these conditions and do not change as long as the conditions do not change. But if the granite is eventually exposed at the earth's surface by uplift and erosion (as in the lower diagram), the rock is exposed to new conditions—lower temperatures and pressures and the presence of atmospheric gases and water. Some of the minerals in granite, such as the ferromagnesian minerals and the feldspars, are *unstable* in the new environment. They gradually weather, forming new chemical combinations—in this case, clay minerals and soluble products carried off in solution—that are stable under surface conditions.

Different minerals weather at different rates. The ferromagnesian minerals in granite, for example, may weather faster than the feldspars. Many rocks exposed at the earth's surface are only partially weathered—some minerals have come into equilibrium with surface conditions, but other minerals have not. The sandy soil surrounding the granite boulders in figure 5.2*B* is composed of quartz and partially weathered feldspar grains.

Role of Oxygen

Oxygen is abundant in the atmosphere and quite active chemically, so it often combines with minerals or with elements within minerals that are exposed at the earth's surface.

The rusting of an iron nail exposed to dampness and air is a simple example of chemical weathering. Oxygen from the atmosphere combines with the iron to form iron oxide, the reaction being expressed as follows:

$$4Fe \quad + \quad 3O_2 \quad \rightarrow \quad 2Fe_2O_3$$
$$\text{iron} \quad + \quad \text{oxygen} \quad \rightarrow \quad \text{iron oxide}$$

Figure 5.15
This cliff of sandstone and shale has been stained red by hematite, released by the chemical weathering of ferromagnesian minerals, western Colorado.

Iron oxide formed in this way is a weathering product of numerous minerals containing iron, such as the ferromagnesian group (pyroxenes, amphiboles, biotite, and olivine). This type of iron oxide (Fe_2O_3) is the mineral **hematite,** which has a brick-red color when powdered. If water is present, as it usually is at the earth's surface, the iron oxide combines with water to form **limonite,** which is the name for a group of amorphous, hydrated iron oxides (often including the mineral *goethite*), which are yellowish-brown when powdered. The general formula for this group is $Fe_2O_3 \cdot nH_2O$ (the *n* representing a variable amount of water). The brown, yellow, or red color of soil and many kinds of sedimentary rock is commonly the result of small amounts of hematite and limonite released by the weathering of iron-containing minerals (figure 5.15).

Role of Acid

The most effective agent of chemical weathering is acid. Acids are chemical compounds that give off hydrogen ions (H^+) when they dissociate, or break down, in water. Strong acids produce a great number of hydrogen ions when they dissociate, and weak acids produce relatively few such ions.

The hydrogen ions given off by natural acids disrupt the orderly arrangement of atoms within most minerals. Because a hydrogen ion has a positive electrical charge and a very small size, it can substitute for other positive ions (such as Ca^{++}, Na^+ or K^+) within minerals. This substitution changes the chemical composition of the mineral

Figure 5.16
A mudpot of boiling mud is created by intense chemical weathering of the surrounding rock by the acid gases dissolved in a hot spring, Yellowstone National Park, Wyoming.

and disrupts its atomic structure. The mineral decomposes, often into a different mineral, when it is exposed to acid.

Some strong acids occur naturally on the earth's surface, but they are relatively rare. Sulfuric acid and hydrofluoric acid are strong acids emitted during many volcanic eruptions. They can kill trees and cause intense chemical weathering of rocks near volcanic vents. The bubbling mud of Yellowstone National Park's mudpots (figure 5.16) is produced by rapid weathering caused by acidic sulfur gases that are given off by some hot springs.

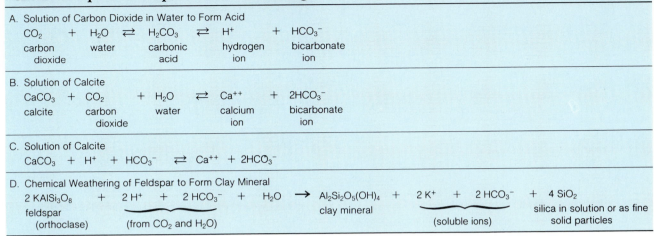

TABLE 5.1

Chemical Equations Important to Weathering

A. Solution of Carbon Dioxide in Water to Form Acid

$$CO_2 + H_2O \rightleftharpoons H_2CO_3 \rightleftharpoons H^+ + HCO_3^-$$

carbon
dioxide   water   carbonic   hydrogen   bicarbonate
                             acid   ion   ion

B. Solution of Calcite

$$CaCO_3 + CO_2 + H_2O \rightleftharpoons Ca^{++} + 2HCO_3^-$$

calcite   carbon   water   calcium   bicarbonate
               dioxide   ion   ion

C. Solution of Calcite

$$CaCO_3 + H^+ + HCO_3^- \rightleftharpoons Ca^{++} + 2HCO_3^-$$

D. Chemical Weathering of Feldspar to Form Clay Mineral

$$2\,KAlSi_3O_8 + 2\,H^+ + 2\,HCO_3^- + H_2O \rightarrow Al_2Si_2O_5(OH)_4 + 2\,K^+ + 2\,HCO_3^- + 4\,SiO_2$$

feldspar   (from CO_2 and H_2O)   clay mineral   (soluble ions)   silica in solution or as fine
(orthoclase)   solid particles

Strong acids also drain from some mines as sulfur-containing minerals such as pyrite oxidize and form acids at the surface. Uncontrolled mine drainage can kill fish and plants downstream and accelerate rock weathering.

The most important natural source of acid for rock weathering at the earth's surface is dissolved carbon dioxide (CO_2) in water. Water and carbon dioxide form *carbonic acid* (H_2CO_3), a weak acid that dissociates into the hydrogen ion and the bicarbonate ion (see equation *A* in table 5.1). Even though carbonic acid is a weak acid, it is so abundant at the earth's surface that it is the single most effective agent of chemical weathering.

The earth's atmosphere (mostly oxygen and nitrogen) contains 0.03% carbon dioxide. Some of this carbon dioxide dissolves in rain as it falls, so most rain is slightly acidic when it hits the ground. Large amounts of carbon dioxide also dissolve in water that percolates through soil. The openings in soil are filled with a gas mixture that differs from air. Soil gas has a much higher content of carbon dioxide than does air, because carbon dioxide is produced by the decay of organic matter and the respiration of soil organisms, such as worms. Rainwater that has trickled through soil is therefore usually acidic and readily attacks minerals in the unweathered rock below the soil (figure 5.17).

Solution Weathering

Some minerals are completely dissolved by chemical weathering. *Calcite,* for instance, goes into solution when exposed to carbon dioxide and water, as shown in equation *B* in table 5.1. The carbon dioxide and water combine to form carbonic acid, which dissociates into the hydrogen ion and the bicarbonate ion, as you have seen, so the equation for the solution of calcite can also be written as equation *C* in table 5.1.

There are no solid products in the last part of the equation, indicating that complete solution of the calcite has occurred. In regions underlain by the sedimentary rock limestone, which is mostly calcite, solution features such

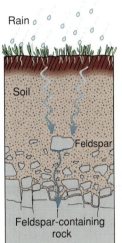

Rain picks up CO_2 from the atmosphere and becomes acidic

Water percolating through the ground picks up more CO_2 from the upper part of the soil, becoming more acidic

A feldspar crystal, loosened from the rock below, slowly alters to a clay mineral as it reacts with the acidic water

The water carries away soluble ions and SiO_2 to the ground-water supply or to a stream

Figure 5.17
Chemical weathering of a feldspar. Water percolating through the soil alters the feldspar to a clay mineral and carries away soluble ions and silica.

as caves can form when flowing water underground removes the soluble calcite. Rain can discolor and dissolve statues and tombstones carved from the metamorphic rock marble, which is also mostly calcite (figure 5.1).

Chemical Weathering of Feldspar

The weathering of feldspar is an example of the alteration of an original mineral to an entirely different type of mineral as the weathered product. When feldspar is attacked by the hydrogen ion of carbonic acid (from carbon dioxide and water), it forms clay minerals. In general, a **clay mineral** is a hydrous aluminum silicate with a sheet-silicate structure like that of mica. Therefore, the entire silicate structure of the feldspar crystal is altered by weathering: feldspar is a framework silicate, but the clay mineral product is a sheet silicate, differing both chemically and physically from feldspar. Partly because of the complexity of the reaction, the chemical weathering of feldspar proceeds at a much slower rate than the solution weathering illustrated by calcite.

BOX 5.1

Acid Rain

The burning of coal, oil, and natural gas (the *fossil fuels*) adds a great deal of carbon dioxide to the atmosphere (box figure 1). As you have seen in table 5.1, this carbon dioxide combines with water to form carbonic acid in rain. Coal and oil can also contain nitrogen and sulfur, which are given off as gases when these fuels are burned, forming nitric acid and sulfuric acid in rain. These two acids are much stronger than carbonic acid.

The strength of an acidic solution is measured on the pH scale from 0 to 14 (box figure 2). A solution of pH 7 is chemically neutral, neither acidic nor alkaline. Values below 7 are acidic; the lower the number the more acidic the solution. Values above 7 are alkaline.

Ordinary rain has a pH of about 5.5 to 6 from the small amount of carbon dioxide given off during respiration (every time we exhale, we add a little CO_2 to the atmosphere). Ordinary rain is about as acidic as milk—hardly a strong acid.

In cities and downwind of industrial smokestacks (often for hundreds of miles) the increased amount of acid gases can reduce the pH of rain to 4, 3, or even 2 (the pH of lemon juice or vinegar). This is the environmental problem termed "acid rain" (although all

Box 5.1 Figure 1
The burning of fossil fuels releases carbon dioxide and other acid gases that combine with water to produce acid rain.

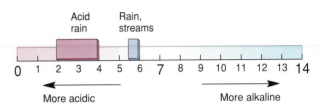

Box 5.1 Figure 2
The pH scale.

rain is really acid). Such low pH values are hard on organisms; fish may die in streams and lakes polluted by acid rain, and forests suffer under acid rain. Chemical weathering is accelerated by such rain. Statues and stone buildings in cities weather many times faster than stone structures in rural areas free of acid rain.

Let us look in more detail at the weathering of feldspar (equation *D* in table 5.1). Rainwater percolates down through soil, picking up carbon dioxide from the atmosphere and the upper part of the soil. The water, now slightly acidic, comes in contact with feldspar in the lower part of the soil (figure 5.17), as shown in the first part of the equation. The acidic water reacts with the feldspar and alters it to a clay mineral.

The hydrogen ion (H^+) attacks the feldspar structure, becoming incorporated into the clay mineral product. When the hydrogen moves into the crystal structure, it releases potassium (K) from the feldspar. The potassium is carried away in solution as a dissolved ion (K^+). The bicarbonate ion from the original carbonic acid does not enter into the reaction; it reappears on the right side of the equation. The soluble potassium and bicarbonate ions are carried away by water (ground water or streams).

All the silicon from the feldspar cannot fit into the clay mineral, so some is left over and is carried away as silica (SiO_2) by the moving water. This excess silica may be carried in solution or as extremely small solid particles.

The weathering process is the same regardless of the type of feldspar: K-feldspar (orthoclase) forms potassium ions; Na-feldspar and Ca-feldspar (plagioclase) form sodium ions and calcium ions, respectively. The ions that result from the weathering of Ca-feldspar are calcium ions (Ca^{++}) and bicarbonate ions (HCO_3^-), both of which are very common in rivers and in underground water, particularly in humid regions.

Chemical Weathering of Other Minerals

The weathering of ferromagnesian or dark minerals is much the same as that of feldspars. Two additional products are found on the right side of the equation—magnesium ions and iron oxides (hematite and limonite). Micas also weather in much the same way (biotite, of course, is both a mica and a ferromagnesian mineral).

In general, the rock-forming minerals weather at rates that are proportional to their positions in Bowen's reaction series (see the chapter on "Intrusive Activity and the Origin of Igneous Rocks"). A high-temperature mineral,

Weathering Products of Common Minerals

Original Mineral	Under Influence of CO_2 and H_2O	Main Solid Product		Other Products (Mostly Soluble)
Feldspar	⟶	Clay mineral	+	Ions, SiO_2
Ferromagnesian minerals (including biotite mica)	⟶	Clay mineral	+	Ions, SiO_2, Fe oxides
Muscovite mica	⟶	Clay mineral	+	Ions, SiO_2
Quartz	⟶	Quartz grains (sand)		
Calcite	⟶	—		Ions

such as pyroxene, forms under conditions very different from surface conditions. When exposed to surface conditions, it is farther out of equilibrium than a low-temperature mineral, such as orthoclase, so pyroxene weathers faster than orthoclase. Similarly, a Ca-rich plagioclase weathers faster than an Na-rich plagioclase.

Quartz is quite stable at the earth's surface. It does not weather chemically, although it is subject to mechanical weathering over very long periods of time. According to Bowen's reaction series, quartz is the last mineral to form, crystallizing at a lower temperature than other minerals. Quartz is resistant to chemical weathering for two reasons: (1) It originally formed at a relatively low temperature; (2) quartz (SiO_2) lacks ions such as Ca^{++}, K^+, and Na^+ that are easily attacked and replaced by H^+. Most other common minerals, which do weather chemically, contain at least one of these ions.

Weathering Products

Table 5.2 summarizes weathering products for the common minerals. Note that quartz and clay minerals commonly are left after complete chemical weathering of a rock. Sometimes other solid products, such as iron oxides, also are left after weathering.

The solution of calcite supplies substantial amounts of calcium ions (Ca^{++}) and bicarbonate ions (HCO_3^-) to underground water. The weathering of Ca-feldspars (plagioclase) into clay minerals can also supply Ca^{++} and HCO_3^- ions, as well as silica (SiO_2), to water. Under ordinary chemical circumstances, the dissolved Ca^{++} and HCO_3^- can combine to form solid $CaCO_3$ (calcium carbonate), the mineral calcite. Dissolved silica can also precipitate as a solid from underground water. This is significant because calcite and silica are the most common materials precipitated as sedimentary rock *cement*, which binds loose particles of sand, silt, and clay into solid rock (see the chapter on "Sediments and Sedimentary Rocks"). The weathering of calcite, feldspars, and other minerals is a likely source for such cement.

If the soluble ions and silica are not precipitated as solids, they remain in solution and may eventually find their way into a stream and then into the ocean. Enormous quantities of dissolved material are carried by rivers into the sea (one estimate is 4 billion tons per year). This is the main reason sea water is salty.

Soil

In engineering and construction, soil is the usual name for any kind of loose, unconsolidated earth material; but most geologists commonly use the term **soil** for a layer of *weathered,* unconsolidated material on top of bed rock (a general term for rock beneath soil). Soil scientists further restrict the term *soil* to layers of weathered, unconsolidated material that contains organic matter and is capable of supporting plant growth. (If this definition is used, then the term *regolith* can be applied to any loose surface sediment; soil would be the upper part of the regolith.) A mature, fertile soil is the product of centuries of mechanical and chemical weathering of rock, combined with the addition and decay of plants and other organic matter.

The common gardening term *loam* refers to a soil of approximately equal amounts of sand, silt, and clay along with a generous amount of organic matter. Such soils are usually well drained and very fertile; they are the best gardening soils. *Topsoil,* which is basically the same thing as loam, is the upper part of the soil and is more fertile than the underlying *subsoil,* which is often stony and lacks organic matter.

Clay minerals and quartz, the two minerals usually remaining after complete weathering of rock (table 5.2), have important roles in soil development and plant growth. Quartz crystals form sand grains that help keep soil loose and aerated, allowing good water drainage. (Partially weathered crystals of feldspar and other minerals can also form sand-sized grains.) Clay minerals help to hold water in a soil. Since most plants need both air and water in a soil for optimum growth, the best soils for plant growth have a balance of clay minerals and sand.

Soil Horizons

As soils mature, a distinct layering becomes apparent (figure 5.18). Layers of soil distinguishable by characteristic physical or chemical properties (including organic-matter content) are termed **soil horizons.** Boundaries between soil horizons are usually transitional rather than sharp.

The **A horizon,** or **zone of leaching,** is the top layer of soil and is characterized by the downward movement of water. Part of the rain falling on the ground percolates downward through the soil. This tends to leach, or wash

BOX 5.2

Clay Minerals and Plant Growth

Clay minerals are hydrous aluminum silicates that occur as microscopic plates with a sheet-silicate structure. Because of ion substitution within the crystal, most clay minerals have a negative electrical charge on the flat faces of the plates.

This negative charge is important to life on earth because it holds water in the soil and helps retain plant nutrients in soil.

The water molecule, made up of two hydrogen atoms and one oxygen atom, is neutral in charge but has a positive end and a negative end. The negative charge on the flat faces of the clay mineral attracts the positive ends of the water molecules to the clay flake (box figure 1). The clay holds the water loosely enough that most of it is available for uptake by plant roots.

Plant nutrients, such as Ca^{++} and K^+, commonly supplied by the weathering of minerals such as feldspar, are also held loosely on the surface of clay minerals. A plant root is able to release H^+ from organic acids and exchange it for the Ca^{++} and K^+ that the plant needs for healthy growth (box figure 2).

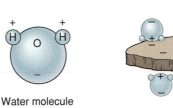

Box 5.2 Figure 1
Negative charges on a clay mineral attract positive ends of water molecules.

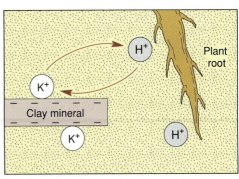

Box 5.2 Figure 2
Ion exchange between plant root and clay mineral.

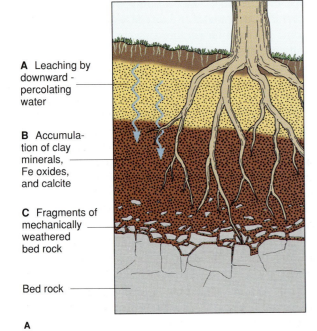

A

A Leaching by downward-percolating water

B Accumulation of clay minerals, Fe oxides, and calcite

C Fragments of mechanically weathered bed rock

Bed rock

B

Figure 5.18
(A) Soil horizons (*A*, *B*, and *C*) that form in a humid climate. **(B)** A pedalfer soil in Illinois. The surface is stained dark by humus. Horizon A is lighter in color, sandy, and crumbly. The hammer rests on the clayey horizon *B*, which is stained red by hematite, leached downward from horizon *A*.

down, some of the soil materials to lower levels. In a humid (wet) climate, clay minerals, iron oxides, and dissolved calcite are most typically leached downward. Leaching may make the A horizon pale and sandy, but the uppermost part is often darkened by humus (decomposed plant material) that collects on the top of the soil. (This dark, upper layer is the loamy topsoil.)

The **B horizon,** or **zone of accumulation,** is a soil layer characterized by the accumulation of material leached downward from the A horizon above. This layer is often quite clayey and stained red or brown by hematite and limonite. Calcite may also build up in B horizons.

The **C horizon** is incompletely weathered parent material, lying below the B horizon. The parent material is commonly the underlying bed rock, which is subjected to mechanical and chemical weathering from frost action, roots, plant acids, and other agents. In such a case, the C horizon is transitional between unweathered bed rock below and developing soil above.

Residual and Transported Soils

A **residual soil** is one that develops from weathering of the rock directly beneath it. Figure 5.18 is a diagram of a residual soil developing in a humid climate from a bedrock source. Although this is a typical situation, a number of important agricultural regions in the United States and elsewhere have developed on **transported soils,** which did not form from the local rock but from parent material brought in from some other region. Transported soils usually form on sediment deposited by running water, wind, or glacial ice. For example, mud deposited by a river during times of flooding can form an excellent agricultural soil next to the river after floodwaters recede. The soil-forming mud was not weathered from the rock beneath its present location but was carried downstream from regions perhaps hundreds of miles away. Transported wind deposits called *loess* (see the chapter on "Deserts and Wind Action") are the parent material for some of the most valuable food-producing soils in the Midwest and the Pacific Northwest.

Soils, Parent Rock, Time, and Slope

The character of a soil depends partly on the parent rock it develops from. A soil developing on weathering granite will be sandy, as sand-sized particles of quartz and partially weathered feldspar are released from the granite. As time passes, the partially weathered feldspar grains weather completely, forming fine-grained clay minerals. The quartz does not weather, so the resulting soil has both sand and clay (and perhaps silt) in it.

A soil forming on basalt may never be sandy, even in its early stages of development (this depends on the relative rates of chemical weathering versus mechanical weathering). The fine-grained feldspars and pyroxenes in the basalt weather to fine-grained clay minerals. Since the parent rock had no coarse-grained minerals and no quartz to start with, the resulting soil may lack sand. Such a soil may not drain well, although it can be quite fertile.

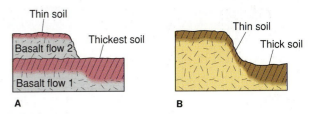

Figure 5.19
Soil thickness. (**A**) Soil thickens with time. Basalt flow 1 was exposed to soil-forming processes for a longer time than flow 2. (**B**) Steep slopes have thin soil.

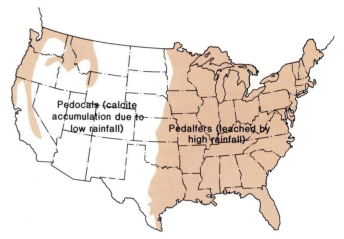

Figure 5.20
Distribution of pedalfer (colored area) and pedocal soils in the United States.
U.S. Department of Agriculture.

Note that the character of a soil changes with time. A soil developing from granite begins as a sandy soil and becomes more clayey with time. Over very long periods, the type of parent rock becomes less and less important. Given enough time, soils forming from many different kinds of igneous, metamorphic, and sedimentary rocks can become quite similar (in the same climate). The presence or absence of coarse grains of quartz in the parent rock becomes the only characteristic of the parent rock to have long-term significance.

With time, soils tend to become thicker (figure 5.19*A*). Another factor controlling soil thickness is the slope of the land surface (figure 5.19*B*). Soils tend to be thick on flat land, and thinner on steep slopes, where gravity pulls soil particles downhill.

Soils and Climate

Climate affects soil thickness and character. Soils in wet climates, as in the eastern United States, tend to be thick and are generally characterized by downward movement of water through the earth materials (figure 5.18 shows such a soil). A general term for such soils is **pedalfer** (figure 5.20) for the high content of aluminum and iron (*Al* and *Fe*) oxides they contain. Pedalfers are marked by effective downward leaching due to high rainfall and to the acids produced by the decay of abundant humus.

In arid (dry) climates, as in many parts of the western United States, soils called **pedocals** form (figures 5.20 and 5.21). These soils tend to be thin and are characterized

A

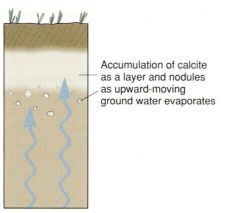

Accumulation of calcite as a layer and nodules as upward-moving ground water evaporates

B

Figure 5.21
Pedocal soils are marked by upward-moving ground water that evaporates underground in a dry climate, precipitating calcium carbonate within the soil, sometimes forming a light-colored layer.
Photo by D. Yost, USDA–Soil Conservation Service.

by little leaching, scant humus, and the *upward* movement of soil water beneath the land surface. The water is drawn up by subsurface evaporation and capillary action.

The evaporation of water beneath the land surface can cause the precipitation of salts within the soil. These salts are usually calcium salts such as calcite (which is the origin of the term pedo*cal*). An extreme example of salt buildup can be found in desert *alkali soils,* in which heavy concentrations of toxic sodium salts may prevent plant growth.

Figure 5.22
Laterite soil develops in very wet climates, where intense downward leaching carries away all but iron and aluminum oxides. Many laterites are deep red.
Photo by D. Yost, USDA–Soil Conservation Service.

Hardpan   "Hardpan" is a term for a hard layer of earth material that is difficult to dig or drill. Geologists restrict the term to a hard, often clayey, layer of cemented soil particles. Such a layer may be too hard for even backhoes to dig through; planting a tree in a lawn with a hardpan layer may require dynamite. Hardpan layers in wet climates are usually formed of clay minerals, silica, and iron compounds that have accumulated in the B horizon. In arid climates a different type of hardpan, called *caliche,* forms from the cementing of soil by calcium carbonate and other salts that precipitate in the soil as water evaporates. Both types of hardpan are really layers of rock within loose soil. A hardpan layer can break plows, prevent water drainage through the soil, and act as a barrier to plant roots. Tree roots may grow laterally along rather than down through hardpan; such shallow-rooted trees are easily uprooted by wind.

Laterites   In tropical regions where temperatures are high and rainfall is abundant, highly leached soils called **laterites** form. Under such conditions weathering is deep and intense. Laterites are usually red and are composed almost entirely of iron and aluminum oxides, generally the least soluble products of rock weathering in tropical climates (figure 5.22). If the soil is rich in hematite, it can be mined as iron ore, but tropical rainfall usually hydrates the hematite to limonite, which is seldom rich enough to mine. However, aluminum is sometimes found in nearly pure layers of *bauxite* ($Al_2O_3 \cdot nH_2O$, the principal ore of aluminum), particularly in laterites formed by the weathering of aluminum-rich igneous rocks. Because bauxite forms under conditions of tropical weathering, the United States has very little aluminum ore and depends almost entirely on tropical countries for its aluminum supply. A small percentage of our aluminum supply has come from bauxite deposits in Arkansas that formed on an igneous rock with a high aluminum content approximately 50 million years ago when the region had a tropical climate.

Laterites are relatively nonproductive soils. This may seem strange when you think of the lush jungle growth that often exists on tropical lateritic soils. Jungle vegetation, though, is nourished largely by a layer of humus on top of the soil. If the jungle and the humus layer are cleared away or burned—an increasingly common practice in tropical regions—the laterite quickly becomes incapable of sustaining plant growth, making tropical agriculture very difficult. Laterite exposed to the sun is apt to bake into a permanent, bricklike layer that makes digging nearly impossible. This hard layer can be quarried, however, and makes a durable building material.

Summary

When rocks that formed deep in the earth become exposed at the earth's surface, they are altered by *mechanical* and *chemical weathering.*

Weathering processes form spheroidally weathered boulders, differentially eroded landforms, and exfoliation domes.

Mechanical weathering, largely caused by *frost action, abrasion* during transportation, and *sheet-jointing* after unloading, disintegrates (breaks) rocks into smaller pieces.

By increasing the exposed surface area of rocks, mechanical weathering helps speed chemical weathering.

Chemical weathering results when a mineral is unstable under surface conditions, which include low temperature and pressure and the presence of water and atmospheric gases. As chemical weathering proceeds, the mineral's components recombine into products that are in equilibrium with surface conditions.

Weak acid, primarily from the solution of carbon dioxide in water, is the most effective agent of chemical weathering.

Calcite dissolves when it is chemically weathered. Most of the silicate minerals form *clay minerals* when they chemically weather. Quartz is very resistant to chemical weathering.

Soil develops by chemical and mechanical weathering of a parent material. Some definitions of soil require that it contain organic matter and be able to support plant growth.

Soils, which can be *residual* or *transported,* usually have distinguishable layers, or *horizons,* caused in part by water movement within the soil.

Climate is the most important factor determining soil type.

Pedalfers are soils characterized by downward leaching. *Pedocals* are soils marked by salt precipitation caused by evaporation of soil water.

Laterites form under conditions of intense tropical weathering. Bauxite, the ore of aluminum, may be found in laterites.

Terms to Remember

abrasion
A horizon (zone of leaching)
B horizon (zone of accumulation)
chemical weathering
C horizon
clay mineral
differential weathering
erosion
exfoliation
exfoliation dome
frost action
frost heaving
frost wedging
hematite
laterite
limonite
mechanical weathering
pedalfer
pedocal
pressure release
residual soil
sheet joints
soil
soil horizon
spheroidally weathered boulder
transportation
transported soil
weathering

Questions for Review

1. Why are some minerals stable several kilometers underground but unstable at the earth's surface?
2. Describe what happens to each mineral within granite during the complete chemical weathering of granite in a humid climate. List the final products for each mineral.
3. Explain what happens chemically when calcite dissolves. Show the reaction in a chemical equation.
4. Why do stone buildings tend to weather more rapidly in cities than in rural areas?
5. Describe at least three processes that mechanically weather rock.
6. How can mechanical weathering speed up chemical weathering?
7. Name at least three natural sources of acid in solution. Which one is most important for chemical weathering?
8. What is the difference between a residual soil and a transported soil?
9. What is a laterite and how does it form?
10. How do pedocals differ from pedalfers?

Questions for Thought

1. Consider Bowen's reaction series. Which mineral weathers at a faster rate—orthoclase or olivine? Why?
2. Why do soils develop internal layers or horizons?
3. In a humid climate, is a soil formed from granite the same as one formed from gabbro? Discuss the similarities and possible differences with particular regard to mineral content and soil color.
4. Discuss the effects of climate on chemical weathering and soil development.

Supplementary Readings

Birkeland, P. W. 1984. *Soils and geomorphology*. New York: Oxford University Press.

Carroll, D. 1970. *Rock weathering*. New York: Plenum Press.

Colman, S. M., and D. P. Dethier. 1986. *Rates of chemical weathering of rocks and minerals*. New York: Academic Press.

Gauri, K. L. The preservation of stone. *Scientific American* 238 (1978): 126–36.

Hunt, C. B. 1972. *The geology of soils*. New York: W. H. Freeman.

Keller, W. D. 1957. *The principles of chemical weathering*. Columbia, Mo.: Lucas Brothers.

Keller, W. D. 1969. *Chemistry in introductory geology*. Columbia, Mo.: Lucas Brothers.

Krauskopf, K. K. 1979. *Introduction to geochemistry*. 2d ed. New York: McGraw-Hill, chapter 4.

Loughnan, F. C. 1969. *Chemical weathering of the silicate minerals*. New York: Elsevier.

McNeil, M. 1964. Lateritic soils. *Scientific American* (November). Offprint #870. New York: W. H. Freeman.

Ollier, C. D. 1969. *Weathering*. New York: American Elsevier.

Singer, M. J., and D. N. Munns. 1986. *Soils*. New York: Macmillan.

The rock cycle (chapter 2) is a theoretical model of the constant recycling of rocks as they form, are destroyed, and then reform. We began our discussion of the rock cycle with igneous rock (chapters 3 and 4), and we now discuss sedimentary rocks. Metamorphic rocks, the third major rock type, are the subject of the next chapter.

You saw in the chapter titled "Weathering and Soil" how weathering produces sediment. In this chapter we explain more about sediment origin, as well as the erosion, transportation, sorting, deposition, and eventual lithification of sediments to form sedimentary rock. Because they have such diverse origins, sedimentary rocks are difficult to classify. We divide them into clastic, chemical, and organic sedimentary rocks, but this classification is not entirely satisfactory. Furthermore, despite their great variety, only three sedimentary rocks are very common—shale, sandstone, and limestone.

Sedimentary rocks contain numerous clues to their origin and the environmental conditions that prevailed at the time of sediment deposition. Geologists find out this information from the size and shape of rock units and from the sediment grains and the sedimentary structures such as fossils, cross-beds, ripple marks, and mud cracks that are contained in the rock.

Sediments and Sedimentary Rocks

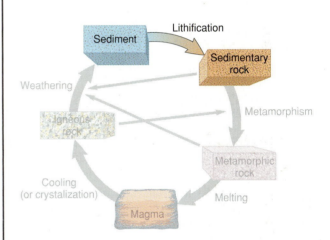

Sedimentary rock layers of limestone (cliff at top), shale, and sandstone (at river level), Grand Canyon National Park, Arizona.

Sedimentary rocks are important because they are wide-spread and because many of them, such as coal and limestone, are economically important. About three-fourths of the surface of continents is blanketed with a thin skin of sedimentary rocks. Concentrated in sedimentary rocks are important resources such as coal, crude oil, natural gas, ground water, salt, and iron ore.

Sediment

Most sedimentary rocks form from loose grains of sediment. Sediment includes such particles as sand on beaches, mud on a lake bottom, boulders frozen into glaciers, pebbles in streams, and dust particles settling out of the air. An accumulation of clam shells on the sea bottom offshore is sediment, as are coral fragments broken from a reef by large storm waves.

Sediment is the collective name for loose, solid particles that originate from:

1. Weathering and erosion of preexisting rocks.
2. Chemical precipitation from solution, including secretion by organisms in water.

These particles usually collect in layers on the earth's surface. An important part of the definition is that the particles are loose. Sediments are said to be *unconsolidated*, which means that the grains are separate, or unattached to one another.

Sediment particles are classified and defined according to the size of individual fragments. Table 6.1 shows the precise definitions of particles by size.

Gravel includes all rounded particles coarser than 2 mm in diameter (angular fragments of this size are called *rubble*). Gravel can be divided into *boulders, cobbles,* and *pebbles,* as you can see from the table. **Sand** grains are from 1/16 to 2 mm in diameter. Grains of this size are visible and feel gritty between the fingers. **Silt** grains are from 1/256 to 1/16 mm. They are too small to see without a magnifying device, such as a geologist's hand lens. Silt does not feel gritty between the fingers, but it does feel gritty between the teeth (geologists often bite sediments to test their grain size). **Clay** is the finest sediment, at less than 1/256 mm, too fine to feel gritty to fingers or teeth. *Mud* is a term loosely used for wet silt and clay.

Note that we now have two different uses of the word *clay*—a *clay-sized particle* (table 6.1) and a *clay mineral.* A clay-sized particle can be composed of any mineral at all provided its diameter is less than 1/256 mm. A clay mineral, on the other hand, is one of a small group of silicate minerals with a sheet-silicate structure. Clay minerals usually fall into the clay-size range.

Quite often the composition of sediment in the clay-size range turns out to be mostly clay minerals, but this is not always the case. Because of its resistance to chemical weathering, quartz may show up in this fine-size grade. (Most silt is quartz.) Intense mechanical weathering can break down a wide variety of minerals to clay size, and these extremely fine particles may retain their mineral

identity for a long time if chemical weathering is slow. The great weight of glaciers is particularly effective at grinding minerals down to the clay-size range, producing "rock flour," which gives a milky appearance to glacial meltwater streams (see the chapter on "Glaciers and Glaciation").

Weathering, erosion, and transportation are some of the processes that affect the character of sediment. Both weathered and unweathered rock and sediment can be eroded, and weathering does not stop after erosion has taken place. Sand being transported by a river also can be actively weathering, as can mud on a lake bottom. The character of sediment can also be altered by *rounding* and *sorting* during transportation, and by eventual *deposition.*

Transportation

Rounding is the grinding away of sharp edges and corners of rock fragments during transportation. Rounding occurs in sand and gravel as rivers, glaciers, or waves cause particles to hit and scrape against one another (figure 6.1) or against a rock surface, such as a rocky stream bed. Boulders in a stream may show substantial rounding in less than one mile of travel. Because rounding during transportation is so rapid, it is a much more important process than spheroidal weathering (see the chapter on "Weathering and Soil"), which also tends to round sharp edges.

Sorting is a process in which sediment grains are selected and separated according to grain size (or grain shape or specific gravity). Because of its high viscosity and manner of flow, a glacier does not sort the sediment it carries. Glaciers deposit all sediment sizes in the same place, so glacial sediment is usually an unsorted mixture of clay, silt, sand, and gravel. A river, however, sorts its sediment, separating sand from gravel, and silt and clay from sand. Sorting takes place because of the greater weight of larger particles. Boulders weigh more than pebbles and are more difficult for the river to transport. Similarly, it takes more of the river's energy to transport pebbles than sand, and more for sand than for silt or clay.

TABLE 6.1

Sediment Particles and Clastic Sedimentary Rocks
(Sandstone and shale are quite common; the others are relatively rare.)

Diameter (mm)	Sediment		Sedimentary Rock
256 —	Boulder	Gravel	Conglomerate (rounded particles) or Breccia (angular particles)
64 —	Cobble		
2 —	Pebble		
1/16 —	Sand		**Sandstone**
1/256 —	Silt	"Mud"	**Shale,** Siltstone, or Mudstone
	Clay		

Figure 6.1
These boulders have been rounded by abrasion as wave action rolled them against one another on this beach.

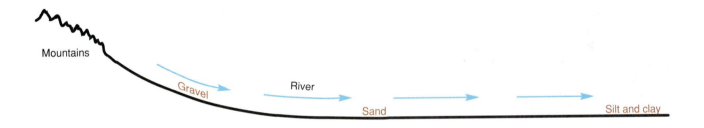

Figure 6.2
Sorting of sediment by a river. The coarse sediment is deposited first, and the finest sediment is carried the farthest.

Figure 6.2 shows a river transporting all sizes of sediment as it flows out of steep mountains onto a gentle plain, where it loses energy and slows down. As the river loses energy, the heaviest particles of sediment are deposited. The boulders come to rest first (figure 6.3). As the river continues to slow down, cobbles and then pebbles are deposited. Sand comes to rest as the river loses still more energy (figure 6.4). Finally, the river is carrying only the finest sediment—silt and clay (figure 6.5). The river has sorted the original sediment mix by grain size.

Deposition

When transported material settles or comes to rest, **deposition** occurs. Sediment is deposited when running water, glacial ice, waves, or wind loses energy and can no longer transport its load.

Deposition also refers to the accumulation of chemical or organic sediment, such as clam shells on the sea floor, or plant material on the floor of a swamp. Such sediments may form as organisms die and their remains accumulate. Deposition of salt crystals can take place as sea water evaporates. A change in the temperature, pressure, or chemistry of a solution may also cause precipitation—hot springs may deposit calcite or silica as the warm water cools.

Figure 6.3
Coarse gravel is deposited first along a river's course as the river sorts out the various sediment sizes. River gravel is usually deposited in or near steep mountains.

Sediments and Sedimentary Rocks **119**

Figure 6.4
A river carries sand farther along its course than it carries gravel.

Figure 6.5
The river on the right is carrying only silt and clay as it enters the clear river on the left. This fine sediment may come to rest only at the mouth of a river where it enters a lake or the sea.
Photo by C. W. Montgomery.

The **environment of deposition** is the location in which deposition occurs. A few examples of environments of deposition are the deep sea floor, a desert valley, a river channel, a coral reef, a lake bottom, a beach, and a sand dune. Each environment is marked by characteristic physical, chemical, and biological conditions. You might expect mud on the sea floor to differ from mud on a lake bottom. Sand on a beach may differ from sand in a river channel. Some differences are due to varying sediment sources and transporting agents, but some are the result of conditions in the environments of deposition themselves.

One of the most important jobs of geologists studying sedimentary rocks is to try to determine the ancient environment of deposition of the sediment that formed the rock. Factors that can help in determining this are the general rock relationships in the field, the features (including fossils) found within the rock, the mineral composition of the rock, and the size, shape, and surface texture of the individual sediment grains. Later in the chapter we give a few examples of interpreting sedimentary rocks.

Lithification

Lithification is the general term for a group of processes that convert loose sediment into sedimentary rock. Most sedimentary rocks are lithified by a combination of *compaction,* which packs loose sediment grains tightly together, and *cementation,* in which the precipitation of cement around sediment grains binds them into a firm, coherent rock. *Crystallization* of minerals from solution, without passing through the loose-sediment stage, is another way that rocks may be lithified.

As sediment grains settle slowly in a quiet environment such as a lake bottom, they form an arrangement with a great deal of open space between the grains (figure 6.6*A*). The open spaces between grains are called *pores,* and in a quiet environment, a deposit of sand may have 40% to 50% of its volume as open **pore space.** (If the grains were traveling rapidly and impacting one another just before deposition, the percentage of pore space will be less.) As more and more sediment grains are deposited on top of the original grains, the increasing weight of this *overburden* packs the original grains together, reducing the

amount of pore space. This shift to a tighter packing, with a resulting decrease in pore space, is called **compaction** (figure 6.6*B*). As pore space decreases, some of the interstitial water that usually fills sediment pores is driven out of the sediment.

As underground water moves through the remaining pore space, solid material called **cement** can precipitate in the pore space and bind the loose sediment grains together to form a solid rock. The cement attaches very tightly to the grains, holding them in a rigid framework. As cement partially or completely fills the pores, the total amount of pore space is further reduced (figure 6.6*C*), and the loose sand forms a hard, coherent sandstone by **cementation.**

Sedimentary rock cement is often composed of the mineral calcite or of other carbonate minerals. Dissolved calcium and bicarbonate ions are common in surface and underground waters, as you saw in the chapter on "Weathering and Soil". If the chemical conditions are right, these ions may recombine to form solid calcite, as shown in the following formula.

$$\underbrace{Ca^{++} + 2HCO_3^-}_{\substack{\text{dissolved} \\ \text{ions}}} \longrightarrow \underset{\text{calcite}}{CaCO_3} + H_2O + CO_2$$

Silica is another common cement. Iron oxides and clay minerals can also act as cement but are less common than calcite and silica. The dissolved ions that precipitate as cement originate from the chemical weathering of minerals such as feldspar and calcite. This weathering may be local, within the sediments being cemented, or very distant, with the ions being transported tens or even hundreds of miles by water.

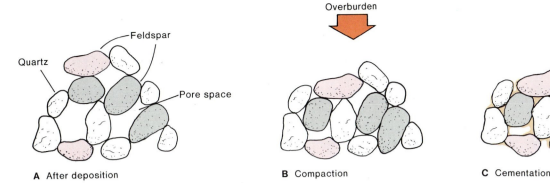

A After deposition   **B** Compaction   **C** Cementation

Figure 6.6
Lithification of sand grains to become sandstone. (**A**) Loose sand grains are deposited with open pore space between the grains. (**B**) The weight of overburden compacts the sand into a tighter arrangement, reducing pore space. (**C**) Precipitation of cement in the pores by ground water binds the sand into the rock sandstone, which has a clastic texture.

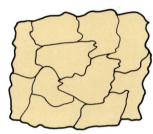

Figure 6.7
Crystalline texture. The rock is held together by interlocking crystals, which grew as they precipitated from solution. Such a rock has no cement or pore space.

A sedimentary rock that consists of sediment grains bound by cement into a rigid framework is said to have a **clastic texture.** Usually such a rock still has some pore space; cement rarely fills the pores completely (figure 6.6*C*).

Some sedimentary rocks form by **crystallization,** the development and growth of crystals by precipitation from solution (the term is also used for igneous rocks that crystallize as magma cools). These rocks have a **crystalline texture,** an arrangement of interlocking crystals that develops as crystals grow and interfere with each other (figure 6.7). Rocks that form by crystallization may never have been sediments. Crystalline rocks lack cement. They are held together by the interlocking of crystals. Such rocks have no pore space because the crystals have grown until they fill all available space.

Types of Sedimentary Rocks

Sedimentary rock is rock that has formed from (1) lithification of sediment, (2) precipitation from solution, or (3) consolidation of the remains of plants or animals. These different types of sedimentary rocks are called, respectively, *clastic, chemical,* and *organic* rocks.

Most sedimentary rocks are **clastic sedimentary rocks,** formed from cemented sediment grains that are fragments of preexisting rocks. The rock fragments can be either identifiable pieces of rock, such as pebbles of granite

or shale, or individual mineral grains, such as sand-sized quartz and feldspar loosened from rocks by weathering and erosion. Clay minerals formed by chemical weathering are also considered fragments of preexisting rocks. In most cases the sediment has been eroded and transported before being deposited. During transportation the grains may have been rounded and sorted. Table 6.1 shows the clastic rocks, such as conglomerate, sandstone, and shale, and shows how these rocks vary in grain size.

Chemical sedimentary rocks are rocks deposited by precipitation of minerals from solution. An example of inorganic precipitation is the formation of *rock salt* as sea water evaporates. Chemical precipitation can also be induced by organisms. The sedimentary rock *limestone,* for instance, can form by the precipitation of calcite within a coral reef by corals and algae. A limestone formed this way has been precipitated directly as a solid rock. Some geologists would classify such a rock as a chemical (or *biochemical*) limestone; others would call it an organic limestone.

Organic sedimentary rocks are rocks that accumulate from the remains of organisms. *Coal* is an organic rock that forms from the compression of plant remains, such as moss, leaves, twigs, roots, and tree trunks. A limestone formed from the accumulation of clam shells on the sea floor might also be called an organic rock.

Appendix B describes and helps you identify the common sedimentary rocks. The geologic symbols for these rocks (such as dots for sandstone, and a "brick-wall" symbol for limestone) are shown in Appendix F and will be used in the remainder of the book.

Clastic Rocks

Breccia and Conglomerate

Sedimentary breccia is a coarse-grained sedimentary rock formed by the cementation of coarse, angular fragments of rubble (figure 6.8). Because grains are rounded so rapidly during transport, it is unlikely that the angular fragments within breccia have moved very far from their source. Sedimentary breccia might form, for example,

Figure 6.8
Breccia is characterized by coarse, angular fragments. The cement in this rock is colored by hematite. The green and white bars on the scale are one centimeter long.

Figure 6.9
An outcrop of conglomerate. Note the rounding of cobbles, which vary in composition.

from fragments that have accumulated at the base of a steep slope of rock that is being mechanically weathered. Landslide deposits also might lithify into sedimentary breccia. This type of rock is not particularly common.

Conglomerate is a coarse-grained sedimentary rock formed by the cementation of rounded gravel. It can be distinguished from breccia by the definite roundness of its particles (figure 6.9). Because conglomerates are coarse-grained, geologists conclude that the original sediment did not travel far; some transport, however, was necessary to round the particles. Angular fragments that fall from a cliff and then are carried a few miles by a river or by waves become rounded. Such fragments could lithify to form conglomerate.

Sandstone

Sandstone is a medium-grained sedimentary rock formed by the cementation of sand grains (figure 6.10). Any deposit of sand can lithify to sandstone. Rivers deposit sand in their channels, and wind piles up sand into dunes. Waves deposit sand on beaches and in shallow water. Deep-sea currents spread sand over the sea floor. As you might

A

B

C

Figure 6.10
Types of sandstone. (*A*) Quartz sandstone; more than 90% of the grains are quartz. (*B*) Arkose; the grains are mostly feldspar and quartz. (*C*) Graywacke; the grains are surrounded by dark, fine-grained matrix. (Scale is in centimeters; most of the sand grains are about 1 millimeter in diameter.)

imagine, sandstones show a great deal of variation in mineral composition, degree of sorting, and degree of rounding.

Quartz sandstone is a sandstone in which more than 90% of the grains are quartz (figure 6.10). Because quartz is not subject to chemical weathering, it tends to concentrate residually in sand as less resistant minerals such as feldspar are weathered away, probably in a low-lying humid region that allows chemical weathering to continue for a long time. The quartz grains in a quartz sandstone are usually well-sorted and well-rounded because they have been transported for great distances. Most quartz sandstone is formed as beach sand or dune sand.

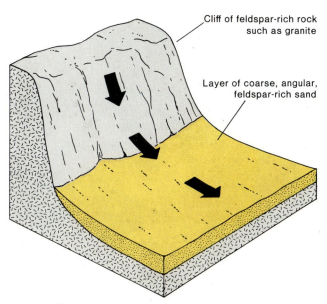

Cliff of feldspar-rich rock
such as granite

Layer of coarse, angular,
feldspar-rich sand

Figure 6.11
**Feldspar-rich sand may accumulate from the rapid erosion of
feldspar-containing rock such as granite. Steep terrain
accelerates erosion rates so that feldspar may be eroded
before it is completely chemically weathered into clay
minerals.**

A sandstone with more than 25% of the grains consisting of feldspar is called *arkose,* (figure 6.10). Because feldspar grains are preserved in the rock, the original sediment obviously did not undergo severe chemical weathering, or the feldspar would have been destroyed. Cliffs of granite are a likely source for such a sediment, for the rapid erosion associated with rugged terrain would allow unweathered feldspar to be eroded. Most arkoses contain coarse, angular grains, so transportation distances were probably short. An arkose may have been deposited at the base of a cliff, as shown in figure 6.11.

Sandstones may contain a substantial amount of **matrix,** fine-grained silt and clay found in the space between larger sand grains (figure 6.12). Matrix usually consists of fine-grained quartz and clay minerals. A matrix-rich sandstone is poorly sorted and often dark in color. It is sometimes called a "dirty sandstone."

Graywacke (pronounced "gray-whacky") is a type of sandstone in which more than 15% of the rock's volume consists of fine-grained matrix (figure 6.10). Graywackes are often tough and dense and are generally dark gray or green. The sand grains may be so coated with matrix that they are hard to see, but they typically consist of quartz, feldspar, and sand-sized fragments of fine-grained sedimentary, volcanic, and metamorphic rocks.

Most graywackes probably were deposited by **turbidity currents,** dense masses of sediment-laden water that flow downslope along the sea floor. The sediment-water mixture is heavier than clear water, so it is pulled downslope by gravity until it comes to rest on the flat sea floor at the base of the slope (figure 6.13). Turbidity currents may be generated by underwater landslides (perhaps triggered by earthquakes) or by violent surface storms such as hurricanes, which stir up bottom sediment. Sediment-laden rivers discharging directly into the sea may also cause turbidity currents.

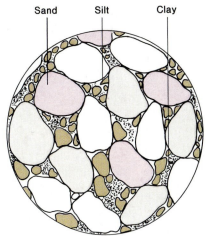

Sand   Silt   Clay

Figure 6.12
**A poorly sorted sediment of sand grains surrounded by a
matrix of silt and clay grains. Lithification of such a sediment
would produce a "dirty sandstone."**

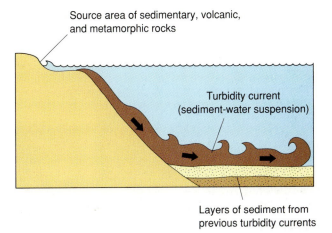

Source area of sedimentary, volcanic,
and metamorphic rocks

Turbidity current
(sediment-water suspension)

Layers of sediment from
previous turbidity currents

Figure 6.13
**A turbidity current flows downslope along the sea floor. Dense
sediment-laden water is heavier than the clear water that it
flows beneath.**

The Fine-Grained Rocks

Rocks consisting of fine-grained silt and clay are called *shale, siltstone,* and *mudstone.*

Shale is a fine-grained sedimentary rock notable for its splitting capability (called *fissility*). Splitting takes place along the surfaces of very thin layers (called *laminations*) within the shale (figure 6.14). Most shales contain both silt and clay and are so fine-grained that the surface of the rock feels very smooth. The silt and clay deposits that lithify as shale accumulate on lake bottoms, at the ends of rivers in deltas, beside rivers in flood, and on quiet parts of the deep ocean floor.

Fine-grained rocks such as shale typically undergo pronounced compaction as they lithify. Figure 6.15 shows the role of compaction in the lithification of shale from wet mud. Before compaction, as much as 80% of the volume of the wet mud may have been pore space, the pores being filled with water. The flakelike clay minerals were randomly arranged within the mud. Pressure from overlying material packs the sediment grains together and reduces the overall volume by squeezing water out of the

Sediments and Sedimentary Rocks   **123**

A

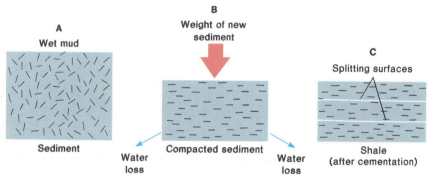

B

Figure 6.14
(*A*) An outcrop of shale in the Grand Canyon. Note how this fine-grained rock tends to break into small, flat pieces.
(*B*) Shale pieces; note the very fine grain (scale in centimeters).

A
Wet mud

B
Weight of new sediment

C
Splitting surfaces

Sediment

Water loss

Compacted sediment

Water loss

Shale (after cementation)

Figure 6.15
Lithification of shale from the compaction and cementation of wet mud. (*A*) Randomly oriented silt and clay particles in wet mud. (*B*) Particles reorient, some water is lost, and pore space decreases during compaction caused by the weight of new sediment deposited on top of the wet mud. (*C*) Splitting surfaces in cemented shale form parallel to oriented mineral grains.

pores. The clay minerals are reoriented perpendicular to the pressure, becoming parallel to one another like a deck of cards. The fissility of shale is due to splitting between these parallel clay flakes.

Compaction by itself does not generally lithify sediment into sedimentary rock. It does help consolidate clayey sediments by pressing the microscopic clay minerals so closely together that attractional forces at the atomic level tend to bind them together. Even in shale, however, the primary method of lithification is cementation.

A rock consisting mostly of silt grains is called *siltstone*. Somewhat coarser-grained than most shales, siltstones lack the fissility and laminations of shale. *Mudstone* contains both silt and clay, having the same grain size and smooth feel of shale but lacking shale's laminations and fissility. Mudstone is massive and blocky, while shale is visibly layered and fissile.

Carbonate Rocks

Limestone

Limestone is a sedimentary rock composed mostly of calcite ($CaCO_3$), usually precipitated in shallow sea water through the actions of organisms. Limestone may be precipitated directly as a solid rock in the core of a reef by corals and encrusting algae (figure 6.16). Such a rock would have a crystalline texture and would contain the fossil remains of organisms still in growth position. Geologists might describe it as a biochemical or organic limestone.

A variety of limestone called *coquina* forms from the cementation of shells that accumulated on the sea floor (figure 6.17). It has a clastic texture and is usually coarse-grained, with easily recognizable shells and shell fragments in it.

Figure 6.16
Corals precipitate calcium carbonate to form limestone in a reef. Water depth about 25 feet (about 8 meters), San Salvador Island, Bahamas.

Figure 6.17
Coquina, a variety of limestone, is formed by the cementation of coarse shells.

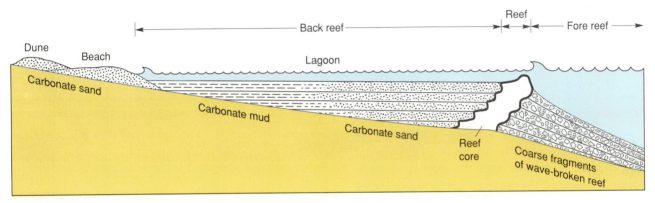

Figure 6.18
A living coral-algal reef sheds bioclastic sediment into the fore-reef and back-reef environments. The fore reef consists of coarse, angular fragments of reef. Coralline algae are the major contributors of carbonate sand and mud in the back reef. Beaches and dunes are often bioclastic sand. The sediments in each environment can lithify to form highly varied limestones.

The great majority of limestones, including coquina, are formed of wave-broken fragments of algae, corals, and shells. The fragments may be of any size (gravel, sand, silt, and clay) and are often sorted and rounded as they are transported by waves and currents across the sea floor (figure 6.18). These *bioclastic* (or *skeletal*) *limestones* take a great variety of appearances. They may be relatively coarse-grained with recognizable fossils (figure 6.19) or uniformly fine-grained and dense from the accumulation of microscopic fragments of coralline algae (figures 6.19 and 6.20). *Chalk* is a light-colored, porous, very fine-grained variety of bioclastic limestone that forms from the sea-floor accumulation of tiny marine organisms that drift near the sea surface (figure 6.21).

Oolitic limestone is a distinctive variety of limestone formed by the cementation of sand-sized *oolites* (or *ooids*), small spheres of calcite inorganically precipitated in warm, shallow sea water (figure 6.22). Strong tidal currents roll the oolites back and forth daily, allowing them to maintain a spherical shape as they grow. Wave action may also contribute to their growth.

Limestones are particularly susceptible to **recrystallization,** the process by which new crystals, often of the same composition as the original grains, develop in a rock. Calcite grains recrystallize easily, particularly in the presence of water and under the weight of overlying sediment. The new crystals that form are often large and can be easily seen in a rock as light reflects off their broad, flat faces. Because recrystallization often destroys the original clastic texture and fossils of a rock, replacing them with a new crystalline texture, the geologic history of such a rock can be very difficult to determine.

Figure 6.19
Bioclastic limestones. The two on the left are coarse-grained and contain visible fossils of corals and shells. The limestone on the right consists of fine-grained carbonate mud formed by coralline algae.

Figure 6.20
Coralline algae on the sea floor in 10 feet (3 meters) of water on the Bahama Banks. The "shaving brush" alga is Penicillus, which produces great quantities of fine-grained carbonate mud.

Figure 6.21
Chalk is a fine-grained variety of bioclastic limestone formed of the remains of microscopic marine organisms that live near the sea surface.

A

B

Figure 6.22
(*A*) Underwater dunes of oolites (ooids) chemically precipitated from sea water on the shallow Bahama Banks, south of Bimini. Tidal currents move the dunes. (*B*) An oolitic limestone formed by the cementation of oolites.

TABLE 6.2

Chemical and Organic Sedimentary Rocks

Rock	Composition	Texture	Origin
Limestone	$CaCO_3$ (Calcite)		
Bioclastic		Clastic	Cementation of fragments of shells, corals, and coralline algae
Oolitic		Clastic	Cementation of oolites (ooids) precipitated chemically from warm, shallow sea water
Dolomite	$CaMg(CO_3)_2$ (Dolomite)	Crystalline	Alteration of limestone by Mg-rich solutions (usually)
Chert	SiO_2 (Silica)	Crystalline (usually)	Cementation of microscopic marine organisms; rock usually recrystallized
Evaporites			Evaporation of sea water or a saline lake
Rock salt	NaCl (Halite)	Crystalline	
Rock gypsum	$CaSO_4 \cdot 2H_2O$ (Gypsum)	Crystalline	
Coal	Plant material rich in organic carbon	Clastic to crystalline	Accumulation and compaction of undecayed plant material (often peat)

Dolomite

The term **dolomite** (table 6.2) is used to refer to both a sedimentary rock and the mineral that composes it, $CaMg(CO_3)_2$. (Some geologists use *dolostone* for the rock.) Dolomite often forms from limestone as the calcium in calcite is partially replaced by magnesium, usually as water solutions move through the limestone.

$$Mg^{++} + 2\,CaCO_3 \rightarrow CaMg(CO_3)_2 + Ca^{++}$$

magnesium calcite   dolomite   calcium

in solution   in solution

Regionally extensive layers of dolomite are thought to form in one of two ways:

1. As magnesium-rich brines created by solar evaporation of sea water trickle through existing layers of limestone.
2. As chemical reactions take place at the boundary between fresh underground water and sea water; this boundary could migrate through layers of limestone as sea level rises or falls.

This replacement process tends to cause recrystallization of the preexisting limestone, so evidence of the rock's origin is often obscured.

A

B

Figure 6.23
(*A*) Chert nodules in Redwall Limestone, Grand Canyon.
(*B*) Bedded chert from the Coast Ranges, California.

Figure 6.24
Salt (and mud) deposited on the floor of a dried-up desert lake, Bonneville salt flats, Utah.

Figure 6.25
A bed of coal near Trinidad, Colorado.

Other Sedimentary Rocks

Chert

A hard, compact, fine-grained sedimentary rock formed almost entirely of silica, **chert** occurs in two principal forms—as irregular, lumpy nodules within other rocks and as layered deposits like other sedimentary rocks (figure 6.23). The nodules, often found in limestone, probably formed from inorganic precipitation as underground water replaced part of the original rock with silica. The layered deposits may also have formed from inorganic precipitation or from the accumulation of hard, shell-like parts of microscopic marine organisms on the sea floor, or from a combination of both.

Microscopic fossils composed of silica are abundant in some cherts. But because chert is susceptible to re-crystallization, the original fossils are easily destroyed, and the origin of many cherts remains doubtful.

Evaporites

Rocks formed from crystals that precipitate during evaporation of water are called **evaporites.** They form from the evaporation of sea water or a saline lake (figure 6.24), such

as Great Salt Lake in Utah. *Rock gypsum,* formed from the mineral gypsum ($CaSO_4 \cdot 2H_2O$), is a common evaporite. *Rock salt,* composed of the mineral halite (NaCl), may also form if evaporation continues. Other, less common evaporites include the borates, potassium salts, and magnesium salts. All evaporites have a crystalline texture.

Coal

A sedimentary rock formed from the consolidation of plant material, **coal** is rich in carbon and is usually black; it burns readily (figure 6.25). Many types of plant material can go into making coal—leaves, roots, woody tree trunks, and stems have been found as fossils in coal. Coal usually develops from *peat,* a brown, lightweight, unconsolidated or semiconsolidated deposit of moss and other plant remains that accumulates in wet bogs. Peat is transformed into coal largely by compaction after it has been buried by sediments. Several varieties of coal are recognized on the basis of the type of original plant material and the degree of compaction (chapter 21).

Valuable Sedimentary Rocks

Many sedimentary rocks have uses that make them valuable. *Limestone* is widely used as building stone and is also the main rock type quarried for crushed rock for road construction. Pulverized limestone is the main ingredient of cement and is also used as a soil conditioner in the humid regions of the United States. *Coal* is a major fuel, used widely for generating electrical power and for heating. Plaster and wallboard for home construction are manufactured from *gypsum,* which is also used to condition soils in some areas. Huge quantities of *rock salt* are consumed by industry, primarily for the manufacture of hydrochloric acid.

More familiar uses of rock salt are for table salt and for melting ice on roads. Some *chalk* is used in the manufacture of blackboard chalk, although most classroom chalk is now made from pulverized limestone. The filtering agent for beer brewing and for swimming pools is likely to be made of *diatomite,* an accumulation of the siliceous remains of microscope diatoms.

Clay from *shale* and other deposits supplies the basic material for ceramics of all sorts, from hand-thrown pottery and fine porcelain to sewer pipe. *Sulfur* is used for matches, fungicides, and sulfuric acid; and *phosphates* and *nitrates* for fertilizers are extracted from natural occurrences of special sedimentary rocks (although other sources also are used). Potassium for soap manufacture comes largely from *evaporites,* as does boron for heat-resistant cookware and fiberglass, and sodium for baking soda, washing soda, and soap. *Quartz sandstone* is used in glass manufacturing. Many *metallic ores,* such as the most common iron ores, have a sedimentary origin. *Crude oil* and *natural gas* are found almost exclusively in the pore spaces of sedimentary rocks. In chapter 21 we take a closer look at these resources and other useful earth materials.

Figure 6.26
Bedding in sandstone and shale, Utah.

Sedimentary Structures

Sedimentary structures are features found within sedimentary rock. They usually form during or shortly after deposition of the sediment, but before lithification.

One of the most prominent structures, seen in most large bodies of sedimentary rock, is **bedding,** a series of visible layers within rock (figure 6.26). Most bedding is horizontal because the sediments from which the sedimentary rocks formed were originally deposited as horizontal layers. The law of **original horizontality,** a general

principle of geology, states that most water-laid sediment is deposited in horizontal or near-horizontal layers that are essentially parallel to the earth's surface. In many cases this is also true for sediments deposited by ice or wind. Sedimentary rocks formed from such sediments preserve the horizontal layering in the form of beds. A **bedding plane** is a nearly flat surface of deposition separating two layers of rock. A change in the grain size or composition of the particles being deposited, or a pause during deposition, can create bedding planes.

Figure 6.27
Cross-bedded sandstone in Zion National Park, Utah. This cross-bedding was probably formed in sand dunes deposited by the wind.
Photo by Frank M. Hanna.

A specialized type of bedding that is not horizontal is **cross-bedding,** a series of thin, inclined layers within a larger bed of rock. The cross-beds form a distinct angle to the horizontal bedding planes of the larger rock unit (figure 6.27). Cross-bedding is found most often in sandstone. It develops as sand is deposited on steep, local slopes (figure 6.28). The distinctive patterns of cross-bedding can form in many places—in sand dunes deposited by wind; in sand ridges deposited by ocean currents on the sea floor; in sediment bars and dunes deposited by rivers in their channels; and in deltas that form at the mouths of rivers.

A **graded bed** is a layer in which particle sizes vary gradually from coarse grains at the bottom of the bed to progressively finer grains toward the top (figure 6.29). A single bed may have gravel at its base and grade upward through sand and silt to fine clay at the top. A graded bed may build up as sediment is deposited by a gradually slowing current. This seems particularly likely to happen during deposition by a *turbidity current* on the deep sea floor (figure 6.13). Figure 6.30 shows the development of a graded bed by turbidity-current deposition.

Mud cracks are a polygonal pattern of cracks formed in very fine-grained sediment as it dries (figure 6.31). Because drying requires air, mud cracks form only in sediment exposed above water. Mud cracks may form in lake-bottom sediment as the lake dries up; in flood-deposited sediment as a river level drops; or in marine sediment exposed to the air, perhaps temporarily by a falling tide.

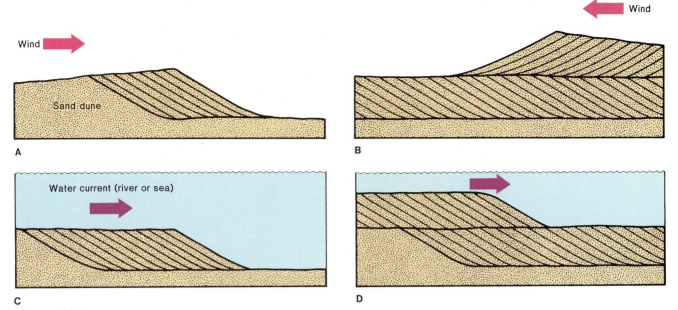

Wind →

Wind ←

Sand dune

A

B

Water current (river or sea) →

→

C

D

Figure 6.28

The development of cross-bedding in wind-blown sand (*A* and *B*) and current-deposited sand (*C* and *D*). (*A*) Sand deposits in inclined layers on the downwind side of a dune. (*B*) Second dune covers first. Cross-bedding may change orientation if wind direction shifts. (*C*) Current fills in a depression on river bottom or sea floor with sediment. (*D*) Continued sedimentation may cover first set of cross-beds with another.

Figure 6.29

A graded bed has coarse grains at the bottom of the bed and progressively finer grains toward the top.

Photo by Frank M. Hanna.

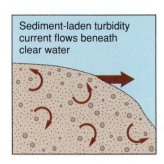
Sediment-laden turbidity current flows beneath clear water

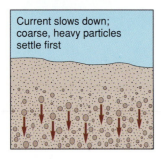
Current slows down; coarse, heavy particles settle first

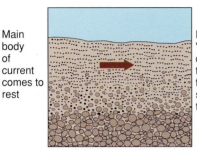
Main body of current comes to rest

Fine-grained "tail" of turbidity current continues to flow, adding fine-grained sediment to top of deposit

Progressively finer sediments settle on top of course particles

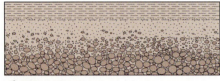
A graded bed

Figure 6.30
Development of a graded bed of sediment deposited by a turbidity current.

Cracked mud can lithify to form shale, preserving the cracks. The filling of mud cracks by sand can form casts of the cracks in an overlying sandstone.

Ripple marks are small ridges formed on the surface of a sediment layer by moving wind or water. The ridges form perpendicular to the motion. Ripple marks, found in gravel, sand, or silt, can be caused by either waves or currents of water or wind (figure 6.32). Wave-caused ripple marks are symmetric ridges; current-caused ripple marks are asymmetric, with steeper sides in the downcurrent direction. Either type can be preserved in rock (figure 6.33).

Fossils, traces of plants or animals preserved in rock, are relatively common sedimentary structures. The hard parts of organisms are most likely to be fossilized. If a bone or shell is covered by sediment before it decays, it may be preserved in the rock that forms from the sediment (figure 6.34). The original bone or shell material is seldom preserved unaltered; calcite or silica may fill the pore spaces of the fossil or may completely replace the original material. Sometimes fossils dissolve entirely, leaving an open mold in a rock. A mold may be filled in later by silica or calcite from underground water, forming a cast of the original fossil. Sometimes flat fossils of fish or leaves are formed as the weight of overburden presses most of the original material out of the fossil, leaving a thin film of black carbon behind (figure 6.35). Preserved footprints, trails, and burrows are also fossils.

Because many limestones are composed of shell and coral debris, fossils are particularly common in limestones, but they also occur in shales and sandstones. Because coal, crude oil, and natural gas form from ancient organic matter, they are sometimes called *fossil fuels.*

A

B

Figure 6.31
(A) Mud cracks in recently dried mud. (B) Mud cracks preserved in shale; they have been partially filled with sediment.

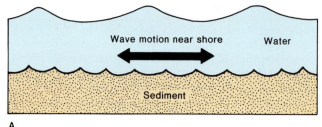

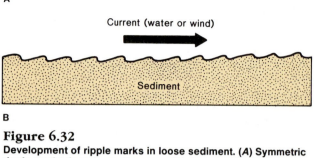

Figure 6.32
Development of ripple marks in loose sediment. (*A*) Symmetric ripple marks form beneath waves. (*B*) Asymmetric ripple marks, forming beneath a current, are steeper on their downcurrent sides.

Figure 6.33
(*A*) Current ripple marks, Alaska. Strong currents flowed to the right. (*B*) Wave ripple marks in sandstone, Australia.
Photo *A* by D. M. Hopkins, U.S. Geological Survey.
Photo *B* by E. D. McKee, U.S. Geological Survey.

Figure 6.34
Fossil shells of clams in sandstone, California. Some of the fossils retain the original white shells; in others the shells have dissolved to form open molds.

Figure 6.35
(*A*) Fossil fish in a rock from western Wyoming. (*B*) Dinosaur footprint in shale, Tuba City, Arizona.
Photo *A* by U.S. Geological Survey.

Sediments and Sedimentary Rocks   **133**

BOX 6.2

Naming of Rock Units

A **formation** is a body of rock of considerable thickness with recognizable lithologic characteristics that make it distinguishable from adjacent rock units. Although a formation is usually composed of one bed or several beds of sedimentary rock, units of metamorphic and igneous rock are also called formations. It is a convenient unit for mapping, describing, or interpreting the geology of a region.

Formations are often based on rock type. A formation may be a single thick bed of sandstone. A sequence of several thin sandstone beds could also be called a formation, as could a sequence of alternating limestone and shale beds.

The main criterion for distinguishing and naming a formation is some visible characteristic that makes it a recognizable unit. This characteristic may be rock type or sedimentary structures or both. For example, a thick sequence of shale may be overlain by basalt flows and underlain by sandstone. The shale, the basalt, and the sandstone are each a different formation. Or a sequence of thin limestone beds, with a total thickness of hundreds of feet, may have recognizable fossils in the lower half and distinctly different fossils in the upper half. The limestone sequence is divided into two formations on the basis of its fossil content.

Formations are given proper names: the first name is often a geographic location where the rock is well exposed, and the second the name of a rock type, such as Navajo Sandstone, Austin Chalk, Baltimore Gneiss, Onondaga Limestone, or Chattanooga Shale. If the formation has a mixture of rock types, so that one rock name does not accurately describe it, it is called simply "formation," as in the Morrison Formation or the Martinsburg Formation.

A **contact** is the boundary surface between two different rock types or ages of rocks. In sedimentary rock formations, the contacts are usually bedding planes.

Box figure 1 shows the three formations that make up the upper part of the canyon walls in Grand Canyon National Park in Arizona. The contacts between formations are also shown.

Box 6.2 Figure 1
The upper three formations in the cliffs of the Grand Canyon, Arizona.

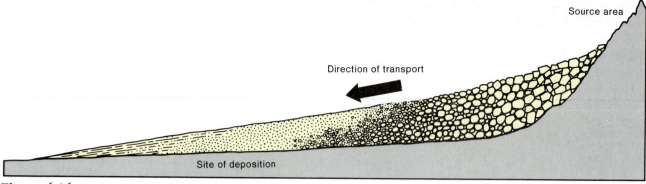

Figure 6.36
Sediment deposits usually become thinner away from the source area, and sediment grains become finer and more rounded. See Appendix F for rock symbols.

Interpretation of Sedimentary Rocks

Sedimentary rocks contain many clues to the character of the sediment's source area and environment of deposition at the time they were deposited.

Source Area

The **source area** of a sediment is the locality that eroded and provided the sediment. The most important things to determine about a source area are the type of rocks that were exposed in it and its location and distance from the site of eventual deposition.

The *rock type* exposed in the source area determines the character of the resulting sediment. The composition of a sediment can indicate the source area rock type, even if the source area has been completely eroded away. A conglomerate may contain cobbles of basalt, granite, and chert; these rock types were obviously in its source area. An arkose containing coarse feldspar, quartz, and biotite may have come from a granitic source area. A quartz sandstone containing well-rounded quartz grains, on the other hand, probably represents the reerosion and redeposition of quartz grains from preexisting sandstone. Quartz is a hard, tough mineral very resistant to rounding by abrasion, so if quartz grains are well-rounded they have undergone many cycles of erosion, transportation, and deposition, probably over tens of millions of years.

Sedimentary rocks are also studied to determine the *direction* and *distance* to the source area. Figure 6.36 shows how several characteristics of sediment vary with distance from a source area. As a general rule, sediment deposits get thinner away from the source, and the sediment grains themselves become finer and (for coarse grains) more rounded.

Sedimentary structures often give clues about the directions of old currents (*paleocurrents*) that deposited sediments. Refer back to figure 6.28 and notice how steep cross-beds slope downward in the direction of current flow. Old current direction can also be determined from asymmetric ripple marks (figure 6.32).

Figure 6.37 shows how three of these characteristics were used to determine the location of the source area for a particular rock unit in the southwestern United States. The unit is the Salt Wash Member of the Morrison Formation. (*Formation* is defined in Box 6.2, *a member* is a subdivision of a formation.) It is an important rock unit, for it contains a great deal of uranium, deposited within the rock by ground water long after the rock formed. The unit thickens and coarsens to the southwest, and crossbeds show that the old currents that deposited the sediment came from the southwest. These three facts strongly suggest that the source area was to the southwest. This information helps prospectors search for uranium within the Salt Wash.

Environment of Deposition

Figure 6.38 shows the common environments in which sedimentary rocks are deposited. Clues to the ancient environments of deposition come from a rock's composition and sedimentary structures (including fossils), as well as the size and shape of the rock unit.

Glacial environments Glacial ice deposits narrow ridges and layers of sediment in valleys and widespread sheets of sediment on plains. Glacial sediment (*till*) is an unsorted mix of unweathered boulders, cobbles, pebbles, sand, silt, and clay. The boulders and cobbles may be scratched from grinding over one another under the great weight of the ice.

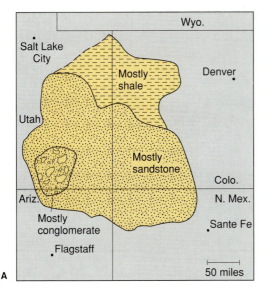

A

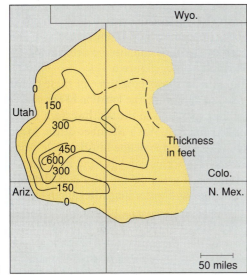

B

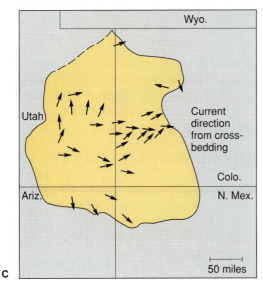

C

Figure 6.37
Characteristics of the Salt Wash Member of the Morrison Formation that help locate its source area. (A) The sediment grains become coarser to the southwest. (B) The deposit becomes thicker to the southwest. The contour lines show the thickness of the Salt Wash Member in feet. (C) Cross-bedding shows that the depositing currents came mostly from the southwest (arrows point downcurrent).
Redrawn and simplified from Craig and others, 1955, *U.S. Geological Survey Bulletin.*

Alluvial fan   As rivers emerge from mountains onto flatter plains, they deposit broad, fan-shaped piles of sediment. The sediment often consists of coarse, arkosic sandstones and conglomerates, marked by coarse cross-bedding and lenslike channel deposits (figure 6.39).

River channel and flood plain   Rivers deposit elongate lenses of conglomerate or sandstone in their channels (figure 6.40). The sandstones may be arkoses or may consist of sand-sized fragments of fine-grained rocks. River channels typically contain cross-beds and current ripple marks. Broad, flat flood plains beside channels are covered by periodic floodwaters, which deposit thin-bedded shales characterized by mud cracks and fossil footprints of animals. Hematite staining may color floodplain deposits red.

Delta   A delta is a large body of sediment deposited when a river flows into standing water, such as the sea or a lake. Deltas contain a great variety of subenvironments but are generally made up of thick sequences of siltstone and shale, marked by low-angle cross-bedding and cut by coarser channel deposits. Delta sequences may contain beds of peat or coal, as well as marine fossils such as clam shells.

Lake   Thin-bedded shale, perhaps containing fish fossils, is deposited on lake bottoms. If the lake periodically dries up, the shales will be mud-cracked and perhaps interbedded with evaporites such as gypsum or rock salt.

Beach, barrier island, dune   A barrier island is an elongate bar of sand built by wave action. Well-sorted quartz sandstone with well-rounded grains is deposited on beaches, barrier islands, and dunes. Beaches and barrier islands are characterized by cross-bedding (often low-angle) and marine fossils. Dunes have both high-angle and low-angle cross-bedding and occasionally contain fossil footprints of land animals such as lizards. All three environments can also contain carbonate sand in tropical regions, thus yielding cross-bedded clastic limestones.

Lagoon   A semienclosed, quiet body of water between a barrier island and the mainland is a lagoon. Fine-grained dark shale, cut by tidal channels of coarse sand and containing fossil oysters and other marine organisms, is formed in lagoons. Limestones may also form in lagoons adjacent to reefs (figure 6.18).

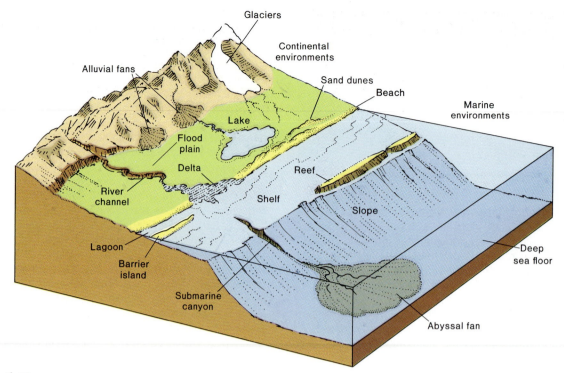

Figure 6.38
The common sedimentary environments of deposition.

Figure 6.39
Alluvial fan deposits, Baja California. A channel deposit of conglomerate occurs within the coarse-grained sequence.

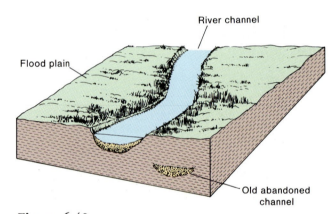

Figure 6.40
A river deposits an elongate lens of sand and gravel in its channel. Fine-grained silt and clay are deposited beside the channel on the river's flood plain.

Shallow marine environments On the broad, shallow shelves adjacent to most shorelines, sediment grain size decreases offshore. Widespread deposits of sandstone, siltstone, and shale can be deposited on such shelves. The sandstone and siltstone contain wave ripple marks, low-angle cross-beds, and marine fossils such as clams and snails. If fine-grained *tidal flats* near shore are alternately covered and exposed by the rise and fall of tides, mud-cracked marine shale will result.

Reefs Massive limestone is deposited in reef cores, with steep beds of limestone breccia forming seaward of the reef, and horizontal beds of sand-sized and finer-grained limestones forming landward (figure 6.18). All these limestones are full of fossil fragments of corals, coralline algae, and numerous other marine organisms.

Deep marine environments On the deep sea floor are deposited shale and graywacke sandstones. The graywackes are deposited by turbidity currents (figure 6.13) and typically contain graded bedding and current ripple marks.

BOX 6.3

Transgressions and Regressions

Sea level is not stable. In the geologic past sea level has risen and fallen many times, flooding and exposing much of the land of the continents as it did so.

On a very broad, shallow, marine shelf several types of sediments may be deposited. On the beach and near shore, waves will deposit sand, which is usually derived from land. Farther from shore, in deeper, quieter water, land-derived silt and clay will be deposited. If the shelf is broad enough and covered with warm water, corals and algae may form carbonate sediments still farther seaward, beyond the reach of land-derived sediment. These sediments can lithify to form a seaward sequence of sandstone, shale, and limestone (box figure 1A).

If sea level rises or the land sinks (subsides), large areas of land will be flooded and these three environments of rock deposition will migrate across the land (box figure 1B). This is a *transgression* of the sea as it moves across the land, and it can result in a bed of sandstone overlain by shale, which in turn is overlain by limestone. Note that different parts of a single rock bed are deposited at different times—the seaward edge of the sandstone bed, for example, is older than the landward edge.

In a *regression* the sea moves off the land and the three rock types are arranged in a new vertical sequence—limestone is overlain by shale, and shale by sandstone (box figure 1C). A drop in sea level alone will not preserve this regression sequence. The land must usually subside rapidly to preserve these rocks, so that they are not destroyed by continental erosion.

The angles shown in the figure are exaggerated—the rocks often appear perfectly horizontal. Geologists use these two contrasting vertical sequences of rock to identify ancient transgressions and regressions.

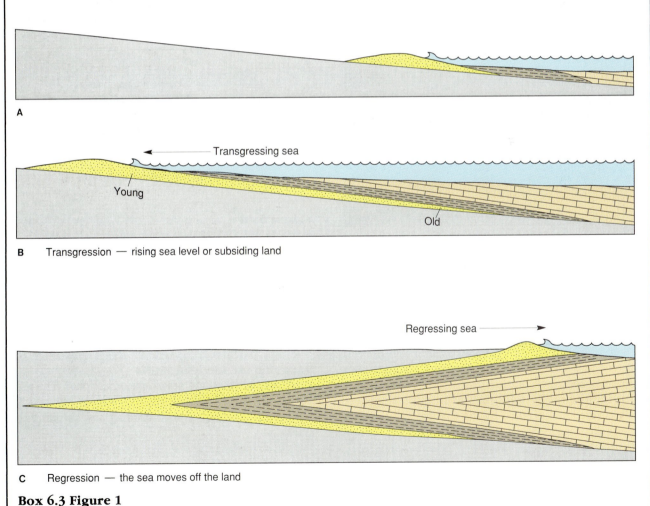

A

B   Transgression — rising sea level or subsiding land

C   Regression — the sea moves off the land

Box 6.3 Figure 1
Transgressions and regressions of the sea can form distinctive sequences of sedimentary rocks.

Summary

Sediment forms by the weathering and erosion of preexisting rocks and by chemical precipitation, sometimes by organisms.

Gravel, sand, silt, and *clay* are sediment particles defined by grain size.

The composition of sediment is governed by the rates of chemical weathering, mechanical weathering, and erosion. During transportation, grains can become rounded and sorted.

Sedimentary rocks form by *lithification* of sediment or by *crystallization* from solution. Sedimentary rocks may be *clastic, chemical,* or *organic.*

Clastic sedimentary rocks form mostly by *compaction* and *cementation* of grains. *Matrix* can partially fill the *pore space* of clastic rocks.

Conglomerate forms from coarse, rounded sediment grains that have been transported only a short distance by a river or waves. *Sandstone* forms from sand deposited by rivers, wind, waves, or turbidity currents. *Shale* forms from river, lake, or ocean mud.

Limestone consists of calcite, formed either as a chemical precipitate in a reef or, more commonly, by the cementation of shell and coral fragments or of oolites. *Dolomite* usually forms from the alteration of limestone by magnesium-rich solutions.

Chert consists of silica and usually forms from the accumulation of microscopic marine organisms. *Recrystallization* often destroys the original texture of chert (and some limestones).

Evaporites, such as rock salt and gypsum, form as water evaporates. *Coal,* a major fuel, is consolidated plant material.

Sedimentary rocks are usually found in *beds* separated by *bedding planes* because the original sediments are deposited in horizontal layers.

Cross-bedding forms where sediment is deposited on a sloping surface in a sand dune, delta, or river bar.

A *graded bed* forms as coarse particles fall from suspension before fine particles, perhaps in a turbidity current.

Mud cracks form in drying mud. *Ripple marks* form beneath waves or currents.

Fossils are the traces of an organism's hard parts or tracks preserved in rock.

A *formation* is a convenient rock unit for mapping and describing rock. Formations are lithologically distinguishable from adjacent rocks; their boundaries are *contacts.*

Geologists try to determine the *source area* of a sedimentary rock by studying its grain size, composition, and sedimentary structures. The source area's rock type and location are important to determine.

The *environment of deposition* of a sedimentary rock is determined by studying the rock unit's size, shape, composition, and sedimentary structures. Typical environments include river channels, flood plains, deltas, lakes, beaches, dunes, shallow marine shelves, and the deep sea floor.

Terms to Remember

bedding	gravel
bedding plane	limestone
cement	lithification
cementation	matrix
chemical sedimentary rock	mud crack
chert	organic sedimentary rock
clastic sedimentary rock	original horizontality
clastic texture	pore space
clay	recrystallization
coal	ripple mark
compaction	rounding
conglomerate	sand
contact	sandstone
cross-bedding	sediment
crystalline texture	sedimentary breccia
crystallization	sedimentary rock
deposition	sedimentary structure
dolomite	shale
environment of deposition	silt
evaporite	sorting
formation	source area
fossil	turbidity current
graded bed	

This chapter on metamorphic rocks, the third major category of rocks in the rock cycle, completes our description of earth materials (rocks and minerals). The information on igneous and sedimentary processes in previous chapters should help you understand metamorphic rocks, which form from *preexisting* rocks.

After reading the chapter on weathering, you know how rocks are altered when exposed at the earth's surface. *Metamorphism* (a word from Latin and Greek that means literally "changing of form") also involves alterations, but the changes are due to deep burial, tectonic forces, and high temperature rather than surface conditions.

As you study this chapter, try to keep clearly in mind how the chemical composition of a rock and the temperature, pressure, and water present each contribute to the metamorphic process and the resultant metamorphic rock.

We also discuss hydrothermally deposited rocks, which are found in association with both igneous and metamorphic rocks. Hydrothermal ore deposits, while not volumetrically significant, are of great importance to the world's supply of metals (chapter 21).

Because nearly all metamorphic rocks form deep within the earth's crust, they provide geologists with many clues about conditions at depth. Therefore understanding metamorphism will help you when we consider geologic processes involving the earth's internal forces. Metamorphic rocks are a feature of the major mountain belts. They are especially important in providing evidence of what happens during subduction.

Appendix B will help you identify the most common metamorphic rocks.

7

Metamorphism, Metamorphic Rocks, and Hydrothermal Rocks

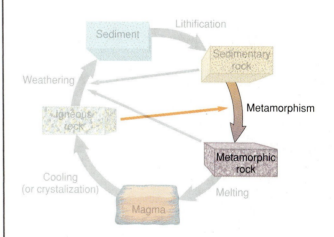

Microfolded mica in the rock shown in figure 9.1 as seen with a polarizing microscope. Width of photo is approximately 1 mm.
Photo by C. C. Plummer.

Figure 7.1
An outcrop of metamorphic rock in Washington's Cascade Mountains. Before metamorphism, the rock was a shale. The bent and deformed rock layers indicate that the rock was plastic during metamorphism.
Photo by C. C. Plummer.

$$CaCO_3 + SiO_2 \rightarrow CaSiO_3 + CO_2$$

calcite　　quartz　　wollastonite　　carbon
　　　　　　　　　　(a mineral)　　　dioxide
　　　　　　　　　　　　　　　　　　　gas

No one has observed metamorphism taking place, just as no one has ever seen a granite pluton form. What, then, leads us to believe that metamorphic rocks form in a solid state (i.e., without melting) at high pressure and temperature? Many metamorphic rocks found on the earth's surface exhibit contorted banding (figure 7.1). These rocks, now hard and brittle, would shatter if smashed with a hammer. But they must have been **plastic** (capable of being bent and molded under stress) to have been folded into such contorted patterns. Because high temperature and pressure are necessary to make rocks plastic, a reasonable conclusion is that these rocks formed at considerable depth, where such conditions exist. Had the rocks melted, however, the banding would have been destroyed.

Laboratory experiments further support the concept that metamorphic rocks form in an environment of high temperature and pressure. Certain minerals found only in metamorphic rocks have been made artificially under the kind of temperature and pressure conditions that exist 5 or more kilometers below the earth's surface.

The role of hot water in forming *hydrothermal rocks* (discussed later in this chapter) furnishes some indirect evidence of metamorphic processes, although hydrothermal rocks are not true metamorphic rocks. Large amounts of hot water that pass through fractures in rocks and deposit material (veins of quartz and other minerals) produce hydrothermal rocks in the process. We can infer from this (and other evidence) that water is important in metamorphism because of its capacity for dissolving elements from one mineral and carrying them to other minerals with which they may react.

Factors Controlling the Characteristics of Metamorphic Rocks

A metamorphic rock owes its characteristic texture and particular mineral content to several factors, the most important being (1) the composition of the parent rock before metamorphism, (2) temperature and pressure during metamorphism, and (3) the effects of fluids, such as water.

Composition of the Parent Rock

Usually no new elements or chemical compounds are added to the rock during metamorphism, except perhaps water. (Metasomatism, discussed later in this chapter, does involve the addition of other elements.) Therefore, the mineral content of the metamorphic rock is controlled by the chemical composition of the parent rock. For example, most metamorphic rocks have a fairly high amount of silica, which indicates that their parent rocks were also high in silica. A basalt always metamorphoses into a rock in which the new minerals can collectively accommodate the approximately 50% silica and relatively high amounts

From your study so far of earth materials and the rock cycle, you know that rocks change, given enough time, when their physical environment changes radically. In the chapter titled "Intrusive Activity and the Origin of Igneous Rocks," you saw how deeply buried rocks melt (or partially melt) to form magma when temperatures are high enough. What happens to rocks that are deeply buried but are not hot enough to melt? They become metamorphosed. **Metamorphism** is the solid-state transformation of preexisting rock into texturally or mineralogically distinct new rock as a result of high temperature, high pressure, or both.

The new rock, the **metamorphic rock,** in nearly all cases has a texture clearly different from that of the *original* rock, or **parent rock.** When limestone is metamorphosed to marble, for example, the fine grains of calcite coalesce and recrystallize into larger calcite crystals. The calcite crystals are interlocked in a mosaic pattern that gives marble a texture distinctly different from that of the parent limestone. If the limestone is composed entirely of calcite (without impurities), then metamorphosis into marble involves no new minerals, only a change in texture.

More commonly, the various elements of a parent rock react chemically and crystallize into new minerals, thus making the metamorphic rock distinct both mineralogically and texturally from the parent rock. This is because the parent rock is unstable in its new environment. The old minerals recrystallize into new ones that are at equilibrium in the new environment. For example, clay minerals form and are stable under low temperature and pressure conditions, such as we find at the earth's surface. When subjected to the temperatures and pressures deep within the earth's crust, the clay minerals of a shale can recrystallize into coarse-grained mica. Another example is that under appropriate temperature and pressure conditions, a quartz sandstone with a calcite cement metamorphoses as follows:

of the oxides of iron, magnesium, calcium, and aluminum in the original rock. On the other hand, a limestone, composed essentially of calcite ($CaCO_3$), cannot metamorphose into a silica-rich rock.

Temperature

A mineral is said to be *stable* if, given enough time, it does not react with or convert to a new mineral or substance. Any mineral is stable only within a given temperature range. The stability temperature range of a mineral varies with factors such as pressure and the presence or absence of other substances. Some minerals are stable over a wide temperature range. Quartz, if not mixed with other minerals, is stable at atmospheric pressure (i.e., at the earth's surface) up to about 800°C. At higher pressures, quartz remains stable to even higher temperatures. Other minerals are stable over a temperature range of only 100 or 200°C. By knowing the particular temperature range in which a mineral is stable, a geologist may be able to deduce the temperature of metamorphism for a rock that includes that mineral.

Minerals stable at higher temperatures tend to be less dense (or have a lower specific gravity) than chemically identical minerals stable at lower temperatures. This is because as temperature increases, the ions vibrate more within their sites in the crystal structure. A more open (less tightly packed) crystal structure, such as high-temperature minerals tend to have, allows greater vibration of ions. (If the heat and resulting vibrations become too great, the crystal breaks apart and the substance becomes liquid.)

Heat is also important because higher temperatures speed chemical reactions. If the temperature in an oven is raised a few degrees, a roast cooks much faster. Similarly, a chemical reaction proceeds much faster at higher temperature, provided a reaction can take place. Because of the slowness of reaction rates, metamorphism rarely takes place below about 200°C, even though minerals may be unstable at these temperatures.

The upper limit on temperature in metamorphism overlaps the temperature of partial melting of a rock. If partial melting takes place, the component that melts becomes a magma; the solid residue remains a metamorphic rock. Temperatures at which the igneous and metamorphic realms can coexist vary considerably. For an ultramafic rock containing only ferromagnesian silicate minerals, the temperature will be well over 1000°C; for a granite or rhyolite under high water pressure, it could be as low as 700°C.

Pressure

Pressure involved in metamorphism is of two types—confining (static) pressure (figure 7.2) and directed (dynamic) pressure (figure 7.3).

Confining pressure Pressure applied equally on all surfaces of a body is **confining** (or static) **pressure.** A diver senses confining pressure proportional to the weight of the water above. The pressure uniformly squeezes the diver's

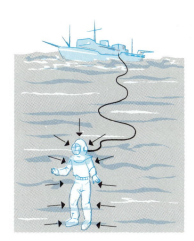

Figure 7.2
Confining pressure.

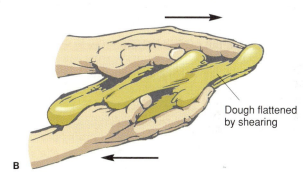

Figure 7.3
(*A*) Compressive directed pressure, and (*B*) shearing.

Dough ball

A

Dough flattened by shearing

B

entire body surface. Likewise, an object buried deep within the earth's crust is compressed by strong confining pressure, which forces grains closer together and eliminates pore space.

Any new mineral that has crystallized under high-pressure conditions tends to occupy less space than did the mineral or minerals from which it formed. The new mineral is denser than its low-pressure counterparts because the pressure forces atoms closer together in the crystal structure.

Directed pressure Pressure applied unequally on the surfaces of a body is **directed** (or dynamic) **pressure.** Directed pressure tends to deform objects into spindle-shaped or flattened forms. If you squeeze a rubber ball between your thumb and forefinger, you are exerting **compressive**

Figure 7.4
Metamorphosed conglomerate in which the pebbles have been flattened and elongated (sometimes called a stretched pebble conglomerate). Compare to figure 6.9.
Photo by C. C. Plummer.

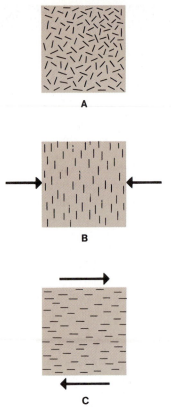

Figure 7.5
Orientation of platy minerals in metamorphic rocks. (**A**) Platy minerals randomly oriented. No directed pressure involved. (**B**) Platy minerals developed under the influence of compressive directed pressure. (**C**) Platy minerals developed under the influence of shearing.

directed pressure on the ball. If you squeeze a ball of dough, it will remain flattened after you stop squeezing, because dough is plastic.

Directed pressure is also caused by **shearing,** when parts of a body move or slide relative to one another and parallel to the forces being exerted. If you flatten a ball of dough by rolling a rolling pin over it, this involves shearing. Note in figure 7.3 that shearing flattens the object *parallel* to the applied pressure, whereas directed compressive pressure flattens the object *perpendicular* to the applied pressure.

A rolling pin exerts shearing force to crush crackers into crumbs. Similarly, intense shearing caused by movements within the earth's crust (tectonic forces) grinds preexisting minerals into finer grains. The increased amount of surface area on which reactions can take place facilitates chemical reactions. Shearing also tends to mix the ground-up fragments by moving grains past one another. Two mineral grains whose elements can react to form a new mineral are likely to be moved into contact with each other.

Foliation

Directed pressure has a very important influence on the texture of a metamorphic rock because it forces the constituents of the rock to become parallel to one another. For instance, the pebbles in the metamorphosed conglomerate shown in figure 7.4 were originally more spherical but have been flattened by directed pressure. The parallel alignment of the textural and structural features of a rock is called **foliation.** Foliation is manifested in various ways. If a platy mineral (such as mica) is crystallizing within a rock that is undergoing directed pressure, the mineral grows in such a way that it remains parallel to the direction of shearing or perpendicular to the direction of directed compression (figure 7.5). Any platy mineral attempting to grow against shearing is either ground up or forced into alignment. Minerals that crystallize in

needlelike shapes (for example, hornblende) behave similarly, growing with their long axes parallel to the plane of shearing or perpendicular to directed compression. The three very different textures described below are all variations of foliation and are important in classifying metamorphic rocks:

1. If the rock splits easily along nearly flat and parallel planes, indicating that preexisting, microscopic, platy minerals were pushed into alignment during metamorphism, we say the rock is **slaty,** or that it possesses **slaty cleavage.**
2. If visible platy or needle-shaped minerals have grown essentially parallel to one another while under the influence of directed pressure, the rock is **schistose** (figure 7.6).
3. If the rock became very plastic and the new minerals separated into distinct (light and dark) layers or lenses, the rock has a layered or **gneissic** texture, such as in figures 7.1 and 7.12.

Engineers and builders must take foliation into account because bed rock moves with relative ease when a force is exerted parallel to foliation planes (see Box 7.1). Rockslides onto highways are a common hazard where roads built on slopes have been cut into foliated rock (see chapter on mass wasting).

Failure of the St. Francis Dam—A Tragic Consequence of Geology Ignored

In 1928 the St. Francis Dam near Los Angeles, California, broke, only a year after it had been completed (box figure 1). The concrete dam was about 60 meters (200 feet) high, and the wall of water that roared down the valley killed about 400 people in two counties.

The eastern edge of the dam had been built against a metamorphic rock with foliation planes parallel to the sides of the valley. Landslide scars in the valley should have been ample warning to the builders that the metamorphic rock moved, even under only the force of gravity. A competent engineer worries as much about the stability of the rock against which a dam is built as about the strength of the dam itself. Water pressure at the base of the dam exerted a force of 5.7 tons per square foot against the dam. With pressure such as this, the dam and part of the bordering foliated rock could easily slide. Movement would be parallel to the weak foliation planes, just as if the dam had been anchored against a giant deck of cards.

Ironically, investigators never found out for sure whether this was what caused the failure of the dam. Many other blunders had been made in construction, and any one of them could have caused the dam to break. The base of the dam was on a fault with ground-up rock; and, incredibly, the other side of the dam was built against rock that disintegrates in water. This is but one of many instances in which ignorance of geology cost lives and money. Had professional geological advice been sought, the dam probably would not have been built in that spot.

A

B

Box 7.1 Figure 1
(*A*) The St. Francis Dam before failure (looking upstream).
(*B*) The St. Francis Dam after failure (looking downstream).
Note pieces of the dam carried downstream and the high-water mark in the standing segment of the dam.
Photos by California Department of Water Resources.

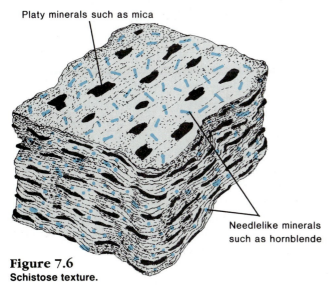

Platy minerals such as mica

Needlelike minerals such as hornblende

Figure 7.6
Schistose texture.

Effects of Fluids

Water (as vapor) is by far the most important fluid involved in metamorphic processes, although other gases, such as carbon dioxide, sometimes play a role.

The only evidence in the rocks to indicate that water was present during metamorphism is that water is part of the crystal structures of many metamorphic minerals, whereas the presumed parent minerals were water-free. Water is thought to help trigger metamorphic chemical reactions. Apparently water under high pressure forces its way between grains, dissolves ions from one mineral, and then carries these ions elsewhere in the rock where they can react with the ions of a second mineral. The new mineral that forms is stable under the existing conditions. Water is a sort of intra-rock rapid transit for ions.

Figure 7.7
Photomicrographs of (A) nonfoliated rock and (B) foliated rock.
Photos by C. C. Plummer.

Time

The effect of time on metamorphism is hard to comprehend. Most metamorphic rocks are composed predominantly of silicate minerals, and silicate compounds are notorious for their sluggish chemical reaction rates. Recently, garnets taken from a metamorphic rock collected in Vermont were analyzed and scientists calculated a growth rate of 1.4 millimeters per million years. The garnets' growth was sustained over a 10.5 million year period. Many laboratory attempts to duplicate metamorphic reactions believed to occur in nature have been frustrated by the time element. The several million years during which a particular combination of temperature and pressure may have prevailed in nature is impossible to duplicate.

Classification of Metamorphic Rocks

As we noted before, the kind of metamorphic rock that forms is determined by the metamorphic environment (primarily the particular combination of pressure and temperature) and by the chemical constituents of the parent rock. Many kinds of metamorphic rocks exist because of the many possible combinations of these factors. These rocks are classified based on similarities. (Appendix B contains a systematic procedure for identifying common metamorphic rocks.)

First consider the texture of a metamorphic rock. Is it *foliated* or *nonfoliated* (figure 7.7)? If the rock is nonfoliated, it is named on the basis of its composition. For example, a nonfoliated quartz-rich metamorphic rock is a *quartzite;* one composed almost entirely of calcite is a *marble.*

If the rock is foliated, you must determine the type of foliation. For example, a schistose rock is called a *schist.* But this name tells us nothing about what minerals are

in this rock; so we add adjectives to describe the composition. Thus, a *garnet-mica schist* (whose parent was shale) is easily distinguishable from a *hornblende schist* (a metamorphic product of basalt). The relationship of texture to rock name is summarized in table 7.1.

Types of Metamorphism

The two most common types of metamorphism are contact, or thermal, metamorphism and regional, or dynamothermal, metamorphism. These are discussed next. Hydrothermal metamorphism, in which hot water plays a major role, is discussed later in this chapter.

Contact (Thermal) Metamorphism

Contact, or **thermal, metamorphism** is metamorphism in which high temperature is the dominant factor. Confining pressure may influence which new minerals crystallize, but directed pressure is not involved. Therefore, these rocks are nonfoliated.

Usually the intense heat required for this process comes from magma intruding into the country rock. Most instances of thermal metamorphism can be thought of as the "cooking" of country rock in contact with an intrusion; hence, the common term *contact metamorphism.* The zone of contact metamorphism (also called an *aureole*) is usually quite narrow—generally from 1 to 10 meters wide.

During contact metamorphism, shale is changed into the very fine-grained metamorphic rock **hornfels.** Characteristically, only microscopically visible micas develop. Sometimes a few minerals grow large enough to be seen with the naked eye; these are minerals that are especially capable of crystallizing under the particular temperature attained during metamorphism. Hornfels can also form from basalt, in which case hornblende or another amphibole, rather than mica, is the predominant fine-grained mineral produced.

Classification and Naming of Metamorphic Rocks (Based Primarily on Texture)

Nonfoliated
Name Based on Mineral Content of Rock

Usual Parent Rock	Rock Name	Predominant Minerals	Identifying Characteristics
Limestone Dolomite	Marble Dolomitic marble	Calcite Dolomite	Coarse interlocking grains of calcite (or, less commonly, dolomite). Calcite (or dolomite) has rhombohedral cleavage; hardness intermediate between glass and fingernail. Calcite effervesces in weak acid.
Quartzose sandstone	Quartzite	Quartz	Rock composed of interlocking small granules of quartz. Has a sugary appearance and vitreous luster; scratches glass.
Shale Basalt	Hornfels Hornfels	Fine-grained micas Fine-grained ferromagnesian minerals, plagioclase	A fine-grained, dark rock that generally will scratch glass. May have a few coarser minerals present.

Foliated
Name Based Principally on Kind of Foliation Regardless of Parent Rock. Adjectives Describe the Composition (e.g., biotite-garnet schist)

Texture	Rock Name	Identifying Characteristics
Slaty	Slate	A fine-grained rock with an earthy luster. Splits easily into thin, flat sheets.
Intermediate between slaty and schistose	Phyllite	Fine-grained rock with a silky luster. Generally splits along wavy surfaces.
Schistose	Schist	Composed of visible platy or elongated minerals that show parallel alignment. A wide variety of minerals can be found in various types of schist (e.g., garnet mica schist, hornblende schist, etc.).
Gneissic	Gneiss	Light and dark minerals are found in separate, parallel layers or lenses. Commonly, the dark layers include biotite and hornblende; the light-colored layers are composed of feldspars and quartz. The layers may be folded or appear contorted.

Limestone recrystallizes during metamorphism into **marble,** a coarse-grained rock composed of interlocking calcite crystals (figure 7.8). Dolomite, less commonly found, recrystallizes into a **dolomitic marble.** Marble has long been valued as a building material and as a material for sculpture, partly because it is easily cut and polished and partly because it reflects light in a shimmering pattern, a result of the excellent cleavage of the individual calcite crystals. Marble is, however, highly susceptible to chemical weathering (see chapter on weathering).

Quartzite is produced when grains of quartz in sandstone are welded together while the rock is subjected to high temperature. This makes it as difficult to break along grain boundaries as through the grains. Therefore quartzite, being as hard as a single quartz crystal, is difficult to crush or break. It is the most durable of common rocks used for construction, both because of its hardness and because quartz is not susceptible to chemical weathering.

(Marble and quartzite can also form under conditions of regional metamorphism. These can be distinguished from their contact metamorphosed counterparts by their foliation, as in a schistose marble.)

Figure 7.8
Marble.

Impact Craters and Shock Metamorphism

Astrogeology Box 7.1 Figure 1
Meteor Crater in Arizona.
© Tom Fridmann/Industrial Visuals.

Very large meteorites have produced large craters when they have collided with Earth's surface. One well-known meteorite crater is Meteor Crater in Arizona, which is a little more than a kilometer in diameter (box figure 1). Many much larger craters are known in Canada, Germany, Australia, and other places. Craters produced by meteorite impact have several identifying characteristics. For example, some rare minerals are produced by the metamorphism of quartz and some other minerals as a result of the brief, very high pressures that occur on impact. The impact often produces shock damage in minerals at the impact site, melts some minerals that then harden into glass, and produces a type of breccia. Beds of rock near the rims of impact craters sometimes have been overturned in all directions away from the crater.

The largest meteorite impacts can throw significant quantities of material into the Earth's atmosphere. Molten blobs become streamlined in their passage through the Earth's atmosphere before they solidify into the objects we call tektites and may be found hundreds of miles from the point of meteorite impact. It has been suggested that large quantities of dust thrown into the Earth's atmosphere would cause a cooling of the Earth of several degrees and might have devastating effects on life, including possible extinctions. There is some evidence that suggests that tektites, along with some elements such as iridium, which are rare on Earth but more abundant in meteorites, are present in large quantities in rocks that were formed at the times of some possible widespread extinctions. However, there is much controversy over the validity and significance of these findings.

Craters are certainly the most common landform on the Moon (box figure 2). There may be as many as 400,000 craters larger than a kilometer in diameter. Approximately two-thirds of the Moon's surface is rugged terrain with thousands of overlapping craters. These rugged regions are called *highlands*. They represent an early period in the Moon's history when intense meteorite bombardment formed craters. Since the initial period of extensive crater formation, occasional meteorites have continued to strike the Moon's surface, producing other craters.

Bright streaks radiate from the most recent craters, giving them the name *rayed craters*. The streaks are apparently composed of dust particles and other debris ejected from the crater at impact. Continued

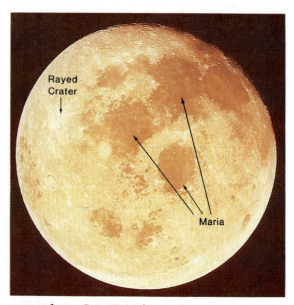

Astrogeology Box 7.1 Figure 2
The Moon. Note rayed craters (left) and dark maria.
NASA.

bombardment by small meteorites will eventually erase the streaks, just as it has erased the streaks from the older craters.

Most lunar craters are circular and surrounded by a blanket of material that was ejected from the crater. A rim higher than the lunar surface encircles a central depressed area. Craters more than about 20 kilometers in diameter often have terraced rim walls and central peaks. The terracing is due to slumping of the crater walls inward, and the formation of central peaks may be the result of the rebounding of surface rock layers—although it is not known whether the

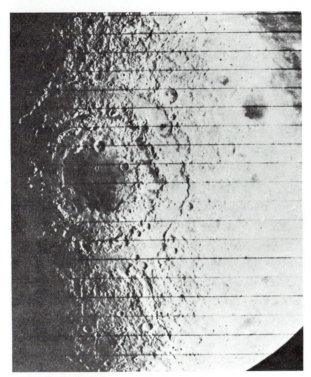

Astrogeology Box 7.1 Figure 3
Mare Orientale.
NASA.

Astrogeology Box 7.1 Figure 4
Fresh impaot crater on Mercury. The diameter of the crater
is 120 kilometers.
NASA.

rebound occurs immediately after the impact or is spread over a long period of time, perhaps with some associated volcanic activity.

Craters range in size from those visible only with a microscope to giant craters over 1,000 kilometers in diameter, such as Oceanus Procellarum, Mare Imbrium, and Mare Australe. These largest craters are usually called *multi-ringed basins* because they are surrounded by a series of concentric rings with intervening lowlands (box figure 3).

The loose, unconsolidated material of the lunar surface, made up of rock fragments, mineral grains, and glassy particles, is called the *regolith*. It has apparently been produced by the continual bombardment of the lunar surface by meteorites, a process that breaks rocks into smaller and smaller fragments and sometimes melts them to produce glass. The regolith varies in composition from place to place because it forms from different rock in different places. In general, the older the surface (that is, the longer it has been exposed to bombardment from meteorites), the thicker the regolith and the more finely crushed its grains.

Lunar breccias are of two principal types: (1) regolith lithified by the shock of meteorite impact and (2) material welded together as it accumulated in a hot mass after being thrown out during the formation of an impact crater. All the particle types found in the regolith occur in the breccias, along with some meteorite fragments. Some of the rock fragments in the breccias are themselves breccias, indicating more than one cycle of breakage and reconsolidation.

Ages for lunar regolith and breccias generally are greater than those obtained for the lunar crystalline rocks; they cluster around 4.6 billion years. This may represent the most common age of the rocks from which the regolith and breccias were derived and therefore the age of the Moon's formation.

Mercury's surface is remarkably similar to that of the Moon. There are densely cratered highlands, with craters ranging in size from barely visible to large, multi-ringed basins up to 1,300 kilometers in diameter (box figure 4). Some of Mercury's craters are highly eroded older features, while others formed recently enough to be still surrounded by extensive systems of rays.

Because Mercury is larger than the Moon and therefore has a larger gravitational pull, material thrown out of craters on Mercury falls closer to the crater rim. The result is that recent cratering on Mercury has not obliterated older surfaces nearly as much as recent cratering on the Moon has covered over traces of older lunar surfaces. Mercury's craters are shallower than those on the Moon. This is probably due to more landsliding of the crater rims because of the greater gravitational pull.

Opposite Caloris, the largest of Mercury's multi-ringed basins, is an area of what is called "peculiar" terrain. This jumbled, chaotic, hilly area is probably composed of material shattered by the convergence of shock waves from the impact that produced the basin. The only other place in the solar system where similar terrain is known to exist is on the Moon, opposite Maria Imbrium and Orientale.

Shallow impact craters, some at least 250 kilometers in diameter and many with central peaks, are known to exist on Venus. Large craters have not been identified in the low areas, and their absence might suggest that these are younger areas.

Astrogeology Box 7.1 Figure 5
Jupiter's moon Callisto.
NASA.

Large areas of Mars, especially in its southern hemisphere, are covered with thousands of craters. These craters are up to 2,000 kilometers in diameter and as much as 3 kilometers deep. Most of the craters were formed by meteorite impact during the first billion years after the formation of Mars; a few are from more recent meteorite impacts, and some are volcanic in origin.

Martian craters are shallower and smoother than those on the Moon. This is the result of more landsliding from the crater rims (because of the greater gravitational pull on Mars) and of erosion of the crater rims and filling of the crater depressions by sediments deposited by agents such as wind. There are few rayed craters, perhaps because the Martian winds blow fine ray material away or deposit fine material from elsewhere on top of the ray material, or maybe because the thickness of the Martian atmosphere prevents ray material from moving through it. Mars has several multi-ringed basins that, except for being more eroded, are very similar to such basins on the Moon. The largest basin, Hellas, is about 2,000 kilometers in diameter.

Many moons throughout the solar system also show evidence of cratering. For example, Jupiter's moon Callisto is one of the most cratered objects in the solar system. One giant impact produced a basin with concentric ridges that look like frozen ripples and extend more than 3,000 kilometers (box figure 5).

TABLE 7.2

Regional Metamorphic Rocks that Form under Approximately Similar Pressure and Temperature Conditions

Parent Rock	Rock Name	Predominant Minerals
Basalt	Amphibole schist (amphibolite)	Hornblende, plagioclase, garnet
Shale	Mica schist	Biotite, muscovite, quartz, garnet
Quartzose sandstone	Schistose quartzite	Quartz
Limestone or dolomite	Schistose marble	Calcite or dolomite
Ultramafic rock	Mafic schist	Magnesium amphibole, talc, olivine

Regional (Dynamothermal) Metamorphism

The great majority of the metamorphic rocks found on the earth's surface are products of **regional,** or **dynamothermal, metamorphism,** which is metamorphism caused by relatively high temperature and pressure (directed and confining). Metamorphic rocks are prevalent in the most intensely deformed portions of mountain ranges. They are visible where once deeply buried cores of mountain ranges are exposed by erosion. Furthermore, large regions of the continents are underlain by metamorphic rocks, thought to be the roots of ancient mountains long since eroded down to plains or rolling hills. Because dynamothermal metamorphic rocks cover large areas, the term *regional metamorphism* is generally regarded as synonymous with dynamothermal metamorphism.

The high temperature associated with regional metamorphism is due to the great depth, the heat generated by the friction of earth movements, and heat radiated from nearby magma bodies. The high confining pressure is due to burial under perhaps 10 or more kilometers of rock. The high directed pressure is a result of tectonism; that is, the constant movement and squeezing of the crust during mountain-building episodes.

Depending on the pressure and temperature conditions during metamorphism, a particular parent rock may recrystallize into one of several metamorphic rocks. For example, if basalt is metamorphosed at relatively low temperatures and pressures, it will recrystallize into a *greenschist,* a schistose rock containing chlorite (a green sheet-silicate), actinolite (a green amphibole), and sodium-rich plagioclase. At higher temperatures and pressures, the same basalt would recrystallize into an *amphibole schist* (also called *amphibolite*), a rock composed of hornblende, plagioclase feldspar, and, perhaps, garnet. Metamorphism of other parent rocks under conditions similar to those that produce amphibole schist from basalt should produce the metamorphic rocks shown in table 7.2.

A

B

Figure 7.9
(*A*) Slate outcrop in Antarctica. (*B*) Slaty cleavage that is nearly
vertical developed at a high angle to bedding. (Light and dark
layers are former sedimentary beds.)
Photo *A* by P. D. Rowley, U.S. Geological Survey.

Intensity of Metamorphism

To show how rocks are changed by regional metamorphism, we look at what happens to shale during *progressive metamorphism*—that is, as progressively greater pressure and temperature act on a rock type with increasing depth in the earth's crust.

The metamorphic rock associated with the lowest pressure and temperature conditions of regional metamorphism is **slate,** a very fine-grained rock that splits easily along flat, parallel planes (figure 7.9). Slate develops under temperatures and pressures only slightly greater than those found in the sedimentary realm. The temperatures are not high enough for the rock to recrystallize significantly. The important controlling factor is intense shearing or compressive directed pressure. Shale, the parent rock, is formed largely of submicroscopic, platy, clay minerals. These minerals partially recrystallize into equally fine-grained micas. The old and new platy minerals are aligned by the directed pressure, creating slaty cleavage in the rock. Sedimentary bedding may still be visible after metamorphism. Figure 7.9*B* shows slaty cleavage developed at an angle to the bedding. A slate indicates that a relatively cool and brittle rock has been subjected to intense tectonic activity.

Because of the ease with which it can be split into thin, flat sheets, slate is used for making chalkboards, pool tables, and roofs.

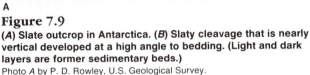

Figure 7.10
Phyllite, exhibiting a crinkled, silky-looking surface.

Phyllite is a rock in which the newly formed micas are larger than in slate, but still cannot be seen with the naked eye. This requires a further increase in temperature and pressure. The extremely fine-grained mica imparts a silky sheen to the rock, which may otherwise closely resemble slate (figure 7.10). However, the slaty cleavage may be destroyed in the process of conversion of slate to phyllite.

Figure 7.11
Garnet-mica schist. Small, subparallel flakes of mica reflect
light. Garnet crystals give the rock a "raisin bread"
appearance.

Figure 7.12
Gneiss.

A **schist** is characterized by megascopically visible,
approximately parallel-oriented minerals. Platy or elongate minerals that crystallize from the parent rock are
clearly visible to the naked eye. Shale may recrystallize
into several mineralogically distinct varieties of schist.
Which minerals form depends on the particular combination of temperature and pressure prevailing during recrystallization. For instance, if the rock is a *mica schist,*
metamorphism probably took place at only slightly higher
temperatures and pressures than those at which a phyllite
forms. A *garnet-mica schist* (figure 7.11) indicates that
the temperature and pressure were somewhat greater than
necessary for a mica schist to form.

In a **gneiss,** a rock consisting of light and dark mineral
layers or lenses, the highest temperatures and pressures
have changed the rock so that minerals have separated
into layers. Platy or elongate minerals (such as mica or
amphibole) in dark layers alternate with layers of light-
colored minerals of no particular shape. In addition, coarse
feldspars have crystallized and form part of the light-
colored layers. In composition, a gneiss may resemble
granite or diorite, but it is distinguishable from those plutonic rocks by its foliation (figure 7.12).

Temperature conditions under which a gneiss develops approach those at which granite solidifies. It is not
surprising, then, that the same minerals are found in gneiss
and in granite. In fact, a previously solidified granite can
be converted to a gneiss under high pressure and temperature conditions.

If the temperature is high enough, partial melting of
rock may take place, and a magma is sweated out into
layers within the foliation planes of the solid rock. After
the magma solidifies, the rock becomes a **migmatite,** a
mixed igneous and metamorphic rock. A migmatite can
be thought of as a "twilight zone" rock that is neither fully
igneous nor entirely metamorphic. Migmatites can also
be formed by metasomatism, as described later in this
chapter.

Plate Tectonics and Metamorphism

The model of the earth's crust and upper mantle that we
get from plate tectonic theory allows us to explain many
of the observed characteristics of metamorphic rocks. To
demonstrate the relationship between regional metamorphism and plate tectonics, we will look at what is believed
to take place at a converging boundary in which oceanic
lithosphere is subducted beneath continental lithosphere,
as shown in figure 7.13.

Directed pressure, which is responsible for foliation,
occurs wherever rocks are being squeezed between the two
plates or wherever rocks are sliding past one another.
Shearing is expected in, for example, the zone where the
oceanic crust slides beneath the continental lithosphere
(figure 7.13*A*). Compressive directed pressure is present
throughout wide zones in which rocks are caught, as if in
a vise.

Confining pressure is directly related to depth. For this
reason, we would expect the same pressure at a depth of
20 kilometers beneath a volcanic area as beneath the relatively cool rocks of a plate's interior.

Metamorphosed rocks indicate wide ranges of temperatures for a given pressure. This is understandable if
we realize that the *geothermal gradient* (the increase in
temperature with depth, as described in the chapter titled
"Intrusive Activity and the Origin of Igneous Rocks") is
not the same everywhere in the world. Each of the three
places (*A, B,* and *C*) in figure 7.13*B* would have a different geothermal gradient. If you were somehow able to
push a thermometer through the lithosphere, you would
find the rock is hotter at shallower depths in areas with
higher geothermal gradients than at places where the geothermal gradient is low. As indicated in figure 7.13*B*, the
geothermal gradient is higher progressing downward
through an active volcanic-plutonic complex (for instance,
the Cascade Mountains of Washington and Oregon) than
it is in the interior of a plate (beneath the Great Plains of
North America, for example).

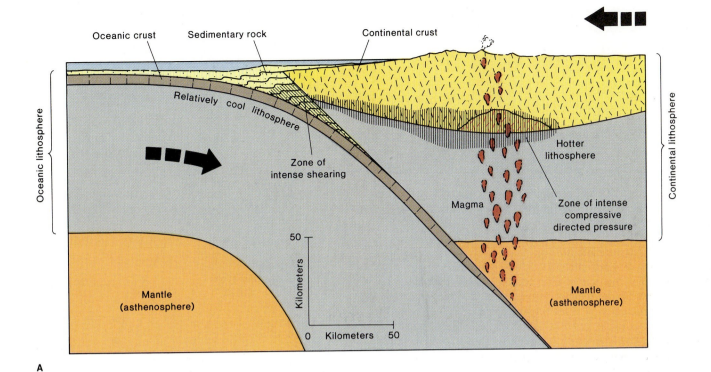

A

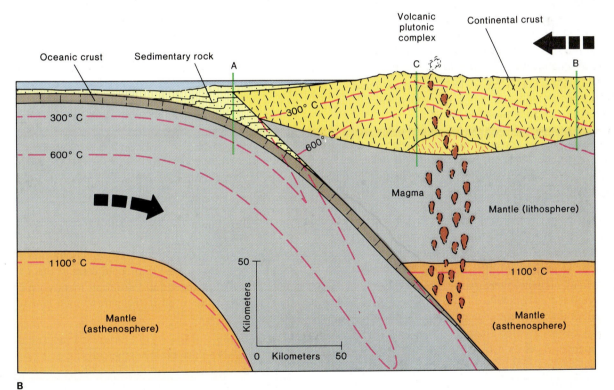

B

Figure 7.13
Metamorphism across a converging plate boundary.
(A) Location of zones of relatively cool and hot lithosphere and
zones of stress where metamorphism is likely to occur.
(B) The 300°, 600°, and 1100° isotherms.

From W. G. Ernst. *Metamorphism and Plate Tectonic Regimes.*
Stroudsberg, Pa.: Dowden, Hutchinson & Ross, 1975; p. 425. Reprinted by
permission of the publisher.

Heat is the most variable of the controlling factors of
metamorphism. Figure 7.13*B* shows the temperatures
calculated for a converging boundary. A line connecting
points that have the same temperature is called an **isotherm.** Note how the 300°C, 600°C, and 1100°C isotherms change depth radically across the subduction zone.

This is because rocks that were cool because they were
relatively near the earth's surface have been transported
rapidly to depth. The oceanic lithosphere along a subduction zone has not had time to heat up and come into
thermal equilibrium with the relatively hot rocks elsewhere at these depths. On the top of the subducted oceanic

BOX 7.2

Metamorphic Facies and the Plate Tectonic Connection

Because rocks such as schist metamorphose over a wide range of conditions, it has been useful to subdivide the various combinations of pressure and temperature into *stability fields,* such as shown in box figure 1. Each such field is called a metamorphic facies. The combination of minerals in a rock is the basis for determining the **metamorphic facies** to which a rock belongs. Each metamorphic facies has a name (for example, *greenschist facies*). The diagram shows the relationships of the various facies to one another and to temperature and pressure. (A beginning geology student does not need to learn the names and compositions of all the facies, but he or she should understand the concept.)

By identifying the metamorphic facies of rocks presently cropping out on the surface, geologists can infer, within broad limits, the depth at which metamorphism took place. They may also (again, within broad limits) be able to determine the corresponding temperature.

The concept of metamorphic facies is analogous to defining climatic zones by the combinations of plants found in each zone. A place where ferns, palm trees, and vines flourish corresponds to a climate with warm temperatures and abundant rainfall. On the other hand,

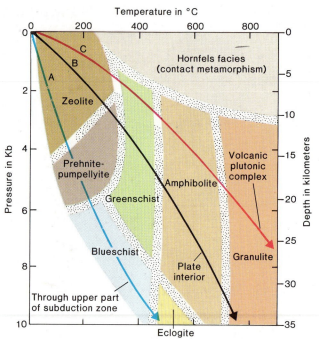

Box 7.2 Figure 1
The metamorphic facies. Facies are named after minerals (prehnite, zeolite, pumpellyite) or rock types (e.g., blueschist, granulite). Boundaries between facies are approximate. The arrows represent increases in temperature with depth for the three lines labeled *A, B, C* in figure 2 and in figure 7.13*B.*
From W. G. Ernst, *Metamorphism and Plate Tectonic Regimes.* Stroudsberg, Pa.: Dowden, Hutchinson & Ross, 1975; p. 425. Reprinted by permission of the publisher.

a combination of palm trees, cactus, and sagebrush implies a hot, dry climate. By way of analogy, assume that we have two specimens of schist, *X* and *Y;* the parent rock of both is basalt. Schist *X* contains the

slab, the extra heat provided by friction along with the heat that is normal for the overlying continental lithosphere and asthenosphere cause the isotherms to rise sharply toward the surface. The isotherms are bowed upward somewhat in the region beneath the volcanic zone because magma created along the lower levels of the subduction zone works its way upward and brings heat from the asthenosphere into the mantle and crust of the continental lithosphere.

To understand why metamorphic rocks can form under such a wide variety of temperatures and pressures, study figure 7.13*B*. You may observe that the bottom of line *A* is at a depth of about 50 kilometers, and if a hypothetical thermometer were here, it would read just over 300° because it would be just below the 300° isotherm. Compare this to vertical line *C* in the volcanic-plutonic complex. The confining pressure at the base of this line would be the same as at the base of line *A*, yet the temperature at the base of line *C* would be well over 600°.

The minerals that could form at the base of line *A* would not be the same as those that could form at line *C*. Therefore, we would expect quite different metamorphic rocks in the two places, even if the parent rock had been the same.

Hydrothermal Processes

Rocks that have precipitated from hot water or have been altered by hot water passing through are hard to classify. As explained earlier, hot water is involved to some extent in most metamorphic processes. However, when rocks have been formed entirely by precipitation of ions derived from hydrothermal solutions, we cannot really call them metamorphic rocks. For this reason we prefer the term **hydrothermal rocks** for rocks formed by precipitation from solution in hot water.

Many gradations exist between hydrothermal rocks and rocks in which water merely assists the metamorphic process. These are summarized in table 7.3. The "dry"

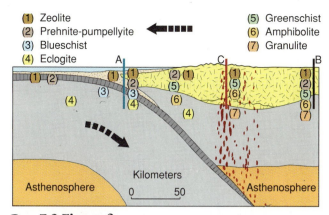

(1) Zeolite
(2) Prehnite-pumpellyite
(3) Blueschist
(4) Eclogite
(5) Greenschist
(6) Amphibolite
(7) Granulite

Box 7.2 Figure 2
Schematic representation of the distribution of facies across a converging plate boundary.
From W. G. Ernst, *Metamorphism and Plate Tectonic Regimes.* Stroudsberg, Pa.: Dowden, Hutchinson & Ross, 1975; p. 426. Reprinted by permission of the publisher.

minerals actinolite, chlorite, and epidote (all green minerals), which indicate (by metamorphic standards) low pressures and temperatures within the range of temperature and pressure combinations under which schist can crystallize. Schist X therefore belongs in the greenschist facies. In schist Y, garnet and hornblende, also derived from basalt, imply a combination of higher pressure and higher temperature within the schist stability range. The presence of these minerals indicates that schist Y belongs in the amphibolite facies.

The concept of metamorphic facies preceded plate tectonic theory by several decades. Although earlier geologists were able to relate the individual facies to pressure and temperature combinations, they had no satisfactory explanation for the environments that produced the various combinations. Figure 7.13, which relates the temperatures of regional metamorphism to plate tectonics, may be used to infer the environment for each of the metamorphic facies shown in box figure 1. Box figure 2 shows the likely distribution of metamorphic facies across the same converging boundary as in figure 7.13. To understand the relationship, study box figures 1 and 2 as well as figure 7.13.

If one were to determine the geothermal gradient represented by the three vertical lines marked A, B, and C on figure 7.13 and box figure 2, the temperatures for particular depths should plot on the corresponding arrows shown in box figure 1. Follow arrow A in box figure 1. This means that if you were able to drill vertically downward along line A, you would find, beneath unmetamorphosed rocks, rocks of the zeolite facies. At a greater depth would be the boundary between the zeolite and prehnite-pumpellyite facies. You would reach this boundary at a depth where the pressure is about 4 kilobars and the temperature is approximately 150°C. Your drill would penetrate rocks of the prehnite-pumpellyite facies until you reached the blueschist facies. In hole B, in the interior of the plate, the progression would be from zeolite facies to prehnite-pumpellyite facies to greenschist facies to amphibolite facies to granulite facies. Hole C, in the volcanic-plutonic complex, would not pass through the prehnite-pumpellyite facies but would go from zeolite facies to greenschist facies to amphibolite facies to granulite facies.

metamorphism referred to is relatively rare. Most of the examples given so far in this chapter probably took place with water present and so would be classed as "wet" metamorphism.

Metasomatism

Metasomatism is metamorphism coupled with the introduction of *ions* from an external source. The ions are brought in by water from outside the immediate environment and are incorporated into the newly crystallizing minerals. At the same time, the hot water may dissolve minerals that were part of the rock and carry them away.

If metasomatism is associated with contact metamorphism, the ions are introduced from a cooling magma. Some important commercial deposits of iron and other materials are of metasomatic origin (figure 7.14). Ions of the metal are transported by water and react with minerals in the host rock. Elements within the host rock are simultaneously dissolved out of the host rock and replaced

TABLE 7.3

Role of Water	Name of Process or Product
No water	"Dry" metamorphism
Water merely transporting ions between grains in a rock	"Wet" metamorphism
Water brings ions from outside the rock, and they are added to the rock during metamorphism. Other ions may be dissolved and removed.	Metasomatism
Water passes through cracks or pore spaces in rock and precipitates minerals on the walls of cracks and within pore spaces.	Hydrothermal rocks

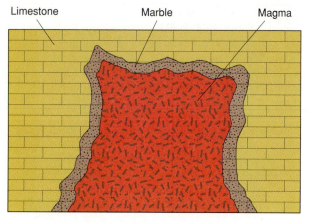

Limestone   Marble   Magma

A   Zone of contact metamorphism (aureole)

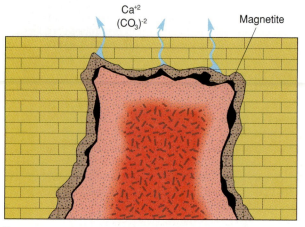

Ca^{+2}
$(CO_3)^{-2}$ Magnetite

B

Figure 7.14
Development of a contact metasomatic deposit of iron
(magnetite). (*A*) Magma intrudes country rock (limestone), and
marble forms along contact. (*B*) As magma solidifies, gases
bearing ions of iron leave the magma, dissolve some of the
marble, and deposit iron as magnetite.

by the metal ions brought in by the fluid. Because of the
solubility of calcite, marble commonly serves as a host for
metasomatic ore deposits.

Metasomatism that takes place during regional metamorphism produces quite different effects. When very hot
water travels through a rock while gneiss or schist is crystallizing, ions (typically K^+, Na^+, Si^{+4}, and O^{-2}) carried
by the water participate in metamorphic reactions. Large
feldspar crystals may develop in schist due to the addition
of potassium or sodium ions. At the same time, magnesium and iron (the mafic components of the rock) may be
removed by solution. Entire layers within a schist or gneiss
may become granitic (composed predominantly of feldspar and quartz), forming a *metasomatic migmatite*
(figure 7.15). The culmination of this process is conversion of the entire rock to predominantly feldspars and
quartz. This process is called **granitization,** the
metasomatic process by which granite is created from other
rock without a melt.

Figure 7.15
Migmatite in the Daniels Range, Antarctica.
Photo by C. C. Plummer.

Hydrothermal Rocks

Hydrothermal rocks are most commonly found in veins.
Veins consisting only of quartz are the most widespread,
although some quartz veins contain other minerals. Veins
with no quartz are less common and are composed of calcite or other minerals.

Quartz veins are especially common where igneous
activity has occurred. These veins can form from hot water
given off by a cooling magma; but probably they are mostly
produced by ground water heated by the pluton and circulated by convection, as shown in figure 7.16. Where the
water is hottest, more material is dissolved. As water vapor
continues upward through increasingly cooler rocks during
its journey toward the earth's surface, pressure is lost, as
is heat. Fewer ions can be carried in solution, and so the
silicon and oxygen leave the water and cake onto the walls
of the crack as silica (SiO_2), forming a quartz vein.

Hydrothermal veins are very important economically.
They are the primary source of zinc, lead, silver, gold, and
other metals on which industrialized nations depend. Ore
minerals containing these metals are found in quartz veins.
Veins containing commercially extractable amounts of
metals are by no means common (and might even be regarded as freaks of nature).

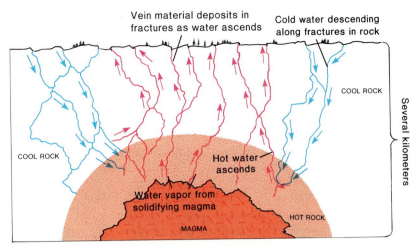

Figure 7.16
How veins form. Cold water descends, is heated, dissolves material, ascends, and deposits material upon ascending.

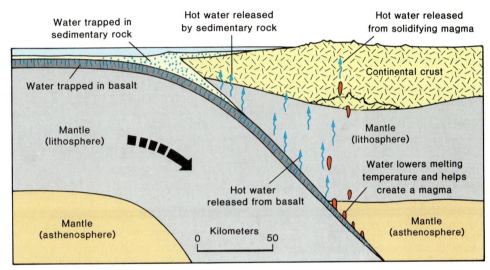

Figure 7.17
Behavior of water in a subduction zone. Sea water trapped in the oceanic crust is carried downward and released upon heating at various depths within the subduction zone.

Some ore-bearing solutions percolate upward between the grains of the rock and deposit very fine grains of ore mineral throughout. These are called *disseminated ore deposits* and account for some of the largest mines in the world. Chapter 21 ("Geologic Resources") contains more information on ores and how they form.

Sources of Water

Where does the water come from? The following is a logical explanation. *Ground water* seeps downward from the earth's surface through pores and fractures in rocks; however, the depth to which surface-derived water can penetrate is quite limited. Plate tectonics can account for water at deeper levels in the lithosphere since sea water trapped in the oceanic crust can be carried to considerable depths through subduction (figure 7.17). Water trapped in sediment and in sedimentary rocks lying on the basaltic oceanic crust is carried down with the descending crust. However, recent studies indicate that most of the water is carried by hydrous minerals (hornblende, for example) in the basalt. When the rocks get hot enough the minerals recrystallize, releasing water. The water vapor works its way upward through the overlying continental lithosphere through fissures. In the process of ascending, water assists in the metamorphism of rocks, dissolves minerals, and carries the ions to interact during metasomatism, or it deposits quartz and other minerals in fissures as veins. The water can also lower the melting points of rocks at depth, allowing magma to form. Later it is boiled off as the magma works its way upward, cools, and solidifies.

Summary

Metamorphic rocks form from other rocks that are subjected to high temperature generally accompanied by high pressure. Recrystallization takes place in the solid state although water, usually present, aids metamorphic reactions. Pressure can be *confining pressure* or *directed pressure* or, more often, both. Directed pressure, either shearing or directed compressive pressure, is implied for foliated rocks (slate, phyllite, schist, gneiss).

Contact (or *thermal*) *metamorphic* rocks are produced during metamorphism without significant directed pressure but with high temperature. In general, contact metamorphism occurs in rocks immediately adjacent to intruded magmas.

Regional (or *dynamothermal*) *metamorphism,* which involves heat, confining pressure, and directed pressure, has created most of the metamorphic rock of the earth's crust. The widely varying combinations of pressure and temperature under which metamorphism can take place are divided into *metamorphic facies*. The combination of minerals in a metamorphic rock indicates the metamorphic facies in which that rock recrystallized.

Hydrothermal rocks form when hot water precipitates material that crystallizes into minerals. During *metasomatism,* hot water introduces ions into a rock being metamorphosed, changing the chemical composition of the metasomatized rock from that of the parent rock. *Migmatites* can be produced by metasomatism or by partial melting of a rock. *Granitization* is the process whereby a rock becomes granitic through metasomatism.

Plate tectonic theory accounts for the features observed in metamorphic rocks and relates their development to other activities in the earth. In particular, plate tectonics explains (1) the deep burial of rocks originally formed at or near the earth's surface; (2) the intense squeezing necessary for the directed pressure, implied by foliated rocks; (3) the presence of water deep within the lithosphere; and (4) the wide variety of pressures and temperatures believed to be present during metamorphism.

Terms to Remember

compressive directed pressure	metamorphic rock
confining pressure	metamorphism
contact (thermal) metamorphism	metasomatism
	migmatite
directed pressure	parent rock
dolomitic marble	phyllite
foliation	plastic
gneiss	quartzite
gneissic	regional (dynamothermal) metamorphism
granitization	
hornfels	schist
hydrothermal rock	schistose
isotherm	shearing
marble	slate
metamorphic facies	slaty
	slaty cleavage

Questions for Review

1. What are the various sources of heat for metamorphism?
2. How do regional metamorphic rocks differ in texture from contact metamorphic rocks?
3. Why is such a variety of combinations of pressure and temperature environments possible during metamorphism?
4. How would you distinguish
   a. schist and gneiss?
   b. slate and phyllite?
   c. quartzite and marble?
   d. granite and gneiss?
5. Why do regional metamorphic rocks present hazards that should be taken into account by builders and engineers?
6. Why is an edifice built with blocks of quartzite more durable than one built of marble blocks?

Questions for Thought

1. What happens to originally horizontal layers of sedimentary rock when they are subjected to the deformation associated with regional metamorphism?
2. What metamorphic and hydrothermal effects would occur in the vicinity of a spreading center?
3. Why is it unlikely for a batholith to form through granitization?
4. Where in the earth's crust would you expect most migmatites to form?

Supplementary Readings

Best, M. G. 1982. *Igneous and metamorphic petrology.* New York: W. H. Freeman.

Ernst, W. G. 1969. *Earth materials.* Englewood Cliffs, N.J.: Prentice-Hall.

Ernst, W. G., ed. 1975. *Metamorphism and plate tectonic regimes.* Stroudsburg, Pa.: Dowden, Hutchinson & Ross, Inc.

Hyndman, D. W. 1985. *Petrology of igneous and metamorphic rocks.* 2d ed. New York: McGraw-Hill.

Mason, R. 1989. *Petrology of the metamorphic rocks.* 2d ed. London: Unwin Hyman.

Yardley, B. W. D. 1989. *An introduction to metamorphic petrology.* Essex, England: Longman.

The immensity of geologic time is hard for humans to perceive. Someone who lives a hundred years is unusual, but a person would have to live 10,000 times that long to observe a geologic process that takes a million years. In this chapter we try to help you develop a sense of the vast amounts of time over which geologic processes have been at work. As mentioned in chapter 1, the principle of *uniformitarianism* radically changed our view of life and the universe. Acceptance of this principle made people aware of the vast amounts of time involved in the earth's history.

To determine what those time spans are—how many years ago a rock formed—requires special techniques of radiometric dating, also described in this chapter. To geologists, however, being able to assign years (absolute age) to a process, significant as it is, is less important than establishing relative time and the sequence in which events have occurred. This chapter explains how to apply several basic principles to decipher a sequence of events responsible for geologic features. These principles can be applied to many aspects of geology—as, for example, in understanding geologic structures (chapter 15). Understanding the complex history of mountain belts (chapter 20) also requires knowing the techniques for determining relative ages of rocks.

Correlation—determining age relationships between geographically widely separated rock units—is necessary for understanding the geologic history of a region, a continent, or the whole earth. Substantiation of the plate tectonics theory depends on intercontinental correlation of rock units and geologic events, piecing together evidence that the continents were once one great body.

Widespread use of fossils for correlation led to the development of the standard geologic time scale. Originally based on relative age relationships, the subdivisions of the standard geologic time scale have now been assigned absolute ages through radiometric dating. Think of the geologic time scale as a sort of calendar to which events and rock units can be referred. Its major subdivisions are referred to frequently in this book.

8

Time and Geology

Grand Canyon, Arizona.
FourbyFive.

The Key to the Past

We who live in the twentieth century are not so resistant to the concept of immense geologic time as were those who lived two or three centuries ago. Then everyone in the Western world believed that the earth was only a few thousand years old (although Chinese and Hindu cultures had a better concept of the age of the universe). In early Christendom, geologic events were placed within a biblical chronology, and catastrophic happenings were blamed for features of the landscape. When rocks several thousand meters above sea level seemed to have been formed of sediment deposited in water (even containing fossils somewhat like modern marine organisms), the explanation was that a worldwide inundation had simply drowned all the earth's mountains in a matter of days. Because no known physical laws could account for such events, they were attributed to divine intervention. In the late 1700s, however, a Scottish naturalist, James Hutton (mentioned in chapter 1), proposed the radical idea that features observed on earth could be explained by natural processes that were still going on in the present. This became known as the principle of **uniformitarianism,** which says that the geological processes operating at present are the same processes that have operated in the past. More succinctly: "The present is the key to the past."

For example, we may see silt and clay settling out of ocean or lake water, and we may also observe the processes of compaction and cementation that turn such sediment into the rock *shale*. If we find a similar layer of shale at a mountain top or exposed near the bottom of Grand Canyon (chapter opening photo), we can, according to the concept of uniformitarianism, presume that the shale was formed through a similar process.

What made the idea of uniformitarianism so difficult for eighteenth-century scholars to accept was the vast amount of time implied. Layers of shale more than 2,000 meters thick are visible at many places on earth, yet our observations tell us that in most places silt and clay settle on ocean floors at rates of less than a millimeter a year. The time required to form a thousand meters of shale, then, would be more than a million years—far longer than the few thousand years Hutton's contemporaries believed the earth had existed.

The principle of uniformitarianism can be applied too rigidly if common sense is ignored. We cannot use uniformitarianism to claim that exactly the same *amount* of volcanic activity went on during the geologic past as at present. But by examining older rocks and comparing them with rocks being formed now by active volcanoes, we can get a reasonably accurate picture of volcanism in the distant geologic past. In short, the principle of uniformitarianism assures physical and chemical laws were operative in the past as well as the present, and, moreover, we need not refer to "magical" processes to explain geological features.

Acceptance of the principle of uniformitarianism led, of course, to the realization that geology involves time periods much greater than a few thousand years. But how long? For instance, were the rocks near the bottom of

Grand Canyon formed closer to 10,000 or 100,000 or 1,000,000 or 1,000,000,000 years ago? What geologists needed was some "clock" that began working when rocks formed. Such a "clock" was found when radioactivity was discovered. Radiometric dating (discussed later in this chapter) allows us to determine a rock's **absolute age** (age given in years or some other unit of time). Geologists, however, usually are more concerned with **relative time,** the *sequence* in which events took place, than with the number of years involved.

These statements show the difference between absolute age and relative time:

"Uncle George planted the plum tree sometime between the time little Oscar was born and the time when the house across the street burned down." This statement gives the time of an event (planting a tree) relative to other events.

But speaking in terms of absolute age: "Let's see, Uncle George planted that plum tree about four and a half, maybe five, years ago." Note that an absolute age does not have to be an *exact* age, merely age given in units of time. Because most geologic problems are concerned with the sequence of events, we discuss relative time first.

Relative Time

The geology of an area may seem, at first glance, to be hopelessly complex. A nongeologist might think it impossible to decipher the sequence of events that created such a geologic pattern. However, a geologist has learned to approach seemingly formidable problems by breaking them down to a number of simple problems. As an example, the geology of Grand Canyon, shown in the chapter opening photo, can be analyzed in four parts: (1) horizontal layers of rock; (2) inclined layers; (3) rock underlying the inclined layers (plutonic and metamorphic rock); and (4) the canyon itself, carved into these rocks. Each part of the problem can be analyzed and interpreted. Ultimately the regional sequence of geologic events can be determined by piecing together the history of the individual parts.

Principles Used to Determine Relative Age

Most of the individual parts of the larger problem are solved by applying several simple principles (or laws, as some prefer to call them) while studying the exposed rock. In this way the sequence of events or the relative time involved can be determined. Contacts are particularly useful for deciphering the geologic history of an area. (**Contacts,** as described in the chapter on sedimentary rocks, are the surfaces separating two different rock types or rocks of different ages.)

Generally we start with the last, or youngest, event to affect the geology of an area of interest and then work toward older events. The block diagram in figure 8.1 represents an area whose geology is somewhat similar to that of Grand Canyon; however, the formation names are fictitious. (**Formations,** as described in the chapter "Sediments and Sedimentary Rocks," are bodies of rock of

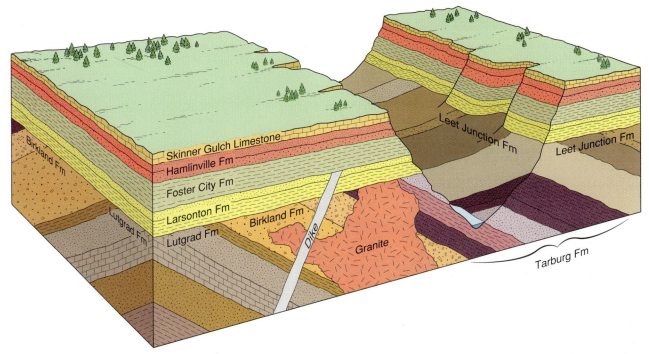

Figure 8.1
Block diagram representing an area somewhat similar to Grand Canyon.

considerable thickness with recognizable characteristics that make each distinguishable from adjacent rock units. They are named after local geographic features, such as towns.) Note the contacts between the tilted formations, the horizontal formations, the granite, and the dike. What sequence of events might be responsible for what is shown in figure 8.1? (You might briefly study the block diagram and see how much of the geologic history of the area you can decipher before reading further.)

To determine the relationship of geologic events to one another, three basic principles are applied one at a time. These are the principles of (1) original horizontality, (2) superposition, and (3) cross-cutting relationships. These three principles will be used in interpreting figure 8.1. (All the layers in the drawing represent sedimentary rock originally deposited in a marine environment.)

Original horizontality The principle of **original horizontality** states that beds of sediment deposited in water formed as horizontal or nearly horizontal layers (as described in the chapter on sedimentary rocks). The few exceptions (such as cross-bedding associated with deltas) are easily identifiable as such by the geologist and can be ignored in this discussion.

Note in figure 8.1 that the Larsonton Formation and overlying rock units (Foster City Formation, Hamlinville Formation, and Skinner Gulch Limestone) are horizontal. Evidently their original horizontal attitude has not changed since they were deposited. However, the Lutgrad, Birkland, Tarburg, and Leet Junction Formations, must have been tilted after they were deposited as horizontal layers. By applying the principle of original horizontality, we have determined that a geologic event—tilting of bed rock—

occurred after the Leet Junction, Tarburg, Birkland, and Lutgrad Formations were deposited on a sea floor. We can also see that the tilting event did not affect the Larsonton and overlying formations. (A reasonable conclusion is that tilting was accompanied by uplift and erosion, all before renewed deposition of younger sediment.)

Superposition The principle of **superposition** states that within a sequence of undisturbed sedimentary rocks, the layers get younger going from bottom to top.

Obviously, if sedimentary rock is formed by sediment settling onto the sea floor, then the first (or bottom) layer must be there before the next layer can be deposited on top of it. The principle of superposition also applies to layers formed by multiple lava flows, where one lava flow is superposed on a previously solidified flow.

Applying the principle of superposition, we can determine that the Skinner Gulch Limestone is the youngest layer of sedimentary rock in figure 8.1. The Hamlinville Formation is the next oldest formation, and the Larsonton Formation is the oldest of the still horizontal sedimentary rock units. Similarly, we assume that the inclined layers were originally horizontal (by the first principle). By mentally restoring them to their horizontal position (or "untilting" them), we can see that the youngest formation of the sequence is the Leet Junction Formation and that the Tarburg, Birkland, and Lutgrad Formations are progressively older.

Cross-cutting relationships The third principle can be applied to determine the remaining age relationships in figure 8.1. The principle of **cross-cutting relationships** states that a disrupted pattern is older than the cause of disruption. A layer cake (the pattern) has to be baked (established) before it can be sliced (the disruption).

TABLE 8.1

Relative Ages of Features in Figure 8.2 Determinable by Cross-cutting Relationships

Feature	Is Younger Than	But Older Than
Valley (canyon)	Skinner Gulch Ls.	
Foster City Fm.	Dike	Hamlinville Fm.
Dike	Larsonton Fm.	Foster City Fm.
Larsonton Fm.	Leet Junction Fm. and granite	Dike
Granite	Tarburg Fm.	Larsonton Fm.

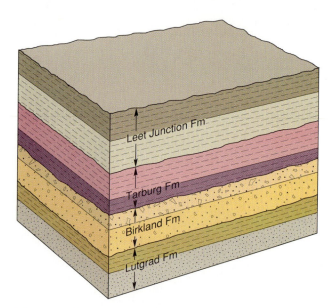

Figure 8.3
The area before intrusion of the granite.

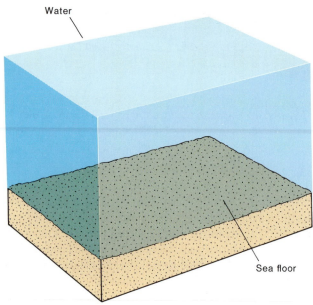

Figure 8.2
The area during deposition of the initial sedimentary layer of the Lutgrad Formation.

To apply this principle, look for disruptions in patterns of rock. Note that the valley in figure 8.1 is carved into the horizontal rocks as well as into the underlying tilted rocks. The sedimentary beds on either side of the valley appear to have been sliced off, or *truncated*, by the valley. (Undisturbed sedimentary beds normally taper [become thinner toward the edges] rather than stop abruptly. This is sometimes called the principle of *lateral continuity*.) So the event that caused the valley must have come after the sedimentation responsible for Skinner Gulch and underlying formations. That is, the valley is younger than these layers. We can apply the principle of cross-cutting relationships to contacts elsewhere in figure 8.1 with the results shown in table 8.1.

We can now describe the geological history of the area represented in figure 8.1 on the basis of what we have learned through applying the three principles. Figures 8.2 through 8.11 show how the area changed over time, progressing from oldest to youngest events.

By *superposition,* we know that the Lutgrad Formation, the lowermost rock unit in the tilted sequence, must be the oldest of the sedimentary rocks as well as the oldest rock unit in the diagram. From the principle of

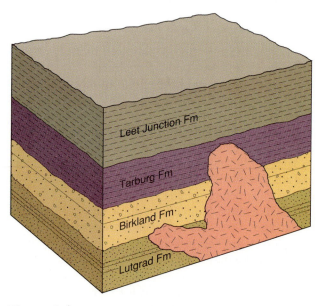

Figure 8.4
The area before layers were tilted.

original horizontality, we infer that these layers must have been tilted after they formed. Figure 8.2 shows initial sedimentation of the Lutgrad Formation taking place.

Superposition indicates that the Birkland Formation was deposited on top of the Lutgrad Formation. Deposition of the Tarburg and Leet Junction Formations followed in turn (figure 8.3).

The truncation of bedding in the Lutgrad, Birkland, and Tarburg Formations by the granite tells us that the granite intruded sometime after the Tarburg Formation was formed (this tells us it is an intrusive contact). Although figure 8.4 shows that the granite was emplaced before tilting of the layered rock, we cannot determine from looking at figure 8.1 whether the granite intruded the sedimentary rocks before or after tilting. We can, however, determine through *cross-cutting relationships*

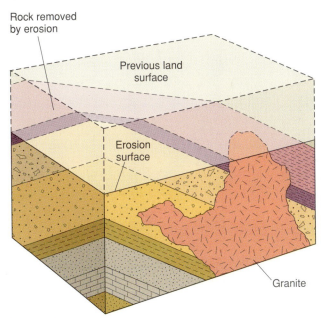

Figure 8.5
The area before deposition of the Larsonton Formation. Dashed lines show rock probably lost through erosion.

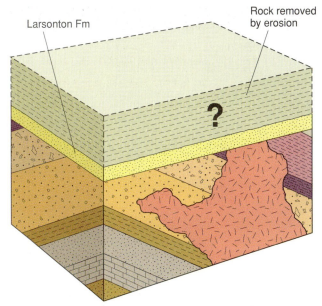

Figure 8.7
Area before intrusion of dike. Thickness of layers above the Larsonton Formation is indeterminable.

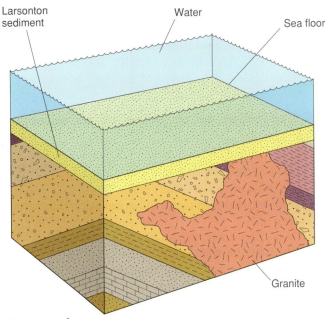

Figure 8.6
The area at the time the Larsonton Formation was being deposited.

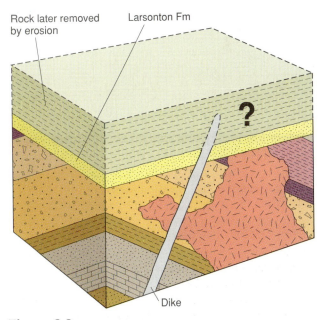

Figure 8.8
Dike intruded into the Larsonton Formation and preexisting overlying layers of indeterminable thickness.

that tilting and intrusion of the granite occurred before deposition of the Larsonton Formation. Figure 8.5 shows the rocks in the area have been tilted and erosion has taken place. Sometime later, sedimentation was renewed and the lowermost layer of the Larsonton Formation was deposited on the erosion surface, as shown in figure 8.6. Contacts representing buried erosion surfaces such as these are called **unconformities** and are discussed in more detail in the chapter on geologic structures.

After the Larsonton Formation was deposited, an unknown additional thickness of sedimentary layers was deposited, as shown in figure 8.7. This can be determined

through application of *cross-cutting relationships*. The dike is truncated by the Foster City Formation; therefore, it must have extended into some rocks that are no longer present, such as shown in figure 8.8. Figure 8.9 shows the area after the erosion that truncated the dike took place.

Once again, sedimentation took place as the lowermost layer of the Foster City Formation blanketed the erosion surface (figure 8.10). Sedimentation continued until the uppermost layer (top of the Skinner Gulch Limestone) was deposited. At some later time, the area was raised above sea level and the stream began to carve the canyon (figure 8.11). Because the valley sides truncated the youngest layers of rock, we can determine from figure 8.1 that the last event was the carving of the valley.

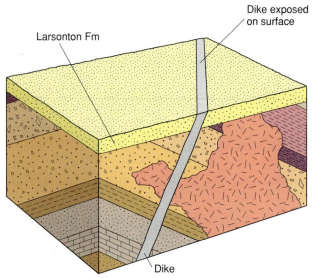

Figure 8.9
The area after rock overlying the Larsonton Formation, along with part of the dike, was removed by erosion.

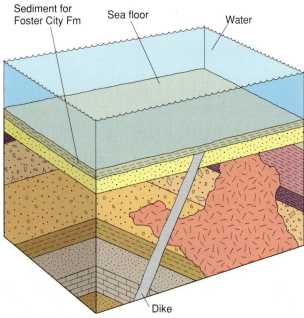

Figure 8.10
Sediment being deposited that will become part of the Foster City Formation.

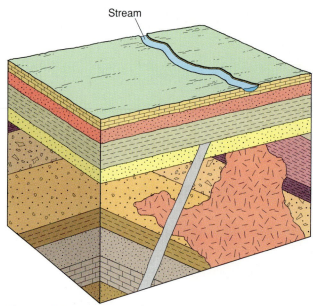

Figure 8.11
Same area as shown in figure 8.1 but before valley was carved into rock.

Note that there are limits on how precisely we can determine the relative age of the granite body. It definitely intruded *before* the Larsonton Formation was deposited and *after* the Tarburg Formation was deposited. As no contacts can be observed between the Leet Junction Formation and the granite, we cannot say whether the granite is younger or older than the Leet Junction Formation. Nor, as mentioned earlier, can we determine whether the granite formed before, during, or after the tilting of the lower sequence of sedimentary rocks.

When combined with common sense and a knowledge of geology, other methods help to determine relative time. For example, we can look for signs of *contact metamorphism* in the rocks of the tilted layers immediately adjacent to the granite body to see if they have been baked by the heat of the granitic magma. (The base of the Larsonton Formation in contact with granite would not be baked because it was deposited after the granite had cooled.)

We must also point out some special circumstances under which these three principles do not apply. For example, intense deformation by tectonic forces can overturn or disrupt beds so much that the principle of superposition cannot be used (see geologic structures chapter). A geologist must avoid being dogmatic in applying principles.

Correlation

In geology, **correlation** usually means determining the *age relationships* between rock units or geologic events in separate areas.

A geologist may want to relate the geology of an area he or she has investigated with adjacent areas studied by others. Or a geologist may want to correlate rocks and events in distant parts of the world to understand better how the earth has evolved. Part of the evidence supporting the theory of plate tectonics (and its predecessor, continental drift) depends on correlating rocks in South America and Africa to see whether rock units of the same age could once have been continuous units. Various methods of correlation are described below.

Physical continuity Finding **physical continuity**—that is, being able to trace physically the course of a rock unit—is one way to correlate rocks between two different places. In figure 8.12, the prominent white layer of cliff-forming rock, the Coconino Sandstone, exposed along the upper part of Grand Canyon can be seen all the way across the photograph. You can physically follow this unit for several tens of kilometers, thus verifying that, wherever it is exposed in Grand Canyon, it is the same rock unit. (Incidentally, the Coconino Sandstone is also shown on box figure 1 of Box 6.2). Grand Canyon is an ideal location

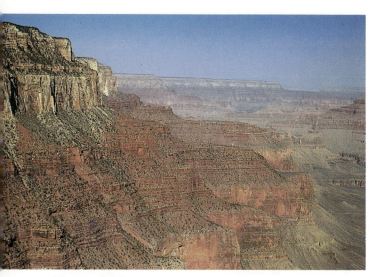

Figure 8.12
The upper layers of sedimentary rock in Grand Canyon. The Coconino sandstone forms the white cliff above the red rocks. Photo by C. C. Plummer.

Figure 8.13
Sandstone in Zion National Park, Utah. Photo by C. C. Plummer.

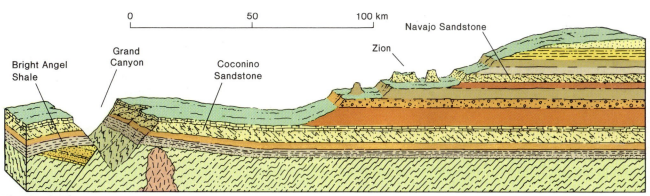

Figure 8.14
Schematic cross section through the Colorado Plateau.

for correlating rock units by physical continuity. However, it is not possible to follow this rock unit from the Grand Canyon into another region because it is not continuously exposed. We usually must use other methods to correlate rock units between regions.

Similarity of rock types   Under some circumstances, correlation between two regions can be made by assuming that similar rock types in two regions formed at the same time. This method must be used with extreme caution, especially if the rocks being correlated are common ones.

To show why correlation by similarity of rock type does not always work, we can try to correlate the white, cliff-forming Coconino Sandstone in Grand Canyon (shown in figure 8.12) with a rock unit of similar appearance in Zion National Park about 100 kilometers away (figure 8.13). Both units are white sandstone. Cross-bedding indicates that both were once a series of sand dunes. It is tempting to correlate them and conclude that both formed at the same time. But if you were to drive or walk from the rim of Grand Canyon (where the Coconino

Sandstone is *below* you), you would get to Zion by ascending a series of layers of sedimentary rock stacked on one another. In other words, you would be getting into progressively younger rock as shown diagrammatically in figure 8.14. In short, you have shown through *superposition* that the sandstone in Zion (called the Navajo Sandstone) is younger than the Coconino Sandstone.

Correlation by similarity of rock types is more reliable if a very unusual sequence of rocks is involved. If you find in one area a layer of green shale on top of a red sandstone that, in turn, overlies basalt of a former lava flow, and then find the same sequence in another area, you probably would be correct in concluding that the two sequences formed at essentially the same time.

Correlation by fossils   Fossils are common in sedimentary rock, and their presence is important for correlation. Plants and animals that lived at the time the rock formed were buried by sediment, and their fossil remains are preserved in sedimentary rock. Most of the fossil species found in rock layers are now extinct. (The concept of *species* in geology is similar to that in biology.)

Time and Geology   **169**

In a thick sequence of sedimentary rock, the fossils nearer the bottom (that is, in the older rock) are more unlike today's plants and animals than are those near the top. As early as the beginning of the nineteenth century, biologists began to notice that fossils in one layer of sedimentary rock often differ markedly from fossils in the layer above (or below). It became clear that many species of plants or animals existed only during the time that certain sedimentary layers were being deposited. Some species survived through the time interval represented by rock layers thousands of meters thick. Other species, confined to thin sequences of rock, apparently became extinct after a relatively short existence on earth. These observations, verified around the world, became formalized into the principle of **faunal succession:** fossil species succeed one another in a definite and recognizable order. In general, fossils in progressively older rock show increasingly greater differences from species now living. Plants and animals have tended to evolve from simpler to more complex species through geologic time. The regularity of changes in species of fossils through successive layers is one of the very strong lines of evidence supporting the theory of evolution.

Historical geology is concerned with the study of life and changes in life forms throughout geologic time as well as with the physical evolution of the earth. *Paleontologists,* specialists in the study of fossils, have patiently and meticulously over the years identified thousands of species of fossils and determined which plant and animal species lived during which subdivision of geologic time. They have also been able to describe the physical environment in which the fossil species lived.

No matter where on earth they are found, individual fossil species always occur in the same sequence relative to one another. By comparing fossils found in a layer of rock in one area with similar fossils in another area, we can correlate the two rock units. More accurately, we can say that both rock units formed during the span of time that the species existed on the earth.

Ideally, a geologist hopes to find an **index fossil,** a fossil from a very short-lived species known to exist during a specific period of geologic time. A single index fossil allows the geologist to correlate the rock in which it is found with all other rock layers in the world containing that fossil.

Many fossils are of little use in time determination because the species thrived during too large a portion of geologic time. Sharks, for instance, have been in the oceans for a long time, so discovering a shark's tooth in a rock is not very helpful in determining the rock's relative age.

A geologist is likely to find several different fossils in a rock. Such a **fossil assemblage** is generally more useful for dating rocks than a single fossil is, because the sediment must have been deposited at a time when all the species represented existed (figure 8.15).

Some fossils are restricted in geographic occurrence, representing organisms adapted to special environments. However, many former organisms apparently lived over

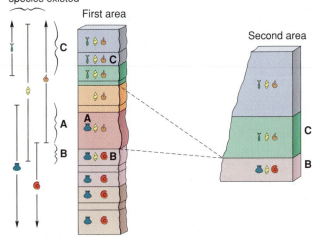

Figure 8.15
The use of fossil assemblages for determining relative ages. Rock *A* contains ⟨symbols⟩ . Therefore, it must have formed during time interval *A*. Rock *B* contains ⟨symbols⟩ . Therefore, it must have formed during time interval *B*. Rock *C* contains ⟨symbols⟩ . Therefore, it must have formed sometime during time interval *C*. In the second area, sequence *A* is missing. Therefore, there is an unconformity between sequence *B* and sequence *C*.

Figure 8.16
Fossil trilobites from the lower horizontal layers of Grand Canyon (the Bright Angel Shale). These trilobites are several centimeters long.
Photo by Tom Bean.

most of the earth, and fossil assemblages from these may be used for worldwide correlation. Fossils in the lowermost horizontal layers of Grand Canyon (figure 8.16) are comparable to ones collected in Wales, Great Britain, and a number of other places in the world. We can therefore correlate these rock units and say they formed during the same general span of geologic time.

BOX 8.1

Demise of the Dinosaurs— Was it Extraterrestrial?

Dinosaurs dominated the continents during the Mesozoic Era. Now they capture the imaginations of children of all ages. It's hard to accept that beings as powerful and varied as dinosaurs existed and were wiped out. But the fossil record is clear—when the Mesozoic came to a close, dinosaurs became extinct. Not a single of the numerous dinosaur species survived into the Cenozoic Era. Not only did the dinosaurs go, but about half of all plant and animal species, marine as well as terrestrial, were extinguished. This was one of the earth's "great dyings"—an even "greater dying" was when the Paleozoic Era ended with the extinction of 90% of all marine species. Most major extinctions have been gradual and scientists usually have attributed them to climate changes.

A decade ago, geologist Walter Alvarez, his father, physicist Luis Alvarez, and two other scientists proposed a hypothesis that the dinosaur extinction was caused by the impact of an asteroid. This was based on the chemical analysis of a thin layer of clay marking the boundary between the Mesozoic and Cenozoic eras (usually referred to as the K-T boundary—it separates the Cretaceous [K] and Tertiary [T] periods). The K-T boundary clay was found to have about 30 times the amount of the rare element iridium as is normal for crustal rocks. Iridium is relatively abundant in meteorites and other extraterrestrial objects such as comets and the scientists suggested that the iridium was brought in by an extraterrestrial body.

The scientists proposed a scenario in which an asteroid 10 kilometers in diameter struck the earth. This generated a gigantic dust cloud, which darkened the earth long enough to disrupt plant life growth and reproduction as well as drop the temperature of the earth worldwide. The dinosaurs perished because of the disruption in their food supply or because of their inability to withstand the sudden climate change.

Other scientists proposed an alternative hypothesis blaming exceptional volcanic activity for the extinction as well as for the high iridium in the K-T boundary layer. (Some Hawaiian eruptions have a high iridium content.) Debate between proponents of the two alternate hypotheses became heated and further evidence to support each was sought. K-T layers throughout the world were found to have grains of quartz that have been subjected to shock metamorphism (see Box 7.2), supporting the asteroid hypothesis. Concentrations of some elements common in volcanic rocks but not in meteorites were found in the K-T boundary layers, supporting the volcanic hypothesis. Critics of the asteroid hypothesis also pointed out that the dinosaurs had been on the decline for millions of years before the end of the Cretaceous. On the other hand, organisms in the oceans apparently died out very rapidly.

The asteroid hypothesis advocates predicted that a large meteorite crater should be found someplace on earth (unless the impact was at sea) that could be dated as having formed around 66 million years ago when the Mesozoic ended. A crater 35 kilometers in diameter was found in Iowa buried beneath glacial deposits and dated as being 67, plus or minus two, million years old—precisely the right age. The only problem is that it is too small to have been formed by a 10-kilometer asteroid (it should have a diameter closer to 150 kilometers).

As the 1980s drew to a close, the original asteroid hypothesis was modified considerably, with new variations. Most agreed that the earth suffered extraterrestrial bombardment. Perhaps it was not just one large asteroid but several smaller ones hitting the earth at about the same time. Or perhaps comets struck the earth, raising the acidity of rain worldwide to deadly levels. The geologic record also shows that it was a time of heavy volcanic activity (flood basalts in India more voluminous than those in the Columbia Plateau erupted at this time). One variation of the hypothesis is that the impact of an asteroid (or asteroids) triggered volcanic activity (as has happened on the moon). Scientists working on specific aspects of the problem are continuing to scrutinize the rocks and fossils formed at and near the K-T boundary. Although the extraterrestrial hypothesis has been widely accepted it has been extensively modified and it will probably be further modified.

A current view is that climates were changing toward the end of the Cretaceous and an extraterrestrial impact (perhaps with extensive volcanic activity triggered by the impact) was perhaps the final unfortunate blow to the dinosaurs. "Unfortunate" is from the perspective of dinosaurs, not humans. The only mammals in the Cretaceous were inconsequential, mouse-sized creatures. They survived the K-T extinction and, with dinosaurs no longer dominating the land, evolved into the many mammal species that populate the earth today, including humans.

Additional Reading   Hsu, K. J. 1986. *The great dying.* New York: Harcourt, Brace, Jovanovich.

Geologic Time Scale

Era	Period	Epoch
	Quaternary	Recent (Holocene) Pleistocene
Cenozoic	Tertiary	Pliocene Miocene Oligocene Eocene Paleocene
Mesozoic	Cretaceous Jurassic Triassic	
Paleozoic	Permian Pennsylvanian Mississippian Devonian Silurian Ordovician Cambrian	
Precambrian Time		

The Standard Geologic Time Scale

Geologists can use fossils in rock to refer the age of the rock to the **standard geologic time scale,** a worldwide relative time scale. Based on fossil assemblages, the geologic time scale subdivides geologic time. On the basis of fossils found, a geologist can say, for instance, that the rocks of the lower portion of horizontal layers in Grand Canyon formed during the *Cambrian Period.* This implicitly correlates these rocks with certain rocks in Wales (in fact, the period takes its name from Cambria, the Latin name for Wales) and elsewhere in the world where similar fossils occur.

The geologic time scale, shown in a somewhat abbreviated form in table 8.2, has had tremendous significance as a unifying concept in the physical and biological sciences. The working out of the evolutionary chronology by successive generations of geologists and other scientists has been a remarkable human achievement. The geologic time scale, representing an extensive fossil record, consists of three **eras,** which are subdivided into **periods,** which are, in turn, divided into **epochs.** (Remember that this is a relative time scale.)

Precambrian denotes the vast amount of time that preceded the Paleozoic Era (which begins with the Cambrian Period). The **Paleozoic Era** (meaning "old life") began with the appearance of abundant and complex life, as indicated by fossils. Rocks older than Paleozoic contain few fossils. This is because creatures with shells or other hard parts, which are easily preserved in fossils, did not evolve until the beginning of the Paleozoic.

The **Mesozoic Era** (meaning "middle life") followed the Paleozoic. On land, reptiles were the dominant animals of the Mesozoic (this was the time when dinosaurs lived). We live in the **Recent (or Holocene) Epoch** of the **Quaternary Period** of the **Cenozoic Era** (meaning "new life"). The Quaternary also includes the most recent ice ages, which were part of the **Pleistocene Epoch.**

Fossils have been used to determine ages of the horizontal rocks in Grand Canyon. All are Paleozoic. The lowermost horizontal formations (figure 8.14) are Cambrian, above which are Mississippian, Pennsylvanian, and Permian rock units. By referring to the geologic time scale (table 8.2), we can see that Ordovician, Silurian, and Devonian rocks are not represented. Thus an unconformity (buried erosion surface) is present within the horizontally layered rocks of Grand Canyon.

Absolute Age

Counting annual growth rings in a tree trunk will tell you how old a tree is. Similarly, layers of sediment deposited annually in glacial lakes can be counted to determine how long those lakes existed (*varves,* as these deposits are called, are explained in the chapter on glaciers). But only within the few decades since the discovery of radioactivity have scientists been able to determine absolute ages of rock units. From these data we have been able to assign absolute values to the geologic time scale and so determine how many years ago the various eras, periods, and epochs began and ended. We can now state with a high degree of confidence that the Cenozoic Era began some 66 million years ago, the Mesozoic Era started about 245 million years ago, and the Precambrian ended (or the Paleozoic began) about 570 million years ago. The Precambrian includes most of geologic time, because the age of the earth is commonly regarded as about 4.5 billion years. (Soviet geologists think that 6 billion years is a better estimate.) The oldest rocks found on earth are from northwestern Canada and have been dated at about 3.964 billion years old. The 4.5 billion-year estimated age of the earth comes from various lines of evidence, mostly worked out by astronomers and planetary geologists who think that all the planets formed about this long ago.

Radiometric Dating

Radioactivity provides a clock that began working when certain types of rocks formed. Under particular conditions radioactivity can be used to determine the age of rocks. We know the rate of radioactive decay or change. By comparing the amount of radioactive elements with the amount of radioactive decay products, we can determine how much time has elapsed since the rocks formed.

Isotopes and radioactive decay   As discussed in the chapter titled "Atoms, Elements, and Minerals," every atom of an element possesses the same number of protons in its nucleus. The number of neutrons, however, need not be the same in all atoms of the same element. The **isotopes** of a given element have different numbers of neutrons, but the same number of protons.

The Origin and History of the Solar System

The solar system began with the condensation of a large volume of interstellar gas and dust called a *nebula.* Much of the material in the nebula collected in the center to form the Sun, but some of it remained farther out. Small groupings of particles formed randomly, and because of the combined gravitational pull of the particles in the group, they attracted other particles. The groupings continued to increase in size until, on the average, they were probably several kilometers in diameter.

These larger bodies sometimes joined together by collision and sometimes became fragmented. The largest bodies tended to increase in size by capturing smaller ones, and the smaller bodies tended to be fragmented and/or captured. The bodies that grew larger became the planets and satellites; those that were fragmented became interstellar dust, asteroids, and the like; and those that were captured were incorporated into the planets and satellites.

Variations in composition among the planets can be partially explained by differences in their sizes. Larger planets were able to attract and hold larger amounts of lighter gases, such as hydrogen and helium. Another important factor in determining planetary composition was the location within the nebula where the planet formed. Those bodies nearest the center (Mercury, Venus, Earth, the Moon, and Mars) would have experienced higher temperatures and so retained smaller percentages of materials that are easily vaporized. The outer planets and their satellites were formed at a great enough distance from the Sun for large quantities of gases and ices to collect around their rocky cores.

Energy from the decay of radioactive elements that were incorporated into the planets would have heated the terrestrial planets, causing partial melting of the planets' interiors. Heavier materials such as iron-nickel alloys would sink to the planets' centers, while lighter materials such as silicate magmas would rise to form crusts. On Earth, the original crust was mostly basalt, with some granitic rocks. The areas of granitic rocks would have formed the first continents. Convection in the mantle would probably have begun at the time of crustal differentiation and would have set in motion plate tectonic processes.

On Venus, which is about the same size as the Earth, the presence of surface rocks such as granite and basalt and possible folded mountains (the parallel ridges that have been observed) and chains of volcanoes suggest that differentiation and plate motions may have occurred in a similar way. Mars, Mercury, and the Moon, all smaller objects, show no evidence of granitic rocks. The absence of folded mountain chains and the presence of large numbers of undeformed craters on Mars indicate that plate tectonic activity has not occurred on Mars, although some suggestions have been made that Mars is in the initial stages of developing a system of convection. It has been noted that the very large size of volcanoes such as Olympus Mons could be explained by assuming a stationary lithosphere over a mantle plume. Mercury and the Moon show no evidence of plate tectonic processes having occurred. Perhaps crustal composition and the extent to which plate tectonic processes occur is related to planetary size.

Space "debris" continued to be swept up by the planets throughout their early histories, producing heavily cratered surfaces and multi-ringed basins. The Moon and Mercury are still heavily cratered. The Earth, too, was subjected to an early period of intensive bombardment. But the Earth is a restless planet, and erosion and tectonic activity have erased most of those early-formed craters. Venus and Mars retain more of their early scars than the Earth does, because of less erosion and tectonic activity.

The heating up of planetary interiors would produce great quantities of magma that would rise to the surface along fractures produced by meteorite impact or by some tectonic process. Vast areas of the Moon, Mercury, and Mars are covered with volcanic rocks, most formed early in the history of the solar system. Similar early extrusive activity must have occurred on Venus and on the Earth. Volcanic activity on a much smaller scale still occurs on the Earth and Venus and may still be occurring on Mars. However, this later volcanic activity is related to tectonic processes.

Gases would escape from planetary interiors as a result of volcanic activity, later condensing to form atmospheres, oceans, ice shells, and polar caps. Lighter elements would quickly have escaped from objects as small as the Moon and Mercury but could have been partially retained by the Earth, Venus, and Mars. The distances of these planets from the Sun might determine whether these light elements exist mostly as atmospheres (as on Venus); as atmospheres, oceans, underground water, and ice caps (as on Earth); or mostly as ice caps and underground ice (as on Mars). On planets with appropriate conditions and temperature ranges, life may have formed, as it did on the Earth. The presence of life on the Earth has certainly extensively modified the Earth's atmosphere (by producing oxygen, for example).

The presence of atmospheres, oceans, underground ice, and so on would have determined the extent to which weathering and erosion modified a planet's surface. Because of extensive erosion on Earth (and because of tectonic activity), other planets must be examined to understand the earliest history of Earth.

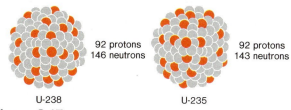

Figure 8.17
Nuclei of isotopes of U-238 and U-235.

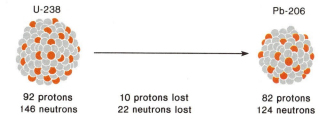

Figure 8.18
U-238 disintegrates radioactively to Pb-206 (intermediate daughter products are not shown).

Uranium, for example, commonly occurs as two isotopes, U-238 and U-235. The former has 238 protons and neutrons in its nucleus, whereas the latter has 235 (figure 8.17). Of these particles, 92 (the atomic number of uranium) must be protons and the rest neutrons. U-238 weighs slightly more than U-235 and may be separated from it by special equipment because of the difference in mass.

Radioactive decay is the spontaneous nuclear disintegration of certain isotopes. As protons and neutrons leave these atoms, energy is produced. The particles departing from radioactive elements can be detected by a Geiger counter or similar device.

When protons are lost during radioactive decay, the atom becomes a different element. When a uranium-238 atom disintegrates radioactively, it loses a total of 32 protons and neutrons, in a complex series of steps, before becoming a nonradioactive, stable atom. Having lost 32 subatomic particles (out of 238), the now-stable atom has 206 neutrons and protons left. Ten of the lost particles are protons, so the atomic number has changed from 92 to 82, and the atom is now the element lead (abbreviated Pb) (figure 8.18). Specifically, it is the Pb-206 isotope. Lead-206 is the **daughter product,** the isotope produced by radioactive decay, of U-238.

An isotope of potassium, K-40, converts to argon, Ar-40, through a different process. From time to time a K-40 nucleus captures one of its orbiting electrons. The negative charge of the electron neutralizes a single positively charged proton. The proton becomes a neutron. Since there is one less proton in the nucleus, that atom no longer is a potassium atom (atomic number 19) but an argon atom (atomic number 18).

Radioactive elements are useful for determining age because in a large number of atoms of a radioactive element, the *rate* of radioactive decay is constant. Even a rock possessing a small fraction of a percent of uranium contains millions of uranium atoms. The rate at which atoms periodically decay is apparently unaffected by the high pressures and high temperatures of the earth's interior.

The decay rate for isotopes is given in terms of **half-life,** the time it takes for a given amount of a radioactive isotope to be reduced by one-half. (The other half disintegrates into daughter products.) The half-lives of some elements are measured in fractions of seconds. K-40, however, has a half-life of 1.3 billion years. If you began with one milligram of K-40, 1.3 billion years later one-half milligram of K-40 would remain. After another 1.3 billion years, there would be one-fourth of a milligram, and after another half-life only one-eighth of a milligram. (Note that two half-lives do not equal a whole life.)

T A B L E 8 . 3

Radioactive Isotopes Commonly Used for Determining Ages of Rocks

Isotope	Half-Life	Daughter Product
K-40	1.3 billion years	Ar-40
U-238	4.5 billion years	Pb-206
U-235	713 million years	Pb-207
Th-232	14.1 billion years	Pb-208
Rb-87	49 billion years	Sr-87
C-14	5,730 years	N-14

To determine the age of a rock by using K-40, the amount of K-40 in that rock must first be determined by chemical analysis. The amount of Ar-40 (the daughter product) must also be determined and then used to calculate how much K-40 was present when the rock formed. By knowing how much K-40 was originally present in the rock and how much is still there, we can calculate the age of the rock on the basis of its half-life by simple mathematics. The graph in figure 8.19 demonstrates the mathematical relationship between a radioactively decaying isotope and time.

Uses of radiometric dating There are some problems in obtaining a meaningful and accurate date from a rock. Laboratory technicians must be highly skilled to operate the sophisticated equipment used in radiometric dating. Most successfully dated rocks are igneous, which, since they solidified from a melt, are unlikely to be contaminated by previously formed daughter products. Sedimentary rock cannot usually be dated by radioactive means. Metamorphic rock can yield inaccurate ages because heat during metamorphism may have driven out some of either the daughter product or the radioactive isotope. In addition, chemically weathered rocks cannot be dated because of the material lost during weathering.

The potassium-argon (K-Ar) method is effective for dating a wide range of ages (from a few hundred thousand years to several billion years) because of its long half-life and because potassium is common in a wide variety of rocks. The disadvantage of the potassium-argon method is that argon is a gas and can be easily lost from rocks.

Table 8.3 shows the commonly used isotopes for dating rocks. U-238/Pb-206, U-235/Pb-207, and Th-232/Pb-208 are used mainly for dating older geologic events (millions or billions of years). The two U-Pb systems are

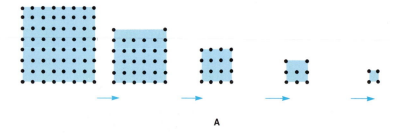

A

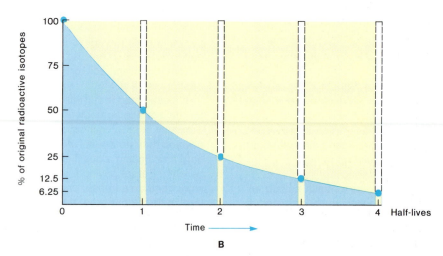

B

Figure 8.19
The curve used to determine the age of a rock by comparing the percentage of radioactive isotope left to the original amount. Heavy bars show the amount left after each half-life. Dashed bars show the amount disintegrated into daughter product and lost nuclear particles. The numbers of dots in the squares above the graph are proportional to the numbers of atoms.

particularly useful if used together, since they provide internal cross-checks on the ages and, therefore, more accurate results.

Combining Relative and Absolute Ages

Radiometric dating can provide absolute time brackets for events whose relative ages are known. Figure 8.20 adds radioactive dates for each of the two igneous bodies to the diagram in figure 8.1. The date obtained for the granite is 540 million years B.P. (before present), while the dike formed 78 million years ago. We can now state that the tilted layers formed before 540 million years ago (though we cannot say how much older they are). We still do not know whether the Leet Junction Formation is older or younger than the granite because of the lack of cross-cutting relationships. The Larsonton Formation's age is bracketed by the age of the granite and the age of the dike. That is, it is between 540 and 78 million years old. The Foster City and overlying formations are younger than 78 million years old; how much younger we cannot say.

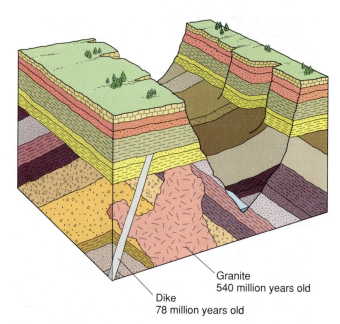

Granite
540 million years old

Dike
78 million years old

Figure 8.20
Same area as shown in figure 8.1 but with radioactive dates for igneous rocks indicated.

Radon, a Radioactive Health Hazard

Radon is an odorless, colorless, gas. The United States Environmental Protection Agency estimates that between 5,000 and 30,000 lung cancer deaths per year in the U.S. are attributable to radon. It is one of the intermediate daughter products in the radioactive disintegration of U-238 to Pb-206. Its half-life is only 3.8 days.

Concentrations of radon are highest in areas where the bed rock is granite, gneiss, limestone, black shale, or phosphate-rich rock—rocks in which uranium is relatively abundant. Even in these areas, radon levels are harmless in open, freely circulating air. Radon may dissolve in ground water or build up to high concentrations in confined air spaces.

Radon was first recognized as a health hazard in mines, where the gas would collect in the confined air spaces. Radon lodges in the respiratory system of an individual and, as it deteriorates into daughter products, the subatomic particles given off cause damage to lung tissue. (Smokers are especially susceptible to radon damage of the lungs.)

During the last few years, high levels of radon were discovered in houses, particularly in parts of Pennsylvania and New Jersey where soil had formed from high-uranium bed rock. In some houses the concentration of radon in the air was greater than the danger level for mines. In most buildings with a high radon level, the gas seeps in from the underlying soil through the building's foundation. If a building's windows are kept open and fresh air circulates freely, radon concentrations cannot build up. However, most houses are kept sealed for air conditioning during the summer and heating during the winter. Air circulation patterns are such that a slight vacuum sucks the gases from the underlying soil into the house. Thus, radon concentrations build up to dangerous levels.

The problem may be solved in several ways (aside from leaving windows open winter and summer). Basements can be made air tight so that gases cannot be sucked into the house from the soil. Air circulation patterns can be altered so that gases are not sucked in from underlying soil, or they are mixed with sufficient fresh, outside air.

It's estimated that it will cost two thousand dollars for the average homeowner to mitigate radon hazards. Currently, a radon mitigation industry has yet to develop (but, no doubt, will); a homeowner may have a hard time finding a qualified person to radon-proof the building. Before paying for having radon mitigation work done on a house, one should have the radon level in the air checked. If you are buying a new house, it would be a good idea to have it tested for radon before buying, particularly if the house is in an area of high-uranium bed rock.

Additional Reading   Nazaroff, W. W. and A. V. Nero Jr. 1988. *Radon and its decay products in indoor air.* New York: Wiley Interscience.

By combining many radioactive dates from sites carefully selected for well-known relative age relationships, geologists have been able to assign absolute ages to the geologic time scale (table 8.4). Radiometric dating has also allowed us to extend the time scale back into the Precambrian. The Proterozoic and the Archean are the two major subdivisions of Precambrian time. There is, of course, a margin of uncertainty in each of the given dates. The beginning of the Paleozoic, for instance, might be off by some 30 million years. There are inherent limitations on the dating techniques as well as problems in selecting the ideal rock for dating. For instance, if you wanted to obtain the date for the end of the Paleozoic Era and the beginning of the Mesozoic Era, the ideal rock would be found where there is no break in deposition of sediments between the two eras, as indicated by fossils in the rocks. But the difficulties in dating sedimentary rock mean you would be unlikely to date such rocks, even if you found them. Therefore, you must find and date igneous rocks that seem to have solidified, according to cross-cutting relationships or other evidence, at *about* the time of transition between the two eras.

Summary

The principle of *uniformitarianism,* a fundamental concept of geology, states that the present is the key to the past.

Relative time, or the sequence in which geologic events occur in an area, can be determined by applying the principles of *original horizontality, superposition,* and *cross-cutting relationships.*

Rocks in separate areas can be correlated by determining the physical continuity of rocks between the two areas (generally this works only for a short distance). A less useful means of correlation is similarity of rock types (which must be used cautiously).

Similarity of *fossil assemblages* is used for worldwide correlation of rocks. Sedimentary rocks are assigned to the various subdivisions of the *geologic time scale* on the basis of fossils they contain, which are arranged according to the principle of *faunal succession.*

Absolute age—how many years ago a geologic event took place—is generally obtained by using *radiometric dating* techniques. Absolute ages have been determined for the subdivisions of the geologic time scale.

Radiocarbon: Dating "Young" Events

Because of its short half-life of 5,730 years, radiocarbon dating is useful only in dating things and events back to about 50,000 years (about nine or ten half-lives). The technique is most useful in archaeological dating and for very young geologic events (Recent, or Holocene, volcanic and glacial features for instance).

Most of the carbon in the air is carbon-12, but a minute fixed amount is radioactive carbon-14. C-14 decays radioactively to nitrogen (N-14). This occurs when a neutron in the C-14 nucleus splits into an electron plus a proton. The electron is emitted from the atom, which now has one more proton than the original carbon atom. Its atomic number is now 7 rather than 6; therefore, it is an isotope of nitrogen.

Living matter incorporates C-12 and C-14 into its tissues; the ratios of C-12 and C-14 in the new tissues are usually the same as in the atmosphere. On dying, the plant or animal ceases to build new tissue. The C-14 disintegrates radioactively at the fixed rate of its half-life (5,730 years). The radioactive emissions per gram of carbon from the plant or animal remains are determined. The greater the radioactivity, the younger the sample. By comparing the radioactivity per gram to previously determined standards, we can indirectly determine the ratio of C-12 to radioactive C-14 in organic remains, and we can determine the time elapsed since the death of the organism.

TABLE 8.4

Geologic Time Scale

Era	Period	Epoch	Approximate Age in Millions of Years Before Present	
Cenozoic	Quaternary	Recent (Holocene)	.01	
		Pleistocene	1.6	
	Tertiary	Pliocene	5.3	
		Miocene	23.7	
		Oligocene	36.6	
		Eocene	57.8	
		Paleocene		
				66.4
Mesozoic	Cretaceous		144	
	Jurassic		208	
	Triassic			245
Paleozoic	Permian		286	
	Pennsylvanian		320	
	Mississippian		360	
	Devonian		408	
	Silurian		438	
	Ordovician		505	
	Cambrian			570
Precambrian Time Proterozoic Eon				2,500
Archean Eon		Formation of Earth		4,500

Source: Decade of North American Geology, 1983 Geologic Time Scale—Geological Society of America

BOX 8.4

The Longest Movie Never Shown—the Earth's Story

One way of trying to visualize and comprehend geologic time is to compare it to a motion picture. A movie is projected at a rate of sixteen frames per second; that is, each image is flashed on the screen for only one-sixteenth of a second, giving the illusion of continuous motion. But suppose that each frame represented 100 years. If you lived 100 years, one frame would represent your whole lifetime.

If we were able to show the movie on a standard projector, each 100 years would flash by in one-sixteenth of a second. It would take only one-eighth of a second to go back to the signing of the Declaration of Independence. The 2,000-year-old Christian era would be viewed for one and a quarter seconds. A section showing all time back to the last major ice age would only be eleven and a quarter seconds long. However, you would have to sit through more than eleven hours of film to view a scene at the close of the Mesozoic Era (perhaps you would see the last dinosaur die). And to give a complete record from the beginning of the Paleozoic Era, this epic film would have to run continuously for four days. You would have to spend over a month (thirty-two and a half days) in the theater, without even a popcorn break between reels, to see a movie entitled "The Complete Story of Earth, from Its Birth to Modern Civilization."

Thinking of our lives as taking less than a frame of such a movie can be very humbling. From the perspective of that one last frame, geologists would like to know what the whole movie is like or, at least, get a synopsis of the most dramatic parts of the film.

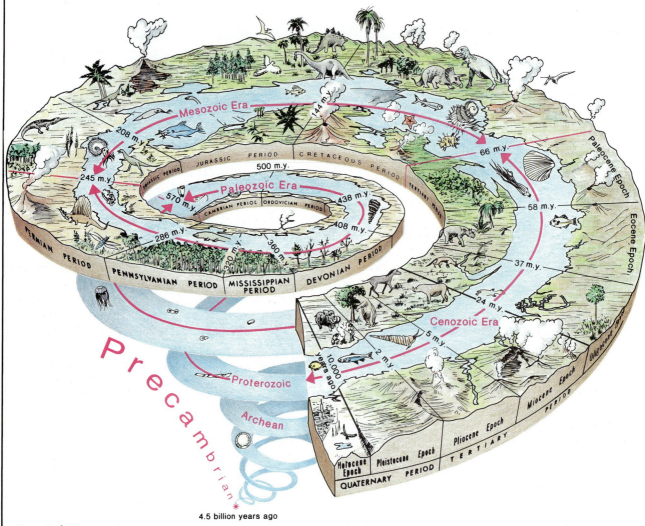

Box 8.4 Figure 1
Representation of geologic time. The linear distance around the whorl representing time is not to scale.
After U.S. Geological Survey publication *Geologic Time*.

Terms to Remember

absolute age	original horizontality
Cenozoic Era	Paleozoic Era
contacts	period
correlation	physical continuity
cross-cutting relationship	Pleistocene Epoch
daughter product	Precambrian
epoch	Quaternary Period
era	radioactive decay
faunal succession	Recent (Holocene) Epoch
formation	relative time
fossil assemblage	standard geologic time scale
half-life	superposition
index fossil	unconformity
isotope	uniformitarianism
Mesozoic Era	

Questions for Review

1. Suppose you had a radioactive isotope X whose half-life in disintegrating to daughter product Y is 10,000 years. By calculating how much it took to make the present amount of Y, you determine that, originally, the rock contained 8 grams of isotope X. At present only ¼ gram of X is in the rock. How old is the rock?
2. What percentage of the earth's history is Precambrian?
3. Why is it desirable to use more than one fossil in a rock unit to determine its relative age?
4. By applying the various principles, draw a cross-section of an area in which the following sequence of events occurred:
   a. Several layers of sedimentary rocks were deposited in the Cambrian through Devonian periods on a much older Precambrian basement of metamorphic rock.
   b. No record exists of Mississippian through Triassic rocks.
   c. A stock intruded in the Jurassic Period.
   d. Tilting and erosion to a flat plain preceded deposition of a sedimentary layer in late Cretaceous.
   e. In the Tertiary Period, more and steeper tilting affected the entire area.
   f. Erosion during the Quaternary Period created slightly hilly terrain.
   g. Following erosion, a volcano with a feeder dike erupted in recent time.
5. How much of the uranium-238 originally in the earth is still present?
6. Name as many types of contacts (e.g., intrusive contact) as you can.

Questions for Thought

1. Moon rocks have been dated that are considerably older than any dated on earth. Give a hypothesis to explain this.
2. As indicated by fossil records, why have some ancient organisms survived through very long periods of time whereas others have been very short-lived?
3. To what extent would a composite volcano (chapter 3) be subject to the three principles described in this chapter?
4. Suppose a sequence of sedimentary rock layers was tilted into a vertical position by tectonic forces. How might you determine (a) which end was originally up and (b) the relative ages of the layers?
5. Note that in table 8.3 the epochs are given only for the Cenozoic Era (as is commonly done in geology textbooks). Why are the epochs for the Mesozoic and Paleozoic considered less important and not given?
6. Why would you not be able to use the principle of superposition to determine the age of a sill (defined in chapter 4)?

Supplementary Readings

Berry, W. B. N. 1987. *Growth of a prehistoric time scale*. Palo Alto, California: Blackwell Scientific Publications.

Cooper, J. D., R. C. Miller, and J. Patterson. 1986. *A trip through time: Principles of historical geology*. Columbus, Ohio: Merrill.

Cowen, R. 1989. *History of life*. Palo Alto, California: Blackwell Scientific Publications.

Faure, G. 1986. *Principles of isotope geology*. 2d ed. New York: John Wiley & Sons.

Gould, S. J. 1988. *Time's arrow/time's cycle*. Boston: Harvard University Press.

Harland, W. B., A. V. Cox, P. G. Llewellyn, C. A. G. Pickton, A. G. Smith, and R. Walters. 1983. *A geologic time scale*. New York: Cambridge University Press.

Thackery, J. 1986. *The age of the earth*. Pamphlet published by Institute of Geological Sciences, London and by Cambridge University Press, New York.

Video:
The earth has a history. VHS. Geological Society of America, Boulder, Colorado.

When material on a hillside has weathered (the process described in the chapter on weathering and soil), it is likely to move downslope because of the pull of gravity. Soil and rock moving in bulk at the earth's surface is called mass wasting. Mass wasting is one of several erosion processes. Other processes of erosion (and deposition)—involving streams, glaciers, wind, and ocean waves—are discussed in the next few chapters.

Landsliding is probably the best known type of mass wasting, for sudden landslides can destroy towns and kill people. While these disasters involve *rapid* movement of debris and rock, mass wasting can also be very slow. A type of mass wasting too slow to be classed as a landslide is called creep.

In this chapter we describe how different types of mass wasting shape the land and alter the environment and what factors control the rapidity or slowness of the process. Understanding mass wasting and its possible hazards is particularly important in hilly or mountainous regions.

Mass Wasting

Landslide in Hong Kong that killed 67 people in 1972.
Supplied by the Geotechnical Control Office, Hong Kong Government.

BOX 9.1

Disaster in the Andes

As a result of a tragic combination of geological conditions, one of the most devastating landslides in history destroyed the town of Yungay in Peru in 1970. Yungay was one of the most picturesque towns in the Santa River Valley, which runs along the base of the highest peaks of the Peruvian Andes. Heavily glaciated Nevado Huascarán, 6,663 meters (21,860 feet) above sea level, rises steeply above the populated narrow plains along the Santa River.

In May 1970 a sharp earthquake occurred. The earthquake was centered offshore from Peru about 100 kilometers from Yungay. Although the tremors in this part of the Andes were no stronger than those that have done only light damage to cities in the United States, many poorly constructed adobe homes collapsed. Because of the steepness of the slopes, thousands of small rockfalls and rockslides were triggered.

The greatest tragedy began when a slab of glacier ice about 800 meters wide, perched near the top of Huascarán, was dislodged by the shaking. (A few years earlier American climbers returning from the peak had warned that the ice looked highly unstable. The Peruvian press briefly noted the danger to the towns below, but the warning was soon forgotten.)

The mass of ice rapidly avalanched down the extremely steep slopes, breaking off large masses of rock debris, scooping out small lakes and loose rock that lay in its path. Eyewitnesses described the mass as a rapidly moving wall the size of a ten story building. The sound was deafening. More than 50 million cubic meters of muddy debris fell 3.7 kilometers (12,000 feet) vertically and traveled 14.5 kilometers (9 miles) horizontally in less than four minutes, attaining speeds between 200 and 435 kilometers per hour (125 to 270 miles per hour). Some geologists believe that

Box 9.1 Figure 1
Air photo showing the 1970 debris avalanche in Peru, which buried Yungay. The main mass of debris destroyed the small village of Ranrahirca.
Photo by Servicio Aerofotografico de Peru, courtesy of U.S. Geological Survey.

for the debris to travel that fast, a cushion of air must have formed under the moving mass to minimize friction between the debris and the land surface. In places the mass shot over ridges into the air like a huge cannonball. The main mass of material traveled down a steep valley until it came to rest blocking the Santa River and burying about 1,800 people in a small village. A relatively small part of the mass of mud and debris overtopped the valley sidewall, or ridge, as it came to a curve in the valley. The mass was airborne for a moment before it fell on the town of Yungay

---

You may recall from previous chapters that mountains are products of tectonic forces. If tectonism were not at work, the surfaces of the continents would long ago have been reduced to featureless plains due to weathering and erosion. We consider the material on mountain slopes or hillsides to be out of equilibrium with respect to gravity.

Because of the force of gravity, the various agents of erosion (moving water, ice, and wind) work to make slopes gentler and therefore increasingly more stable. The process of erosion discussed in this chapter is mass wasting.

Mass wasting (also called mass movement) is movement in which bed rock, rock debris, or soil moves downslope in bulk, or as a mass, because of the pull of gravity. Mass wasting includes movement so slow that it is almost imperceptible (called *creep*) as well as **landslides,** a general term for the slow-to-very-rapid descent of rock or soil.

B

A

Box 9.1 Figure 2
(*A*) Yungay is completely buried, except for the cemetery and a few houses on the small hills in the lower right of the photograph. (*B*) Beneath the palm trees is the top of a church at Yungay's central plaza, buried under five meters of debris. (*C*) Three years later.
Photos *A* and *B* by George Pflafker, U.S. Geological Survey. Photo *C* by C. C. Plummer.

below, completely burying it under several meters of mud and loose rock. Only the church steeple and the tops of palm trees could still be seen.

It was estimated that 17,000 people were killed in Yungay. This was far more than the town's normal population, for it was Sunday, a market day, and many families had come in from the country.

C

For several days after the slide the debris was too muddy for people to walk on, but within three years grass had grown over the site. Except for the church steeple and the tops of palm trees that still protrude above the ground, and the crosses erected by families of those buried in the landslide, the former site of Yungay appears to be a scenic meadow overlooking the Santa River. The U.S. Geological Survey and Peruvian geologists found evidence that Yungay itself had been built on top of debris left by an even bigger slide in the recent geologic past. More slides will almost surely occur here in the future. Forewarned, the Peruvian government will not allow the building of a new town in the danger area.

Mass wasting affects humans in many ways. Its effects range from the devastation of a killer landslide (see Box 9.1) to the nuisance of having a fence slowly pulled apart by soil creep. The cost in lives and property from landslides is surprisingly high. According to the U.S. Geological Survey, more people in the United States died from landslides during the last three months of 1985 than were killed during the last twenty years by all other geologic hazards, such as earthquakes and volcanic eruptions. The damage to property from landslides each year exceeds the cost of earthquake damage for the last twenty years. In many cases of mass wasting, a little knowledge of geology, along with appropriate preventive action, could have averted destruction.

Whether slow, fast, or somewhere in between, the various kinds of mass wasting play an important role in the wearing away of the earth's surface.

Classification of Mass Wasting

A number of systems are used by geologists, engineers, and others for classifying mass wasting, but none has been universally accepted. Some are very complex and useful only to the specialist.

The classification system used here and summarized in table 9.1 is based on (1) rate of movement, (2) type of material, and (3) nature of the movement.

Rate of Movement

A landslide like the one in Peru clearly involved rapid movement. Just as clearly, movement of soil at a rate of less than a centimeter a year is slow movement. Between these extremes is a wide range. Some of the factors controlling the speed of mass movement are discussed later in this chapter.

Type of Material

Mass wasting processes are usually distinguished on the basis of whether the descending mass started as bed rock (as in a rockslide) or as debris. The term **debris,** as applied to mass wasting processes, means any unconsolidated material at the earth's surface, such as soil and rock fragments (weathered or unweathered) of any size.

The amount of water (or ice and snow) in a descending mass strongly influences the rate and type of movement. This influence is explained later in this chapter.

Type of Movement

In general, the type of movement in mass wasting can be classified as mainly flow, slip, or fall (figure 9.1). A **flow** implies that the descending mass is moving downslope as a viscous fluid. **Slip** means the descending mass remains relatively coherent, moving along one or more well-defined surfaces. A **fall** occurs when material free-falls or bounces down a cliff.

Two kinds of slips are shown in figure 9.1. In a **slide,** the descending mass moves along a plane approximately parallel to the slope of the surface. A **slump** involves movement along a curved surface, the upper part moving downward while the lower part moves outward.

Controlling Factors in Mass Wasting

Table 9.2 summarizes the factors that influence the likelihood and the rate of movement of mass wasting. The table makes apparent some of the reasons why the landslide in Peru (Box 9.1) occurred and why it moved so rapidly.

TABLE 9.1

Some Types of Mass Wasting[1]

	Slowest ———→	Increasing Velocities ————→		Fastest
Type of Movement	Less than 1 cm/year	1 mm/day to 10 km/hour	1 to 5 km/hour	Velocities generally greater than 4 km/hour
Flow	Creep (Debris)	Earthflow without slumping	Mudflow (Water-saturated debris)	Rock avalanche (Bed rock) / Debris avalanche (Debris)
Slip		Earthflow (Debris) / Earthflow with slumping	Rockslide (Bed rock) / Debris slide (Debris)	
Fall				Rockfall (Bed rock) / Debris fall (Debris)
	←———————————— "Landslides" ————————————→			

1. The type of material at the start of movement is shown in parentheses. Rates given are typical velocities for each type of movement.

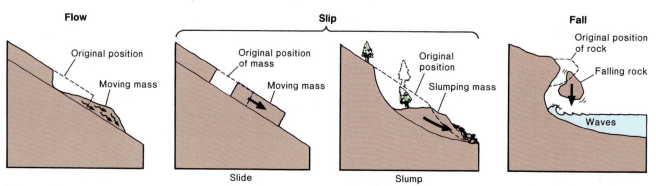

Figure 9.1
Flow, slip (slide and slump), and fall.

(1) The slopes were exceptionally steep, and (2) the **relief** (the vertical distance) between valley floor and mountain summit was great, allowing the mass to pick up speed and momentum. (3) Water and ice not only added weight to the mass of debris but also acted as lubricants. (4) Much loose rock and debris were available in the course of the slide. (5) Where the slide began, there were no plants with roots to anchor loose material on the slope. Finally, (6) the area was and is earthquake prone. Although the slide would have occurred eventually even without an earthquake, the earthquake acted as a trigger.

Other factors influence susceptibility to mass wasting as well as its rate of movement. The orientation of planes of weakness in bed rock (bedding planes, foliation planes, etc.) is important if the movement involves bed rock rather than debris. Fractures or bedding planes oriented so that slabs of rock can slide easily along these surfaces greatly increase the likelihood of mass wasting.

Table 9.2 shows that climatic controls inhibit some types of mass wasting and aid others. Climate influences how much and what kinds of vegetation grow in an area and what type of weathering occurs. Infrequent but heavy rainfall aids mass wasting because it quickly saturates debris that lacks the protective vegetation found in wetter climates. By contrast, rain that drizzles intermittently much of the year tends to inhibit mass wasting. In cold climates, freezing and thawing contribute to downslope movement.

Water

Water is a critical factor in downslope movement. When debris is saturated with water (as from heavy rain or melting snow), it becomes heavier and is more likely to flow downslope. The added gravitational force from the increased weight, however, is probably less important than the effect of increased *pore pressure* in which water forces grains of debris apart.

Paradoxically, a small amount of water in soil can actually prevent downslope movement. When water does not completely fill the pore spaces between the grains of soil, it forms a thin film around each grain (as shown in figure 9.2). Loose grains adhere to one another because of the *surface tension* created by the film of water. Surface tension of water between sand grains is what allows you to build a sand castle. The sides of the castle can even be vertical because surface tension holds the moist sand grains in place. Dry sand cannot be shaped into a sand castle because the sand grains slide back into a pile that generally slopes at an angle of about 30° to 35° from the horizontal. On the other hand, an experienced sand castle builder also knows that it is impossible to build anything with sand that is too wet. In this case the water completely occupies the pore space between sand grains, forcing them apart and allowing them to slide easily past one another. When the tide comes in, or someone pours a pail of water on your sand castle, all you have is a puddle of wet sand.

Similarly, as the amount of water in debris increases, rate of movement also increases. Damp debris may not move at all, whereas moderately wet debris moves slowly downslope. Slow types of mass wasting, such as creep, are generally characterized by a relatively low ratio of water to debris. Earthflows tend to have, and mudflows always have, high ratios of water to debris. A mudflow that continues to gain water eventually becomes a muddy stream.

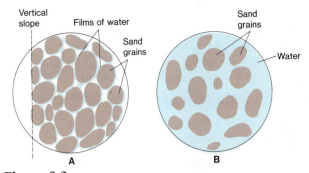

Figure 9.2
The effect of water in sand. (A) Unsaturated sand held together by surface tension of water. (B) Saturated sand grains forced apart by water; mixture flows easily.

TABLE 9.2

Summary of Controls of Mass Wasting

Driving Force: Gravity

Contributing Factors	Most Stable Situation	Most Unstable Situation
Slope angle	Gentle slopes or horizontal surface	Steep or vertical
Local relief	Low	High
Thickness of debris over bed rock	Slight thickness (usually)	Great thickness
Orientation of planes of weakness in bed rock	Planes at right angles to hillside slopes	Planes parallel to hillside slopes
Climatic factors		
Ice	Temperature stays above freezing	Freezing and thawing for much of the year
Water in soil or debris	Film of water around debris particles	Oversaturation of debris with water
Precipitation	Frequent but light rainfall or snow	Long periods of drought with rare episodes of heavy precipitation
Vegetation	Heavily vegetated	Sparsely vegetated

Triggering Mechanisms: (1) earthquakes; (2) weight added to upper part of a slope.

BOX 9.2

Preventing Downslope Movement of Soil

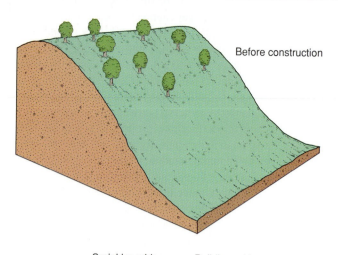

Before construction

Usually mass movements of debris can be prevented. Proper engineering is essential when the natural environment of a hillside is altered by construction. As shown in box figure 1, construction generally makes a slope more susceptible in several ways to mass wasting of debris. (1) The base of the slope is undercut, removing the natural support for the upper part of the slope. (2) Vegetation is removed during construction. (3) Buildings constructed on the upper part of a slope add weight to the potential slide. (4) Extra water may be allowed to seep into the debris.

Some preventive measures can be taken during construction. A retaining wall is usually built where a cut has been made in the slope, but this alone is seldom as effective a deterrent to down-slope movement as people hope. If, in addition, drain pipes are put through the retaining wall and into the hillside, water can percolate through and drain away rather than collecting in the debris behind the wall (box figure 2). Without drains, excess water may overcome the surface tension between grains. In this case, the trapped water forces the grains apart as well as adds weight to the debris, and eventually the whole soggy mass bursts through the wall.

Another practical preventive measure is to avoid oversteepening the slope. The hillside can be cut back in a series of terraces rather than in a single steep cut. Besides removing much of the material that could add to the weight of a potential slide, this procedure prevents loose material (such as boulders dislodged from the top of the cut) from rolling to the bottom, knocking off more stones and debris during descent. Road cuts constructed in this way are usually reseeded with rapidly growing grass or plants whose roots help anchor the slope. A vegetation cover also minimizes erosion from running water.

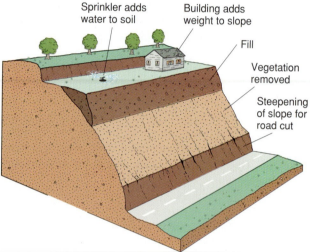

Sprinkler adds water to soil

Building adds weight to slope

Fill

Vegetation removed

Steepening of slope for road cut

Box 9.2 Figure 1
A hillside becomes vulnerable to mass wasting due to construction activities.

Building a heavy structure high on a slope demands special precautions. To prevent movement of both the slope and the building, pilings may have to be sunk through the debris, perhaps even into bed rock. Developers may have to settle for fewer buildings than planned because too many structures will make the slope unsafe. If a thorough geologic study of conditions indicates that safe construction is not possible, the builders should be required by law to abandon the planned project. Even better would be zoning laws that prohibit construction in hazardous areas.

The common types of mass wasting are shown in table 9.1. Here we will describe each type in detail.

Creep

Creep is very slow, continuous, downslope movement of soil or unconsolidated debris. The rate of movement is usually less than a centimeter per year and can be detected only by observations taken over months or years.

When conditions are right, creep can take place along nearly horizontal slopes. Some indicators of creep are illustrated in figures 9.3 and 9.4.

Two factors that contribute significantly to creep are water in the soil and daily cycles of freezing and thawing. As we have said, water-saturated ground facilitates movement of soil downhill. What keeps downslope movement from becoming more rapid in most areas is the presence of abundant grass or other plants that anchor the soil. Understandably, overgrazing can severely damage sloping pastures.

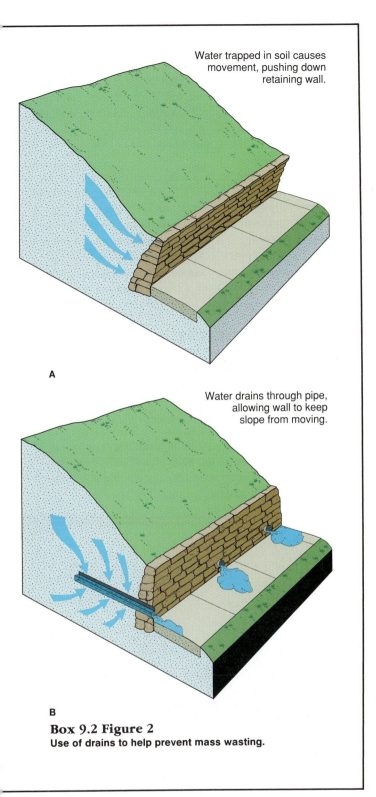

Water trapped in soil causes movement, pushing down retaining wall.

A

Water drains through pipe, allowing wall to keep slope from moving.

B

Box 9.2 Figure 2
Use of drains to help prevent mass wasting.

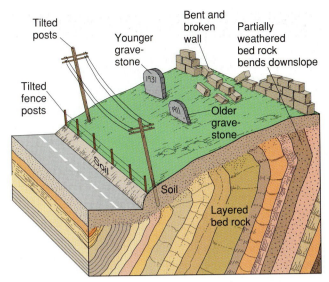

Tilted posts

Bent and broken wall

Younger grave-stone

Partially weathered bed rock bends downslope

Tilted fence posts

1931

1911

Older grave-stone

Soil

Soil

Layered bed rock

Figure 9.3
Indicators of creep.
After C. F. S. Sharpe.

A

B

Figure 9.4
(*A*) Tilted gravestones in a churchyard at Lyme Regis, England.
(*B*) Soil developed from nearly vertical rock sedimentary strata has crept downslope.
Photo *A* by C. C. Plummer. Photo *B* by Frank M. Hanna.

Although creep does take place in year-round warm climates, the process is more active where the soil freezes and thaws during part of the year. During the winter in regions like the northeastern United States, the temperature may rise above and fall below freezing once a day or several times daily. When there is moisture in the soil, each freeze-thaw cycle moves soil particles a minute amount downhill, as shown in figure 9.5.

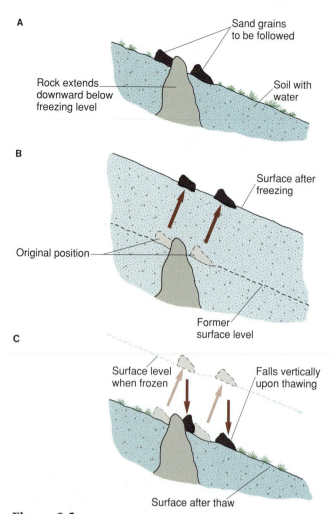

A

Sand grains
to be followed

Rock extends
downward below
freezing level

Soil with
water

B

Surface after
freezing

Original position

Former
surface level

C

Surface level
when frozen

Falls vertically
upon thawing

Surface after thaw

Figure 9.5
Downslope movement of soil, illustrated by following two sand grains during a freeze-thaw cycle.

Earthflow

In an **earthflow,** debris moves downslope as a viscous fluid; the process can be slow or rapid. Earthflows usually occur on hillsides that have a thick cover of debris, often after heavy rains have saturated the soil. Typically, the flowing mass remains covered by a blanket of vegetation, with a *scarp* (steep cut) developing where the moving debris has pulled away from the upper slope, which remains stationary.

An earthflow can be only a flow, with soil movement roughly parallel to the slope. Most earthflows, however, involve both slip and flow. Figure 9.6*B* shows how the upper part of an earthflow slumps downward as a relatively coherent block, shoving the lower part outward and causing the debris to flow downhill. Meanwhile, a hummocky lobe forms at the toe or front of the earthflow (figure 9.7). An earthflow can be active over a period of hours, days, or months; in some earthflows intermittent slow movement continues for years.

Humans can trigger earthflows by adding too much water to soil from septic tank systems or by overwatering lawns. In one case, in Los Angeles, a man departing on a long trip forgot to turn off the sprinkler system for his hillside lawn. The soil became saturated, and both house and lawn were carried downward on an earthflow whose lobe spread out over the highway below.

Earthflows, like other kinds of landslides, can be triggered by undercutting at the base of a slope. The undercutting can be caused by waves breaking along shorelines or streams eroding and steepening the base of a slope. Along coastlines, mass wasting commonly destroys buildings (figure 9.8). Entire housing developments and expensive homes built for a view of the ocean are lost. A

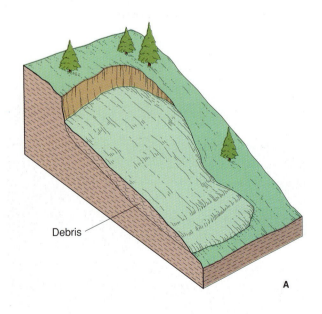

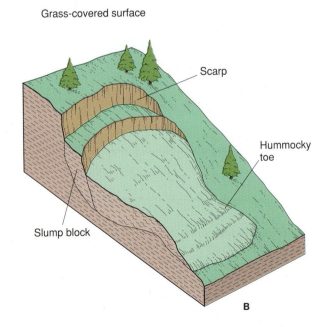

Grass-covered surface

Scarp

Hummocky
toe

Slump block

Debris

A

B

Figure 9.6
Earthflow. (*A*) Earthflow involving only flow. (*B*) Earthflow combining slip and flow.

home buyer who knows nothing of geology may not realize that the sea cliff is there entirely because of the relentless erosion of waves along the shoreline. Nor is the person likely to be aware that a steepened slope creates the potential for landslides.

Bulldozers can undercut the base of a slope more rapidly than wave erosion, and such oversteepening of slopes by human activity has caused many landslides. Unless careful engineering measures are taken at the time a cut is made, roadcuts or platforms carved into hillsides for houses may bring about disaster.

Solifluction and permafrost   One variety of earthflow is usually associated with colder climates. **Solifluction** is the flow of water-saturated debris over impermeable material. Because the impermeable material beneath the debris prevents water from draining freely, the debris between the vegetation cover and the impermeable material becomes saturated (figure 9.9). Even a gentle slope is susceptible to movement under these conditions (figure 9.10).

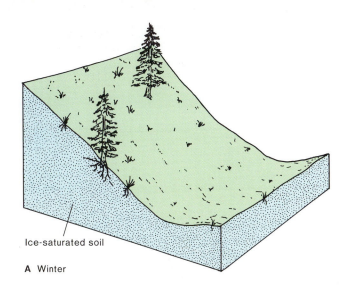

Ice-saturated soil

A Winter

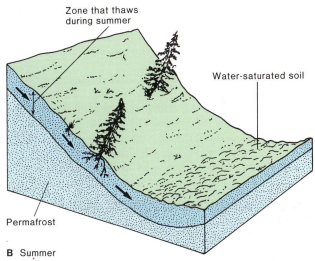

Zone that thaws during summer

Water-saturated soil

Permafrost

B Summer

Figure 9.9
Solifluction due to thawing of ice-saturated soil.

Figure 9.7
Earthflows in the Horse Heaven Hills, Washington. Lobes are encroaching on the highway.
Photo by D. A. Rahm, courtesy Rahm Memorial Collection, Western Washington University.

Figure 9.8
Landslide with houses and roads moved on slump blocks. Portuguese Bend, Los Angeles, California.
Photo by Frank M. Hanna.

Figure 9.10
A railroad built on permafrost terrain in Alaska.
Photo by Lynn A. Yehle, U.S. Geological Survey.

BOX 9.3

Los Angeles, A Mobile Society

The following newspaper column was written by Art Buchwald during a year in which southern California had many landslides because of unusually wet weather.

Los Angeles—I came to Los Angeles last week for rest and recreation, only to discover that it had become a rain forest.

I didn't realize how bad it was until I went to dinner at a friend's house. I had the right address, but when I arrived there was nothing there. I went to a neighboring house where I found a man bailing out his swimming pool.

I beg your pardon, I said. Could you tell me where the Cables live?

"They used to live above us on the hill. Then, about two years ago, their house slid down in the mud, and they lived next door to us. I think it was last Monday, during the storm, that their house slid again, and now they live two streets below us, down there. We were sorry to see them go—they were really nice neighbors."

I thanked him and slid straight down the hill to the new location of the Cables' house. Cable was clearing out the mud from his car. He apologized for not giving me the new address and explained, "Frankly, I didn't know until this morning whether the house would stay here or continue sliding down a few more blocks."

Cable, I said, you and your wife are intelligent people, why do you build your house on the top of a canyon, when you know that during a rainstorm it has a good chance of sliding away?

"We did it for the view. It really was fantastic on a clear night up there. We could sit in our Jacuzzi and see all of Los Angeles, except of course when there were brush fires.

"Even when our house slid down two years ago, we still had a great sight of the airport. Now I'm not too sure what kind of view we'll have because of the house in front of us, which slid down with ours at the same time."

But why don't you move to safe ground so that you don't have to worry about rainstorms?

"We've thought about it. But once you live high in a canyon, it's hard to move to the plains. Besides, this house is built solid and has about three more good mudslides in it."

Still, it must be kind of hairy to sit in your home during a deluge and wonder where you'll wind up next. Don't you ever have the desire to just settle down in one place?

"It's hard for people who don't live in California to understand how we people out here think. Sure we have floods, and fire and drought, but that's the price you have to pay for living the good life. When Esther and I saw this house, we knew it was a dream come true. It was located right on the tippy top of the hill, way up there. We would wake up in the morning and listen to the birds, and eat breakfast out on the patio and look down on all the smog.

"Then, after the first mudslide, we found ourselves living next to people. It was an entirely different experience. But by that time we were ready for a change. Now we've slid again and we're in a whole new neighborhood. You can't do that if you live on solid ground. Once you move into a house below Sunset Boulevard, you're stuck there for the rest of your life.

"When you live on the side of a hill in Los Angeles, you at least know it's not going to last forever."

Then, in spite of what's happened, you don't plan to move out?

"Are you crazy? You couldn't replace a house like this in L.A. for $500,000."

What happens if it keeps raining and you slide down the hill again?

"It's no problem. Esther and I figure if we slide down too far, we'll just pick up and go back to the top of the hill, and start all over again; that is, if the hill is still there after the earthquake."

Reprinted by permission of the author.

The impermeable material beneath the saturated soil can be either impenetrable bed rock or, as is more common, **permafrost,** ground that remains frozen for many years. Most solifluction takes place in areas of permanently frozen ground, such as in Alaska and northern Canada. Permafrost occurs at depths ranging from a few centimeters to a few meters beneath the surface. The ice in permafrost acts as a cementing agent for the debris. When all the pores are filled with ice, the ground becomes a concretelike mass.

Above the permafrost is a zone that, if the debris is saturated, is frozen during the winter and indistinguishable from the underlying permafrost. When this zone thaws during the summer, the water, along with water from rain and runoff, cannot seep downward through the permafrost, and so the slopes become susceptible to solifluction.

Since solifluction movement is not rapid enough to break up the overlying vegetation cover into blocks, the water-saturated debris flows downslope, pulling vegetation along with it and forming a wrinkled surface. Gradually the debris collects at the base of the slope, where the vegetated surface bulges into a hummocky lobe.

Permafrost Ice on Mars

On Earth most underground water is in the liquid form, with the only abundant underground ice occurring in the polar regions. In these areas underground ice is responsible for producing several distinctive landforms, including surfaces with polygonal cracks called patterned ground (box figure 1*A*), soil-covered ice mounds called pingos, and various collapse pits and linear depressions caused by the melting or sublimation of the ice.

On Mars, because of its cooler temperatures, ice is the normal state of underground water and is apparently quite abundant. Many features of the Martian landscape have been attributed to this underground ice, including polygonally patterned ground (box figure 1*B*), rampart craters, and chaotic and fretted terrain.

Rampart craters (box figure 2) are Martian meteorite craters that are surrounded by material that appears to have flowed from the point of impact. It has been suggested that the underground ice in the surface layers of Mars could have melted as a result of the impact and, along with rock fragments, sloshed outward from the impact area.

Patches of jumbled and broken angular slabs and blocks called *chaotic terrain* occur in some places on Mars (box figure 3). Some channels originate in these areas, and it is believed that this terrain may be caused by melting of ground ice and consequent collapse of the ground. Subsidence due to withdrawal of magma has also been suggested as the cause of chaotic terrain.

Melting of ground ice may also be responsible for producing another kind of terrain. *Fretted terrains* are flat lowlands with some scattered high plateaus. The plateaus are bounded by high cliffs and scarps, some more than a kilometer high and 800 kilometers long. Ground ice exposed in the cliffs may melt, undermining the cliff tops and causing them to landslide, converting plateau regions into lowlands.

A

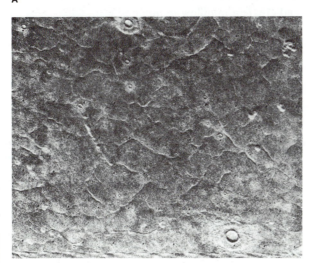

B

Astrogeology Box 9.1 Figure 1
(*A*) Patterned ground on Earth. (*B*) Patterned ground on Mars. The polygons in photo (*B*) are about 100 times larger than those in photo (*A*).
Photo *A* courtesy T. L. Pewl, U.S. Geological Survey. Photo *B* NASA.

Astrogeology Box 9.1 Figure 2
The rampart crater Arandas on Mars. Material around it appears to have flowed along the surface.
NASA.

Astrogeology Box 9.1 Figure 3
Chaotic and fretted terrain with associated channels on Mars.
NASA.

Figure 9.11
A dried mudflow in the Peruvian Andes.
Photo by C. C. Plummer.

Figure 9.12
Man examining a 75-meter-long bridge on Washington state highway 504, across the North Fork of the Toutle River. The bridge was washed out by mudflow during the May 18, 1980, eruption of Mount St. Helens. The steel structure was carried about 0.5 kilometers downstream and partially buried by the mudflow.
Photo by R. L. Schuster, U.S. Geological Survey.

Solifluction is not the only hazard associated with permafrost. Great expanses of flat terrain in arctic and subarctic climates become swampy during the summer because of permafrost, making overland travel very difficult. Building and maintaining roads is an engineering headache. In the preliminary stages of planning the Alaska pipeline, a road was bulldozed across permafrost terrain during the winter, removing the vegetation from the rock-hard ground. It was an excellent truck route during the winter, but when summer came, the road became a quagmire several hundred kilometers long (see Box 1.1, figure 1). The strip can never be used by vehicles as planned, nor will the vegetation return for many decades. Building structures on permafrost terrain presents serious problems. Unless preventive measures are taken, water-saturated soil trapped between a building's foundation and permafrost is likely to freeze during winter months. When water freezes, its volume increases by 9%. This results in an upward heaving that, all too frequently, will break a building apart.

Mudflow

A **mudflow** is a flowing mixture of debris and water, usually moving down a channel (figure 9.11). It can be visualized as a stream with the consistency of a thick milkshake. Usually after a heavy rainfall a slurry of soil and water forms and begins moving down a slope. Most mudflows quickly become channeled into valleys. They then move downvalley like a stream except that, because of the heavy load of debris, they are more viscous. Mud moves more slowly than a stream but, because of its high

viscosity, can transport boulders, automobiles, and even locomotives. Houses in the path of a mudflow will be filled with mud, if not broken apart and carried away.

Mudflows are more likely to occur in arid regions than in wet climates where a dense cover of vegetation protects soil and debris. A hillside in a desert environment, where it may not have rained for many years, may be covered with a blanket of loose material. With sparse desert vegetation offering little protection, a sudden thunderstorm with drenching rain can rapidly saturate the loose debris and create a mudflow in minutes.

Lack of vegetation can be the cause of mudflows in other situations. Mudflows frequently occur on young volcanoes that are littered with ash. Heavy rains or rapid melting of snow provides the water. Alternatively, a mudflow can be triggered by renewed volcanic activity on a glaciated volcano, as occurred at Mount St. Helens in 1980 (figure 9.12) and Nevado del Ruiz in 1985 (which, as described in chapter 1, cost 23,000 lives). The heat from fresh, hot ash or a new lava flow melts glacial ice, mixing meltwater with loose pyroclastic debris and starting a mudflow. Mudflows also occur after forest fires have destroyed slope vegetation that normally anchors soil in place. Burned-over slopes are extremely vulnerable to mudflow if heavy rains fall before the vegetation is restored.

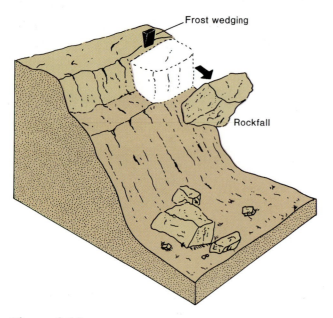

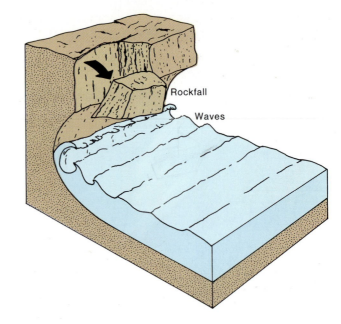

Figure 9.13
Two examples of rockfall.

Rockfall

In the movement called **rockfall,** rock falls freely or bounces down a cliff (figure 9.13). When a steep slope or cliff is being undercut, as by a river or wave erosion or highway construction, large blocks of rock tend to break off along cracks in the bed rock or along planes of weakness such as bedding planes. Blocks of rock commonly break off through frost wedging (a type of mechanical weathering). Most cliffs have an apron of fallen rock at their bases that are the result of rockfall. An accumulation of broken rock (from rockfalls) at the base of a cliff is called **talus** (figure 9.14).

Rockslide

A **rockslide** is, as the term suggests, the rapid sliding of a mass of bed rock along an inclined surface of weakness, such as a bedding plane, a major fracture in the rock, or a foliation plane (figures 9.15 and 9.16). Once sliding begins, a rock slab usually breaks up into rubble. Like rockfalls, rockslides can be caused by undercutting at the base of the slope from erosion or construction.

Some rockslides travel only a few meters before halting at the base of a slope. In country with high relief, however, a rockslide may travel hundreds or thousands of meters before reaching a valley floor. If movement becomes very rapid, the rockslide may break up and become a rock avalanche. A **rock avalanche** is a very rapidly moving, turbulent mass of broken-up bed rock. Movement in a rock avalanche is flowage on a grand scale.

Figure 9.14
Talus.
Photo by C. C. Plummer.

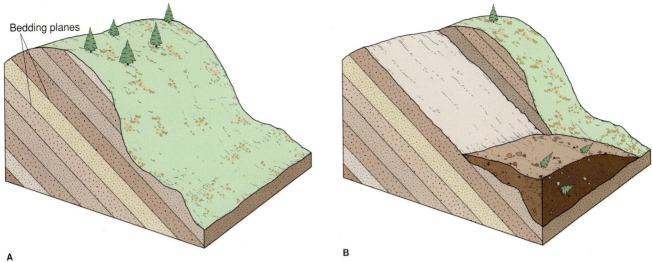

A

B

Figure 9.15
A rockslide along bedding planes. (*A*) Before slide, a steep slope with bedding planes undercut. (*B*) After slide.

Figure 9.16
A rockslide in southern California. Sliding took place along steep fractures in the rock.
Photo by California Division of Mines and Geology.

Some geologists have suggested that in very rapidly moving rock avalanches, air trapped under the rock mass creates an air cushion that reduces friction. This could explain why some landslides reach speeds of several hundred kilometers per hour. But other geologists have contended that the rock mass is too turbulent to permit such an air cushion to form.

Ultimately, a rockslide or rock avalanche comes to rest as the terrain becomes less steep. Sometimes the mass of rock fills the bottom of a valley and creates a natural dam. If the rock mass suddenly enters a lake or bay, it can create a huge wave that destroys lives and property far beyond the area of the original landslide.

As in slower mass movements, water can play an important role in causing a rockslide. In 1925, exceptionally heavy rains in the Gros Ventre Mountains of Wyoming caused water to seep into a layer of shale (figure 9.17). The shale, along with other layers of sedimentary rock above and below it, was inclined roughly parallel to the

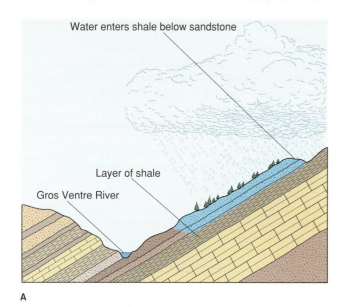

Water enters shale below sandstone

Layer of shale

Gros Ventre River

A

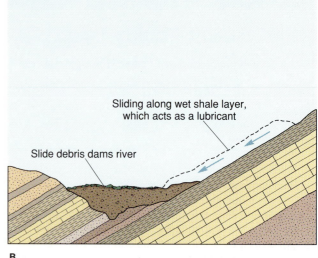

Sliding along wet shale layer, which acts as a lubricant

Slide debris dams river

B

C

Figure 9.17
(*A* and *B*) **Diagram of the Gros Ventre, Wyoming, slide.**
(*C*) **Photo of Gros Ventre slide.**
A and *B* after W. C. Alden, U.S. Geological Survey. Photo *C* by John S. Shelton.

hillside. With the wet shale acting as a lubricant, the overlying sedimentary rock and its soil cover slid into the valley, blocking the river. The slide itself merely created a lake, but the subsequent breaking of the natural dam resulted in a flood that destroyed the small town of Kelly several kilometers downstream and killed several residents who were standing on a bridge watching the floodwaters come down the valley.

Debris Slides, Falls, and Avalanches

Landslides composed of debris or rock fragments rather than bed rock have names preceded by the word *debris* rather than *rock*. A **debris slide** is a rapidly moving coherent mass of debris. A **debris fall** is a free-falling mass of debris. A **debris avalanche** is debris moving very rapidly and turbulently downslope.

BOX 9.4

Preventing Rockfalls and Rockslides on Highways

Rockslides and rockfalls are a major problem on highways built through mountainous country. Steep slopes and cliffs are created when road cuts are blasted and bulldozed into mountain sides. These slopes and cliffs are susceptible to rockslides or rockfalls. If the bed rock has planes of weakness (such as joints, bedding planes, or foliation planes), the orientation of these planes relative to the road cut determines whether there is a rockslide hazard (as shown in box figure 1). If the planes of weakness are inclined away from the road cut, there is little likelihood of a rockslide. On the other hand, where the planes of weakness are approximately parallel to the slope of the hillside, a rockslide may occur.

Various techniques are used to prevent rockslides. By doing a detailed geologic study of an area before a road is built, builders might avoid a hazard by choosing the least dangerous route for the road. If a road cut must be made through bed rock that appears prone to sliding, all of the rock that might slide could be removed (sometimes at great expense), as shown in box figure 2.

In some instances, slopes prone to rock sliding have been "stitched" in place by the technique shown in box figure 3.

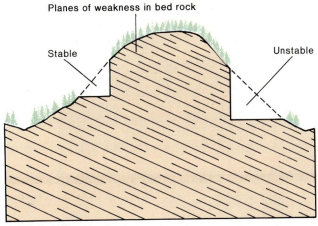

Box 9.4 Figure 1
Cross section of a hill showing a relatively safe road cut on the left and a hazardous road cut on the right.

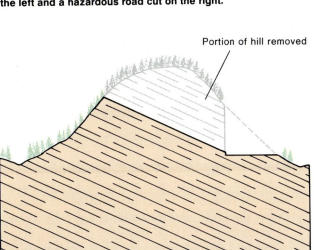

Box 9.4 Figure 2
The same hazardous road cut shown in figure 1, but after removal of rock that might slide.

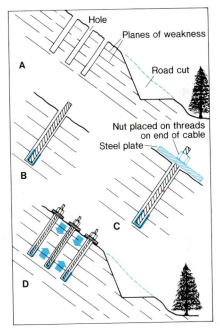

Box 9.4 Figure 3
"Stitching" a slope to keep bed rock from sliding along planes of weakness. (A) Holes are drilled through unstable layers into stable rock. (B) Expanded view of one hole. A cable is fed into the hole and cement is pumped into the bottom of the hole and allowed to harden. (C) A steel plate is placed over the cable and a nut tightened. (D) Tightening all the nuts pulls unstable layers together and anchors them in stable bed rock.

Summary

Mass wasting is the movement of a mass of debris (soil and loose rock fragments) or bed rock toward the base of a slope. Movement can take place as flow, slip, or fall. Gravity is the driving force. A number of factors determine whether movement will occur and, if it does, the rate of movement.

The slowest type of movement, *creep,* occurs mostly on relatively gentle slopes, usually aided by water in the soil. In colder climates, repeated freezing and thawing of water within the soil contributes to creep.

Landsliding is a general term for more rapid mass wasting of rock, debris, or both. Some types of landslides are earthflow, mudflow, rockfall, rockslide, debris fall, debris slide, rock avalanche, and debris avalanche.

Earthflows vary greatly in velocity. Debris *flows,* and often *slips* as well. Usually the upper part of the moving mass slumps downward while the lower part flows outward. Water filling the pore spaces of the debris contributes to earthflow. *Solifluction,* a special variety of earthflow, usually takes place in arctic or subarctic climates where the ground is permanently frozen (*permafrost*).

A *mudflow* is a slurry of debris and water. Most mudflows flow in channels much as streams do.

Rockfall is the fall of broken rock down a vertical or near-vertical slope. A *rockslide* is a slab of rock sliding down a less-than-vertical surface. *Debris falls* and *debris slides* involve unconsolidated material rather than bed rock. A *rock* or *debris avalanche* is a turbulent mass of rock fragments or soil moving very rapidly downslope.

Terms to Remember

creep	permafrost
debris	relief
debris avalanche	rock avalanche
debris fall	rockfall
debris slide	rockslide
earthflow	slide
fall	slip
flow	slump
landslide	solifluction
mass wasting	talus
mudflow	

Questions for Review

1. How does a slump differ from a slide?
2. What is the distinction between solifluction and creep?
3. What role does water play in each of the types of mass wasting?
4. Why is solifluction more common in colder climates than in temperate climates?
5. List and explain the key factors that control mass wasting.

Questions for Thought

1. How would you classify the types of mass wasting if the amount of water involved were a criterion?
2. If you were building a house on a cliff, what would you look for to ensure that your house would not be destroyed through mass wasting?
3. Why isn't the land surface of the earth flat after millions of years of erosion by mass wasting as well as by other erosional agents?
4. Can any of the indicators of creep be explained by processes other than mass wasting?

Supplementary Readings

Brabb, E. E. and B. L. Harrod. 1989. *Landslides: extent and economic significance.* Brookfield, Vermont: A. A. Balkema.

Dennen, W. H., and B. R. Moore. 1986. *Geology and engineering.* Dubuque, Iowa: Wm. C. Brown Publishers.

Ericksen, G. E., G. Plafker, and J. Fernandez Concha. 1970. *Preliminary report on the geologic events associated with the May 31, 1970, Peru earthquake.* U.S. Geological Survey Circular 639.

Legget, R. F. 1973. *Cities and geology.* New York: McGraw-Hill.

McPhee, J. 1989. *The control of nature.* New York: Farrar, Strauss, and Giroux.

Montgomery, C. W. 1989. Environmental geology. Dubuque, Iowa: Wm. C. Brown.

Ritter, D. F. 1986. *Process geomorphology.* 2d ed. Dubuque, Iowa: Wm. C. Brown Publishers.

Schuster, R. L. and R. J. Krizek, eds. 1978. *Landslides, analysis and control.* Transport Research Board Special Report 176. Washington, D.C.: National Academy of Sciences.

Washburn, A. L. 1973. *Periglacial processes and environments.* London: Edward Arnold Ltd.

Running water, aided by mass wasting, is the most important geologic agent in eroding, transporting, and depositing sediment. Almost every landscape on earth shows the results of stream erosion or deposition. Although other agents—ground water, glaciers, wind, and waves—can be locally important in sculpturing the land, stream action and mass wasting are the dominant processes of landscape development.

The first part of this chapter deals with the various ways that streams erode, transport, and deposit sediment. The second part describes landforms produced by stream action, such as valleys, flood plains, deltas, alluvial fans, and terraces, and shows how each of these is related to changes in stream characteristics. The chapter concludes with a discussion of the factors that control slope angle and the appearance of stream-eroded landscapes.

CHAPTER

10

Streams and Landscapes

Stream erosion has formed this valley in limestone. Havasu Canyon, Grand Canyon National Park, Arizona.

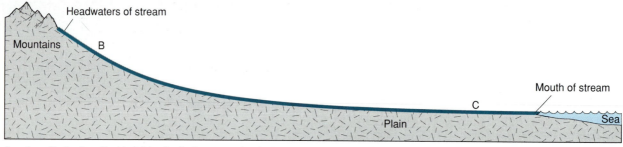

A Longitudinal profile (dark blue line) of a stream beginning in mountains and flowing across a plain into the sea.

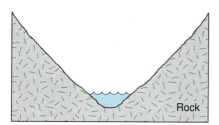

B Cross profile of the stream at point B. The channel is at the bottom of a V-shaped valley cut into rock.

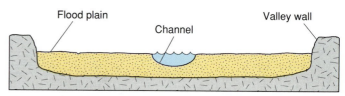

C Cross profile at point C. The channel is surrounded by a broad flood plain of sediment.

Figure 10.1
Longitudinal and cross profiles of a typical stream.

A **stream** is a body of running water that is confined in a channel and moves downhill under the influence of gravity. In some parts of the country, *stream* implies size: rivers are large, streams somewhat smaller, and brooks or creeks even smaller. Geologists, however, use *stream* for any body of running water, from a small trickle to a huge river.

Channel Flow and Sheet Flow

Figure 10.1*A* shows a *longitudinal profile* of a typical stream viewed from the side. The stream begins in steep mountains and flows out across a gentle plain into the sea. The *headwaters* of a stream are the upper part of the stream near its source in the mountains. The *mouth* is the place where a stream enters the sea, a lake, or a larger stream. The *cross profile* of a stream in steep mountains is usually a V-shaped valley cut into solid rock, with the stream channel occupying the narrow bottom of the valley; there is no flat land next to the stream on the valley bottom (figure 10.1*B*). Near its mouth a stream usually flows within a broad, flat-floored valley. The stream channel is surrounded by a flat *flood plain* of sediment deposited by the stream (figure 10.1*C*).

A stream normally stays in its **stream channel,** a long narrow depression eroded by the stream into rock or sediment. The stream *banks* are the sides of the channel; the

stream *bed* is the bottom of the channel. During a flood the waters of a stream may rise and spill over the banks onto the flat flood plain of the valley floor (figure 10.2).

Not all water that moves over the land surface is confined to channels. Sometimes, particularly during heavy rains, water runs off as **sheetwash,** a thin layer of unchanneled water flowing downhill. Sheetwash is particularly common in deserts, where the lack of vegetation allows rainwater to spread quickly over the land surface. It also occurs in humid regions during heavy thunderstorms when water falls faster than it can soak into the ground. A series of closely spaced storms can also promote sheetwash; as the ground becomes saturated, more water runs over the surface.

Sheetwash, along with the violent impact of raindrops on the land surface, can produce considerable *sheet erosion,* in which a thin layer of surface material, usually topsoil, is removed by the flowing sheet of water. This gravity-driven movement of sediment is a process intermediate between mass wasting and stream erosion.

Overland sheetwash becomes concentrated in small channels, forming tiny streams called *rills*. Rills merge to form small streams, and small streams join to form larger streams. Most regions are drained by networks of coalescing streams.

Figure 10.2
(*A*) A stream normally stays in its channel, but during a flood it can spill over its banks onto the adjacent flat land.
Photo by Frank M. Hanna.

(*B*) This flooding river in northern California has spilled out of its channel, marked by lines of trees in the right half of the photo, onto the adjacent flat valley floor.
Photo by U.S. Army Corps of Engineers.

Preventing Sheet Erosion on Farms

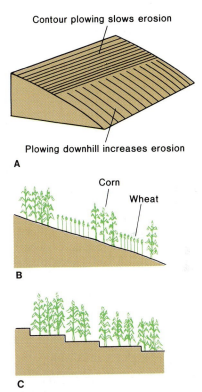

Box 10.1 Figure 1
**Methods of preventing sheet erosion. (*A*) Contour plowing.
(*B*) Strip planting. (*C*) Terracing.**

Tilled fields are particularly susceptible to sheet erosion because plowing and cultivating during the planting season remove the vegetation cover. A rapid loss of topsoil is a very serious problem if it exceeds the rate at which new soil is forming. A loss of 10 tons of soil per acre in a single year results in a loss of one inch (2.5 cm) of soil each twenty-five years. Even in wet regions, soil seldom develops this rapidly. Sheet erosion can remove as much as 100 tons of topsoil per acre in a single year if the land slopes steeply. A field can be destroyed in just a few years, losing the soil that took centuries to form. Farmers use several techniques to slow sheet erosion in their fields (box figure 1).

In *contour plowing,* the furrows are plowed along contours (lines of equal elevation), rather than up and down the slopes. Furrows that run directly down a slope channelize sheet flow into small rivulets. They increase the velocity of the water and as a result cause more soil erosion. If the furrows run along contour lines, however, they are perpendicular to the direction of sheet flow. Each ridge and groove tends to retard the water, thereby reducing sheet erosion of topsoil.

In *strip planting,* two different crops are grown in alternating bands along contour lines. For example, a strip of hay might be planted downhill from a strip of corn. The dense growth and root systems of the hay slow any sheet flow moving downhill from the more widely spaced corn.

Terracing the land involves cutting or building flat surfaces, again parallel to the contours. The terraces can be so narrow as to hold only one row of plants, or wide enough for an entire field. The flat surfaces of the terraces slow the downhill flow of water and decrease soil erosion.

Combinations of contour plowing, strip planting, terracing, and other techniques such as shallow and infrequent plowing can very effectively slow sheet erosion of topsoil. A recent promising development in further controlling sheet erosion is to mulch fields with organic matter to reduce raindrop impact and sheet flow. Mulching sometimes is done on untilled land into which seeds have been drilled. Such protection can essentially eliminate topsoil loss by sheet erosion.

Figure 10.3
The drainage basin of the Mississippi River is the land area drained by the river and all its tributaries, including the Ohio and Missouri Rivers; it covers more than 1 million square miles. Heavy rain in any part of the basin can cause flooding on the lower Mississippi River in the states of Mississippi and Louisiana. The Continental Divide separates rivers that flow into the Pacific from rivers that flow into the Atlantic and the Gulf of Mexico.

Drainage Basins

A **drainage basin** is the total area drained by a stream and its tributaries (a **tributary** is a small stream flowing into a larger one). A drainage basin can be outlined on a map by drawing a line around the region drained by all the tributaries to a river (figure 10.3). The Mississippi River's drainage basin, for example, includes all the land area drained by the Mississippi River itself and by all its tributaries, including the Ohio and Missouri Rivers. This great drainage system includes approximately half the land area of the contiguous 48 states.

A ridge or strip of high ground dividing one drainage basin from another is termed a **drainage divide** (figure 10.3). The best known in the United States is the Continental Divide, an imaginary line separating streams that flow to the Pacific Ocean from those that flow to the Atlantic and the Gulf of Mexico. The Continental Divide, which extends from the Yukon Territory down into Mexico, crosses Montana, Wyoming, Colorado, and New Mexico in the United States. Signs indicating the crossing of the Continental Divide have been placed at numerous points where major highways intersect the divide.

Flooding

Only about 25% of rainfall normally ends up as surface runoff in rivers; the rest soaks into the ground or returns to the atmosphere by evaporation and transpiration from plants. Steady, continuous rains, however, can saturate the ground and the atmosphere and lead to greatly increased runoff and floods.

In January 1937, a twenty-five-day series of storms dumped huge amounts of rain on the Ohio River drainage basin, which is aligned parallel to the path of eastward-moving storms. The vast quantity of water mostly flowed over the surface, filling the Ohio River until it spilled over its banks onto the flat flood plain adjacent to the river channel. The river level rose at Cairo, Illinois, where the Ohio flows into the Mississippi, until it was almost 60 feet (18 meters) deep, 40 feet above its normal low-water level and 13 feet (4 meters) above its level during a disastrous flood ten years earlier. As the surging floodwaters flowed into the Mississippi, gates were opened and low spots were dynamited open in the *levees* (artificial banks) along the river to deliberately flood low-lying areas beside the river in order to keep the river from cresting still higher. Flood-control gates from Missouri to Louisiana were opened, lowering the level of the entire lower Mississippi. Despite these actions, 13,000 homes were destroyed and more than 130 people died.

Some floods occur rapidly and die out just as quickly. *Flash floods* are local, sudden floods of large volume and short duration, often triggered by heavy thunderstorms.

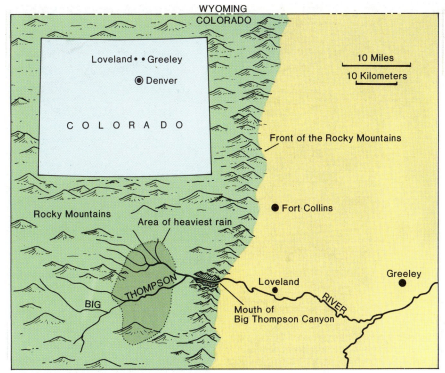

Figure 10.4
Location map of the 1976 flash flood on the Big Thompson
River in Colorado. The worst flood damage occurred in the
steep, narrow canyon near the front of the Rocky Mountains.

Figure 10.5
(*A*) A cabin sits crushed against a bridge following the Big
Thompson Canyon flash flood of 1976.

(*B*) Boulder bar deposited by the Big Thompson River during
the 1976 flash flood.
Photo *A* by W. R. Hansen, U.S. Geological Survey.

A startling example occurred in 1976 in north-central
Colorado along the Big Thompson River (figure 10.4).
Strong winds from the east pushed moist air up the front
of the Colorado Rockies, causing thunderstorms in the
steep mountains. The storms were unusually stationary,
allowing as much as 12 inches (30 cm) of rain to fall in
two days. Some areas received 7.5 inches (19 cm) in just
over an hour. Little of this torrential rainfall could soak
into the ground. The volume of water in the Big Thompson

River swelled to four times the previously recorded maximum, and the river's velocity rose to an impressive 15
miles per hour (about 25 kilometers per hour) for a few
hours on the night of July 31. By the next morning the
flood was over, and the appalling toll became apparent—
139 people dead, 5 missing, and more than $35 million in
damages (figure 10.5).

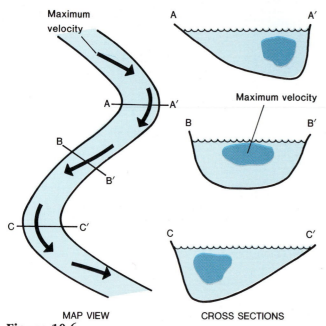

Figure 10.6
Regions of maximum velocity in a stream. Arrows on the map show how the maximum velocity shifts to the outside of curves. Sections show maximum velocity on outside of curves and in the center of the channel on a straight stretch of stream.

Factors Affecting Stream Erosion and Deposition

Stream erosion and deposition are controlled primarily by a river's *velocity* and, to a lesser extent, by its *discharge.* Velocity is largely controlled by the stream *gradient,* channel shape, and channel roughness.

Velocity

The speed at which water in a stream travels is called the **stream velocity.** A moderately fast river flows at about 3 miles per hour (5 kilometers per hour). Rivers flow much faster during flood, sometimes exceeding 15 miles per hour (25 kilometers per hour).

The cross-sectional views of a stream in figure 10.6 show that a stream reaches its maximum velocity near the middle of the channel. Near the stream's banks and bed, friction between the water and the stream channel slows the water. When a stream goes around a curve, the region of maximum velocity is displaced by centrifugal force toward the outside of the curve. Velocity is the key factor in a stream's ability to erode, transport, and deposit. High velocity generally results in erosion and transportation; deposition occurs when a river slows down. Slight changes in velocity can cause great changes in the sediment load carried by the river.

Gradient

One factor that controls a stream's velocity is the **stream gradient,** the downhill slope of the bed (or of the water surface, if the stream is very large). A stream gradient is usually measured in feet per mile in the United States, because these units are used on U.S. maps (elsewhere,

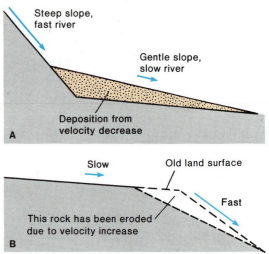

Figure 10.7
Changes in gradient can cause deposition and erosion.

gradients are expressed in meters per kilometer). A gradient of 5 feet per mile means that the river drops 5 feet vertically for every mile that it travels horizontally. Mountain streams may have gradients as steep as 50 to 200 feet per mile (about 10 to 40 m/km). The lower Mississippi River has a very gentle gradient, 0.5 feet/mile (0.1 m/km) or less. A stream's gradient usually decreases downstream (figure 10.1).

Two streams of equal size (and with similar channel shape and cross-sectional area) will have different velocities if they have different gradients. A stream with a steep gradient has a high velocity, while a similar stream on a gentle slope travels more slowly. A very slight change in gradient can significantly change velocity, which in turn can greatly affect the stream's ability to erode, transport, and deposit sediment.

Figure 10.7 shows two examples of a change in gradient along a river's course. If a river flows out of steep mountains onto a flatter plain, the river's gradient may change suddenly from steep to gentle (figure 10.7*A*). Such a change in gradient causes the river to slow down, and the sudden loss of velocity can cause sediment deposition at the base of the steep slope. If the gradient becomes steeper in a downstream direction (figure 10.7*B*), the velocity increases on the steep slope, and the river erodes rapidly there, deepening its valley as a result.

Slow uplift of the land may steepen a stream's gradient, causing a once-sluggish stream to flow faster and cut its valley downward (figure 10.8).

A river's gradient can be steepened artificially by dredging to straighten its channel. This process, called *channelization,* is used to improve navigation on a river, or to help drain a flat-lying region, or to increase the runoff capacity of a river to help control floods. Figure 10.9 shows a once-curving river that has been straightened. The difference in elevation between points *A* and *B* is the same in the new channel as it was in the old, but the river now travels a shorter distance between the two points. The gradient is therefore steeper and the velocity increases.

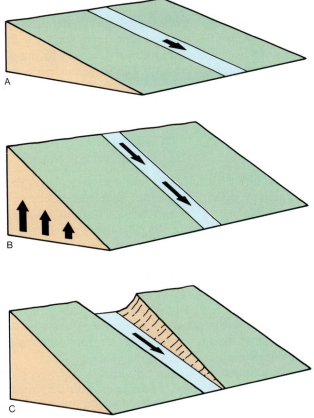

Figure 10.8
Uplift of the land can increase the rate of erosion. (**A**) River flows slowly on a gentle gradient, with little erosion. (**B**) Differential uplift increases the gradient, which increases the river's velocity. (**C**) The valley is deepened by rapid erosion resulting from the increase in velocity.

This can be the intent of channelization, but sometimes the resulting erosion between *A* and *B* is unanticipated. The newly eroded material will be deposited downstream from *B,* where the river reencounters its original gentle gradient.

Channel Shape and Roughness

Another factor that controls velocity is the cross-sectional *shape of the channel.* As the water travels in a stream, it drags against the stream banks and stream bed (the *wetted perimeter*), and the resulting friction tends to slow down the water. The more friction between water and channel, the more slowly the river travels. For a given cross-sectional area, different channel shapes can vary considerably in the amount of wetted perimeter. A wide, shallow channel slows a river because this channel shape has a greater wetted perimeter for the water to drag against (figure 10.10).

A stream may change its channel width as it flows across different rock types. Hard, resistant rock is difficult to erode, so a stream may have a relatively narrow channel in such rock. As a result it flows rapidly. If the stream flows onto a softer rock that is easier to erode, the channel may widen, and the river will slow down because of the increased wetted perimeter and friction. Sediment may be deposited as the velocity decreases (figure 10.11*A*).

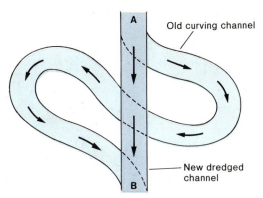

Figure 10.9
Channelization can increase a river's gradient (and velocity) by shortening the distance between two points.

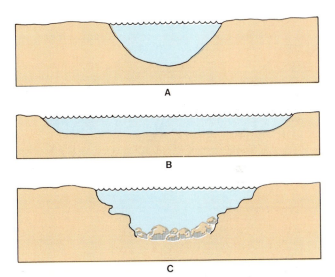

Figure 10.10
Channel shape and roughness influence stream velocity. (**A**) Narrow, deep channel with semicircular cross section allows stream to flow rapidly. (**B**) Wide, shallow channel increases friction, slowing river down. (**C**) Rough, boulder-strewn channel slows river.

The width of a stream may be controlled by factors external to the stream. A landslide may carry debris onto a valley floor, partially blocking a stream's channel. The constriction causes the stream to speed up as it flows past the slide, and the increased velocity may quickly erode the landslide debris, carrying it away downstream (figure 10.11*B*). Human interference with a river can promote erosion and deposition. Construction of a culvert or bridge can partially block a channel, increasing the stream's velocity (figure 10.11*C*). If the bridge was poorly designed, it may increase velocity to the point where erosion widens the stream and perhaps even causes the bridge to collapse.

The *roughness of the channel* also controls velocity. A stream can flow rapidly over a smooth channel, but a rough, boulder-strewn channel floor creates more friction and slows the flow (figure 10.10). Coarse particles increase the roughness more than fine particles, and a rippled or wavy sand bottom is rougher than a smooth sand bottom.

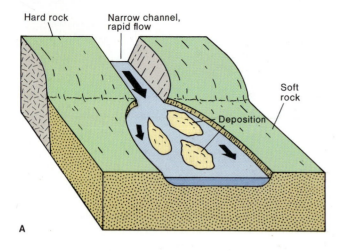

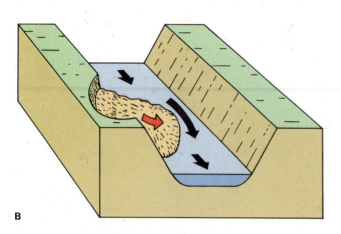

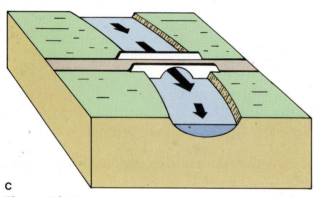

Figure 10.11
Channel width variations caused by rock type and obstructions. Length of arrow indicates velocity. (*A*) A channel may widen in soft rock. Deposition may result as stream velocity drops. (*B*) Landslide may narrow a channel, increasing stream velocity. Resulting erosion usually removes landslide debris. (*C*) Bridge piers or other obstructions will increase velocity and sometimes erosion.

Discharge

The **discharge** of a stream is the volume of water that flows past a given point in a unit of time. It is found by multiplying the cross-sectional area of a stream by its velocity (or width × depth × velocity). Discharge can be reported in cubic feet per second (cfs), which is standard in the United States, or in cubic meters per second (m³/sec).

$$\text{Discharge (cfs)} = \text{channel width (ft)}$$
$$\times \text{ channel depth (ft)}$$
$$\times \text{ average velocity (ft/sec)}$$

A river with a channel 100 feet wide and 15 feet deep flowing at 4 miles per hour (6 ft/sec) has a discharge of 9,000 cubic feet per second. In most streams, discharge increases downstream for two reasons: (1) water flows out of the ground into the river through the stream bed; and (2) small tributary streams flow into a larger stream along its length, adding water to the stream as it travels.

To handle the increased discharge, streams generally increase in width and depth downstream. Some rivers also increase slightly in velocity downstream, as a result of the increased discharge (the effect of discharge apparently overrides the effect of a lessening gradient).

During floods a stream's discharge and velocity increase, usually as a result of heavy rains over the stream's drainage basin. Flood discharge may be 50 to 100 times normal flow. Stream erosion and transportation generally increase enormously as a result of a flood's velocity and discharge. Flooded land may be intensely scoured, with river banks and adjacent lawns and fields washed away. As floodwaters recede, both velocity and discharge decrease, leading to the deposition of a blanket of sediment, usually mud, over the flooded area.

In a dry climate, a river's discharge can decrease in a downstream direction as river water evaporates into the air and soaks into the dry ground (or is used for irrigation). As the discharge decreases, the load of sediment is gradually deposited.

Stream Erosion

A stream usually erodes the rock and sediment over which it flows. In fact, streams are one of the most effective sculptors of the land. Streams cut their own valleys, deepening and widening them over long periods of time and carrying away the sediment that mass wasting delivers to valley floors. The particles of rock and sediment that a stream picks up are carried along to be deposited farther downstream. Streams erode rock and sediment in three ways—*hydraulic action, solution,* and *abrasion.*

Hydraulic action refers to the ability of the water itself to pick up and move rock and sediment (figure 10.12). The force of water swirling into a crevice in a rock can eventually crack the rock and break loose a fragment to be carried away by the stream. A loose sediment grain on the stream bed may tumble or slide along the stream bottom, pushed by the pressure of flowing water. A swirling

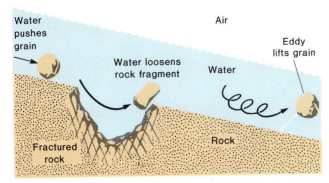

Figure 10.12
Hydraulic action.

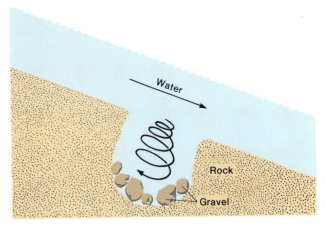

Figure 10.13
Sediment moved by swirling water can scour potholes in the rock of a stream bed.

eddy of water may exert enough force to lift a rock fragment above the stream bed into the main body of flowing water. Standing beside a swiftly flowing mountain stream, you may be able to hear cobbles and boulders tumbling along the stream bed and hitting one another. Clearly, hydraulic action is effective in moving rock and sediment.

From what you have learned about weathering (see the chapter on "Weathering and Soil"), you know that some rocks can be dissolved by water. **Solution,** although ordinarily slow, can be an effective process of weathering and erosion (weathering because it is a response to surface chemical conditions; erosion because it removes material). A stream flowing over limestone, for example, gradually dissolves it, deepening the stream channel. The solution of calcite cement within sedimentary rocks (see the chapter on "Sediments and Sedimentary Rocks") loosens sediment grains that were once tightly bound together. The loosened grains can then be picked up by hydraulic action.

The erosive process that is usually most effective on a rocky stream bed is **abrasion,** the grinding away of the stream channel by the friction and impact of the sediment load. Sand and gravel tumbling along near the bottom of a stream wear away the stream bed much as moving sandpaper wears away wood. The abrasion of sediment on the stream bed is generally much more effective in wearing away the rock than hydraulic action alone. The more sediment a stream carries, the faster it is likely to wear away its bed.

The coarsest sediment is the most effective in stream erosion. Sand and gravel strike the stream bed frequently and with great force, while the finer-grained silt and clay weigh so little that they are easily suspended throughout the stream and have little impact when they hit the channel.

Potholes are depressions that are eroded into the hard rock of a stream bed by the abrasive action of the sediment load. During high water when a stream is full, the swirling water can cause sand and pebbles to scour out smooth, bowl-shaped depressions in hard rock (figure 10.13). Potholes tend to form in spots where the rock is a little weaker than the surrounding rock. Although potholes are fairly uncommon, you can see them on the beds

Figure 10.14
A pothole about 2 feet (0.6 meter) wide scoured into granite, Yosemite National Park, California. Photo was taken through the clear water of a stream. Note rounded gravel in the pothole.
Photo by B. Amundson.

of some streams at low water level. Potholes may contain sand or an assortment of beautifully rounded pebbles (figure 10.14).

Stream Transportation of Sediment

The sediment load transported by a stream can be subdivided into *bed load, suspended load,* and *dissolved load.* Most of a stream's load is carried in suspension and in solution.

The **bed load** is the large or heavy sediment particles that travel near or on the stream bed. Sand and gravel, which form the usual bed load of streams, move by either *traction* or *saltation.*

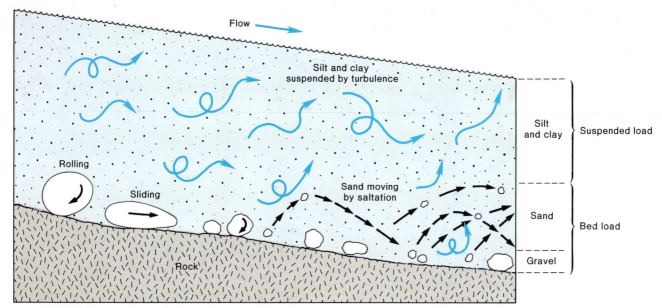

Figure 10.15
A stream's bed load consists of sand and gravel moving on or near the stream bed by traction and saltation. Finer silt and clay form the suspended load of the stream.

Large, heavy particles of sediment, such as cobbles and boulders, may never lose contact with the stream bed as they move along in the flowing water. They roll or slide along the stream bottom, eroding the stream bed and each other by abrasion. Movement by rolling, sliding, or dragging is called **traction.**

Sand grains move by traction, but they also move downstream by **saltation,** a series of short leaps or bounces off the bottom (figure 10.15). Saltation begins when sand grains are momentarily lifted off the bottom by *turbulent* water (eddying, swirling flow). The force of the eddying water temporarily counteracts the downward force of gravity, suspending the grains in water above the stream bed. But because the velocity of water in an eddy is not constant, the water soon slows down; then gravity overcomes the lift of the water, and the sand grain once again falls to the bed of the stream. While it is suspended, the grain moves downstream with the flowing water. After it lands on the bottom, it may be picked up again if turbulence increases, or it may be thrown up into the water by the impact of another falling sand grain. In this way sand grains saltate downstream in leaps and jumps, partly in contact with the bottom and partly suspended in the water.

The **suspended load** is sediment that is light enough to remain lifted indefinitely above the bottom by water turbulence. The muddy appearance of a stream during a flood or after a heavy rain is due to a large suspended load. Silt and clay usually are suspended throughout the water, while the coarser bed load moves on or near the stream bottom. Suspended load has less effect on erosion than the less visible bed load, which causes most of the abrasion of the stream bed. Vast quantities of sediment, however, are transported in suspension.

Figure 10.16
These large boulders of granite in a mountain stream are moved only during flood. Note the rounding of the boulders and the scoured high-water mark of floods on the valley walls.

Soluble products of chemical weathering processes can make up a substantial **dissolved load** in a stream. Most streams contain numerous ions in solution, such as sodium, calcium, potassium, bicarbonate, chloride, and sulfate. The ions may precipitate out of water as evaporite minerals if the stream dries up, or they may eventually reach the ocean. Very clear water may in fact be carrying a large load of material in solution, for the dissolved load is invisible. Only if the water evaporates does the material become visible as crystals begin to form.

One estimate is that rivers in the United States carry about 250 million tons of solid load and 300 million tons of dissolved load each year. (It would take a freight train eight times as long as the distance from Boston to Los Angeles to carry 250 million tons.)

BOX 10.2

Stream Velocity and Competence

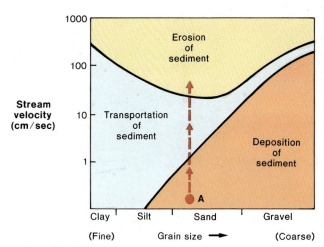

Box 10.2 Figure 1
Curves showing the stream velocities at which erosion and deposition of sediment occur. These velocities vary with the grain size of the sediment. See text for a discussion of point A and the dashed red line above it.

Box figure 1 shows the stream velocities at which sediments are eroded, transported, and deposited. For each grain size, these velocities are different. The upper curve represents the minimum velocity needed to erode sediment grains. This curve shows the velocity at which previously stationary grains are first picked up by moving water. The lower curve represents the velocity below which deposition occurs, when moving grains come to rest. Between the two curves the water is moving fast enough to transport grains that have already been eroded. Note that it takes a higher stream velocity to erode grains (set them in motion) than to transport grains (keep them in motion).

Let's assume that fine sand covers the bed of a stream that is barely flowing. No sand is moving, because the stream velocity is so low. These conditions are represented by point A on the graph. The vertical line drawn upward from point A represents a gradually increasing stream velocity, perhaps as the stream starts to flood. No sediment moves as the velocity rises through the lower curve and through the region marked "transportation of sediment." Only when the velocity is high enough to intersect the *upper* curve do sand grains begin to move. As the flood recedes and the velocity decreases, the velocity falls below the upper curve and into the transportation region. Under these conditions the sand that was already eroded continues to be transported, but no new sand is eroded. As the velocity falls below the lower curve, all the sand is deposited again, coming to rest on the stream bed.

The right half of the diagram shows that coarser particles require progressively higher velocities for erosion and transportation, as you might expect (boulders are harder to move than sand grains). The erosion curve also rises toward the left of the diagram, however. This shows that fine-grained silt and clay are actually harder to erode than sand. The reason is that molecular forces tend to bind silt and clay into a cohesive mass that resists erosion. Once silt and clay are eroded, however, they are easily transported. As you can see from the lower curve, the silt and clay in a river's suspended load are not deposited until the river virtually stops flowing.

Two terms are used to describe a river's ability to carry sediment. The *capacity* of a stream is the total load of sediment that the stream can carry. A stream's capacity increases as its discharge increases—the more water flowing in a stream, the more sediment it can carry. The *competence* of a stream is measured by the largest particle size that the stream can carry. The competence of a stream increases as its velocity increases—the faster a stream flows, the larger the particles that it can move. The impact force of water varies with the square of its velocity, so if a stream's velocity doubles, the impact force on sediment particles increases by a factor of four. This is why swift mountain streams in flood can sometimes move boulders the size of automobiles (figure 10.16). During floods, a river's velocity and discharge both increase dramatically, so the amount of sediment (the capacity) and the size of the sediment (the competence) moved by the river both increase. A single flood may cause more erosion than 50 or 100 years of normal stream flow.

Stream Deposition

The sediments transported by a stream are often deposited temporarily along the stream's course (particularly the bed-load sediments). Such sediments move sporadically downstream in repeated cycles of erosion and deposition, forming *bars* and *flood-plain deposits*. At or near the end of a stream, sediments may be deposited more permanently in a *delta* or an *alluvial fan*.

Bars

Stream deposits may take the form of a **bar,** a ridge of sediment, usually sand or gravel, deposited in the middle or along the banks of a stream (figure 10.17). Bars may be formed by deposition when a stream's discharge or velocity decreases. During flood, a river can move all sizes of sediment, from silt and clay up to huge boulders, because the greatly increased volume of water is moving very

Figure 10.17
Gravel bars along the banks and in the middle of a stream.
Note the sand deposited on the boulders at the right.

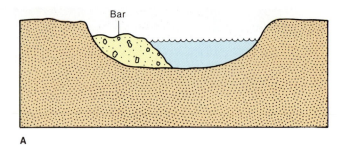

A

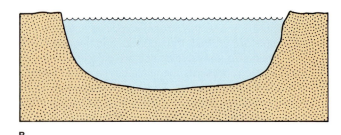

B

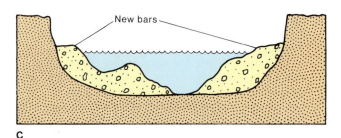

C

Figure 10.18
A flood can wash away bars in a stream, depositing new bars
as the water recedes. (*A*) Normal water flow with sand and
gravel bar. (*B*) Increased discharge and velocity during flood
moves all sediment downstream. Channel deepens and widens
if banks erode easily. (*C*) New bars are deposited as water
level drops and stream slows down.

rapidly. As the flood begins to recede, the water level in
the stream falls and the velocity drops. With the stream
no longer able to carry all its sediment load, the larger
boulders drop down on the stream bed, slowing the water
locally even more. Finer gravel and sand are deposited between the boulders and downstream from them. In this
way, deposition builds up a sand or gravel bar that may
become exposed as the water level falls.

The next flood on the river may erode all the sediment
in this bar and move it farther downstream. But as the
flood slows, it may deposit new gravel in approximately
the same place, forming a new bar (figure 10.18). After
each flood, river fishermen and boat operators must re-
learn the size and position of the bars. Sometimes gold
panners discover fresh gold in a mined-out river bar after
a flood has shifted sediment downstream.

Braided Streams

Deposition of a bar in the center of a stream (a *midchannel bar*) diverts the water toward the sides where it
washes against the stream banks with greater force,
eroding the banks and widening the stream. A stream
heavily loaded with sediment may deposit many bars in
its channel, causing the stream to widen continually as
more bars are deposited. Such a stream typically goes
through many stages of deposition, erosion, redeposition,
and reerosion, especially if its discharge fluctuates. The
stream may lose its main channel and become a **braided
stream,** flowing in a network of interconnected rivulets
around numerous bars (figures 10.19 and 10.20). A
braided stream characteristically has a wide, shallow
channel.

A stream tends to become braided when it is heavily
loaded with sediment (particularly bed load) and has banks
that are easily eroded. The braided pattern develops in
deserts as a sediment-laden stream loses water through
evaporation and through percolation into the ground. In
meltwater streams flowing off glaciers, braided patterns
tend to develop when the discharge from the melting glaciers is low relative to the great amount of sediment the
stream has to carry.

Meandering Streams and Point Bars

Rivers that carry fine-grained silt and clay in suspension
tend to be narrow and deep and to develop pronounced,
sinuous curves called **meanders** (figure 10.21). In a long
river, sediment tends to become finer downstream, so
meandering is common in the lower reaches of a river.

You have seen in figure 10.6 that a river's velocity is
higher on the outside of a curve than on the inside. This
high velocity can erode the river bank on the outside of a
curve (figure 10.22).

The low velocity on the inside of a curve promotes
sediment deposition. The sand bars in figure 10.21 have
been deposited on the inside of curves because of the lower
velocity there. Such a bar is called a **point bar** and usually
consists of a series of arcuate ridges of sand or gravel.

The simultaneous erosion on the outside of a curve
and deposition on the inside can deepen a gentle curve into
a hairpin-like meander (figure 10.23). Meanders are rarely
fixed in position. Continued erosion and deposition cause
them to migrate back and forth across a flat valley floor,
leaving scars and arcuate point bars to mark their former
positions (figure 10.24).

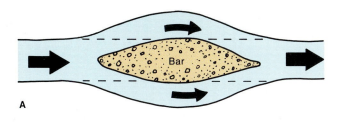

A

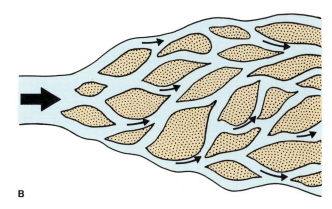

B

Figure 10.19
(*A*) A midchannel bar can divert a stream around it, widening the stream. (*B*) Braided stream. Bars split main channel into many smaller channels, greatly widening the stream.

A

B

Figure 10.20
(*A*) A braided stream in coarse desert sand. (*B*) A braided stream in Alaska. Water flows toward the bottom of the photo from a melting glacier in the background. Large amounts of coarse bed load cause the braiding.
Photo *A* by Frank M. Hanna. Photo *B* by D. A. Rahm, courtesy Rahm Memorial Collection, Western Washington University.

A

B

Figure 10.21
(*A*) Meanders on the Mississippi River, New Madrid, Missouri. Point bars of sand have been deposited on the inside of meander curves. (*B*) A point bar of gravel has been deposited on the inside of this curve.
Photo *A* by Frank M. Hanna.

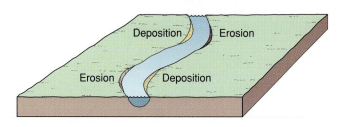

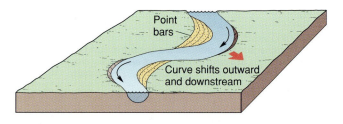

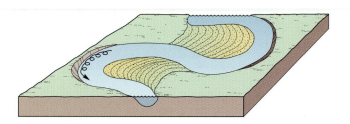

Figure 10.22
River erosion on the outside of a curve, Newaukum River, Washington. Pictures were taken in January and March 1965.
Photos by P. A. Glancy, U.S. Geological Survey.

At times, particularly during floods, a river may form a **meander cutoff,** a new, shorter channel across the narrow neck of a meander (figure 10.25). The old meander may be abandoned as sediment separates it from the new, shorter channel. The cutoff meander becomes a crescent-shaped **oxbow lake** (figure 10.26). With time, an oxbow lake may fill with sediment and vegetation.

Flood Plains

A **flood plain** is a broad strip of land built up by sedimentation on either side of a stream channel (figure 10.27). As meanders shift back and forth laterally over the valley floor, they leave point bar deposits on the insides of curves. Some flood plains are constructed almost entirely of such bar deposits, usually made of sand and gravel (figure 10.28*A*). Other flood plains consist of horizontal layers of fine-grained sediment deposited during floods when the river overflows its banks (figure 10.28*B*). During floods, flood plains may be covered with water carrying suspended silt and clay (figure 10.27*B*). When the floodwaters recede, these fine-grained sediments are left behind as a horizontal deposit on the flood plain. In some cases coarse bar deposits may be covered by fine-grained flood deposits (figure 10.28*C*).

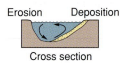
Erosion   Deposition
Cross section

Corkscrew water motion on a curve helps cause erosion and deposition.

Figure 10.23
Development of river meanders and point bars by erosion and deposition on curves.

Figure 10.24
The migration of river meanders across this valley has left arcuate point bar deposits and visible scars marking old meander positions, Panama City, Florida.
Photo by Frank M. Hanna.

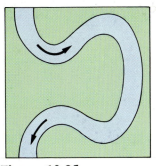

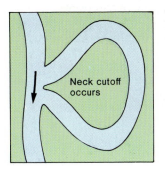
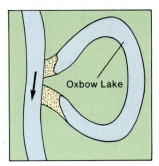

Figure 10.25
Creation of an oxbow lake by a meander neck cutoff. Old channel is separated from river by sediment deposition.

Figure 10.26
An oxbow lake marks the old position of a river meander, Sweetwater River, Wyoming.
Photo by W. R. Hansen, U.S. Geological Survey.

A

B

C

Figure 10.27
River flood plains. (*A*) Flat flood plain adjacent to a stream channel, northwestern Wyoming. (*B*) Flooded flood plain of the Animas River, Colorado. (*C*) Sediment deposited on a flood plain by the Trinity River in California during a 1964 flood.
Photo *A* by B. Amundson. Photo *B* by D. A. Rahm, courtesy Rahm Memorial Collection, Western Washington University. Photo *C* by A. O. Waananen, U.S. Geological Survey.

River Cities and Floods— Do You Live on a Flood Plain?

Why are many of the great cities of the world built on the banks of rivers? In the United States, cities such as New York, Philadelphia, New Orleans, St. Louis, Detroit, Minneapolis, and Portland (Oregon) are located beside major rivers. Kansas City and Pittsburgh were built where two major rivers come together.

Transportation and trade are the usual reasons for the location of these cities. Some cities began as river landings in the days when rivers, rather than highways, were the primary routes of transportation. Even today, river cities rely heavily on ships and barges for transporting crops and supplies. Some river cities were founded near bridges or fords where roads crossed rivers. Others are located at major obstructions, such as rapids, that stopped boats from going upstream.

While a city may owe much of its importance to its river location, there are several geologic hazards associated with rivers. *Floods,* which may cause little or no damage in uninhabited flood plains, can bring death and destruction to cities built on valley floors. Buildings, railroads, and highways along river banks are easily washed out by floods.

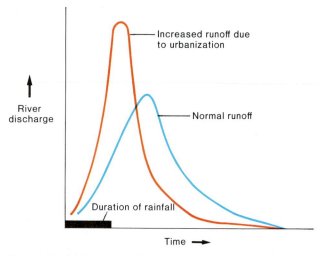

Box 10.3 Figure 1
The presence of a city can increase the chance of floods. The blue curve shows the normal increase in a river's discharge following a rainstorm (black bar). The red curve shows the great increase in runoff rate and amount caused by pavement and storm sewers in a city.

Although flooding is not likely to occur every year on every river, flooding is a natural process on all rivers and must be prepared for. Heavy seasonal rains and rapidly melting snow in springtime are the usual causes of floods. Variations in rainfall rate and volume and the geographic path of rainstorms determine to a large extent whether flooding will occur.

Cities themselves enhance flooding. Paved areas and storm sewers increase the amount and rate of surface runoff of water, making river levels higher during storms (box figure 1). Bridges, docks and

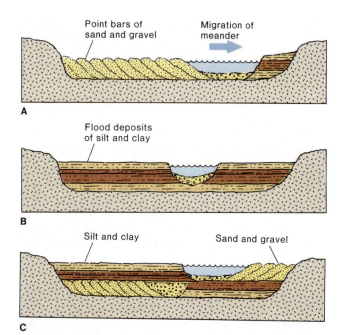

Figure 10.28
Three types of flood plains. (*A*) Point bar deposits of sand and gravel formed on the inside banks of shifting meanders. (*B*) Horizontal layers of fine-grained flood deposits. Each flood deposits a new layer on the top. (*C*) Combination of point bar deposits and flood deposits.

As a flooding river spreads over a flood plain, it slows down. The velocity of the water is abruptly decreased by the increased wetted perimeter and friction as the water leaves the deep channel and moves in a thin sheet over the flat valley floor. The sudden decrease in velocity of the water causes the river to deposit most of its sediment near the main channel, with progressively less sediment deposited away from the channel. A series of floods may build up **natural levees**—low ridges of flood-deposited sediment that form on either side of a stream channel and thin away from the channel (figure 10.29). The sediment near the river is coarsest, often sand and silt, while the finer clay is carried farther from the river into the flat, lowland area (the backswamp).

Deltas

Most streams ultimately flow into the sea or into large lakes. A stream flowing into standing water usually builds a **delta,** a body of sediment deposited at the mouth of a river when the river's velocity decreases (figure 10.30).

Box 10.3 Figure 2
Mud deposited by flooding Potomac River during Hurricane Agnes, 1972, Alexandria, Virginia.
Photo by USDA–Soil Conservation Service.

buildings built on flood plains can constrict the flow of floods, increasing the water height and velocity and promoting erosion.

Another flood problem is *flood deposits.* Silt and clay deposited on flood plains in an agricultural region can be beneficial, renewing the fields with topsoil from upstream. On the other hand, a layer of wet mud blanketing a city can destroy lawns, furniture, and machinery. Cleaning up after a flood is slow and very expensive.

Flood-control structures can partially reduce the dangers of floodwaters and sedimentation to river cities. Upstream dams can trap water and release it slowly after the storm. (A dam also catches sediment, which eventually fills its reservoir and ends its life as a flood-control structure.) Artificial levees are embankments built along the banks of a river channel, sometimes on top of natural levees, to contain floodwaters within the channel.

Flood-control structures are generally built to control *most,* not *all,* floods. Exceptional floods can fill reservoirs and overtop dams and levees. Floods are described by *recurrence interval,* the average time interval between floods of a given size. A "50-year flood" is one that occurs, on the *average,* every 50 years. A 50-year flood has a 1-in-50 chance of occurring in any one year. It is perfectly possible to have two 50-year floods in two successive years—or even in the same year. If a 50-year flood occurs this year on the river you live beside, you should not assume that there will be a 49-year period of safety before the next one.

Dams and levees are designed to control certain specified floods. If the flood-control structures on your river were designed for 75-year floods, then a much larger 100-year flood will likely overtop these structures, and may destroy your home. The disastrous floods in Pennsylvania and New York from Hurricane Agnes in 1972 resulted from many such failures in flood control. Wise land-use plans for flood plains should go hand in hand with flood control. Wherever possible, buildings should be kept out of areas that might someday be flooded by 100-year floods.

The *migration of river channels* can be a hazard to riverfront property and bridges. As a river undercuts its banks on the outside of curves, buildings and piers may fall into the river. Protective walls of stone (riprap) or concrete are often constructed to slow bank erosion.

The surface of most deltas is marked by **distributaries**—small, shifting channels that carry water away from the main river channel and distribute it over the surface of the delta (figure 10.31). Sediment deposited at the end of a distributary tends to block the water flow, causing distributaries and their sites of sediment deposition to shift periodically.

The shape of a delta in map view depends on the balance between sediment supply from the stream and the erosive power of waves and currents in the sea or lake (figure 10.31). Some deltas, like that of the Nile River, are broadly triangular; this delta's resemblance to the Greek letter *delta* (Δ) is the origin of the name. Other deltas, including that of the Mississippi River, are created when very large amounts of sediment are carried into relatively quiet water. Partly because dredging has kept the major distributary channels (locally called "passes") fixed in position for many decades, the Mississippi's distributaries have built long fingers of sediment out into the sea. The resulting shape has been termed a *birdfoot delta.* The deltas of the Amazon and the Niger rivers are examples of what happens when a river empties on a coast where vigorous waves and currents sweep much of the sediment along the shoreline. The St. Lawrence is an example of a major river that does not carry enough sediment to form a delta. This is because the St. Lawrence is the outlet for the Great Lakes, which tend to trap sediment.

Many deltas, particularly small ones in freshwater lakes, are built up from three types of deposits, shown in the diagram in figure 10.32. *Foreset beds* form the main body of the delta. They are deposited at an angle to the horizontal. This angle can be as great as 20° to 25° in a small delta where the foreset beds are sandy, or less than 5° in large deltas with fine-grained sediment. On top of the foreset beds are the *topset beds,* nearly horizontal beds of varying grain size formed by distributaries shifting across the delta surface. Out in front of the foreset beds are the *bottomset beds,* deposits of the finest silt and clay that are carried out to sea by the fresh water flow or by sediments sliding downhill on the sea floor. Many of the world's great deltas are far more complex than the simplified diagram shown in the figure. Shifting river mouths, wave energy, currents, and other factors produce many different internal structures.

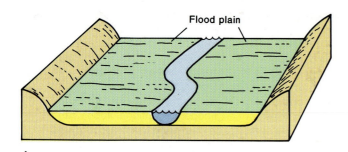

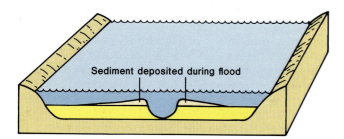

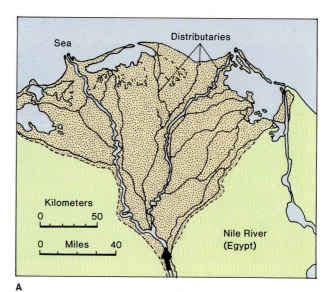

Figure 10.29
Natural levee deposition during a flood. Levees are thickest
and coarsest next to the river channel and build up from many
floods, not just one. (Relief of levees is exaggerated.)
(*A*) Normal flow. (*B*) Flood. (*C*) After flood.

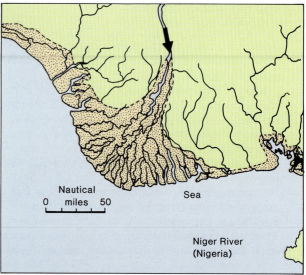

Figure 10.30
Delta of Nooksack River, Washington. Note the sediment-laden
water, and how the land is being built outward by river
sedimentation. A river typically divides into several channels
(distributaries) on a delta.
Photo by D. A. Rahm, courtesy Rahm Memorial Collection, Western
Washington University.

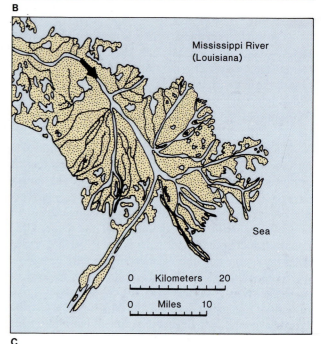

Figure 10.31
The shape of a delta depends on the amount of sediment
being carried by the river and on the vigor of waves and
currents in the sea.

D

Figure 10.31 continued
Landsat image of Mississippi River delta. Note how sediment
(light yellow) is carried by both river and ocean currents.
Photo by NASA.

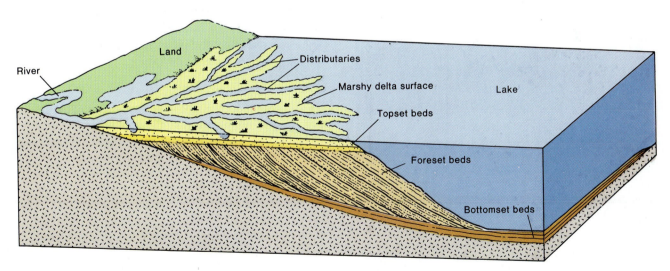

Figure 10.32
Internal construction of a small delta. Most large deltas are
more complicated than this.

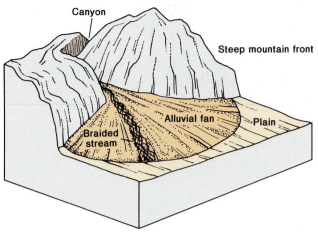

Figure 10.33
An alluvial fan at the mouth of a desert canyon.

A

B

Figure 10.34
(*A*) This alluvial fan formed on the salt-encrusted floor of Death Valley, California, as sediment was washed out of the canyon by thunderstorms. (*B*) Giant alluvial fans in the distance joined together as they grew at the base of the mountains. The surface of another fan forms the foreground. Death Valley.
Photo (*A*) by Frank M. Hanna.

Alluvial Fans

Some streams, particularly in dry climates, do not reach the sea or any other body of water. They build alluvial fans instead of deltas. An **alluvial fan** is a large, fan- or cone-shaped pile of sediment that usually forms where a stream's velocity decreases as it emerges from a narrow mountain canyon onto a flat plain (figures 10.33 and 10.34). Alluvial fans are particularly well developed and exposed in the southwestern desert of the United States and in other desert regions; but they are by no means limited to arid regions.

An alluvial fan builds up its characteristic cone shape gradually as streams and mudflows shift back and forth across the fan surface and deposit sediment, usually in a braided pattern. Deposition on an alluvial fan in the desert is discontinuous because streams typically flow for only a short time after the infrequent rainstorms. When rain does come, the amount of sediment to be moved is apt to be greater than the available water. When the rain stops, the stream ends up as a mudflow spreading out over the fan.

The sudden loss of velocity when a stream flows from steep mountains onto a plain causes the sediment to deposit on an alluvial fan. The loss of velocity is due to both the decrease in gradient and the widening or branching of the channel as it leaves the mountains. The gradual loss of water as it infiltrates into the fan also promotes sediment deposition. On large fans, deposits are graded in size within the fan, with the coarsest sediment dropped nearest the mountains and the finer material deposited progressively farther away. Small fans do not usually show such grading.

Valley Development

Valleys, the most common landforms on the earth's surface, are usually cut by streams. By removing rock and sediment from the stream channel, a stream deepens, widens, and lengthens its own valley.

Downcutting and Base Level

The process of deepening a valley by erosion of the stream bed is called **downcutting.** If a stream removes rock from its bed, it can cut a narrow *slot canyon* down through rock. Such narrow canyons do not commonly form because mass wasting and sheet erosion remove rock from the valley walls. These processes widen the valley from a narrow, vertical-walled canyon to a broader, open, V-shaped canyon (figure 10.35). Slot canyons persist, however, in very resistant rock or in regions where downcutting is rapid.

Downcutting cannot continue indefinitely because the headwaters of a stream cannot cut below the level of the stream bed at the mouth. If a river flows into the ocean, sea level becomes the lower limit of downcutting. The river cannot cut below sea level, or it would have to flow uphill to get to the sea.

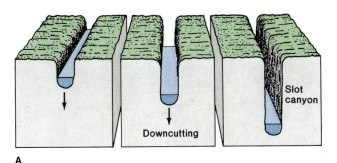

A

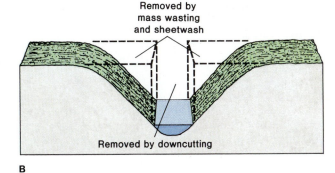

B

Figure 10.35
Downcutting, mass wasting, and sheetwash shape canyons and valleys. (*A*) Downcutting can create slot canyons in resistant rock, particularly where downcutting is rapid. (*B*) Downslope movement of rock and soil on valley walls widens most canyons into V-shaped valleys.

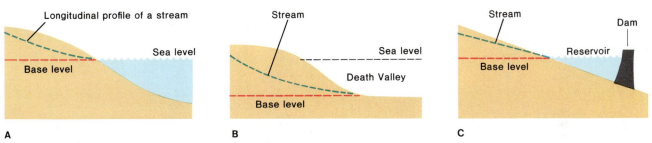

A   B   C

Figure 10.36
Base level is the lowest level of downcutting.

The limit of downcutting is known as **base level;** it is a theoretical limit for erosion of the earth's surface (figure 10.36). Downcutting will proceed until the stream bed reaches base level. If the stream is well above base level, downcutting can be quite rapid; but as the stream approaches base level, the rate of downcutting slows down. For streams that reach the ocean, base level is close to sea level, but since streams need at least a gentle gradient in order to flow, base level slopes gently upward in an inland direction.

During the glacial ages of the Pleistocene Epoch (see the chapter on "Glaciers and Glaciation"), sea level rose and fell as water was removed from the sea to form the glaciers and returned to the sea when the glaciers melted. This means that base level rose and fell for streams flowing into the sea. As a result, the lower reaches of such rivers alternated between erosion (caused by low sea level) and deposition (caused by high sea level). Since the glaciers advanced and retreated several times, the cycle of erosion and deposition was repeated many times, resulting in a complex history of cutting and filling near the mouths of most old rivers.

Base levels for streams that do not flow into the ocean are not related to sea level. In Death Valley in California (figure 10.36*B*), base level for in-flowing streams corresponds to the lowest point in the valley, 282 feet (86 meters) *below* sea level (the valley has been dropped below sea level by tectonic movement along faults). On the other hand, base level for a stream above a high reservoir or a mountain lake can be hundreds of feet above sea level. The surface of the lake or reservoir serves as base level for all the water upstream (figure 10.36*C*). The base level of a tributary stream is governed by the level of its junction with the main stream. A ledge of resistant rock may act as a temporary base level if a stream has difficulty eroding through it.

The Concept of a Graded Stream

As a stream begins downcutting into the land, its longitudinal profile is usually irregular, with rapids and waterfalls along its course (figure 10.37). Such a stream, termed *ungraded,* is using most of its erosional energy in downcutting to smooth out these irregularities in gradient.

As the stream smooths out its longitudinal profile to a characteristic concave-upward shape (figure 10.38), it becomes graded. A **graded stream** is one that exhibits a delicate balance between its transporting capacity and the sediment load available to it. This balance is maintained by cutting and filling any irregularities in the smooth longitudinal profile of the stream.

Earlier in this chapter you learned how changes in a stream's gradient can cause changes in its sediment load. An increase in gradient causes an increase in a stream's velocity, allowing the stream to erode and carry more sediment. A balance is maintained—the greater load is a result of the greater transporting capacity caused by the steeper gradient.

Figure 10.37
The waterfall and rapids on the Yellowstone River in Wyoming indicate that the river is actively downcutting. Note the V-shaped cross-profile and lack of flood plain.
Photo by B. Amundson.

Headwaters

Ungraded profile on irregular land surface; waterfalls and rapids

Mouth

Concave upward-graded profile

Figure 10.38
The longitudinal profile of a stream is a side view of the stream's gradient from its headwaters to its mouth. An ungraded stream has an irregular profile with many waterfalls and rapids. A graded stream has smoothed out its longitudinal profile to a smooth, concave-upward curve.

The relationship also works in reverse—a change in sediment load can cause a change in gradient. For example, a decrease in sediment load may bring about erosion of the stream's channel, thus lowering the gradient. Because dams trap sediment in the calm reservoirs behind them, most streams are sediment-free just downstream from dams. In some streams this loss of sediment has caused severe channel erosion below a dam, as the stream adjusts to its new, reduced load.

A river's energy is used for two things—transporting sediment and overcoming resistance to flow. If the sediment load decreases, the river has more energy for other things. It may use this energy to erode more sediment, deepening its valley. Or it may change its channel shape or length, increasing resistance to flow, so that the excess energy is used to overcome friction. Or the river may increase the roughness of its channel, also increasing friction. The response of a river is not always predictable, and construction of a dam or channelization of a river can sometimes have unexpected and perhaps harmful results.

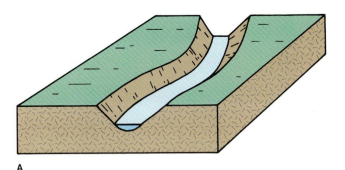

A

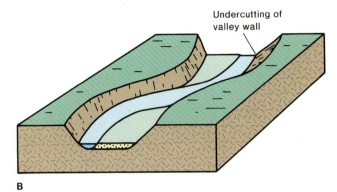

Undercutting of
valley wall

B

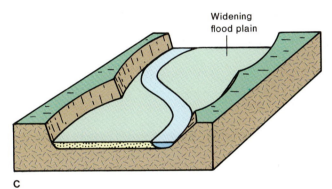

Widening
flood plain

C

Figure 10.39
Lateral erosion can widen a valley by undercutting and eroding valley walls.

Lateral Erosion

A graded stream can be deepening its channel by downcutting while part of its energy is also widening the valley by **lateral erosion,** the erosion and undercutting of a stream's banks and valley walls as the stream swings from side to side across its valley floor. The stream channel remains the same width as it moves across the flood plain, but the valley widens by erosion, particularly on the outside of curves and meanders where the stream impinges against the valley walls (figure 10.39). The valley widens as its walls are eroded by the stream and as its walls retreat by mass wasting triggered by stream undercutting. As a valley widens, the stream's flood plain increases in width also.

Figure 10.40
Headward erosion is lengthening this gully and destroying the field. The head of the gully is extending uphill, largely by mass wasting.
Courtesy Ward's Natural Science Est., Inc., Rochester, N.Y.

Headward Erosion

Building a delta or alluvial fan at its mouth is one way a river can extend its length. A stream can also lengthen its valley by **headward erosion,** the slow uphill growth of a valley above its original source through gullying, mass wasting, and sheet erosion (figure 10.40). This type of erosion is particularly difficult to stop. When farmland is being lost to gullies that are eroding headward into fields and pastures, farmers must divert sheet flow and fill the gully heads with brush and other debris to stop, or at least retard, the loss of topsoil.

Regional Erosion

Geologists have two contrasting ideas about how landscape develops as a region is eroded. The first idea is that rugged mountains are gradually worn flat by erosion, eventually producing a level plain (figure 10.41). The second idea is that slopes retreat laterally across a region, maintaining a constant angle of slope as they do so (*parallel retreat* of slopes—see figure 10.42).

The first idea, that mountains are gradually worn flat to produce plains, dominated geologic thought in the first half of the twentieth century. According to this concept, steep slopes form early in a region's erosional history and become gentler as erosion progresses (figure 10.41). As the valley sides are worn away, the upland areas are eroded downward, and the eroded material is carried away by streams. The end result is a nearly flat, featureless plain (a *peneplain*) produced as a region is lowered nearly to base level by stream erosion, sheet erosion, and mass wasting.

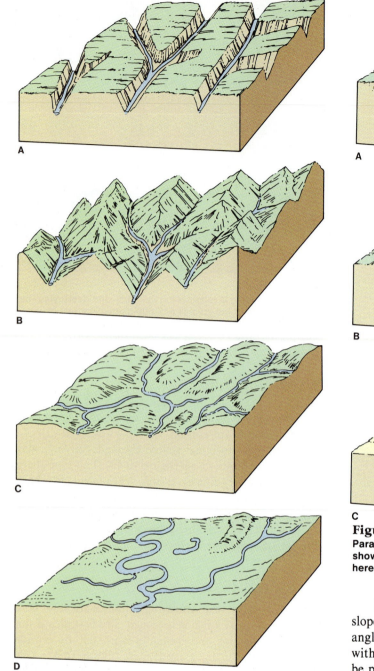

Figure 10.41
An early idea of landscape development. Streams cut into a flat plain to form rugged mountains, which are gradually worn flat to produce another nearly flat plain. Notice how the steep slopes become gentler as erosion proceeds.

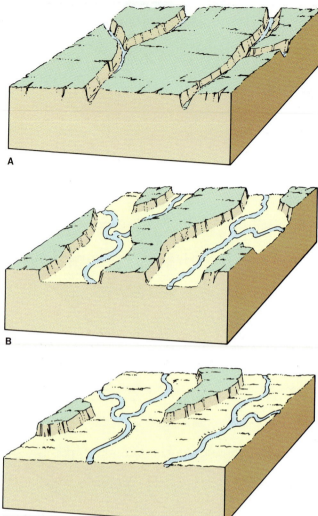

Figure 10.42
Parallel retreat of slopes, which contrasts with the scheme shown in figure 10.41. The land is worn flat to form a plain here, but the slope angles remain constant with time.

More recently, field studies have suggested that many slopes undergo *parallel retreat*. They maintain a constant angle as they are worn away; they do not become gentler with time (figure 10.42). Although a broad, flat plain can be produced by either of these mechanisms, the key difference between them is whether slope angles remain constant or become gentler with time.

Modern workers on slope erosion believe that the role of *time* in landscape development was overemphasized by early geologists. Many workers today feel that slopes are controlled by *rock type, climate,* and *rock structure*. A slope angle may represent an equilibrium among these factors and remain constant for centuries.

It is clear that *rock type* influences slope angle. Figure 10.43 shows how slopes can vary on different geologic materials. Some solid rocks can hold vertical cliffs. Coarse, angular talus often holds an angle of about 45°, and loose, dry sand in dunes generally piles up at an angle of about 34°. Clayey soil may be stable at an angle of only a few

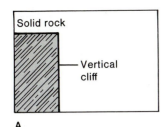
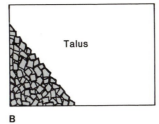
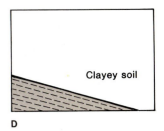

A
B
C
D

Figure 10.43
Maximum slope angles for different types of geologic materials.

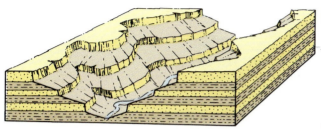

Figure 10.44
Cliffs and slopes caused by the erosion of horizontal sedimentary rock layers. Resistant sandstone layers form cliffs. Shale beds are less resistant to weathering and erosion and form gentle slopes.

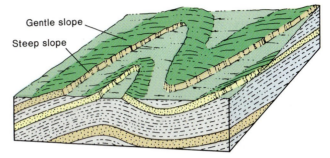

Figure 10.45
The erosion of gently folded sedimentary rocks can produce arcuate ridges with unequal slopes.

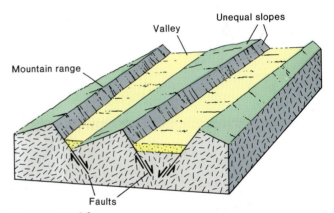

Figure 10.46
Rock movements along faults can form valleys and mountain ranges with unequal slopes.

degrees; the angle becomes lower as the water content of the soil increases. Figure 10.44 shows how different rock types maintain different slope angles. This situation is common in the deserts of the southwestern United States, where horizontal sandstone beds stand as vertical cliffs and shale weathers and erodes to much gentler slopes. Similar slope differences show up in wetter climates as well.

Climate can also influence the slope angle. Many geologists think that a humid (wet) climate results in smoothly rounded topography due to a rapid rate of soil creep and other processes. An arid (dry) climate is more apt to result in sharp, angular topography, due to low precipitation, slow chemical weathering and soil creep, and sparse vegetation. (There is by no means universal agreement among geologists on these points. Some believe that climate does not influence slope, and others think that *humid* climates yield angular topography.)

One problem with studying hillslopes is the recent climatic fluctuations associated with the glacial ages of the Pleistocene Epoch. As the huge glaciers advanced and retreated, the climate for a region became alternately warmer and colder as well as wetter and drier. These climatic fluctuations may mean that the present landscape of a region formed a short time ago under a very different climate than the present one. Therefore, any measurable changes in slope today may be a response to climate change rather than an indication of a progressive change with time.

Rock structure also contributes to slope angle and local elevation. *Structure* refers to features such as *folds* (bends) and *faults* (breaks) that deform rocks (see the

chapter on "Geologic Structures"). Undeformed, horizontal sedimentary rocks of varying resistances can lead to a staircase-like topography of cliffs and slopes (figure 10.44), as is well developed in the Grand Canyon in Arizona. Gently folded sedimentary rocks can weather and erode as shown in figure 10.45, resulting in resistant beds forming arcuate ridges on the land surface. If the folding is gentle, the ridges will have unequal slope angles on their sides. Some types of faulting produce mountain ranges with unequal slopes on opposite sides (figure 10.46).

A series of folded sedimentary rocks with varying resistances to erosion might result in a landscape like that in figure 10.47. The landscape has three different levels of land surface, one connecting the highest summits, one

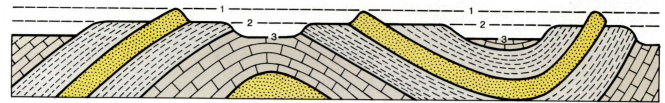

Figure 10.47
Folded sedimentary rock layers with three different levels of
land surface, resulting from the differential removal of rocks of
varying resistances to weathering and erosion.

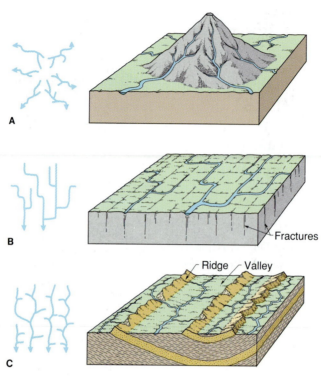

Figure 10.48
Drainage patterns can reveal something about the rocks
underneath. (*A*) A radial pattern develops on a conical
mountain or dome. (*B*) A rectangular pattern develops on
regularly fractured rock. (*C*) A trellis pattern develops on
alternating ridges and valleys caused by the erosion of
resistant and nonresistant tilted rock layers.

connecting the valley floors, and one in between. Such a
landscape can form by the differential removal of resistant and nonresistant rocks. Resistant sandstones form the
high ridges; nonresistant limestone, which dissolves in a
humid climate, forms the valley floors. Shale, with an intermediate resistance, forms the intermediate surface.

Drainage Patterns

The arrangement, in map view, of a river and its tributaries is a **drainage pattern.** A drainage pattern can in
many cases reveal the nature and structure of the rocks
underneath it.

Most tributaries join the main stream at an acute
angle, forming a V pointing downstream. If the pattern
resembles branches of a tree or veins in a leaf, it is called

dendritic (figure 10.3). Dendritic drainage patterns develop on uniformly erodible rock, usually horizontal sedimentary rocks or unfractured crystalline rock, such as
igneous rock in a batholith. A **radial pattern,** in which
streams diverge outward like spokes of a wheel, forms on
high conical mountains, such as composite volcanoes and
domes (figure 10.48*A*). A **rectangular pattern,** in which
tributaries have frequent 90° bends and tend to join other
streams at right angles, develops on regularly fractured
rock (figure 10.48*B*). A network of fractures meeting at
right angles forms pathways for streams because fractures are eroded more easily than unbroken rock. A **trellis
pattern** consists of parallel main streams with short tributaries meeting them at right angles (figure 10.48*C*). A
trellis pattern forms in a region where tilted layers of resistant rock such as sandstone alternate with nonresistant
rock such as shale. Erosion of such a region results in a
surface topography of parallel ridges and valleys.

How Can a River Slice Through a Mountain Range?

Some narrow mountain ranges have steep-sided river valleys slicing directly across them (box figure 1). The origin of these valleys is often puzzling. Why did the river erode *through* the mountain range rather than flow parallel to it or perhaps around the end of it?

The usual reason is that the river has been *superimposed* (or superposed) on the range from a gently sloping plain that used to exist above it (box figure 2). If sediment once buried the range, a river could develop on the smooth surface of the sediment layer. As the sediment gradually was eroded away by the river, the river itself would be let down onto the once-buried range and could cut a canyon through the range. The thick sedimentary layer might have been deposited by glaciers or as giant alluvial fans. It could even be marine.

A less common possibility is that the river is *antecedent,* existing before the mountain range even formed, perhaps as a fold. Some rivers are thought to be many millions of years old and could predate geologic structures in a tectonically active region (box figure 3). As a fold forms, a large river might be able to keep up with deformation, cutting through the rising fold.

Another relatively rare origin for such a valley is by *stream piracy* (or *stream capture*). If two streams flow downhill in opposite directions on either side of a range, one stream may capture the other. One of the streams may erode faster than the other, perhaps because it is on the rainy side of the mountain and has more water in it or because it is flowing over softer rock than the other stream. The rapidly eroding, larger stream will cut through most of the mountain and may eventually capture the drainage basin of the smaller, slower stream by headward erosion (box figure 4). The larger stream grows at the expense of the smaller one, and the water that used to flow in one direction in the small stream now flows in the opposite direction in the new, larger stream.

Box 10.4 Figure 1
The Bighorn River cuts a steep canyon directly across the folded rocks of Sheep Mountain, Wyoming.
Photo by D. A. Rahm, courtesy Rahm Memorial Collection, Western Washington University.

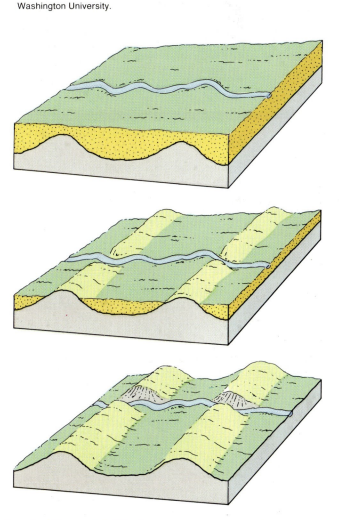

Box 10.4 Figure 2
A superimposed stream initially flows over a sedimentary plain and gradually cuts through once-buried ridges as the sediments are eroded away.

Box 10.4   *continued*

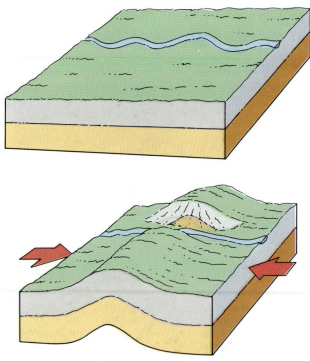

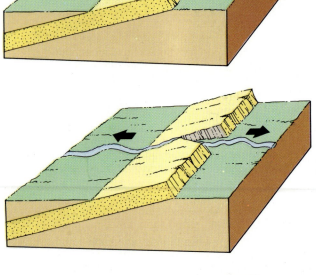

Box 10.4 Figure 3
An antecedent stream existed before a ridge forms. The stream is able to erode its way through a gradually rising fold.

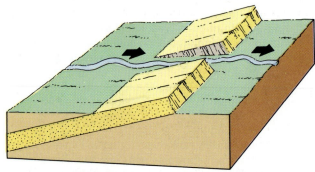

Box 10.4 Figure 4
Stream piracy is the capture of one river by another. Here two streams flow in opposite directions on different sides of a ridge. One stream erodes more rapidly than the other and captures the slower stream. In this case, the captured stream reverses direction after joining the larger stream.

Figure 10.49
River terraces in Alberta, Canada. The river has cut downward into its old flood plain.

Stream Terraces

Stream terraces are steplike landforms found above a stream and its flood plain (figure 10.49). Terraces may be benches cut in rock (sometimes sediment-covered), or they may be steps formed in sediment by deposition and subsequent erosion.

Paired terraces are terraces found at the same elevation on each side of a river. Figure 10.50 shows how one type of paired terrace forms as a river cuts downward into a thick sequence of its own flood-plain deposits. A stretch of river with characteristics favorable to deposition can build up a thick section of flood-plain deposits. Later the river may change from deposition to erosion and cut into its old flood plain, parts of which remain as terraces above the river.

Why might a river change from deposition to erosion? One reason might be regional uplift, raising a river that was once meandering near base level to an elevation well above base level. Uplift would steepen a river's gradient, causing the river to speed up and begin erosion (figure 10.8). But there are several other reasons why a river might change from deposition to erosion. A change from a dry to a wet climate may increase discharge and cause a river to begin eroding. A drop in base level (such as lowering of sea level) can have the same effect. A situation like that shown in figure 10.50 can develop in a recently glaciated region. Thick valley fill such as glacial outwash (see the chapter on "Glaciers and Glaciation") may be deposited in a stream valley and later, after the glacier stops producing large amounts of sediment, be dissected into terraces by the river.

Paired terraces can also develop from erosion of a bedrock valley floor (figure 10.51). Bedrock benches are usually capped by a relatively thin veneer of flood-plain deposits.

Unpaired terraces are terraces with unequal elevation on opposite sides of a river (figure 10.52). A river that is downcutting and laterally eroding at the same time shifts back and forth across its valley as it cuts downward. This movement can cut a series of rock benches or terraces at progressively lower levels on either side of a valley.

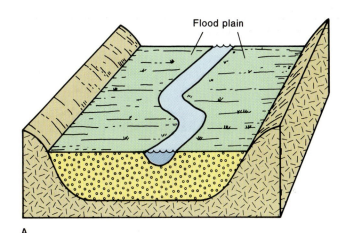

A

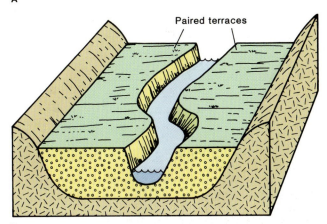

B

C

Figure 10.50
Paired terraces formed by a river cutting downward into its own flood-plain deposits. (*A*) River deposits thick, coarse, flood-plain deposits. (*B*) River erodes its flood plain by downcutting. Old flood-plain surface forms paired terraces. (*C*) Lateral erosion forms new flood plain below terraces.

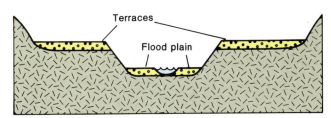

Figure 10.51
Paired terraces may be erosional benches cut in rock.

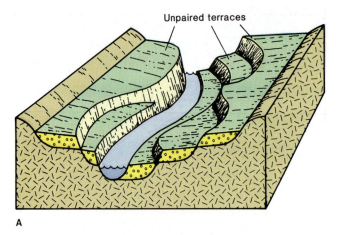

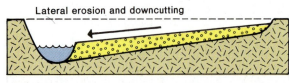

A

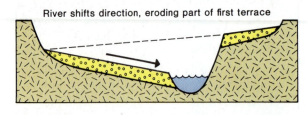

Lateral erosion and downcutting

River shifts direction, eroding part of first terrace

Repeated shifting produces new terraces

B

Figure 10.52
Unpaired terraces do not match across a river. They form by simultaneous downcutting and lateral erosion.

Figure 10.53
Incised meanders of the Green River, near Vernal, Utah.
Photo by C. M. Wentworth, U.S. Geological Survey.

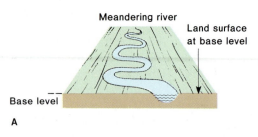

A

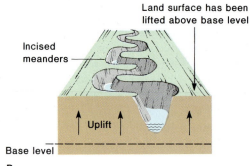

B

Figure 10.54
Incised meanders can form by uplift.

Incised Meanders

Incised meanders are meanders that retain their sinuous curves as they cut vertically downward below the level at which they originally formed. The result is a meandering *valley* with essentially no flood plain, cut into the land as a steep-sided canyon (figure 10.53).

Some incised meanders may be due to the profound effects of a change in base level. They may originally have been formed as meanders in a laterally eroding river flowing over a flat flood plain, perhaps near base level. If regional uplift elevated the land high above base level, the river would begin downcutting and might be able to maintain its characteristic meander pattern while deepening its valley (figure 10.54). A drop in base level without land uplift (possibly because of a lowering of sea level) could bring about the same result.

Although uplift may be a key factor in the formation of many incised meanders, it may not be *required* to produce them. Lateral erosion certainly seems to become more prominent as a river approaches base level, but some meandering can occur as soon as a river develops a graded profile. A river flowing on a flat surface high above base level may develop meanders early in its erosional history, and these meanders may become incised by subsequent downcutting. In such a case uplift is not necessary.

Stream Features on the Planet Mars

There is no liquid water on the surface of Mars today. With the present surface temperatures, atmospheric pressures, and water content in the Martian atmosphere, any liquid water would immediately evaporate. There are some indications, however, that conditions may have been different in the past and that liquid water existed on Mars, at least temporarily. Certain features on Mars, called *channels,* closely resemble certain types of stream channels on Earth. They have tributary systems and meanders and are sometimes braided. The channels trend downslope and tend to get wider toward their mouths. These Martian channels are restricted to certain areas and appear to have been formed by intermittent episodes of erosion.

Martian channels of one type (box figure 1*A*) are similar in appearance to channels in the Channeled Scablands of Washington state (box figure 1*B*). The Scablands channels were formed by extensive flooding during the Pleistocene glacial ages when a naturally formed ice dam broke and released the waters from a large lake. Several flooding events may have been responsible for the Martian channels, but the source of the water is not known. It may have been produced by the melting of underground ice, perhaps caused by climatic changes or nearby volcanic activity.

A second kind of Martian channel (box figure 2) occurs mostly on the sides of some volcanoes and some meteorite impact craters. Here the channels form systems of dendritic canyons and gullies. Channels of this type occur on the older surfaces of Mars (more than 3.5 billion years old) and may indicate that early in the history of Mars temperature and atmospheric conditions were such that rainfall could have occurred and long-lived river systems could have existed. Possibly as the climate of Mars cooled, the water froze and formed the extensive deposit of underground ice known to exist in the surface layers of Mars. (See Astrogeology Box 9.1 for a look at underground ice.)

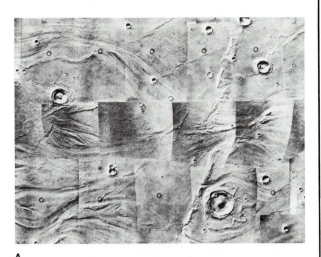

A

B

Astrogeology Box 10.1 Figure 1
(*A*) Viking Orbiter photograph of channel Maja Vallis. (*B*) Landsat photograph of the channel scablands of Washington.

Photo *A* by NASA. Photo *B* by Dr. J. C. Boothroyd, University of Rhode Island.

Astrogeology Box 10.1 Figure 2
Viking Orbiter View of a portion of channel Nirgal Vallis and its tributary system.

Photo by Dr. Michael H. Carr/National Space Science Data Center.

Summary

Normally stream *channels* are eroded and shaped by the streams that flow in them. Unconfined sheet flow can cause significant erosion.

Drainage basins are separated by *drainage divides*.

Stream *velocity* is the key factor controlling sediment erosion, transportation, and deposition. Velocity is in turn controlled by several factors.

An increase in a stream's *gradient* increases the stream's velocity. *Channel shape* and *roughness* affect velocity by increasing or lessening friction. As tributaries join a stream, the stream's *discharge* increases downstream.

Streams erode by *hydraulic action, abrasion,* and *solution.* They carry coarse sediment by *traction* and *saltation* as *bed load.* Finer-grained sediment is carried in *suspension.* A stream can also have a substantial *dissolved load.*

Streams create features by erosion and deposition. *Potholes* form by abrasion of hard rock on a stream bed. *Bars* form in the middle of streams or on stream banks, particularly on the inside of curves where velocity is low (*point bars*). A *braided pattern* can develop in streams with a large amount of bed load.

Meanders are created when a laterally eroding stream shifts across the flood plain, sometimes creating cutoffs and oxbow lakes.

A *flood plain* develops by both lateral and vertical deposition. *Natural levees* are built up beside streams by flood deposition.

A *delta* forms when a stream flows into standing water. The shape and internal structure of deltas are governed by river deposition and wave and current erosion. *Alluvial fans* form, particularly in dry climates, at the base of mountains as a stream's channel widens and its velocity decreases.

Rivers deepen their valleys by *downcutting* until they reach *base level,* which is either sea level or a local base level.

A *graded stream* is one with a delicate balance between its transporting capacity and its available load.

Lateral erosion widens a valley after the stream has become graded.

A valley is lengthened by both *headward erosion* and sediment deposition at the mouth.

Regional erosion may proceed (1) as slope angles decrease with time until a region is reduced to base level; or (2) as slope angles remain constant during *parallel retreat* of slopes. Climate, rock type, and structure control slope angle as well.

A river and its tributaries form a *drainage pattern.* A *dendritic* drainage pattern develops on uniform rock, a *rectangular* pattern on regularly jointed rock. A *radial* pattern forms on conical mountains, while a *trellis* pattern usually indicates erosion of folded sedimentary rock.

Stream terraces can form by erosion of rock benches or by dissection of thick valley deposits during downcutting.

Incised meanders form as (1) river meanders are cut vertically downward following uplift or (2) lateral erosion and downcutting proceed simultaneously.

Terms to Remember

abrasion
alluvial fan
bar
base level
bed load
braided stream
delta
dendritic pattern
discharge
dissolved load
distributary
downcutting
drainage basin
drainage divide
drainage pattern
flood plain
graded stream
headward erosion
hydraulic action
incised meander
lateral erosion

meander
meander cutoff
natural levee
oxbow lake
point bar
pothole
radial pattern
rectangular pattern
saltation
sheetwash
solution
stream
stream channel
stream gradient
stream terrace
stream velocity
suspended load
traction
trellis pattern
tributary

Questions for Review

1. What factors control a stream's velocity?
2. Describe how bar deposition creates a braided stream.
3. In what part of a large alluvial fan is the sediment the coarsest? Why?
4. What does a trellis drainage pattern tell about the rocks underneath it?
5. Describe one way that incised meanders form.
6. Compare and contrast an alluvial fan and a delta.
7. How does a meander neck cutoff form an oxbow lake?
8. How does a natural levee form?
9. Describe two ways in which river terraces form.
10. Describe three ways in which a river erodes its channel.
11. Name and describe the three main ways in which a stream transports sediment.
12. How does a stream widen its valley?
13. What is base level?

Questions for Thought

1. Over the last few thousand years, sea level has been rising because of the melting of large glaciers on land. What effects does such a rise in sea level have on rivers that flow into the sea?
2. Several rivers have recently been set aside as "wild rivers" on which dams cannot be built. Give at least four arguments against building dams on rivers. Give at least four arguments in favor of building dams.

Supplementary Readings

Bloom, A. L. 1978. *Geomorphology: A systematic analysis of late Cenozoic landforms.* Englewood Cliffs, N.J.: Prentice-Hall.

Clark, C. 1982. *Planet earth: Flood.* Alexandria, Va.: Time-Life Books.

Czaya, E. 1981. *Rivers of the World.* New York: Van Nostrand Reinhold.

Hoyt, W. B., and W. B. Langbein. 1955. *Floods.* Princeton, N.J.: Princeton University Press.

Leopold, L. B. 1974. *Water, a primer.* New York: W. H. Freeman.

Leopold, L. B., and W. B. Langbein. 1966. River meanders. *Scientific American* (June). Offprint #869. New York: W. H. Freeman.

Leopold, L. B., and W. B. Langbein. 1966. River meanders. *Scientific American* (June). New York: W. H. Freeman.

Morisawa, M. 1968. *Streams: Their dynamics and morphology.* New York: McGraw-Hill.

Ritter, D. F. 1986. *Process geomorphology.* 2d ed. Dubuque, Iowa: Wm. C. Brown Publishers.

Schumm, S. A. 1977. *The fluvial system.* New York: John Wiley & Sons.

Thornbury, W. D. 1969. *Principles of geomorphology.* 2d ed. New York: John Wiley & Sons.

Tuttle, S. D. 1980. *Landforms and landscapes.* 3d ed. Dubuque, Iowa: Wm. C. Brown Publishers.

Surprisingly, water underground is 30 to 40 times as plentiful as fresh water on the land surface (not including water stored as ice in glaciers). Ground water is a tremendously important resource. How it gets underground, how it is stored, how it moves while underground, how we look for it, and, perhaps most important of all, why we need to protect it are the main topics of this chapter.

Also discussed here are the relationships among springs, rivers, and ground water. Ground water forms distinctive geologic features, such as caves, sinkholes, and petrified wood. It also can appear as hot springs and geysers. Hot ground water can be used to generate power.

11

Ground Water

Ground water gushes from a large spring in the Redwall limestone, Grand Canyon National Park, Arizona.

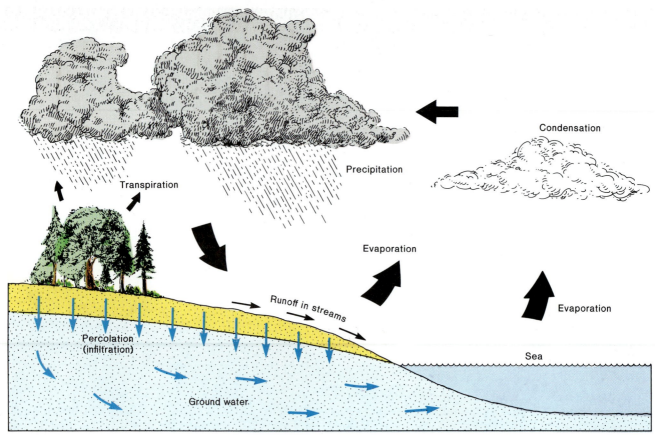

Figure 11.1
The hydrologic cycle. Water vapor evaporates from the land and sea, condenses to form clouds, and falls as precipitation (rain and snow). Water falling on land runs off over the surface as streams, or percolates into the ground to become ground water. It returns to the atmosphere again by evaporation and transpiration (the loss of water to the air by plants).

Many communities obtain the water they need from rivers, lakes, or reservoirs, sometimes using aqueducts or canals to bring water from distant surface sources. Another source of water lies directly beneath most towns. This resource is **ground water,** the water that lies beneath the ground surface, filling cracks, crevices, and the pore space between grains in clastic sedimentary rocks.

Ground water is a major economic resource, particularly in the dry western areas of the United States and Canada where surface water is scarce. Many towns and farms pump great quantities of ground water from drilled wells. Even cities next to large rivers may pump their water from the ground, however, for ground water is often less polluted and more economical to use than surface water.

The Hydrologic Cycle

The movement of water and water vapor from the sea to the atmosphere, to the land, and back to the sea and atmosphere again is called the **hydrologic cycle** (figure 11.1). When rain (or snow) falls on the land surface, more than half the water returns rather rapidly to the atmosphere by evaporation or by transpiration from plants. The remainder either flows over the land surface as runoff in streams or percolates down into the ground to become

ground water. The source of ground water is therefore rain and snow that falls onto the land. How much precipitation soaks into the ground is influenced by climate, land slope, soil and rock type, and vegetation. In general, perhaps 15% to 20% of the total precipitation ends up as ground water, but that varies locally and regionally.

Porosity and Permeability

Porosity, the percentage of a rock's volume that is openings, is a measurement of a rock's ability to hold water. Most rocks can hold some water. Some sedimentary rocks, such as sandstone, conglomerate, and many limestones, tend to have a high porosity and therefore can hold a considerable amount of water. A deposit of loose sand may have a porosity of 30% to 50%, but this may be reduced to 10% to 20% by compaction and cementation as the sand lithifies (table 11.1). A sandstone in which pores are nearly filled with cement and fine-grained matrix may have a porosity of 5% or less. Crystalline rocks, such as granite, schist, and some limestones, do not have pores but may hold water in joints and other openings.

Even though most rocks can hold some water, they vary a great deal in their ability to allow water to pass through them. **Permeability** refers to the capacity of a rock

TABLE 11.1

Porosity and Permeability of Sediments and Rocks

Sediment	Porosity (%)	Permeability
Gravel	25 to 40	excellent
Sand (clean)	30 to 50	good to excellent
Silt	35 to 50	moderate
Clay	35 to 80	poor (impermeable)
Glacial till	10 to 20	poor to moderate
Rock		
Conglomerate	10 to 30	moderate to excellent
Sandstone		
Well-sorted, little cement	20 to 30	good to very good
Average	10 to 20	moderate to good
Poorly sorted, well-cemented	0 to 10	poor to moderate
Shale	0 to 30	very poor to poor
Limestone, dolomite	0 to 20	poor to good
Cavernous limestone	up to 50	excellent
Crystalline rock		
Unfractured	0 to 5	very poor
Fractured	5 to 10	poor
Volcanic rocks	0 to 50	poor to excellent

to transmit a fluid such as water or petroleum. In other words, permeability measures the relative ease of water flow and indicates the interconnection of the openings in a rock. The distinction between porosity and permeability is important. A rock that holds much water is called *porous;* a rock that allows water to flow easily through it is described as *permeable.* Most sandstones and conglomerates are both porous and permeable. An *impermeable* rock is one that does not allow water to flow through it easily. Unjointed granite and schist are impermeable. Shale can have a substantial porosity but is usually impermeable because its pores are too small to permit easy passage of water.

The Water Table

Responding to the pull of gravity, water percolates down into the ground through the soil and through cracks and pores in the rock. Several kilometers down in the crust percolation stops. With increasing depth, sedimentary rock pores tend to be closed by increasing amounts of cement and by the plastic flow of grains. Sedimentary rock also typically lies over igneous and metamorphic crystalline basement rock, which usually contains no pores. The depth where downward percolation stops is generally about 5 kilometers below the surface, although in thick wedges of sedimentary rock, such as those found along the Gulf Coast of the United States, percolation may reach a depth of 10 kilometers or more.

The subsurface zone in which all rock openings are filled with water is called the **saturated zone** (figure 11.2). If a well were drilled downward into this zone, ground water would fill the lower part of the well. The water level inside the well marks the upper surface of the saturated zone; this surface is the **water table.**

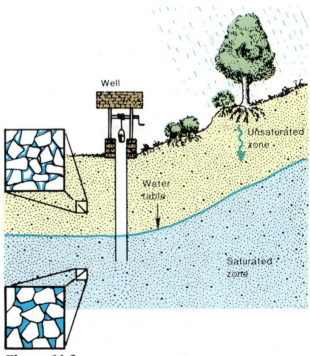

Figure 11.2
The water table marks the top of the saturated zone, in which water completely fills the rock pore space (inset figure). Above the water table is the unsaturated zone, in which rock openings contain both air and water.

Above the saturated zone, the rock openings are filled partly with air and partly with water. This zone is termed the **unsaturated zone** (or *zone of aeration*). The force of *surface tension* holds water droplets on the sides of rock openings. Surface tension also causes water to rise upward in small openings by *capillary action* if the openings are

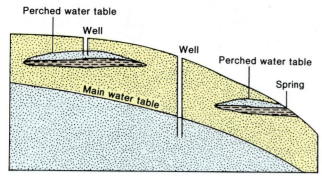

Figure 11.3
Perched water tables above lenses of impermeable shale within a large body of sandstone. Downward percolation of water is stopped by the impermeable shale.

not completely saturated with water. (If you dip the corner of a paper towel in water, water rises up the towel by capillary action.) In rocks, capillary action causes water to rise above the saturated zone in a *capillary fringe,* a transitional zone a few feet thick between the saturated rocks and the unsaturated zone above.

Plant roots generally obtain their water from the belt of soil moisture near the top of the unsaturated zone, where fine-grained clay minerals hold water and make it available for plant growth. Most plants "drown" if their roots are covered by water in the saturated zone; plants need both water and air in soil pores to survive. (The water-loving plants of swamps and marshes are an exception.)

A **perched water table** is the top of a body of ground water separated from the main water table beneath it by a zone that is not saturated (figure 11.3). It may form as ground water collects above a lens of relatively impermeable shale within a more permeable rock, such as sandstone. If the perched water table intersects the land surface, a line of springs can form along the upper contact of the shale lens.

The Movement of Ground Water

Compared to the rapid flow of water in surface streams, most ground water moves relatively slowly through underground rock. The water moves in response to differences in water pressure. This usually means that water within the upper part of the saturated zone moves down the slope of the water table (figure 11.4).

The circulation of ground water in the saturated zone is not confined to a shallow layer beneath the water table. Ground water may move hundreds of meters vertically downward before rising again to discharge as a spring or to seep into the beds of rivers and lakes at the surface (figure 11.4).

The *slope of the water table* strongly influences ground-water velocity. The steeper the slope of the water table, the faster ground water moves. Water-table slope is controlled largely by topography—the water table roughly parallels the land surface (particularly in humid regions). Even in highly permeable rock, ground water will not move if the water table is flat.

How fast ground water flows also depends on the *permeability* of the rock or other materials through which it passes. If rock pores are small and poorly connected, water moves slowly. When openings are large and well connected, the flow of water is more rapid. One way of measuring ground-water velocity is to introduce a tracer, such as a dye, into the water and then watch for the color to appear in a well or spring some distance away. Such experiments have shown that the velocity of ground water varies widely, averaging several centimeters to several meters a day. Nearly impermeable rocks may allow water to move only a few centimeters a *year,* but highly permeable materials, such as unconsolidated gravel or cavernous limestone, may permit flow rates well over 100 meters per day.

Springs and Rivers

A **spring** is a place where water flows naturally from rock onto the land surface (figure 11.5). Some springs discharge where the water table intersects the land surface, but they also occur where water flows out from caverns or along fractures, faults, or rock contacts that come to the surface (figures 11.6 and 11.7).

Climate determines the relationship between stream flow and the water table. In rainy regions most streams are **gaining streams;** that is, they are receiving water from the saturated zone (figure 11.8). The surface of these streams coincides with the water table. Water from the saturated zone flows into the stream through the stream bed and banks that lie below the water table. Because of the added ground water, the discharge of these streams increases downstream. Where the water table intersects the land surface over a broad area, ponds, lakes, and swamps are found.

In drier climates rivers tend to be **losing streams;** that is, they are losing water to the saturated zone (figure 11.8). The channels of losing streams lie above the water table. The water percolating into the ground beneath a losing stream may cause the water table below the stream to rise. This ground-water mound remains beneath the stream even when the stream bed is dry. In a desert the nearest source of water may be a short distance under a dry stream bed.

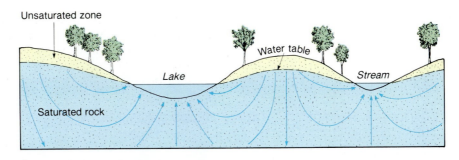

A

B

Figure 11.4
(*A*) Movement of ground water beneath a sloping water table in uniformly permeable rock. Near the surface the ground water tends to flow parallel to the sloping water table. (*B*) Ground water fills this abandoned limestone quarry in Indiana that is cut below the water table.

Labels in figure A: Unsaturated zone, Water table, Lake, Stream, Saturated rock

Figure 11.5
A large spring issuing from limestone, Jasper National Park, Alberta, Canada.

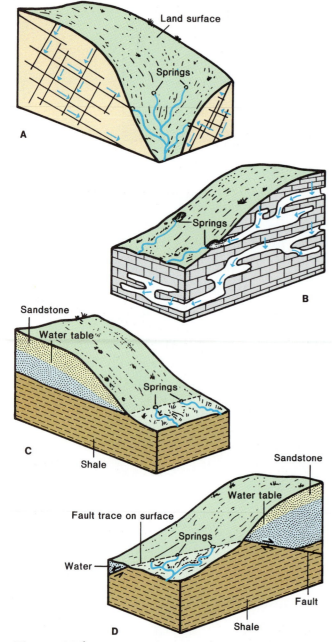

Figure 11.6
Springs can form in many ways. (*A*) Water moves along fractures in crystalline rock and forms springs where fractures intersect the land surface. (*B*) Water enters caves along joints in limestone and forms springs at the mouths of caves. (*C*) Springs form at the contact between a permeable rock such as sandstone and an underlying impermeable rock such as shale. (*D*) Springs can form along faults when permeable rock has been moved against impermeable rock. Fault motion shown by arrows.

Aquifers

An **aquifer** is a body of saturated rock or sediment through which water can move easily. Aquifers are both porous and permeable. A well must be drilled into an aquifer to reach an adequate supply of water (figure 11.9). Good aquifers include sandstone, conglomerate, well-jointed limestone, bodies of sand and gravel, and some fragmental

Figure 11.7
Numerous springs issuing from a cliff of fractured volcanic rock (note fractures upper right) beneath a double waterfall. **Burney Falls, California.**
Photo by B. Amundson.

or fractured volcanic rocks such as columnar basalt (table 11.1). These favorable geologic materials are sought in "prospecting" for ground water or looking for good sites to drill water wells.

Wells drilled in shale beds are not usually very successful because shale, although sometimes quite porous, is relatively impermeable (figure 11.9). Wet mud may have a porosity of 80% to 90% and even when compacted to form shale may still have a high porosity of 30%. Yet the extremely small size of the pores, together with the electrostatic attraction that clay minerals have for water molecules (see the chapter on "Weathering and Soil"), prevents water from moving through the shale into a well.

Because they are not very porous, crystalline rocks such as granite, gabbro, gneiss, schist, and some types of limestone are not good aquifers. The porosity of such rocks may be 1% or less. (Shale and crystalline rocks are sometimes called *aquitards* because they retard the flow of ground water.) Crystalline rocks that are highly fractured, however, may be porous and permeable enough to provide a dependable water supply to wells (figure 11.10).

Wells

A **well** is a deep hole, generally cylindrical, that is dug or drilled into the ground to penetrate an aquifer within the saturated zone (figures 11.9 and 11.10). Water that flows into the well from the saturated rock usually must be lifted or pumped to the surface. As figure 11.11 shows, a well dug in a valley usually has to go down a shorter distance

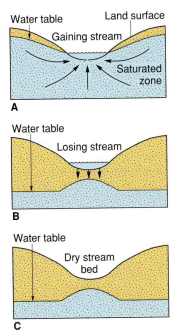

Figure 11.8
Gaining and losing streams. (*A*) Stream gaining water from saturated zone. (*B*) Stream losing water through stream bed to saturated zone. (*C*) Water table can be close to land surface beneath dry stream bed.

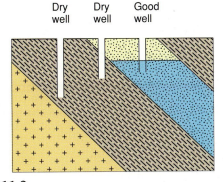

Figure 11.9
A well must hit an aquifer to obtain water. The saturated part of the sandstone is an aquifer, but the shale is not.

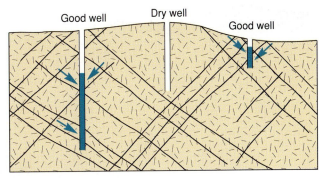

Figure 11.10
Wells can obtain some water from fractures in crystalline rock. Well must intersect fractures to obtain water.

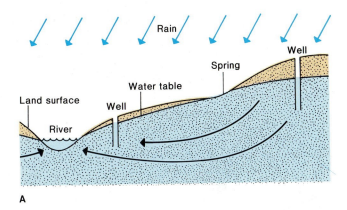

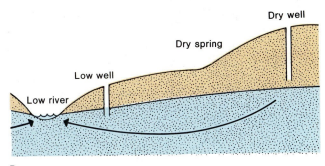

Figure 11.11
The water table rises in wet seasons and falls in dry seasons as water drains out of the saturated zone into rivers. (*A*) Wet season: water table and rivers high; springs and wells flow readily. (*B*) Dry season: water table and rivers low; some springs and wells dry up.

to hit water than a well dug on a hilltop. During dry seasons the water table falls as water flows out of the saturated zone into springs and rivers. Wells not deep enough to intersect the lowered water table go dry, but the rise of the water table during the next rainy season normally returns water to the dry wells. The addition of new water to the saturated zone is called **recharge.**

When water is pumped from a well, the water table is usually drawn down around the well into a depression shaped like an inverted cone known as a **cone of depression** (figure 11.12). This local lowering of the water table, called **drawdown,** tends to change the direction of flow of ground

water by changing the slope of the water table. In lightly used wells that are not pumped, drawdown does not occur and a cone of depression does not form. In a simple, rural well with a bucket lowered on the end of a rope, water cannot be extracted rapidly enough to significantly lower the water table. Wells of this type are shown in the early figures of this chapter.

Prospecting for Ground Water

Many wells are drilled or dug without any effort to locate a promising area of ground-water flow. Many of these wells are successful (especially if only small amounts of water are needed) because most rocks hold some water, which flows into wells that intersect the water table. However, if a large and dependable supply of water is needed—as for a city water system—specialists in ground-water geology may be called in to locate a promising well site. Geologists use many methods to locate aquifers. A detailed knowledge of the local rocks is necessary, so a geologist may map the rocks and use electrical, magnetic, and seismic surveys to study subsurface rocks to try to determine the depth and type of possible aquifers. Sometimes a small-diameter test well is drilled before the larger, more expensive supply well is sunk. The geologist looks for potentially high-producing aquifers rather than searching for water directly, which would be much more difficult. In some regions, however, the presence of certain plants is sometimes a useful guide to locating water, particularly the depth of the water table.

Some people search for water by water witching, or *dowsing,* with a divining rod (also sometimes used to search for metals or lost objects). Usually the dowser holds a forked stick horizontally in the hands while walking over an area. The stick is supposed to deflect or twist downward of its own accord when the dowser passes over water. This method has been tried for centuries, the only modification being that a twisted metal rod, often made of a coat hanger, now may be substituted for the stick. Carefully controlled tests conducted by workers in psychic research have shown that water witchers' "success" is equal to or less than pure chance, while geologists' results are superior both to witching and to chance. Records kept on thousands of wells in Australia in the early 1900s show that of wells that were not divined, more than 83% produced flows of 100 gallons per hour and 7.4% were failures, finding no water at all. Of wells that were divined by water witchers, only about 70% produced more than 100 gallons per hour, and 14.7% were dry. In the early part of this century the U.S. Geological Survey concluded that any future testing of the results of water witching would be a misuse of public funds.

Despite such findings, many people believe strongly in dowsing. Water witchers themselves devoutly believe they can find water, and in some regions of the United States almost no wells are drilled without a witcher's advice. Dowsers are helped in locating water by the fact that most rocks hold some water, and dowsers often have a long-standing knowledge of a particular region and its potential water resources. This is not to say that dowsers are deliberate frauds. Many are convinced that they perform a valuable public service, and some do not charge for their services. Scientists see no reason for dowsing to work and are skeptical about dowsers' "success." Geologists would almost unanimously urge you not to pay for a dowser's service if a fee were charged.

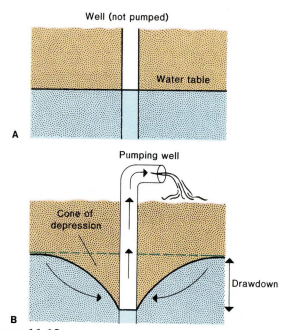

Figure 11.12
Pumping well lowers the water table into a cone of depression.

Artesian Aquifers and Wells

An **artesian aquifer** is an aquifer confined above, and sometimes below, by less permeable rocks. If recharge is to replenish such an aquifer, part of the aquifer must be exposed at (or be close to) the surface. Figure 11.13 shows an artesian system with a sandstone bed between two shale beds. The sandstone has been bent by tectonic action and subsequently exposed by erosion in a highland area where recharge can take place. Water from rain or snow gradually fills up the pores in the sandstone until it is completely saturated, almost up to the recharge area.

Most of the aquifer (figure 11.13) is *below* the level of recharge. Therefore, the water in distant parts of the aquifer is under pressure from the weight of the water above. When a well is drilled into the aquifer, this water pressure forces water up the well. A well is known as an **artesian well** if the water in it rises naturally above the level of the aquifer, without pumping. In an artesian well, the water may rise above the land surface, producing a flowing well that spouts continuously into the air unless it is capped. Flowing wells used to exist in South Dakota when the extensive artesian aquifer in that state was first

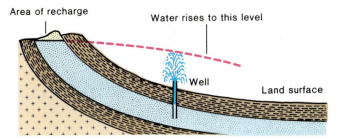

Figure 11.13
Artesian aquifer of sandstone between two shale beds. Water rises above the land surface when a well taps the artesian aquifer.

A

Figure 11.14
Artesian well spouts water above land surface in South Dakota, early 1900s. Heavy use of this aquifer has reduced water pressure so much that spouts do not occur today.
Photo by N. H. Darton, U.S. Geological Survey.

B

C

D

tapped (figure 11.14), but continued use has lowered the water pressure level below the ground surface in most parts of the state. Water still rises above the aquifer but does not reach the land surface.

Pollution of Ground Water

Ground water in its natural state tends to be relatively free of contaminants. Because it is a widely used source of drinking water, pollution of ground water can be a very serious problem.

Pesticides and *herbicides* (such as DDT and 2,4-D) applied to agricultural crops (figure 11.15A) can find their way into ground water when rain or irrigation water leaches the poisons downward into the soil. *Fertilizers* are also a concern. Nitrate, one of the most widely used fertilizers, is harmful in even small quantities in drinking water.

Figure 11.15
Some sources of ground-water pollution. (A) Pesticides. (B) Household garbage. (C) Animal waste. (D) Industrial toxic waste.
Photo A from Media Services.
Photo B by Frank M. Hanna. Photos C and D from USDA–Soil Conservation Service.

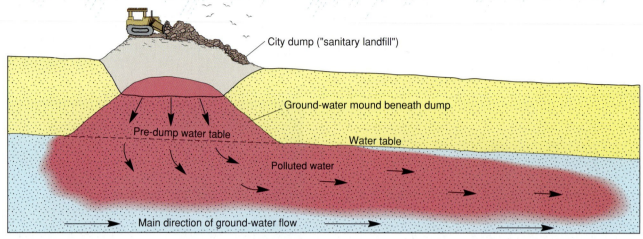

Rain

City dump ("sanitary landfill")

Ground-water mound beneath dump

Pre-dump water table

Water table

Polluted water

Main direction of ground-water flow

A Cross section

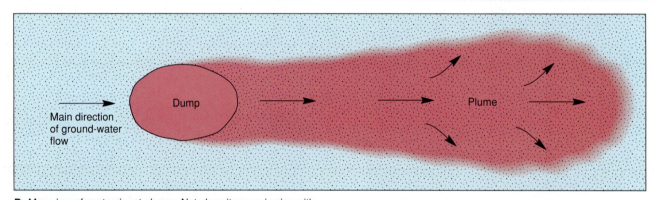

Main direction of ground-water flow

Dump

Plume

B Map view of contaminant plume. Note how it grows in size with distance from the pollution source.

Figure 11.16
Dump waste piled on the land surface creates a ground-water mound beneath it because the dump forms a hill, and because the waste material is more porous and permeable than the surrounding soil and rock. Rain leaches pollutants into the saturated zone. A plume of polluted water will spread out in the direction of ground-water flow.

Rain can also leach pollutants from city dumps into ground-water supplies (figure 11.15B). Consider for a moment some of the things you threw away last year. A partially empty aerosol can of ant poison? The can will rust through in the dump, releasing the poison into the ground and into the saturated zone below. A broken thermometer? The toxic mercury may eventually find its way to the ground-water supply. A half-used can of oven cleaner? The dried-out remains of a can of lead-base paint? *Heavy metals* such as mercury, lead, chromium,

copper, and cadmium, together with household chemicals and poisons, can all be concentrated in ground-water supplies beneath dumps (figure 11.16).

Liquid and solid wastes from septic tanks, sewage plants, and animal feedlots and slaughterhouses may contain *bacteria, viruses,* and *parasites* that can contaminate ground water (figure 11.15C). Liquid wastes from industries (figure 11.15D) and military bases can be highly toxic, containing high concentrations of heavy metals and compounds such as cyanide and *PCBs* (polychlorinated biphenyls), which are widely used in industry. A degreaser called *TCE* (trichloroethylene) has been increasingly found to pollute both surface and underground water in numerous regions. Toxic liquid wastes are often held in surface ponds or pumped down deep disposal wells. If the ponds leak, ground water can become polluted. Deep wells are relatively safe for liquid waste disposal if they are deep enough, but contamination of drinking water supplies and even surface water has resulted in some localities from improper design of the disposal wells.

Acid mine drainage from coal and metal mines can contaminate both surface and ground water. It is usually caused by sulfuric acid formed by the oxidation of sulfur in pyrite and other sulfide minerals when they are exposed to air by mining activity. Fish and plants are often killed by the acid waters draining from long-abandoned mines.

Radioactive waste is both an existing and a very serious potential source of ground-water pollution. The shallow burial of low-level solid and liquid radioactive wastes from the nuclear power industry has caused contamination of ground water, particularly as liquid waste containers leak into the saturated zone and as the seasonal rise and fall of the water table at some sites periodically covers the waste with ground water. The search for a permanent disposal site for solid, high-level radioactive waste (now stored temporarily on the surface) is a major national concern for the United States. The permanent site will be deep underground and must be isolated from ground-water circulation for thousands of years. Salt beds, shale, glassy tuffs, and crystalline rock deep beneath the surface have all been studied, as well as arid regions where the water table is hundreds of meters below the land surface. The likely site for disposal of high-level waste, primarily spent fuel from nuclear reactors, is Yucca Mountain, Nevada, 110 miles (180 km) northwest of Las Vegas. The site would be deep underground in volcanic tuff well above the current (or predicted future) water table, and in a region of very low rainfall. The U.S. Congress essentially chose the site in late 1988 by eliminating the funding for the study of all alternative sites, but the final decision regarding the safety of Yucca Mountain will not be made until about 1995 after much additional study. Even if the site is deemed safe, it could not open before the year 2010.

Not all ground-water pollutants form plumes within the saturated zone as shown in figure 11.16. *Gasoline*, which leaks from gas station storage tanks at thousands of U.S. locations, is less dense than water, and floats upon the water table (figure 11.17). Some liquids are heavier than water and sink to the bottom of the saturated zone, perhaps traveling in unpredicted directions upon the surface of an impermeable layer (figure 11.17). Determining the extent and flow direction of ground-water pollution is a lengthy process requiring the drilling of tens, or even hundreds, of costly wells for each pollution site.

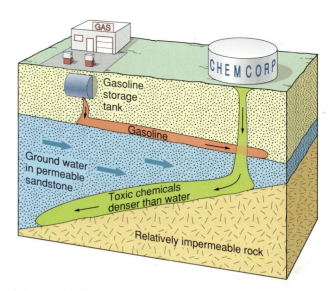

Figure 11.17
Not all pollutants move within the saturated zone as shown in figure 11.16. Gasoline floats on water; many dense chemicals move along impermeable rock surfaces below the saturated zone.

Not all sources of ground-water pollution are man-made. Naturally occurring *minerals within rock and soil* may contain elements such as arsenic, selenium, mercury, and other toxic metals. Circulating ground water can leach these elements out of the minerals and raise their concentrations to harmful levels within the water. Not all spring water is safe to drink. Like a "bad waterhole" depicted in a Western movie, some springs contain such high levels of toxic elements that the water can sicken or kill humans and animals that drink it. Many desert springs contain such high concentrations of sodium chloride or other salts that their water is undrinkable.

Soil and rock filter some contaminants out of ground water. This filtering ability depends on the permeability and mineral composition of rock and soil. Under ideal conditions, human sewage can be purified by only 100 to 150 feet (30 to 45 meters) of travel through a sandy loam soil (a mixture of clay minerals, sand, and organic humus). The sewage is purified by filtration, ion absorption by clay minerals and humus, and decomposition by soil organisms (figure 11.18). On the other hand, extremely permeable rock, such as highly fractured granite or cavernous limestone, has little purifying effect on sewage. Ground water flows so rapidly through such rocks that it is not purified even after hundreds of meters of travel. Some pesticides and toxic chemicals are not purified by passage through rock and soil at all, even soil rich in humus and clay minerals.

Polluted ground water is extremely difficult to clean up. Networks of expensive wells are needed to pump contaminated water out of the ground and replace it with clean water. Because of the slow movement and large volume of ground water, the clean-up process can take years to complete.

BOX 11.2

Hard Water and Soapsuds

"Hard water" is water that contains relatively large amounts of dissolved calcium (often from the chemical weathering of calcite or dolomite) or magnesium (from the ferromagnesian minerals). Water taken from the ground-water supply or from a stream for home use may contain enough of these ions to prevent soap from lathering. Calcium ions in hard water form gray curds with soap. The curd continues to form until all the calcium ions are removed from the water and bound up in the curd. Only then will soap lather and clean laundry. Cleaning laundry in hard water therefore takes an excessively large amount of soap.

Hard water may also precipitate a scaly deposit inside teakettles and hot-water tanks and pipes (box figure 1). The entire hot-water piping system of a home in a hard-water area eventually can become so clogged that the pipes must be replaced.

"Soft water" may carry a substantial amount of ions in solution but not the ions that prevent soap from lathering. Water softeners in homes replace calcium ions with sodium ions, which do not affect lathering or cause scale. However, water containing a large amount of sodium ions, whether from a softener or

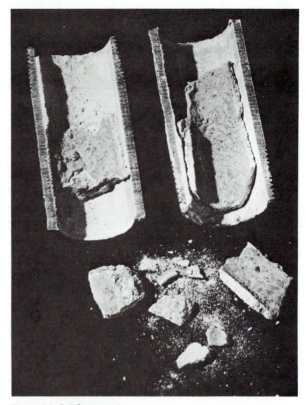

Box 11.2 Figure 1
Scale caused by hard water coats the inside of this hot-water pipe.
Photo by Hauser Water Conditioning, Inc.

from natural sources, may be harmful if used as drinking water by persons on a "salt-free" (low sodium) diet for health reasons.

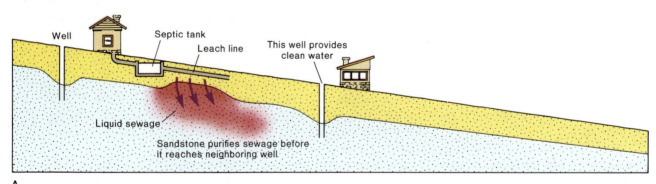

A

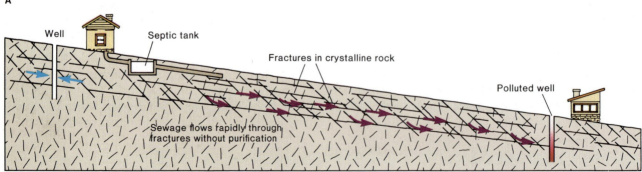

B

Figure 11.18
Rock type and distance control possible sewage contamination of neighboring wells. (A) As little as 30 meters of movement can effectively filter human sewage in sandstone and some other rocks and sediments. (B) If the rock has large open fractures, contamination can occur many hundreds of meters away.

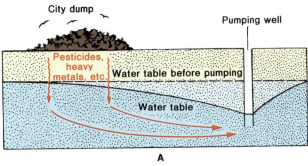

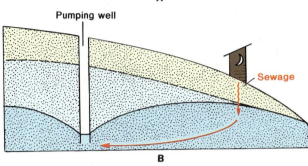

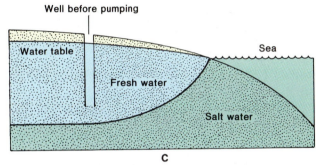

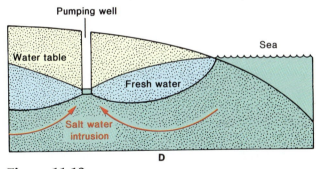

Figure 11.19
Ground-water pollution problems caused or aggravated by pumping wells. (*A*) Water table steepens near a dump, increasing the velocity of ground-water flow and drawing pollutants into a well. (*B*) Water-table slope is reversed by pumping, changing direction of the ground-water flow, and polluting the well. (*C*) Well near a coast (before pumping). Fresh water floats on salt water. (*D*) Well in *C* begins pumping, thinning the fresh water lens and drawing salt water into the well.

Pumping wells can cause or aggravate ground-water pollution (figure 11.19). Well drawdown can increase the slope of the water table locally, thus increasing the rate of ground-water flow and giving the water less time to be purified underground before it is used (figure 11.19*A*). Drawdown can even reverse the original slope of the water

table, perhaps contaminating wells that were pure before pumping began (figure 11.19*B*). Heavily pumped wells near a coast can be contaminated by *saltwater intrusion* (figures 11.19*C* and *D*).

Balancing Withdrawal and Recharge

A local supply of ground water will last indefinitely if it is withdrawn for use at a rate equal to or less than the rate of recharge to the aquifer. If ground water is withdrawn faster than it is being recharged, however, the supply is being reduced and will one day be gone.

Heavy use of ground water causes a regional water table to drop. In parts of western Texas and eastern New Mexico the pumping of ground water has caused the water table to drop 100 feet (30 meters) over the past few decades. The lowering of the water table means that wells must be deepened and more electricity must be used to pump the water to the surface. Moreover, as water is withdrawn, the ground surface may settle because the water no longer supports the rock and sediment. Mexico City has subsided more than 20 feet (6 meters) and portions of California's Central Valley 30 feet (9 meters) because of extraction of ground water. Such *subsidence* can crack building foundations, roads, and pipelines.

To avoid the problems of falling water tables and subsidence, many towns use *artificial recharge* to increase the natural rate of recharge. Natural floodwaters or industrial wastewaters are stored in infiltration ponds on the surface to increase the rate of water percolation into the ground. Reclaimed, clean water from sewage treatment plants is often used for this purpose. In some cases water is actively pumped down into the ground to replenish the ground-water supply. This is more expensive than filling surface ponds, but it reduces the amount of water lost through evaporation.

Effects of Ground-Water Action

Caves, Sinkholes, and Karst Topography
Caves (or **caverns**) are naturally formed underground chambers. Most caves develop when slightly acidic ground water dissolves limestone along joints and bedding planes, opening up cavern systems as the rock components are carried away in solution (figure 11.20). Natural ground water is often slightly acidic because of the solution of carbon dioxide (CO_2) from the atmosphere or from soil gases (as discussed in the chapter titled "Weathering and Soil").

Geologists disagree whether limestone caves form above, below, or at the water table. Some caves probably are formed by ground water circulating below the water table, as shown in figure 11.20. If the water table drops or the land is elevated above the water table, the cave may

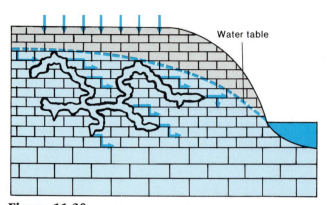

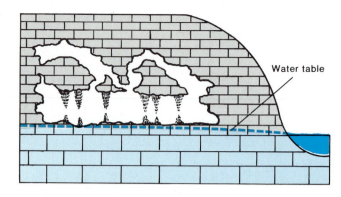

Figure 11.20
Solution of limestone to form caves. **(A)** Water moves along fractures and bedding planes in limestone, dissolving the limestone to form caves below the water table.

(B) Falling water table allows cave system, now greatly enlarged, to fill with air. Calcite precipitation forms stalactites and stalagmites above the water table.

begin to fill in again by calcite precipitation. Read the equation below from left to right for calcite solution, and from right to left for the calcite precipitation reaction (see also table 5.1).

$$H_2O + CO_2 + CaCO_3 \rightleftharpoons Ca^{++} + 2HCO_3^-$$

water / carbon dioxide / calcite in limestone / calcium ion / bicarbonate ion

$\longrightarrow$ development of caves (solution)

$\longleftarrow$ development of flowstone and dripstone (precipitation)

Ground water with a high concentration of calcium (Ca^{++}) and bicarbonate (HCO_3^-) ions may drip slowly from the ceiling of an air-filled cave. As a water drop hangs on the ceiling of the cave, some of the dissolved carbon dioxide (CO_2) may be lost into the cave's atmosphere. The CO_2 loss causes a small amount of calcite to precipitate out of the water onto the cave ceiling. When the water drop falls to the cave floor, the impact may cause more CO_2 loss, and another small amount of calcite may precipitate on the cave floor. A falling water drop, therefore, can precipitate small amounts of calcite on both the cave ceiling and the cave floor. The next drop precipitates a bit more calcite over the first deposits, with each subsequent drop adding still more calcite.

Deposits of calcite (and, rarely, other minerals) built up in caves by dripping water are called *dripstone*. Figure 11.21 shows some of the intriguing features formed by dripstone. **Stalactites** are icicle-like pendants of dripstone hanging from cave ceilings. They are generally slender and often are aligned along cracks in the ceiling, which act as conduits for ground water. **Stalagmites** are cone-shaped masses of dripstone formed on cave floors, generally directly below stalactites. Splashing water precipitates calcite over a large area on the cave floor, so stalagmites are usually thicker than the stalactites above them. As a stalactite grows downward and a stalagmite grows upward, they may eventually join to form a *column* (figure 11.22).

In parts of some caves, water flows in a thin film over the cave surfaces rather than dripping from the ceiling. Sheetlike or ribbonlike *flowstone* deposits develop from calcite that is precipitated by flowing water on cave walls and floors.

The floors of most caves are covered with sediment, much of which is *residual clay*, the fine-grained particles left behind as insoluble residue when a limestone containing clay dissolves. (Some limestone contains only about 50% calcite.) Other sediment, including most of the coarse-grained material found on cave floors, may be carried into the cave by streams, particularly when surface water drains into a cave system from openings on the land surface.

Solution of limestone underground can sometimes produce features that are visible on the surface. Extensive cavern systems can undermine a region so that roofs collapse and form depressions in the land surface above. **Sinkholes** are closed depressions found on land surfaces underlain by limestone (figure 11.23). They form either by the collapse of a cave roof or by solution as descending water enlarges a crack in limestone. Limestone regions in Florida, Missouri, Indiana, and Kentucky are heavily dotted with sinkholes. Sinkholes can also form in regions underlain by gypsum or rock salt, which are also soluble in water.

Figure 11.21
Stalactites and stalagmites in Lehman Caves, Nevada. A few have joined to form slender columns near the right center. Note the alignment of stalactites along a ceiling fracture.
Photo by O. M. Terry, Great Basin Natural History Association.

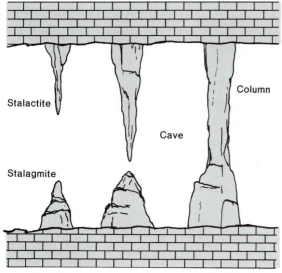

Figure 11.22
Stalactites grow downward from cave ceilings, and stalagmites build upward from cave floors. If they meet, they form columns.

An area with many sinkholes and with cave systems beneath the land surface is said to have **karst topography** (figure 11.24). Karst areas are characterized by a lack of surface streams, although one major river may flow at a level lower than the karst area.

Streams sometimes disappear down sinkholes to flow through caves beneath the surface. In this specialized instance, a true *underground stream* exists. Such streams are quite rare, however, as most ground water flows very slowly through pores and cracks in sediment or rock. You may hear people with wells describe the "underground stream" that their well penetrates, but this is almost never the case. Wells tap ground water in the rock pores and crevices, not underground streams. If a well did tap a true underground river in a karst region, the water would

A

B

Figure 11.23
(*A*) Trees grow in a sinkhole formed in limestone near Mammoth Cave, Kentucky. (*B*) A collapse sinkhole that formed suddenly in Winter Park, Florida, in 1981.
Photo *A* courtesy Ward's Natural Science. Photo *B* by Richard A. Davis, Jr.

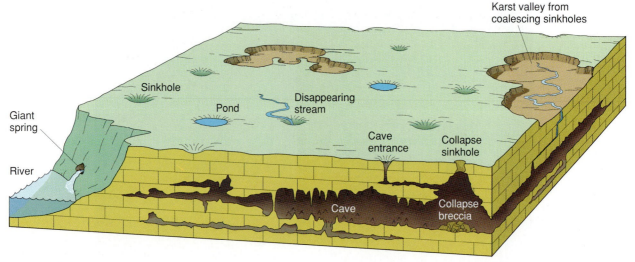

Figure 11.24
Karst topography is marked by underground caves and numerous surface sinkholes. A major river may cross the region, but small surface streams generally disappear down sinkholes.

Figure 11.25
Petrified logs in the Petrified Forest National Park, Arizona. Small amounts of iron and other elements color the silica in the logs.
Photo by E. D. McKee, U.S. Geological Survey.

Figure 11.26
Concretions that have weathered out of shale. Concretions contain more cement than the surrounding rock and therefore are very resistant to weathering.

probably be too polluted to drink, especially if it had washed down from the surface into a cavern without being filtered through soil and rock.

Other Effects

Ground water is important in the preservation of *fossils* such as **petrified wood,** a material that forms when the porous organic matter of buried wood is either filled in or replaced by inorganic silica carried in by ground water (figure 11.25). The result is a hard, permanent rock, often preserving the growth lines and other details of the wood. Calcite or silica carried by ground water can also replace the original material in marine shells and animal bones.

Sedimentary rock *cement,* usually silica or calcite, is carried into place by ground water. When a considerable amount of cementing material precipitates locally in a rock, a hard rounded mass called a **concretion** develops, often around an organic nucleus such as a leaf, tooth, or other fossil (figure 11.26).

Geodes are partly hollow, globe-shaped bodies found in some limestones and occasionally in other rocks. The outer shell is of silica, and well-formed crystals of quartz or calcite project inward toward a central cavity (figure 11.27). The origin of geodes is complex but clearly related to ground water. Crystals in geodes may have filled original cavities or have replaced fossils or other crystals.

In arid and semiarid climates, *alkali soil* may develop because of the precipitation of great quantities of sodium salts by evaporating ground water. Such soil is generally unfit for plant growth. Alkali soil generally forms at the ground surface in low-lying areas. (See also the chapter on "Weathering and Soil.")

Figure 11.27
Geodes. Concentric layers of silica are lined with well-formed crystals growing inward toward a central cavity. (Scale is in centimeters.)

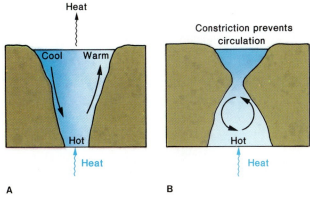

Figure 11.28
(*A*) Hot springs and (*B*) geysers. Many hot springs are funnel-shaped pools that allow water to rise and cool at the surface. Geysers may have constrictions that prevent such circulation, so that the water continues to rise in temperature. The figures are simplified.

Hot Water Underground

Hot springs are springs in which the water is warmer than human body temperature. Water can gain heat in two ways while it is underground. First, and more commonly, ground water may circulate near a magma chamber or a body of cooling igneous rock. In the United States most hot springs are found in the western states associated with relatively recent volcanism. The hot springs and pools of Yellowstone National Park in Wyoming are of this type.

Ground water can also gain heat if it circulates unusually deeply in the earth, perhaps along joints or a fault. As discussed in a previous chapter, the normal geothermal gradient (the increase in temperature with depth) is 25° C/kilometer (about 75° F/mile). Water circulating to a depth of 2 or 3 kilometers is warmed substantially above normal surface water temperature. The famous springs at Warm Springs, Georgia, have been warmed by deep circulation. Warm water, regardless of its origin, is lighter than cold water and readily rises to the surface.

A **geyser** is a type of hot spring that periodically erupts hot water and steam. The water is generally near boiling (100°C). Eruptions may be caused by a constriction in the underground "plumbing" of a geyser, which prevents the water from rising and cooling (figure 11.28). The

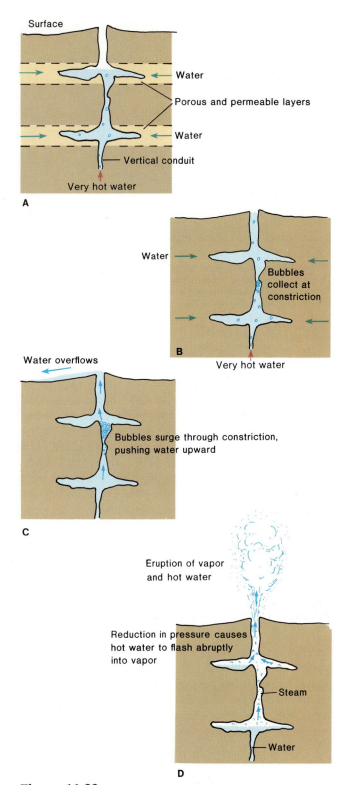

Figure 11.29
Eruption of a geyser. See text for explanation.
Modified from W. R. Keefer, 1971, *U.S. Geological Survey Bulletin,* 1347.

events thought to lead to a geyser eruption are illustrated in figure 11.29. Water gradually seeps into a partially emptied geyser chamber. Heat supplied from below slowly warms the water. Bubbles of water vapor and other gases begin to form as the temperature of the water rises. The bubbles may clog the constricted part of the chamber. The

Figure 11.30
Precipitation of calcite in the form of travertine terraces around a hot spring (Mammoth Hot Springs, Yellowstone National Park). Algae living in the hot water provide the color.
Photo by B. Amundson.

upward pressure of the bubbles pushes out some of the water above in a gentle surge, thus lowering the pressure on the water in the lower part of the chamber. This drop in pressure causes the chamber water, now very hot, to flash into vapor. The expanding vapor blasts upward out of the chamber, driving hot water with it and condensing into visible steam. The chamber, now nearly empty, begins to fill again and the cycle is repeated. The entire cycle may be quite regular, as it is in Yellowstone's Old Faithful Geyser, which averages about 65 minutes between eruptions (though it varies from about 30 to 95 minutes). Many geysers, however, erupt irregularly, some with weeks or months between eruptions.

As hot ground water comes to the surface and cools, it may precipitate some of its dissolved ions as minerals. *Travertine* is a deposit of *calcite* that often forms around hot springs (figure 11.30), while dissolved *silica* precipitates as *sinter* (called *geyserite* when deposited by a geyser, as shown in figure 11.31). The composition of the subsurface rocks generally determines which type of deposit forms, although sinter can indicate higher subsurface temperatures than travertine because silica is harder to dissolve than calcite. Both deposits can be stained by the pigments of algae living in the hot water. The algae can be used to estimate water temperature—they change from green to brown to orange to yellow as the temperature rises.

A *mudpot* is a special type of hot spring that contains thick, boiling mud. Mudpots are usually marked by a small amount of water and strongly sulfurous gases, which combine to form strongly acidic solutions. The mud probably results from intense chemical weathering of the surrounding rocks by these strong acids (figure 5.16).

Figure 11.31
Geyserite deposits around the vent of Castle Geyser, Yellowstone National Park.

Geothermal Energy

Electricity is generated in most power plants by directing a jet of steam against the curved blades of a turbine (hydroelectric plants at dams use falling water rather than steam). The steam jet causes the turbine to rotate, and the moving turbine spins a generator that produces electricity. Steam to turn the turbine can be produced in many ways. Coal, oil, or natural gas can be burned to boil water and create steam. These fuels are expensive, and add to consumers' electric bills. In addition the burning of coal and oil releases sulfur gases and other pollutants that cause acid rain. All three fossil fuels create carbon dioxide when they burn, perhaps contributing to global warming by the "greenhouse effect."

Nuclear power plants produce steam using the heat given off during nuclear reactions. Although the acid rain and carbon dioxide problems of the fossil fuels are eliminated, uranium is expensive and nuclear wastes present a very serious disposal problem, as you have seen earlier in the chapter. There is also a danger of radioactive contamination if a nuclear power plant is damaged, as at Three Mile Island near Harrisburg, Pennsylvania in 1979 and at Chernobyl near Kiev in the Soviet Union in 1986.

Figure 11.32
Geothermal power plant at The Geysers, California.
Underground steam, piped from wells to the power plant, is
being discharged from the cooling towers in the background.
Photo by M. Smith, U.S. Geological Survey.

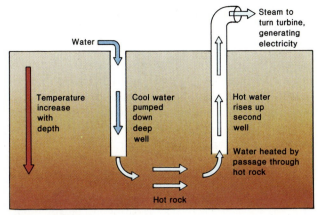

Figure 11.33
Proposed system to use the earth's heat to generate power.
Rock is artificially fractured between wells.

Geothermal energy is energy produced by harnessing
naturally occurring steam and hot water in areas that are
exceptionally hot underground. A well drilled into a geothermal area taps steam or superheated water that is piped
to a powerhouse to run a generator. A geothermal area
need not be marked by surface features such as hot springs
and geysers (just as oil fields are seldom marked by surface oil and tar seeps).

The advantages of geothermal energy production are
that no costly fuel in short supply need be burned to generate steam, air pollution problems are minimal, and the
radiation hazards of nuclear plants are eliminated. Geothermal energy is therefore a relatively clean and non-
polluting alternative to other steam-using power plants.

Tapping geothermal power has some environmental
effects. Hot water brought to the surface often contains
appreciable quantities of hydrogen sulfide gas and dissolved salts and metals, such as mercury and lead, which
can harm plant and fish life if allowed to enter surface
streams. (Geothermal workers often need protective
clothing and breathing devices.) The salts can also cor-
rode generating equipment. The land surface may subside
as water is withdrawn from underground. Pumping
wastewater back underground can help reduce these environmental effects, however, and prolong the life of the
field (geothermal fields can be depleted by pumping; they
are not inexhaustible sources of energy).

At present, geothermal power is generated in relatively few locations. The largest field in the world is at The
Geysers in California (figure 11.32), 80 miles northeast
of San Francisco. The Geysers field is capable of generating nearly 2,000 megawatts of electricity (enough to
serve 2 million people). Electrical production at The Geysers is declining however; the field may soon run out of

steam. Another commercial site currently generating geothermal power in the United States is a hot-water plant
near the Salton Sea in southern California. Other geothermal areas in the west are being intensively explored.
Large geothermal plants are also in operation in Italy,
Mexico, El Salvador, New Zealand, Japan, the Philippines, and Iceland.

Geothermal energy's greatest potential may be in
nonelectric uses such as space heating and water heating.
Iceland's capital, Reykjavik, with a population of more
than 100,000, is heated almost entirely by geothermal
energy. Geothermal heating systems also exist in Klamath
Falls, Oregon, and Boise, Idaho. Specialized uses such as
heating greenhouses, manufacturing paper, processing ore,
and preparing food are being investigated.

Will geothermal energy ever become a major source
of power generation? Areas containing substantial geothermal resources are not common, but their power potential is great. One method of geothermal energy
production may be widely used in the future in areas
without hot fluids underground. Figure 11.33 shows this
proposed system, called the hot dry rock system. Since
deep rock in all areas is hot, cold water could be pumped
down a deep well and allowed to heat through contact with
deep, hot rock. The hot water could then be raised up another well to run a surface power plant. The rock usually
must be fractured to allow the water to travel from one
well to another, and this creates added costs and problems
that are not yet solved. New technology may make geothermal energy a source of as much as 20% of our future
energy needs, about the same percentage now supplied by
natural gas, but other estimates of geothermal potential
are much more conservative.

Summary

Twenty percent or less of the water that falls on land percolates underground to become ground water. Ground water fills pores and joints in rock, creating a large reservoir of usable water in most regions.

Porous rocks can hold water. *Permeable* rocks permit water to move through them.

The *water table* is the top surface of the *saturated zone* and is overlain by the *unsaturated zone* (or *zone of aeration*).

Local variations in rock permeability may develop a *perched water table* above the main water table.

Ground-water velocity depends on rock permeability and the slope of the water table.

Gaining streams, springs, and lakes form where the water table intersects the land surface. *Losing streams* contribute to the ground water in dry regions.

An *aquifer* is porous and permeable and can supply water to wells. An *artesian aquifer* holds water under pressure, which can create flowing or spouting wells.

Ground water can be polluted by city dumps, agriculture, industry, or sewage disposal. Some pollutants can be filtered out by passage of the water through moderately permeable geologic materials.

A pumped well causes a *cone of depression* that in turn can cause or aggravate ground-water pollution. Near a coast, it can cause *saltwater intrusion.*

Artificial *recharge* can help create a balance between withdrawal and recharge of ground-water supplies.

Solution of limestone by ground water forms *caves, sinkholes,* and *karst topography.* Calcite precipitating out of ground water forms *stalactites* and *stalagmites* in caves.

Precipitation of material out of solution by ground water helps form petrified wood, other fossils, sedimentary rock cement, concretions, geodes, and alkali soils.

Geysers and *hot springs* occur in regions of hot ground water. *Geothermal energy* can be tapped and may generate much power in the future.

Terms to Remember

aquifer	losing stream
artesian aquifer	perched water table
artesian well	permeability
cave (cavern)	petrified wood
concretion	porosity
cone of depression	recharge
drawdown	saturated zone
gaining stream	sinkhole
geode	spring
geothermal energy	stalactite
geyser	stalagmite
ground water	unsaturated zone
hot spring	water table
hydrologic cycle	well
karst topography	

Questions for Review

1. What conditions are necessary for an artesian aquifer system?
2. What distinguishes a geyser from a hot spring? What conditions determine which of these will form?
3. What is karst topography? How does it form?
4. What chemical conditions are necessary for caves to develop in limestone? For stalactites to develop in a cave?
5. What causes a perched water table?
6. Describe several ways in which ground water can become polluted.
7. Discuss the difference between porosity and permeability.
8. What is the water table? Is it fixed in position?
9. Sketch four different origins for springs.
10. What controls the velocity of ground-water flow?
11. Name several geologic materials that make good aquifers. Define *aquifer.*
12. How does petrified wood form?
13. What happens to the water table near a pumped well?

Question for Thought

1. Describe any difference between the amounts of water that would percolate downward to the saturated zone beneath a flat meadow in northern New York and beneath a rocky hillside in southern Nevada. Discuss the factors that control the amount of percolation in each case.

Supplementary Readings

Baldwin, H. L., and C. I. McGuinness. 1963. *A primer on ground water.* Washington, D.C.: U.S. Geological Survey.

Bouwer, H. 1978. *Groundwater hydrology.* New York: McGraw-Hill.

Davis, S. N., and R. J. M. De Wiest. 1966. *Hydrogeology.* New York: John Wiley & Sons.

Fetter, C. W. 1988. *Applied hydrogeology.* 2d ed. Columbus, Ohio: Merrill.

Freeze, R. A., and J. A. Cherry. 1979. *Groundwater.* Englewood Cliffs, N.J.: Prentice-Hall.

Heath, R. C. 1983. *Basic ground-water hydrology.* Washington, D.C.: U.S. Geological Survey Water-Supply Paper 2220.

Heath, R. C. 1984. *Ground-water regions of the United States.* Washington, D.C.: U.S. Geological Survey Water-Supply Paper 2242.

Keefer, W. R. 1971. *The geologic story of Yellowstone National Park.* U.S. Geological Survey Bulletin 1347.

Leopold, L. B., and W. B. Langbein. 1960. *A primer on water.* Washington, D.C.: U.S. Geological Survey.

Moore, G. W., and G. Nicholas. 1964. *Speleology: The study of caves.* Boston: D. C. Heath.

Ritter, D. F. 1986. *Process geomorphology.* 2d ed. Dubuque, Iowa: Wm. C. Brown Publishers.

Swenson, H. A., and H. L. Baldwin. 1965. *A primer on water quality.* Washington, D.C.: U.S. Geological Survey.

Todd, D. K. 1980. *Ground water hydrology.* 2d ed. New York: John Wiley & Sons.

Waller, R. M. 1988. *Ground water and the rural homeowner.* Washington, D.C.: U.S. Geological Survey General Interest Publication.

Waltham, T. 1975. *Caves.* New York: Crown Publishers.

In the preceding three chapters you have seen how the surface of the land is shaped by mass wasting, running water, and, to some extent, ground water. Running water is regarded as the erosional agent most responsible for shaping the earth's land surface. Where glaciers exist, however, they are far more effective agents of erosion, transportation, and deposition. Geologic features characteristic of glaciation are distinctly different from the features formed by running water. Once recognized, they lead one to appreciate the great extent of glaciation during the recent geologic past (that age popularly known as the Ice Age).

Immense and extensive glaciers, covering as much as a third of the earth's land surface, had a profound effect on the landscape and on our present civilization. Moreover, worldwide climatic changes during the glacial ages distinctively altered landscapes in areas far from the glacial boundaries. For instance, water stored as ice in glaciers lowered the levels of the world's oceans, exposing more land than is presently exposed.

These episodes of glaciation took place within only the last few million years, ending about 10,000 years ago. Preserved in the rock record, however, is evidence of extensive older glaciations. The chapter on plate tectonics shows how the record of these ancient glaciations supports the theory of plate tectonics.

To understand how glacial erosion and deposition could have created the features regarded as evidence for past glaciation, you must first appreciate how present-day glaciers erode, transport, and deposit material. In other words, you must apply the principle of uniformitarianism to your study of glaciation.

CHAPTER

12

Glaciers and Glaciation

A valley glacier in Antarctica.
Photo by C. C. Plummer.

Figure 12.1
Yosemite Valley, Yosemite National Park, California.
Photo by C. C. Plummer.

A **glacier** is a large, long-lasting mass of ice formed on the land that moves because of gravity. It is formed as snow is compacted and recrystallized. A glacier can develop anyplace where over a period of years more snow accumulates than melts away or is otherwise lost.

There are two types of *glaciated* terrain on the earth's surface. **Alpine glaciation** is found in mountainous regions, while **continental glaciation** exists where a large part of a continent (thousands of square kilometers) is covered by glacial ice. In both cases the moving masses of ice profoundly and distinctively change the landscape.

The Theory of Glacial Ages

In the early 1800s the hypothesis of past extensive continental glaciation of Europe was proposed. Among the many people who regarded the hypothesis as outrageous was the Swiss naturalist Louis Agassiz. But, after studying the evidence in Switzerland, he changed his mind. In 1837, he published a discourse that eventually led to wide acceptance of the theory. Later, Agassiz traveled widely throughout Europe and North America promoting and extending the theory. Agassiz and his colleagues had observed the characteristic erosional and depositional features of present glaciers in the Alps and had compared these with similar features found in northern Europe and the British Isles, well beyond the farthest extent of the Alpine glaciers. Based on these observations, they proposed that very large glaciers had covered most of Europe while a colder climate prevailed during the past. Agassiz later came to North America and worked with American geologists who had found similar indications of large-scale past glaciation on this continent.

As more evidence accumulated, the hypothesis became accepted as a theory that today is seldom questioned. The **theory of glacial ages** states that at times in the past, colder climates prevailed during which much more of the land surface of the earth was glaciated than at present.

Because the last episode of glaciation was at its peak only about 18,000 years ago, its record has remained largely undestroyed by subsequent erosion and so provides abundant evidence to support the theory. This most recent glacial episode was the last of several glacial ages that alternated with periods of warmer climate (only slightly warmer than today's climate) around the world.

The glacial ages are not just a scientific curiosity. Our lives and environment today have been profoundly influenced by their effects. For example, much of the fertile soil of the northern Great Plains of the United States developed on the loose debris transported and dumped by glaciers that moved southward from Canada. The spectacularly scenic areas in many North American national parks owe much of their beauty to glacial action. Yosemite Valley in California might have been another nondescript valley if glaciers had not carved it into its present shape (figure 12.1).

Before we can understand how a continental glacier was responsible for much of the soil of the Midwest, or how a glacier confined to a valley could carve a Yosemite, we must learn something about present-day glaciers.

BOX 12.1

Glaciers as a Water Resource

Few persons think of glaciers as frozen reservoirs supplying water for irrigation, hydroelectric power, recreation, and industrial and domestic use. In the state of Washington, however, streamflow from the approximately 800 glaciers there amounts to about 470 billion gallons of water during a summer, according to the U.S. Geological Survey. More water is stored in glacier ice in Washington than in all of the state's lakes, reservoirs, and rivers.

One important aspect of glacier-derived water is that it is available when needed most. Snow accumulates on glaciers during the wet winter months. During the winter, streams at lower elevations, where rain falls rather than snow, are full and provide plenty of water. During the summer, however, the climate in the Pacific Northwest is hotter and drier. Because of

this, demand for water increases, especially for irrigation of crops. Streams that were fed by rainwater may have dried up. Yet, in the heat of summer, the period of peak demand, snow and ice on glaciers are melting, and streams draining glaciers are at their highest level.

Paradoxically, the greater the snowfall on a glacier during a winter, the smaller the amount of meltwater during the summer. A larger blanket of white snow reflects the sun's radiation more effectively than the darker, bare glacier ice, which absorbs more of the heat of the sun. Experiments have shown that melting can be greatly increased by darkening the snow surface, for instance, by sprinkling coal dust on it. Similarly, the melting of a glacier can be slowed artificially by covering it with highly reflective material. Such means of controlling glacial meltwater have been proposed to benefit power generating stations or to provide additional irrigation.

These ideas are appealing from a short-sighted point of view. However, the long-term effect of tampering with a glacier's natural regime can adversely affect the overall environment. It is conceivable, for example, that we could melt a glacier out of existence. Because most of the United States' glaciers are in national parks or wilderness areas, glaciers have not been tampered with to control the output of meltwater.

Glaciers—Where They Are, How They Form and Move

Distribution of Glaciers

Glaciers are found both in polar regions, where there is little melting during the summer, and in temperate climates that have heavy snowfall during the winter months. Certain climatic conditions are necessary for glaciation.

The coastal mountains of Washington have more glaciers than any state other than Alaska, even though the climate of Washington is considerably warmer than that of the Rocky Mountain states. Glaciers can flourish in these coastal mountains because Washington has very high precipitation during the winter months, and more snow accumulates in higher elevations than melts away during the summer. Glaciers are common even near the equator in the very high mountains of South America because of the low temperatures at high altitudes.

Glaciation is most extensive in polar regions, where little melting takes place at any time of the year. At present about one-tenth of the land surface of the earth is covered by glaciers (compared with about one-third during the peak of the glacial ages). About 85% of the present-day glacier ice is on the Antarctic continent, covering an area larger than the combined areas of western Europe and the United States; 10% is in Greenland. All the remaining glaciers of the world amount to only about 5% of the earth's freshwater ice. This means that Antarctica is in fact storing most of the earth's fresh water in the form of ice.

Some people have suggested that ice from the Antarctic could be brought to areas of dry climate to alleviate water shortages. While nobody has suggested melting the entire Antarctic ice sheet, it is worth noting that if all that continent's ice were to melt, the sea level around the world would rise about 60 meters (200 feet). This would flood the world's coastal cities and significantly decrease the land surface available for humans to live on.

Types of Glaciers

A simple criterion—whether or not a glacier is restricted to a valley—is the basis for classifying glaciers by form. A **valley glacier** is a glacier that is confined to a valley and flows from a higher to a lower elevation. Like streams, small valley glaciers may be tributaries to a larger trunk system. Valley glaciers are prevalent in areas of alpine glaciation. As might be expected, most glaciers in the United States and Canada, being in mountains, are of the valley type (figure 12.2).

In contrast, an **ice sheet** is a mass of ice that is not restricted to a valley but covers a large area of land (over 50,000 square kilometers). Ice sheets are associated with continental glaciation. Only two ice sheets exist on the earth now, one in Greenland and one in Antarctica. A similar but smaller body is called an **ice cap.** Ice caps (and valley glaciers as well) are found in a few mountainous regions and on islands in the Arctic Ocean, off Canada, Russia, and Siberia. An ice cap or ice sheet flows downward and outward from a central high point, as figure 12.3 shows.

Figure 12.2
Valley glacier on Mount Logan, Canada's highest mountain.
Photo by C. C. Plummer.

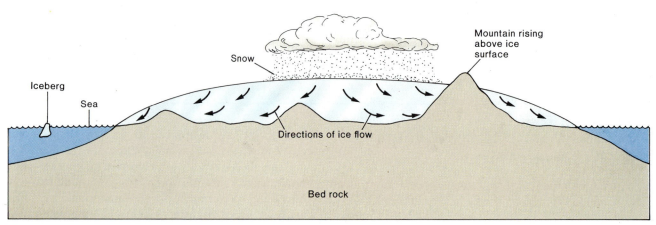

Figure 12.3
Diagrammatic cross section of an ice sheet. Vertical scale is highly exaggerated.

Formation and Growth of Glaciers

Snow converts to glacier ice in somewhat the same way that sediment turns into a sedimentary rock and then into metamorphic rock; figure 12.4 shows the process. A snowfall can be compared to sediment settling out of water. A new snowfall may be in the form of light "powder snow," which consists mostly of air trapped between many six-pointed snowflakes. In a short time the snowflakes settle by compaction under their own weight and much of the air between them is driven out. Meanwhile, the sharp points of the snowflakes are destroyed as flakes reconsolidate into granules—the "corn snow" of spring skiing. The compacted mass of granular snow, transitional between snow and glacier ice, is called **firn.** The granules are weakly "cemented" together by ice. Firn, then, is analogous to a sedimentary rock such as sandstone.

As fresh snow piles on top of firn, and as time passes, air is expelled, reducing the remaining pore space and forcing the granules into a tighter pattern. Individual grains, with little melting, recrystallize to fill the voids and become "welded" together, developing a mosaic-like texture similar to that of quartzite, a metamorphic rock.

Under the influence of gravity, glacier ice moves downward and is eventually **wasted,** or lost. For glaciers in all but the coldest parts of the world, wastage is due mostly to melting, although some ice evaporates directly into the atmosphere. If a moving glacier reaches a body of water, blocks of ice break off (or *calve*) and float free as **icebergs** (figure 12.5). In the Antarctic, wastage of the ice sheet is due largely to calving of icebergs and to direct evaporation. Only along the coast does melting take place, and there for only a few weeks of the year.

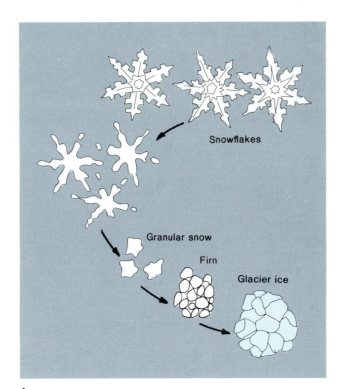

A

B

Figure 12.4
(*A*) Conversion of snow to glacier ice. (*B*) Thin slice of glacier
ice in polarized light.
Photo by C. C. Plummer.

Figure 12.5
An iceberg off the coast of Antarctica.
Photo by C. C. Plummer.

Glacial budgets   If, over a period of time, the amount
of snow a glacier gains is greater than the amount of ice
and water it loses, the glacier's budget is *positive* and it
expands. If the opposite occurs, the glacier decreases in
volume and is said to have a *negative budget*. Glaciers with
positive budgets push outward and downward at their
edges; they are called **advancing glaciers.** Those with negative budgets grow smaller and their edges melt back; they
are **receding glaciers.** Bear in mind that the glacial ice
moves downvalley, as shown in figure 12.6, whether the
glacier is advancing or receding. In a receding glacier,
however, the rate of flow of ice is insufficient to replace all
of the ice lost in the lower part of the glacier. If the amount
of snow retained by the glacier equals the amount of ice
and water lost, the glacier has a *balanced budget* and is
neither advancing nor receding.

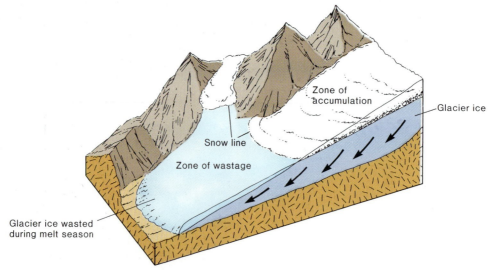

Figure 12.6
A valley glacier as it would appear at the end of a melt season. Below the snow line, glacier ice and snow have been lost during the melting season. In the zone of accumulation above that line, firn is added to the glacier from the previous winter snowfall.

A

B

Figure 12.7
South Cascade Glacier, Washington. If the photos were taken at the end of the melt season, the firn limit would be the boundary between white snow and darker glacier ice. These two photos were taken 23 years apart. (*A*) was taken in 1957; note that the glacier extended into the lake and that small

icebergs calved from it. (*B*) was taken in 1980; notice that the glacier has shrunk as well as retreated. During the period between photos, the glacier lost approximately 7.5 meters of water averaged over its surface or the equivalent of 18.7 million cubic meters of water for the entire glacier.
Photos by U.S. Geological Survey.

The upper part of a glacier, called the **zone of accumulation,** is the part of the glacier with a perennial snow cover (figure 12.6). The lower part is the **zone of wastage,** for there ice is lost, or wasted, by melting, evaporation, and calving.

The boundary between these two altitudinal zones of a glacier is an irregular line called the **snow line,** which marks the highest point at which the glacier's winter snow cover is lost during a melt season (figure 12.7).

The snow line may shift up or down from year to year, depending on whether there has been more accumulation or more wastage. Its location therefore indicates whether a glacier has a positive or negative budget. A snow line migrating upglacier over a period of years is a sign of a negative budget, whereas a snow line migrating downglacier indicates that the glacier has a positive budget. If a snow line remains essentially in the same place year after year, the glacier has a balanced budget.

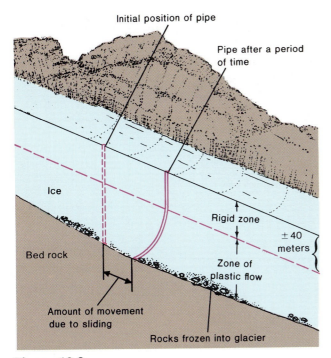

Figure 12.8
Movement of a glacier. Cross-sectional view.

The **terminus,** the lower edge of a glacier, moves farther downvalley when a valley glacier has a positive budget. In a receding glacier the terminus melts back upvalley. Because most glaciers move slowly, migration of the terminus tends to lag several years behind a change in the budget.

An ice sheet with a positive budget increases in volume, advancing its outer margins. If the expanding ice sheet extends into the ocean, an increasing number of icebergs break off and float away in the open sea.

Advancing or receding glaciers are significant and sensitive indicators of climatic change. However, an advancing glacier does not necessarily indicate that the climate is getting colder. It may mean that the climate is getting wetter, or that more precipitation is falling during the winter months, or that the summers are cloudier. It is estimated that a worldwide decrease in the mean annual temperature of only 4° or 5°C could bring about a new ice age.

Movement of Valley Glaciers

Valley glaciers move downslope under the influence of gravity and their own weight, the rate being very variable, ranging from less than a few millimeters a day to more than 15 meters a day. The upper part of a glacier—where the volume of ice is greater and slopes tend to be steeper—generally moves faster than ice farther down or on gentler slopes. In this way ice from the higher altitudes keeps replenishing ice lost in the zone of wastage. Glaciers in temperate climates—where the temperature of the glacier is at or near the melting point for ice—tend to move faster than those in colder regions—where the ice temperature stays well below freezing.

Velocity also varies within the glacier itself. The central portion of a valley glacier moves faster than the sides, and the surface moves faster than the base. How ice moves within a valley glacier has been demonstrated by studies in which holes are drilled through the glacier ice and flexible pipes inserted. Changes in the shape and position of the pipes are measured periodically. The results of these studies are shown diagrammatically in figure 12.8.

Note in the diagram that the base of the pipe has moved downglacier. This indicates **basal sliding,** which is the sliding of the glacier as a single body over the underlying rock. A thin film of meltwater that develops along the base from the pressure of the overlying glacier facilitates basal sliding. Think of a large bar of wet soap sliding down an inclined board.

Note that the lower portion of the pipe is bent in a downglacier direction. The pipe is bent more sharply near the base of the glacier, indicating that greater pressure from the overlying ice permits greater motion within the glacier ice near the bottom. These effects are caused by **plastic flow** of ice, movement that occurs within the glacier due to the plastic or "bendable" nature of the ice itself.

In the **rigid zone,** or upper part of the glacier, the pipe has been moved downglacier; however, it has remained unbent. In all glaciers the ice nearer the top apparently rides along passively on the plastically moving ice closer to the base.

Crevasses   Along its length, a valley glacier moves at different rates in response to changes in the steepness of the underlying rock. Typically a valley glacier rides over a series of rock steps. Where the glacier is passing over a steep part of the bed, it moves faster. The upper rigid zone of ice, however, cannot stretch to move as rapidly as the underlying plastic-flowing ice. Being brittle, the ice of the rigid zone is broken by the tensional forces. Open fissures, or **crevasses,** develop (figures 12.9 and 12.10). Theoretically, a crevasse should be no deeper than about 40 meters, the usual limit of the rigid zone.

After the ice has passed over a steep portion of its course, it slows down, and compressive forces close the crevasses. If a glacier descends a long, very steep slope without slowing down, the ice splits into pinnacles and blocks forming a chaotic jumble of crevasses called an *icefall* (figure 12.11).

Figure 12.9
Crevasses on a glacier, looking down from Mount Logan,
Yukon Territory, Canada.
Photo by C. C. Plummer.

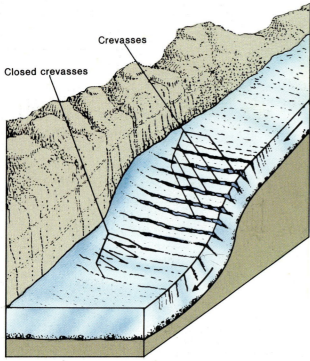

Figure 12.10
Crevasses along the course of a glacier.

Figure 12.11
An icefall in the Peruvian Andes.
Photo by C. C. Plummer.

BOX 12.2

Mars on a Glacier

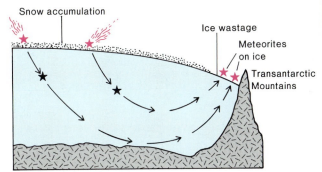

Box 12.2 Figure 1
Diagram showing the way in which meteorites are concentrated in a narrow zone of wastage along the Transantarctic Mountains. Two meteorites are shown as well as the paths they would have taken from the time they hit the ice sheet until they reached the zone of wastage. The vertical scale is greatly exaggerated.
Source: *Antarctic Journal of the United States.*

Meteorites are extraterrestrial rocks—fragments of material from space that have managed to penetrate the earth's atmosphere and land on the earth's surface (see chapter 4). They are of interest not only to astronomers but to geologists, for they may give us clues to what the earth's interior is like (see the chapter on the earth's interior) because many of the meteorites are thought to represent fragments of destroyed minor planets. Meteorites are rarely found; for one reason, they usually do not look very different from the earth's rocks with which they are mixed.

In recent years, a bonanza of meteorites has been found on the Antarctic ice sheet. Over a thousand meteorites have been collected from one small area where the ice sheet terminates against the Transantarctic Mountains. The reason for this heavy concentration is that meteorites landing on the surface of the ice over a vast area have been incorporated into the glacier and transported to where wastage takes place. The process is illustrated in box figure 1.

Two of the meteorites are especially intriguing. One almost certainly is a rock from the moon. There is considerable evidence that the other one came from Mars. Its chemistry and physical properties match what we would expect of a Martian rock. But how could a rock escape Mars and travel to Earth? Scientists suggest that a meteorite hit Mars with such force that fragments of that planet were launched into space.

If the meteorite is indeed a Martian rock, think of how much more economical it is to collect samples of the planet from a glacier rather than sending a space mission to Mars.

Movement of Ice Sheets

An ice sheet or ice cap moves like a valley glacier except that it moves downward and outward from a central high area toward the edges of the glacier (as shown in figure 12.3).

Glaciological studies in Antarctica have determined how an ice sheet grows and moves. Nearly all the Antarctic ice sheet is a zone of accumulation, because so little melting takes place and because occasional snowfalls nourish its high central part. Most of the ice sheet lies over interior lowlands, but it also completely buries several mountain ranges. At the South Pole (figure 12.12)—neither the thickest part nor the center of that vast ice sheet—the ice is 2,700 meters thick.

Some of the movement of the Antarctic ice sheet is due to sliding along its base, but most is by plastic flow. This is because the rigid zone is quite thin compared with the great thickness of most of the ice sheet.

Figure 12.12
The South Pole. Actually, the true South Pole is several kilometers from here. The moving ice sheet has carried the striped pole away from the site of the true South Pole, where the pole was erected in 1956.
Photo by C. C. Plummer.

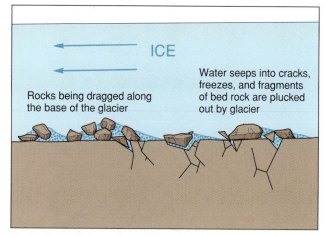

Figure 12.13
Rock fragments and bed rock being plucked out and abraded by movement of a glacier.

Figure 12.14
Striated and polished rock surface in South Australia. Unlike glacial striations commonly found in North America, this was caused by late Paleozoic glaciation.
Photo by C. C. Plummer.

Figure 12.15
An avalanche on Mount Logan, Yukon Territory, Canada.
Photo by C. C. Plummer.

Glacial Erosion

Wherever basal sliding takes place, the rock beneath the glacier is abraded and modified. As meltwater works into cracks in bed rock and refreezes, pieces of the rock are broken loose and frozen into the base of the moving glacier. While being dragged along by the moving ice, the rock within the glacier grinds away at the rock beneath (figure 12.13). The thicker the glacier, the more pressure on the rocks and the more effective the grinding and crushing.

Small pebbles and good-sized boulders that are dragged along are **faceted,** that is, given a flat surface by erosion. The bed rock, as well as the ice-carried rocks, is *polished* by the grinding. Sharp corners of rock fragments dragged along make grooves and **striations,** or scratches, in the rock, usually in the direction of ice movement (figure 12.14).

The grinding of rock across rock produces a powder of fine fragments called **rock flour.** Rock flour is composed largely of very fine (silt- and clay-sized) particles of unaltered minerals (pulverized from chemically unweathered bed rock). When *meltwater* washes rock flour from a glacier, the streams draining the glacier turn milky white.

Not all glacier-associated erosion is caused directly by glaciers. Frost wedging breaks up bedrock ridges and cliffs above a glacier, causing frequent rockfalls. Snow avalanches (figure 12.15) bring down loose rocks and other materials onto the glacier surface, where they ride on top of the ice. The important effects of meltwater washing out of glaciers are discussed later in this chapter.

Erosional Landscapes Associated with Alpine Glaciation

Mountain ranges and highland regions that have been subjected to alpine glaciation are noted for their rugged terrain and spectacular scenery. Figure 12.16 shows how a previously unglaciated mountainous region may have been altered by erosional effects of valley glaciers and frost wedging on exposed rock, producing some of the striking and unique features associated with mountain glaciation.

Glacial valleys Glacially carved valleys are usually easy to recognize. A **U-shaped valley** (in cross profile) is characteristic of glacial erosion, just as a V-shaped valley is characteristic of stream erosion.

Valley glaciers, which usually occupy valleys formerly carved by streams, tend to straighten the curves formed by running water. This is because the mass of ice of a glacier is too sluggish and inflexible to move easily around the curves. In the process of carving the sides of its valley, a glacier wears away or "truncates" any ridges that extended to the valley. **Truncated spurs** are the lower parts of ridges that have been carved into triangular facets by glacial erosion (figure 12.16B).

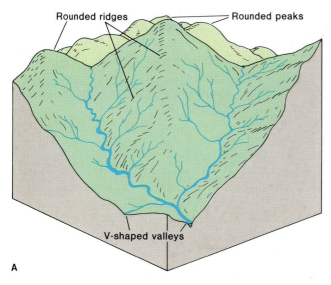

A

B

C

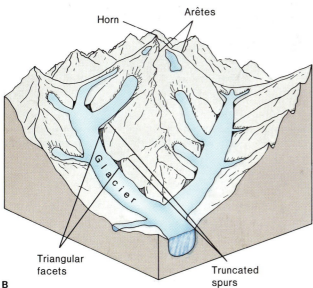

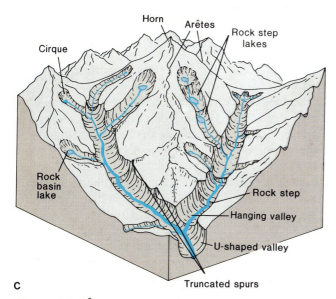

Figure 12.16
(*A*) A stream-carved mountain landscape before glaciation.
(*B*) The same area during glaciation. Ridges and peaks
become sharper due to frost wedging. (*C*) The same area after
glaciation.

Figure 12.17
A hanging valley in Yosemite National Park, California.
Photo by C. C. Plummer.

The thicker a glacier is, the more erosive force it exerts on the valley floor beneath, and the more bed rock is ground away. For this reason, a large trunk glacier erodes downward more rapidly and carves a deeper valley than do the smaller tributaries that join it. After the glaciers disappear, these tributaries remain as **hanging valleys** high above the main valley (figure 12.17).

Although a glacier tends to straighten and smooth the side walls of its valley, ice action often leaves the surface of the underlying bed rock carved into a series of steps. This is due to the variable resistance of bed rock to glacial erosion. Figure 12.18 shows what happens when a glacier abrades a relatively weak rock with closely spaced fractures. Water seeps into cracks in the bed rock, freezes there, and enlarges fractures or makes new ones. Rock frozen into the base of the glacier grinds and loosens more pieces. After the ice has melted back, a chain of **rock-basin lakes** may occupy the depressions carved out of the weaker rock. On stronger, less fractured rock, glacial erosion works in a similar way but is less effective because of the bed rock's resistance to grinding and crushing.

Cirques, horns, and arêtes A glacial **cirque** is a steep-sided, rounded hollow carved into a mountain at the head of a glacial valley (figure 12.19). In this unique, often spectacular, topographic feature, a large percentage of the snow accumulates that eventually converts to glacier ice and spills over the threshold as the valley glacier starts its downward course.

A cirque is not entirely carved by the glacier itself but is also shaped by the weathering and erosion of the rock walls above the surface of the ice. Frost wedging and avalanches break up the rock and steepen the slopes above the glacier. Broken rock tumbles onto the valley glacier and becomes part of its load, and some rock may fall into a crevasse, called a *bergschrund,* that develops where the glacier is pulling away from the cirque wall (figure 12.20).

Glaciers and Glaciation   **265**

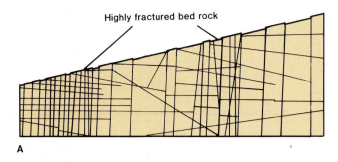

Highly fractured bed rock

A

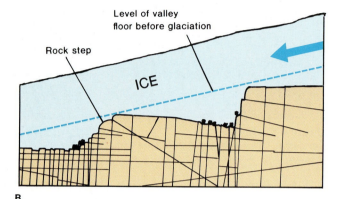

Level of valley
floor before glaciation

Rock step

ICE

B

Figure 12.18
Development of rock steps. (*A*) Valley floor before glaciation.
(*B*) During glaciation. (*C*) Rock steps and rock-basin lakes,
Sierra Nevada, California.
A and *B* after F. E. Matthes, 1930, U.S. Geological Survey. Photo by C. C.
Plummer.

C

Figure 12.19
A cirque occupied by a small glacier in the Canadian Rocky
Mountains.
Photo by C. C. Plummer.

The headward erosional processes that enlarge a
cirque also create the sharp peaks and ridges characteristic of glaciated mountain ranges. A **horn** is the sharp
peak that remains after cirques have cut back into a
mountain on several sides (figure 12.21).

Frost wedging works on the rock exposed above the
glacier, steepening and cutting back the side walls of the
valley. Sharp ridges called **arêtes** eventually separate adjacent glacially carved valleys (figure 12.22).

Erosional Landscapes Associated
with Continental Glaciation

The rock underneath an ice sheet is eroded in much the
same way as the rock beneath a valley glacier. However,
the weight and thickness of the ice sheet may produce more
pronounced effects. Grooved and striated bed rock is
common. Some grooves are actually channels several
meters deep and many kilometers long. The orientation of
grooves and striations indicates the direction of movement
of a former ice sheet.

An ice sheet may be thick enough to bury mountain
ranges, rounding off the ridges and summits and perhaps
streamlining them in the direction of ice movement. Much
of northeastern Canada, with its rounded mountains and
grooved and striated bedrock surface, shows the erosional
effects of ice sheets that formerly covered that part of
North America (figure 12.23).

A

Figure 12.20
(*A*) A bergschrund (partially filled with snow) in the Peruvian Andes. Glacier is pulling away to the right. (*B*) Cutaway view of a bergschrund in a cirque.
Photo by C. C. Plummer.

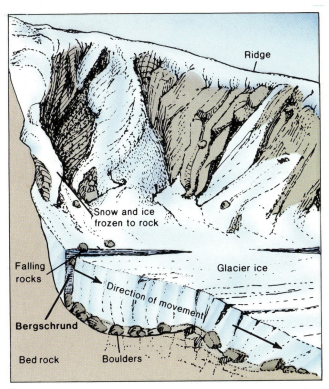

B

Figure 12.21
The Matterhorn in Switzerland. Engraving by Edward Whymper, who, in 1865, became the first person to climb the Matterhorn.

Figure 12.22
This arête in the Peruvian Andes extends downward from a horn and separates two glaciers, parts of which are visible in the lower part of the picture.
Photo by C. C. Plummer.

Figure 12.23
Air view of glacially scoured terrain in Canada. Ice moved from upper right to lower left.
National Air Photo Library of Canada.

Figure 12.24
Unsorted debris, including boulders, transported on top of and alongside a glacier in Peru. View is downglacier.
Photo by C. C. Plummer.

The rock fragments scraped and plucked from the underlying bed rock and carried along at the base of the ice make up most of the load carried by an ice sheet, but only part of a valley glacier's load. Much of a valley glacier's load comes from rocks broken from the valley walls.

Most of the rock fragments carried by glaciers are angular, as the pieces have not been tumbled around enough for the edges and corners to be rounded. The debris is unsorted, and clay-sized to boulder-sized particles are mixed together (figure 12.24). The unsorted and unlayered rock debris carried or deposited by a glacier is called **till** (figure 12.25).

Glaciers are capable of carrying virtually any size of rock fragment, even boulders as large as a house. An **erratic** is an ice-transported boulder that has not been derived from the nearby bed rock. If its bedrock source can be found, the erratic indicates the direction of movement of the glacier that carried it.

Moraines

A **moraine** is a body of till either carried on or within a glacier or left behind after the glacier has receded. Most of the loose material that falls from the steep cliffs along the course of a valley glacier accumulates along the edges of the ice. These ridgelike piles of till along the sides of a glacier are called **lateral moraines** (figures 12.26 and 12.27).

Figure 12.25
Till.
Photo by C. C. Plummer.

Lateral moraines Medial moraines End moraines

Terminus of glacier Recessional moraine Ground moraine Terminal moraine

Figure 12.26
Moraines associated with valley glaciers.

Figure 12.27
Lateral moraine in Switzerland. Moraine formed when the glacier to the left was higher. The Matterhorn is on the skyline.
Photo by B. Amundson.

Figure 12.28
Medial moraines on valley glaciers, Yukon Territory, Canada.
Ice is flowing toward viewer and to lower right.
Photo by C. C. Plummer.

Figure 12.29
End moraines in front of and lateral moraines on the sides of
Wallowa Lake in northeast Oregon. The lake is dammed up by
one of several recessional moraines.
Photo by D. A. Rahm, courtesy Rahm Memorial Collection, Western
Washington University.

Where tributary glaciers come together, the adjacent lateral moraines join and are carried downglacier as a single long ridge of till known as a **medial moraine.** In a large trunk glacier that has formed from many tributaries, the numerous medial moraines give the glacier the appearance from the air of a multilane highway (figures 12.26 and 12.28).

An actively flowing glacier brings debris to its terminus. If the terminus remains stationary for a few years, a distinct **end moraine,** or ridge of till, piles up along the front edge of the ice. Valley glaciers build end moraines that are crescent-shaped or sometimes horseshoe-shaped (figures 12.26 and 12.29). The end moraine of an ice sheet takes a similar lobate form but is much larger and more irregular than that of a valley glacier.

Geologists distinguish two special kinds of end moraines. A **terminal moraine** is the end moraine marking the farthest advance of a glacier. A **recessional moraine** is an end moraine built while the terminus of a receding glacier remains temporarily stationary. A single receding glacier can build several recessional moraines (figures 12.26 and 12.29).

As ice melts, rock debris that has been dragged along by a glacier is deposited to form a **ground moraine,** a fairly thin, extensive layer or blanket of till (figure 12.26). Very large areas that were once covered by an ice sheet now have the gently rolling surface characteristic of ground moraine deposits.

In some areas ground moraine has been reshaped into streamlined hills of till called **drumlins** (figure 12.30). A drumlin is shaped like an inverted spoon with the long axis parallel to the direction of ice movement (drumlins are sometimes called "whalebacks"). Drumlins were probably produced by an ice sheet overriding and reshaping a deposit of till left by an earlier glacial advance. Their gently sloping ends point in the direction the former glacial ice traveled.

Figure 12.30
Drumlins in Washington.
Photo by D. A. Rahm, courtesy Rahm Memorial Collection, Western
Washington University.

Outwash

In the zone of wastage, large quantities of meltwater usually run over, beneath, and away from the ice. The material deposited by the debris-laden meltwater is called **outwash.** Because it has the characteristic layering and sorting of stream-deposited sediment, outwash can be easily distinguished from the unsorted and unlayered deposits of till. Because outwash is fairly well sorted and the

Figure 12.31
An esker in northeastern Washington.
Photo by D. A. Rahm, courtesy Rahm Memorial Collection, Western Washington University.

Figure 12.32
A kettle (*foreground*) and outwash (*background* and *left*) from a glacier. Stagnant ice underlies much of the till. Yukon Territory. Canada.
Photo by C. C. Plummer.

Figure 12.33
Varves from a former glacial lake. Each pair of light and dark layers represents a year's deposition. Gradations on ruler are centimeters.
Photo by Brian Atwater, U.S. Geological Survey.

particles generally are not chemically weathered, it is an excellent source of aggregate for building roadways and for mixing with cement to make concrete.

An outwash feature of unusual shape associated with former ice sheets and some very large valley glaciers is an **esker,** a long, sinuous ridge of water-deposited sediment (figure 12.31). Eskers can be up to 10 meters in height and are formed of cross-bedded and well-sorted sediment. Evidently eskers are deposited in tunnels within or under glaciers, where meltwater loaded with sediment flows under and out of the ice.

As meltwater builds thick deposits of outwash alongside and in front of a retreating glacier, blocks of stagnant ice may be surrounded and buried by sediment. When the ice block finally melts (sometimes years later), a depression called a **kettle** forms (figure 12.32). Many of the small scenic lakes in the upper Middle West of the United States are kettle lakes.

The streams that drain glaciers tend to be very heavily loaded with sediment, particularly during the melt season. As they come off the glacial ice and spread out over the outwash deposits, the streams form a braided pattern (see the chapter on streams and landscapes).

The large amount of rock flour that these streams carry in suspension settles out in quieter waters—in bars, mud flats, shallow lake bottoms, and the like. In dry seasons or drought, the water may dry up and the rock flour deposits be picked up by the wind and carried long distances. Some of the best agricultural soil in the United States has been formed by rock flour that has been redeposited by wind. Such fine-grained, wind-blown deposits of dust are called **loess** (see chapter "Deserts and Wind Action").

Glacial Lakes and Varves

Lakes often form in depressions carved by glacial erosion but can also be the result of dams built by glacial deposition. Commonly a lake forms between a retreating glacier and an earlier end moraine.

In the still water of the lake, clay and silt settle on the bottom in two thin layers—one light-colored, one dark—that are characteristic of glacial lakes. Two layers of sediment representing one year's deposition in a lake are called a **varve** (figure 12.33). The light-colored layer consists of slightly coarser sediment (silt) deposited during the warmer part of the year when the glacier is melting and sediment is being dumped in the lake. The silt settles within a few weeks or so after being washed into the lake. The dark layer is finer sediment (clay)—material that sinks down more slowly during the winter after the lake surface freezes and the supply of fresh, coarser sediment stops because of lack of meltwater. The dark color is attributed to fine organic matter mixed with the clay.

Because each varve represents a year's deposit, varves are like tree rings and indicate how long a glacial lake was in existence.

Bodies of Ice

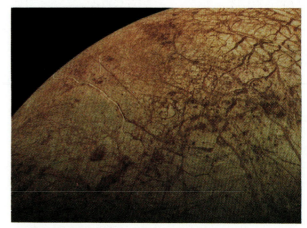

Astrogeology Box 12.1 Figure 2
Jupiter's satellite Europa.
NASA.

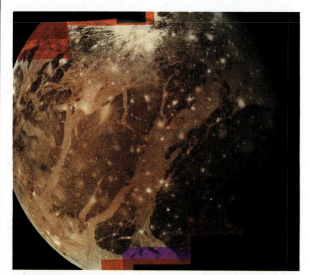

Astrogeology Box 12.1 Figure 1
Jupiter's satellite Ganymede.
NASA.

Some bodies in the solar system are mostly made of ice. These include satellites of the large outer planets and comets.

Geologists are particularly interested in the larger satellites of Jupiter, Saturn, Uranus, and Neptune because they are similar in size to the terrestrial planets and because they have solid surfaces. Several of these have been studied extensively by spacecraft.

Jupiter's satellites Ganymede, Europa, and Callisto are all believed to have rocky cores covered by shells of ice and possibly liquid water. Ganymede and Callisto may be at least half water ice. These three satellites have vastly different surface markings, however. Ganymede (box figure 1) has heavily cratered polygon-shaped darker regions separated by lighter zones made up of closely spaced parallel ridges and troughs. In some zones at least twenty ridges and troughs can be counted, each several kilometers wide and hundreds of kilometers long. This "grooved" terrain may have been caused by liquid water rising to the surface along cracks and then freezing. Europa (box figure 2) is crisscrossed by thousands of dark intersecting stripes tens of kilometers wide and thousands of kilometers long. The dark stripes may have originated as cracks that were then filled by some kind of material from below. Callisto's surface is heavily cratered but is otherwise uniform.

Saturn's Titan hides its surface beneath an atmosphere composed mainly of nitrogen but appears to have a polar ice cap and may be covered with an ocean of liquid nitrogen. Saturn's remaining satellites are covered with water ice, and some are composed mostly of water ice. Tethys has a 750-kilometer-long valley, perhaps caused by cracking as the ice of which the satellite is composed formed and expanded. Dione (box figure 3) and Rhea exhibit bright wispy areas, perhaps where large fractures have allowed water to escape and form frost. Rhea also has linear grooves and troughs.

Uranus's largest satellites seem to be predominantly made of water ice with some silicates and perhaps mixed with small amounts of ammonia and methane. All are cratered. Extensive systems of troughlike features can be seen on Miranda (box figure 4), Aerial, and Titania. Faulting perhaps occurred as

Effects of Past Glaciation

As the glacial theory gained general acceptance during the latter part of the nineteenth century, it became clear that much of northern Europe and the northern United States as well as most of Canada had been covered by great ice sheets during the so-called Ice Age. It also became evident that even areas not covered by ice had been affected because of the changes in climate and the redistribution of large amounts of water.

We now know that the last of the great North American ice sheets melted away from Canada less than 10,000 years ago. In many places, however, till from that ice sheet overlies older tills, deposited by earlier glaciations. The older till was deeply weathered during times of warmer climate between glacial episodes and is distinguishable from the newer till.

Astrogeology Box 12.1 Figure 3
Saturn's satellite Dione.
NASA.

Astrogeology Box 12.1 Figure 4
Uranus's satellite Miranda.
NASA.

the icy crust expanded in response to the freezing of ice in the satellites' interiors. Other fractures may have been caused by impacts. On Aerial some troughs show evidence of fluid flow along their floors, perhaps indicating filling by ice lavas from the satellite's interior. Oberon does not exhibit obvious fractures, but its craters appear to have been flooded with a dark material (carbon-rich ice?), and the surface of Umbriel appears to have been coated with similarly dark material.

Neptune has two large moons, Triton (box figure 5) and Nereid, and at least six others. These moons were closely studied for the first time in September of 1989 by the Voyager 2 Spacecraft. Triton is of unusual beauty, covered with slushy oceans and heavily cratered pink ice and criss-crossed by large fractures which have been interpreted as faults.

Comets are small objects, with nuclei no more than 1 or 2 kilometers in diameter, composed of dry ice (frozen carbon dioxide), frozen methane, frozen ammonia, and water ice, with small solid particles and dust imbedded in the ices. It was probably a small comet that landed in Siberia in 1908, rattling windows 450 kilometers away, knocking down trees 30 kilometers away, and causing much speculation about the event. It is believed that most comets orbit on the fringes of the solar system and only come within the central part of the solar system when their orbits are disturbed by the gravitational pulls of passing stars. Some comets orbit the Sun in highly elliptical orbits; at

Astrogeology Box 12.1 Figure 5
Neptune's moon, Triton (note faults).

their nearest approach to the Sun they sometimes come close enough that some of their ices evaporate and, along with the released dust particles, are swept out into a long tail. Some of these tails are more than 150 million kilometers long. Comets' tails are mostly empty space. In 1910 the Earth passed through the tail of Halley's Comet, and it could not even be detected.

The Glacial Ages

Geologists can reconstruct with considerable accuracy the last episode of extensive glaciation, which covered large parts of North America and Europe and was at its peak about 18,000 years ago. There has not been enough time for weathering and erosion to alter significantly the effects of glaciation. Less evidence is preserved for each successively older glacial episode, because (1) weathering and erosion occurred during warm interglacial periods and (2) later ice sheets and valley glaciers overrode and obliterated many of the features of earlier glaciation. However, from piecing together the evidence, geologists can see that earlier glaciers covered approximately the same region as the more recent ones.

BOX 12.3

Causes of Glacial Ages

What caused the glacial ages? This question has been asked by scientists since the theory of glacial ages was accepted over a century ago. Only in the last few years have climatologists felt they are beginning to provide acceptable answers.

The question is of more than academic importance. Understanding climate changes could provide the key to accurate, long-range weather predictions. Even minor climate changes affect crops and the problem of feeding humanity. If we knew the causes of fluctuations between glacial and interglacial episodes, we would probably know whether another glacial age is imminent or whether we are in a cooling or warming stage in the earth's history.

A number of theories and hypotheses have been proposed to account for glacial ages. Each explains some of our observations, but none alone satisfactorily accounts for all of the data. We shall review some of the more plausible ideas.

Variations in the earth's orbit and inclination to the sun The amount of heat from solar radiation received by any particular portion of the earth is related to the angle of the incoming sun's rays and, to a lesser degree, the distance to the sun. The dimensions of the earth's orbit around the sun change slightly over a period of thousands of years. The angle of the earth's poles relative to the plane of the earth's orbit about the sun also changes periodically. Recent analysis of data has added strong support to the theory (now known as the *Milankovitch theory*) that variations in orbital relationships and "wobble" of the earth's axis are largely responsible for the glacial and

interglacial episodes. Milutin Milankovitch, a Yugoslavian mathematician, calculated that the incoming solar radiation varies in cycles of 21,000, 41,000 and 100,000 years. Proof for the Milankovitch cycles came from oceanographic data. Cores of deep sea sediment have provided a fairly precise record of climatic variations over the past few hundred thousand years. The cycles of cooling and warming determined from the marine sediments are 23,000, 42,000 and 100,000 years, thus closely matching the times predicted by Milankovitch.

The orbital variation theory fails to explain the absence of glaciation over most of geologic time. One or more of the other mechanisms described below must have contributed to glacial ages.

Changes in the atmosphere A set of hypotheses seeks to explain glacial ages by changes in the ability of the atmosphere to filter solar radiation. Much of the solar energy reaching our planet is either reflected back out to space or absorbed by the atmosphere. If the composition of the atmosphere changes, the amount of the sun's heat reaching and retained by the earth's surface also changes.

One hypothesis regards carbon dioxide as responsible for major climate changes. According to this proposal, if there is an increase in the carbon dioxide content in the atmosphere, the earth warms up. This is because of a "greenhouse effect" in which solar energy penetrates the atmosphere and heat is retained or trapped at the earth's surface. Carbon dioxide in the atmosphere reduces the amount of heat that can radiate from earth back out into space. This results in a warming trend until, for some reason, the carbon dioxide in the atmosphere decreases. High carbon dioxide content would coincide with warm episodes between glacial ages, low carbon dioxide content with periods of glaciation.

A major problem in the carbon dioxide hypothesis is explaining a cyclic change in the concentration of carbon dioxide in the atmosphere. One suggestion has been that when vegetation is abundant, there is less carbon dioxide in the atmosphere. The cooling this causes results in an ice age, and much of the earth's vegetation dies, releasing carbon dioxide to the

Until a few years ago geologists thought the Pleistocene Epoch (see chapter 8) included all the glacial ages, but recent work indicates that worldwide climate changes necessary for continental glaciation probably began late in the Cenozoic Era, at least a million years before the Pleistocene. Antarctica has been glaciated for at least 20 million years, and continental glaciers elsewhere probably existed at least 3 million years ago.

The significance of this is that the earth has undergone episodic changes in climate during the last 2 to 3 million years. Actually, the climatic changes necessary for a glacial age to occur are not so great as one might imagine. During the height of a glacial age, the worldwide average of annual temperatures was probably only about 5°C cooler than at present. The intervening interglacial periods were probably a bit warmer worldwide than present-day average annual temperatures.

atmosphere to begin the warming trend. However, there is no evidence of profound changes of vegetation. An additional difficulty comes in explaining why the process is effective for such a small percentage of geologic time.

Another hypothesis involves volcanoes. Worldwide temperature dropped about 2°C following the major volcanic eruption of Krakatoa in Indonesia in 1883. The dust carried by high atmospheric winds reduced solar energy penetrating the atmosphere for a few years. Presumably a closely spaced series of large eruptions could cause enough of a temperature drop to trigger glaciation and begin a glacial age.

Changing of the positions of continents

Another hypothesis is that glacial ages occur when land masses move closer to polar regions. Plate tectonics provides the mechanism for motion of land masses. But the fluctuations in late Cenozoic climates are not explained by plate tectonics. All evidence for continental drift indicates that the positions of the continents have not changed significantly with respect to the poles during the late Cenozoic Era. Certainly the continents have not shuffled back and forth during the recent glacial and interglacial ages.

However, the earlier movement of continents from positions closer to the equator into more northerly latitudes might have placed the present northern continents in a position favorable to glaciation.

Changes in circulation of sea water   Our

present climates are very much affected by patterns of circulation of sea water. Land masses block the worldwide free circulation of ocean water, so some oceans are warmer than others.

According to one hypothesis, a glacial episode begins when Atlantic water circulates freely with Arctic Ocean water. At present the warmer waters of the Atlantic Ocean cannot mix freely with water in the Arctic Ocean. For this reason the surface of the Arctic Ocean is frozen for much of the year. Continental glaciation begins, according to this hypothesis, when warmer Atlantic water flows through a shallow channel between Greenland and Canada. This would keep the

Arctic Ocean from completely freezing over during the winter. The moisture picked up by winds blowing over the Arctic Ocean would precipitate heavy amounts of snow on the northern continental landmasses. Ice sheets would grow so long as ice-free seas supplied the moisture. But as glaciation continued, sea level would drop due to the loss of water to the ice sheets. Sea level would eventually drop below the floor of the shallow channel between Greenland and Canada, shutting off the supply of warm Atlantic waters to the Arctic Ocean. Because of this, the surface of the Arctic Ocean would freeze, reducing the moist air supplied to the ice sheets. The ice sheets would recede and disappear.

Some scientists question the adequacy of this mechanism to explain large-scale glaciation. Others point out that this hypothesis cannot explain the very warm interglacial climates. Also, snow and ice reflect much solar radiation. Therefore, a frozen Arctic Ocean coupled with glacial ice covering up to one-third of the world's land mass should tend to perpetuate a glacial age by reflecting more solar radiation into space.

Sliding of the Antarctic ice sheet over the

ocean   One intriguing although highly speculative hypothesis regards changes in the Antarctic ice sheet as responsible for glacial ages. A large segment of the ice sheet, lubricated by water along its base, slides as a mass onto the ocean surface. Before the large floating slab of ice can be broken up by wave action, the ice reflects a significant amount of solar heat back into space. This results in a worldwide cooling sufficient to trigger a glacial age.

There is some evidence that parts of the Antarctic ice sheet have slid rapidly and suddenly in the past. Whether the mechanism is capable of cooling the world's climate so much has yet to be demonstrated.

Summary   Scientists do not fully understand what causes glacial ages and intervening warm episodes. Only recently has the Milankovitch theory become widely accepted. This, at least, seems to explain what controls the cycles of climatic variation. But one or more of the other postulated mechanisms must also contribute significantly.

Direct Effects of Past Glaciation in North America

Moving ice abraded vast areas of northern and eastern Canada during the growth of the North American ice sheet (figure 12.34). Most of the soil was scraped off and bed rock was scoured. Many thousands of future lake basins were gouged out of the bed rock.

The directions of ice flow can be determined from the striations and grooves in the bed rock. The ice sheet moved outward from the general area now occupied by Hudson Bay, which is where the highest part of the North American ice sheet was located. The present generally barren surface of the Hudson Bay area contrasts markedly with the Great Plains surface of southern Canada and northern United States, where vast amounts of till were deposited. Why is this so? The ice sheet's erosive ability may have been greater in northern Canada, where it was thickest. On the other hand, the erodibility of the bed rock may

Figure 12.34
Extent of glaciation in North America during the Pleistocene.
Arrows show direction of ice movement.
After C. S. Denny, U.S. Geological Survey National Atlas of the United States.

have been a more significant factor. The scoured bed rock of northern Canada is igneous and metamorphic rock, resistant to erosion. The bed rock beneath most of the till in midwestern Canada and northern United States is easily eroded sedimentary rock.

Most of the till was deposited as ground moraine, which, along with outwash deposits (figure 12.35), has partially weathered to give excellent soil for agriculture. In many areas along the southern boundaries of land covered by ground moraines, broad and complex end moraines extend for many kilometers, indicating that the ice margin must have been virtually stationary for a long period of time. Numerous drumlins are preserved in areas such as New England and upstate New York. Kettle lakes dot the landscape in large sections of Wisconsin and Minnesota. New York's Long Island was built up by morainal debris, most of it probably scoured out of New England.

The Great Lakes are, at least in part, a legacy of continental glaciation. Former stream valleys were widened by the ice sheet into the present lake basins. End moraines border the Great Lakes, as shown in figure 12.36. Large regions of Manitoba, Saskatchewan, North Dakota, and Minnesota were covered by ice-dammed lakes (Lake Agassiz is the name for the largest of these). The former lake beds are now rich farmland.

Alpine glaciation was much more extensive throughout the world during the glacial ages than it is now. For example, small glaciers in the Rocky Mountains that now barely extend beyond their cirques were then valley glaciers 10, 50, or 100 kilometers in length. Yosemite Valley, which is no longer glaciated, was filled by a glacier about a kilometer thick. Its terminus was at an elevation of about 1,300 meters above sea level. Furthermore, cirques and other features typical of valley glaciers can be found in regions that at present have no glaciers, such as

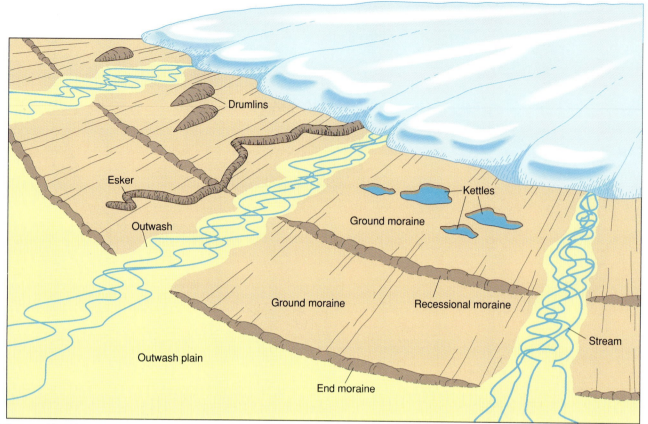

Figure 12.35
Depositional features in front of a receding ice sheet.

the northern Appalachians—notably in the White Mountains of New Hampshire. Apparently alpine glaciers were present in New England for some time after the main ice sheet retreated.

Indirect Effects of Past Glaciation

As the last continental ice sheet wasted away, what effects did the tremendous volume of meltwater have on American rivers? Rivers that now contain only a trickle of water were huge in the glacial ages. Other river courses were blocked by the ice sheet or clogged with morainal debris. Large dry stream channels have been found that were preglacial tributaries to the Mississippi and other river systems.

Pluvial lakes   During the glacial ages the climate in North America, even beyond the glaciated parts, was more humid than it is now. Most of the presently arid regions of the western United States had moderate rainfall, as traces or remnants of numerous lakes indicate. These **pluvial lakes** (formed in a period of abundant rainfall) once existed in Nevada, Utah, and eastern California (figure 12.36). Some may have been fed by meltwater from mountain glaciers, but most were simply the result of a wetter climate.

Great Salt Lake in Utah is but a small remnant of a much larger body of fresh water called Lake Bonneville, which was, at its maximum size, nearly as large as Lake

Michigan is today. (Geologists have mapped ancient beaches and wave-cut terraces that indicate Lake Bonneville's former shorelines.) As the climate became more arid, lake levels lowered, outlets were cut off, and the water became salty, eventually leaving behind the Bonneville salt flats and the present very saline Great Salt Lake.

Even Death Valley in California—now the driest place in the United States—was occupied by a deep lake during the Pleistocene. The salt flats that were left when this lake dried (see figure 6.24) include rare boron salts that were mined during the pioneer days of the American West.

Lowering of sea level   All of the water for the great glaciers had to come from somewhere. It is reasonable to assume that the water was "borrowed" from the oceans and that sea level worldwide was lower than it is today—at least 130 meters lower, according to scientific estimates.

What is the evidence for this? Stream channels have been charted in the present continental shelves, the gently inclined submerged edges of the continents (described in the chapter on the sea floor). These submerged channels are continuations of today's major rivers and could not have been eroded unless the continental shelf surfaces were above sea level during the glacial ages. Bones and teeth from now-extinct mammoths and mastodons have been

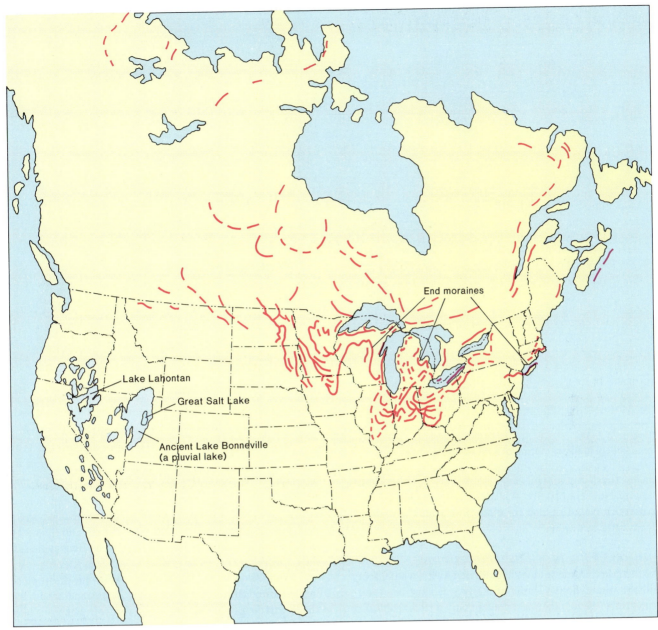

Figure 12.36
**End moraines in northern United States and southern Canada
(shown by dark lines) and pluvial lakes in the southwestern
United States.**
After C. S. Denny, U.S. Geological Survey, and the Geological Map of
North America, Geological Society of America, and The Geological Survey
of Canada.

dredged up from the Atlantic continental shelf, indicating
that these relatives of elephants roamed over what must
have been dry land at the time.

A **fiord** is a coastal inlet that is a drowned glacially
carved valley (figure 12.37). Fiords are common along the
mountainous coastlines of Alaska, British Columbia, New
Zealand, and Norway. They are evidence that valleys
eroded by past glaciers were later partly submerged by
the rising seas.

Crustal rebound The weight of an ice sheet several
thousand meters thick depresses the crust of the earth
much as the weight of a person depresses a mattress. A
land surface bearing the weight of a continental ice sheet
may be depressed several hundred meters.

Once the glacier is gone, the land begins to rebound
slowly to its previous height (see figures 17.13 and 17.14).
Uplifted and tilted shorelines along lakes are an indication of this process. The Great Lakes region is still rebounding as the crust slowly adjusts to the removal of the
last ice sheet.

Martian Ice Caps and Glaciers

The polar regions of Mars are covered with ice caps (box figure 1), which are only a few meters thick and composed mostly of frozen carbon dioxide (dry ice). During the summer on each hemisphere, the ice caps shrink markedly as the carbon dioxide vaporizes; however, a small cap remains. This small residual cap (400 kilometers in diameter) is probably composed of water ice.

Two distinctive types of terrain can be observed in the Martian polar regions. *Laminated terrain* is the name given to areas where series of alternating light and dark layers can be seen. The layers are essentially horizontal, and each is about 15 to 35 meters thick. As many as fifty layers have been counted in one location. The layers are thought to represent alternating beds of high dust content (loess deposits?) and high ice content, and their alternation may be due to some kind of climatic change. That the layers are stratified outwash from glaciers has also been suggested. Near the margins of the polar caps are large troughs and ridges that could be glacial valleys and moraines.

Underlying the laminated terrain is another terrain, which is characterized by small pits. The pits of this *etch-pitted terrain* may be due to wind erosion, and, if so, they would be deflation basins (blowouts). They may also be glacial kettles or collapse pits caused by the loss of undergound ice.

Astrogeology Box 12.2 Figure 1
Mars, with Olympus Mons, Valles Marineris & Argyre Basin visible.

Figure 12.37
Fiord in Alaska.
Photo by D. A. Rahm, courtesy Rahm Memorial Collection, Western Washington University.

Evidence for Older Glaciation

Throughout most of geologic time, the climate has been warmer and more uniform than it is today. We think that the late Cenozoic Era is unusual because of the periodic fluctuations of climate and the widespread glaciations. However, glacial ages are not restricted to the late Cenozoic.

The evidence of older glaciation comes from rocks called tillites. A **tillite** is lithified till. Unsorted rock particles, including angular, striated, and faceted boulders, have been consolidated into a sedimentary rock. In some places, the tillite layers overlie surfaces of older rock that have been polished and striated. Tillites of the late Paleozoic and tillites representing a minor part of the late Precambrian crop out in parts of the southern continents. (The striated surface in Australia, shown in figure 12.14, is overlain in places by late Paleozoic tillite.)

Late Paleozoic tillites in the southern continents (South Africa, Australia, Antarctica, South America) have been used as evidence that these land masses were once joined (see the chapter on plate tectonics). Directions of striations indicate that an ice sheet flowed onto South America from what is now the South Atlantic Ocean. Because an ice sheet can build up only on land, it is reasonable to conclude that the former ice sheet was centered on the ancient supercontinent before it broke up into the present continents.

Summary

A *glacier* is a large, long-lasting mass of ice that forms on land and moves by its own weight. A glacier can form wherever more snow accumulates than is lost. *Ice sheets* and *valley glaciers* are the two most important types of glaciers. Glaciers move downward from where the most snow accumulates toward where the most ice is wasted.

A glacier moves both by basal sliding and by internal flow. The upper portion of a glacier tends to stay rigid and is carried along by the ice moving beneath it.

Glaciers advance and recede in response to changes in the climate. A glacier recedes if it has a *negative budget* and advances if it has a *positive budget*. A glacier's budget for the year can be determined by noting the relative position of the *snow line*.

Snow recrystallizes into firn, which eventually becomes converted to glacier ice. Glacier ice is lost (or wasted) by melting, by breaking off as icebergs, and by direct evaporation of the ice into the air.

A glacier erodes by the grinding action of the rock it carries. The grinding produces *rock flour* and faceted and polished rock fragments. Bed rock over which a glacier moves is generally polished and grooved.

A mountain area showing the erosional effects of alpine glaciation possesses relatively straight valleys with U-shaped cross-profiles. The floor of a glacial valley usually has a *cirque* at its head and descends as a series of rock steps. Small *rock-basin lakes* are commonly found along the steps and in cirques. *Hanging valleys* indicate that smaller tributaries joined the main glacier. A *horn* is a peak between several cirques. *Arêtes* usually separate adjacent glacial valleys.

A glacier deposits unsorted rock debris or *till*. Till contrasts sharply with the sorted and layered deposits of glacial *outwash*. Till forms various types of *moraines*.

Fine silt and clay may settle as *varves* in a lake in front of a glacier, each pair of layers representing a year's accumulation.

Multiple till deposits and other glacial features indicate several major episodes of glaciation during the late Cenozoic Era. During each of these episodes, large ice sheets covered most of northern Europe and northern North America, and glaciation in mountain areas of the world was much more extensive than at present. At the peak of glaciation about a third of the earth's land surface was glaciated (in contrast to the 10% of the land surface presently under glaciers). Warmer climates prevailed during interglacial episodes.

The glacial ages also affected regions never covered by ice. Because of the wetter climate in the past, large lakes formed in now-arid regions of the United States. Sea level was considerably lower.

Glacial ages also occurred in the more distant geologic past. Evidence for late Paleozoic and late Precambrian glaciation is found in the southern continents.

Terms to Remember

advancing glacier	medial moraine
alpine glaciation	moraine
arête	outwash
basal sliding	plastic flow
cirque	pluvial lake
continental glaciation	receding glacier
crevasse	recessional moraine
drumlin	rigid zone
end moraine	rock-basin lake
erratic	rock flour
esker	snow line
faceted	striations
fiord	terminal moraine
firn	terminus
glacier	theory of glacial ages
ground moraine	till
hanging valley	tillite
horn	truncated spur
iceberg	U-shaped valley
ice cap	valley glacier
ice sheet	varve
kettle	wasted (or wastage)
lateral moraine	zone of accumulation
loess	zone of wastage

Questions for Review

1. How do features caused by stream erosion differ from features caused by glacial erosion?
2. How does material deposited by glaciers differ from material deposited by streams?
3. Why is the North Pole not glaciated?
4. How would you distinguish a sample of glacial rock flour from nonglacial silt and clay deposited in a sea?
5. How do arêtes, cirques, and horns form?
6. What differences are there in the way ice moves in a valley glacier and the way it moves in an ice sheet?
7. How does the glacial budget control the migration of the firn limit?
8. How do recessional moraines differ from terminal moraines?

Questions for Thought

1. How might a warming trend cause increased glaciation?
2. How do or do not the Pleistocene glacial ages fit in with the principle of uniformitarianism?
3. Is ice within a glacier a mineral? Is a glacier a rock?
4. Could tillites be deposited by any agent other than an ice sheet?
5. What is the likelihood of a future glacial age? What effect might human activity have on causing or preventing a glacial age?

Supplementary Readings

Dyson, J. L. 1962. *The world of ice.* New York: Alfred A. Knopf.

Flint, R. F. 1971. *Glacial and quaternary geology.* New York: John Wiley & Sons.

Imbrie, J. I. and K. P. Imbrie. 1986. *Ice Ages: Solving the mystery.* Boston: Harvard University Press.

Matsch, C. L. 1976. *North America and the great Ice Age.* New York: McGraw-Hill.

Matthes, F. E. 1930. *The geologic history of Yosemite Valley.* U.S. Geological Survey Professional Paper 160.

Post, A., and E. R. LaChappelle. 1971. *Glacier ice.* Seattle: University of Washington Press.

Ritter, D. F. 1986. *Process geomorphology.* 2d ed. Dubuque, Iowa: Wm. C. Brown Publishers.

Sharp, R. P. 1989. *Living Ice: Understanding glaciers and glaciation.* New York: Cambridge University Press.

Tuttle, S. D. 1980. *Landforms and landscapes.* 3d ed. Dubuque, Iowa: Wm. C. Brown Publishers.

U.S. Geological Survey. 1973. *Glaciers, a water resource.* U.S. Geological Survey Information Pamphlet.

Deserts have a distinctive appearance because a dry climate controls erosive and depositional processes and the rates at which they operate. Although it seldom rains in the desert, running water is the dominant agent of land sculpture. Flash floods cause most desert erosion and deposition, even though they are rare events.

In chapters 9 through 12 you have seen how the land is sculptured by mass wasting, streams, ground water, and glaciers. Here we discuss the fifth agent of erosion and deposition: wind. Deserts and wind action are discussed together because of the wind's particular intensity in dry regions. But wind erosion and deposition can be very significant in many other climates as well.

13

Deserts and Wind Action

Mudcracks and butte near Canesville, Utah.
© Doug Sherman.

Figure 13.1
Typical appearance of the southwestern United States desert. Widely spaced plants have adapted to less than 10 inches (25 centimeters) of rain per year.

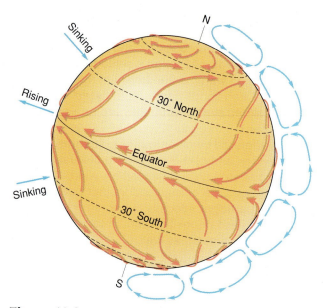

Figure 13.2
Global air circulation. Red arrows show surface winds. Blue arrows show vertical circulation of air. Air sinks at 30° N and 30° S latitude.

The word *desert* may evoke for you a region of shifting sand dunes. Although moviemakers usually film sand dunes to represent deserts, only small portions of most deserts are covered with dunes. Actually a **desert** is any region with low rainfall. A region is usually classified as a desert if it has a dry or *arid climate* with less than 10 inches (25 centimeters) of rain per year. Few plants can tolerate low rainfall, so most deserts have a barren look.

Some specialized types of plants, however, grow well in desert climates despite the dryness. These plants are generally salt-tolerant; and they have extensive root systems to conserve water, so they often are widely spaced (figure 13.1). The leaves are usually very small, minimizing water loss by transpiration; they may even drop off the plants between rainstorms. During much of the year many desert plants look like dead, dry sticks. When rain does fall on the desert, the plants become green, and many will bloom.

Distribution of Deserts

The location of most deserts is related to descending air. The global pattern of air circulation is shown in simplified form in figure 13.2. The equator receives the sun's heat more directly than the rest of the earth. Air warms and rises at the equator, then moves both northward and southward to sink near 30° North latitude and 30° South latitude. The world's largest deserts lie beneath these two belts of sinking air (figure 13.3).

Air sinking down through the atmosphere is compressed by the weight of the air above it. As air compresses, it warms up; and as it warms, it is able to hold more water vapor. Evaporation of water from the land

surface into the warm, dry air is so great under belts of sinking air that moisture seldom falls back to earth in the form of rain. The two belts at 30° North and South latitude characteristically have clear skies, much sunshine, little rain, and high evaporation.

In contrast to the belts at 30°, the equator is marked by rising air masses. The rising air expands and cools as it rises. In cooling, the air loses its moisture, causing cloudy skies and heavy precipitation. Thus a belt of high rainfall at the equator separates the two major belts of deserts.

Not all deserts lie on the 30° latitude belts. Some of the world's deserts are the result of the **rain shadow** effect of mountain ranges (figure 13.4). As moist air is forced up to pass over a mountain range, it expands and cools, losing moisture as it rises. The dry air coming down the other side of the mountain compresses and warms, bringing high evaporation with little or no rainfall to the downwind side of the range. This dry region downwind of mountains is the *rain shadow zone.* Parts of the southwestern United States desert in Nevada and northern Arizona are largely the result of the rain shadow effect of the Sierra Nevada range in eastern California.

Great distance from the ocean is another factor that can create deserts, since most rainfall comes from water evaporated from the sea. The dry climate of the large arid regions in China, well north of 30° North latitude, is due to their location in a continental interior and to the rain shadow effect of mountains such as the Himalayas.

Deserts also tend to develop on tropical coasts next to *cold ocean currents.* Cold currents run along the western edges of continents, cooling the air above them. The cold marine air warms up as it moves over land, causing high evaporation and little rain on the coasts. This effect is particularly pronounced on the Pacific coast of South America and the Atlantic coast of Africa.

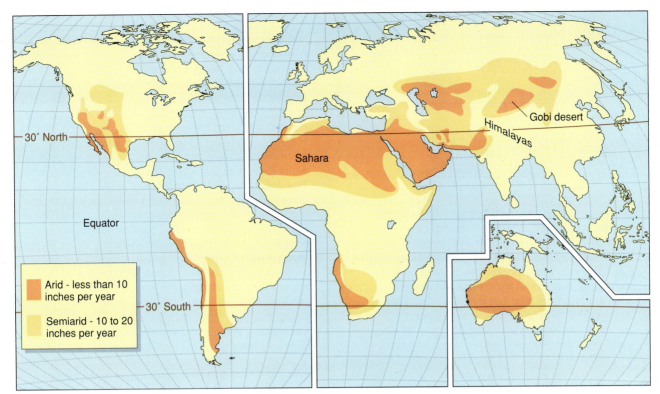

Figure 13.3
World distribution of deserts. Most deserts lie in two bands near 30° N and 30° S.
From map by U.S. Department of Agriculture.

Arid - less than 10 inches per year

Semiarid - 10 to 20 inches per year

Figure 13.4
Rain shadow causes deserts on the downwind side of mountain ranges. Prevailing winds are from left to right.

Some Characteristics of Deserts

Because of their low rainfall, deserts have characteristic drainage and topography, which differ from those of humid regions. Desert streams usually flow intermittently. Water runs over the surface after storms, but during most of the year stream beds are dry. As a result, most deserts *lack through-flowing streams.* The Colorado River in the southwestern United States and the Nile River in Egypt are notable exceptions. Both are fed by heavy rainfall in distant mountains. The runoff is great enough to sustain stream flow across dry regions with high evaporation.

Many desert regions have *internal drainage;* the streams drain toward landlocked basins instead of toward the sea. The surface of an enclosed basin acts as a local base level. Because each basin is generally filled to a different level than the neighboring basins, desert erosion may be controlled by many different *local base levels.* As a basin fills with sediment, its surface rises, leading to a *rising base level,* which is a rare situation in humid (wet) regions.

The limited rainfall that does occur in deserts often comes from violent thunderstorms, with a high volume of rain falling in a very short time. Desert thunderstorms may dump more than 5 inches (13 centimeters) of rain in one hour. Such a large amount of rain cannot soak readily into the sun-baked ground, so the water runs rapidly over the land surface, particularly where vegetation is sparse. This high runoff can create sudden local floods of high discharge and short duration called **flash floods.** Flash floods are more common in arid regions than in humid regions. They can turn normally dry stream beds into raging torrents for a short time after a thunderstorm. Because soil particles are not held in place by plant roots, these occasional floods can effectively erode the land surface in a desert region. As a result, desert streams normally are very heavily laden with sediment. Flash floods can easily erode enough sediment to become *mudflows* (see chapter on "Mass Wasting").

BOX 13.1

Expanding Deserts

Many geologists and geographers use a two-part definition of *desert*. A desert must not only have less than 10 inches (25 centimeters) of rain per year, but it also must be so devoid of vegetation that few people can live there. Many dry regions have supported marginally successful agriculture and moderate human populations in the past but are being degraded into barren deserts today by overgrazing and overpopulation. The expansion of barren deserts into once-populated regions is called *desertification.*

Limited numbers of people can exist in dry regions through careful agricultural practices that protect water sources and limit grazing of sparse vegetation. Overuse of the land by livestock and humans, however, can strip it bare and make it uninhabitable. The large desert in northern Africa, shown in figure 13.3, is the Sahara Desert, and the semiarid region (10 to 20 inches of rain per year) to the south of it is the

Sahel (box figure 1). In the early 1960s a series of abnormally wet years encouraged farmers in the Sahel to expand their herds and grazing lands.

A severe drought in the late 1960s and early 1970s caused devastation of the plant life of the region as starving livestock searched desperately for food, and humans gathered the last remaining sticks for firewood. Vast areas that were once covered with trees and sparse grass became totally barren, and an acute famine began, killing more than 100,000 people. The desert expanded southward, accelerating its advance in some places from 4 miles per year to 30 (6 to 50 kilometers). The denuded soil in many regions became susceptible to wind erosion, leading to choking dust storms and new, advancing dune fields (some even migrating into cities).

Some of the same problems afflicted the midwestern United States in the 1930s, as intense land cultivation coupled with a prolonged drought produced the barren Dust Bowl during the time of the Great Depression. Renewed rains and improved soil-conservation practices have reversed the trend in the United States, but the area is still vulnerable to a future drought, and the possibility of future dust storms in prairie states is very real.

Drought accelerates desertification but is not necessary for it to occur. Overloading the land with livestock and humans can strip marginal regions of vegetation even in wet years.

Box 13.1 Figure 1
The Sahel, south of the Sahara Desert, has been undergoing a severe drought.

Desert stream channels are distinctive in appearance because of the great erosive power of flash floods and the intermittent nature of streamflow. Most stream channels are normally dry and covered with sand or gravel that is moved only during occasional flash floods. Rapid down-cutting by sediment-laden floodwaters tends to produce narrow canyons with vertical walls and flat, gravel-strewn floors (figure 13.5).

Newcomers to deserts sometimes get into serious trouble in desert canyons in rainy weather. Imagine for a moment that you have camped on the canyon floor in figure 13.5 to get out of strong desert winds. Later that night a towering thunderhead cloud forms, and heavy rain falls on the mountains several miles away from you. Although no rain has fallen on you, you are awakened several minutes later by a distant roar. The roar grows louder until a 10-foot (3-meter) "wall" of water rounds a bend in the

Figure 13.5
Desert stream channel showing dry, gravel-covered floor and
steep, vertical sides (cut in sandstone and conglomerate) in
Death Valley, California.

canyon, heading straight for you at the speed of a galloping horse. Boulders, brush, and tree trunks are being swept along in this raging flash flood. The walls of the canyon are too steep to climb. Stay out of desert canyons if there is any sign of rain; sleeping in such canyons is particularly dangerous.

The resistance of some rocks to weathering and erosion is partly controlled by climate. In a humid (wet) climate limestone dissolves easily, forming low places on the earth's surface. In a desert climate the lack of water makes limestone resistant, so it stands up as ridges and cliffs in the desert just as sandstone and conglomerate do. Lava flows and most igneous and metamorphic rock are also resistant. Shale is the least resistant rock in a desert so it usually erodes deeper than other rock types and forms gentler slopes (figure 10.44).

Desert topography characteristically looks more angular than the gently rounded hills and valleys of a humid region. This may be due indirectly to the low rainfall. Shortage of water slows chemical weathering processes to the point where few minerals break down to form fine-grained clay minerals. Soils are coarse and rocky with few chemically weathered products. Plants, which help bind soil into a cohesive layer in humid climates, are rare in deserts, and so desert soils are easily eroded by wind and rainstorms. Downhill creep of thick, fine-grained soil is partly responsible for softening the appearance of topography in humid climates. With thin, rocky soil and slow rates of creep, desert topography should remain steep and angular.

As mentioned in the chapter on streams, not all geologists are sure that the angular topography common to many deserts was formed by the dry climate. The recent fluctuations of climate caused by the glacial ages of the Pleistocene Epoch (see the chapter on glaciers) mean that most modern deserts probably had a much wetter climate a short time ago. It is possible that the angular topography formed then and is well exposed in present deserts only because the lack of vegetation allows us to see it.

As you learned in the chapter on streams, climate is only one of many things that determine the shape and appearance of the land. Rock structure is another. As an example, in the next section we will look closely at two different structural regions within the desert of the southwestern United States.

Desert Features in the Southwestern United States

Much of the southwestern United States has an arid (or semiarid) climate, partly because it is close to 30° North latitude and partly because of the rain shadow effect of the Sierra Nevada and other mountain ranges. Within this region of low rainfall are two areas of markedly different geologic structure. One area is the Colorado Plateau and the other is the Basin and Range province, a mountainous region centered on the state of Nevada. The boundaries of these two areas are shown in figure 13.6.

The *Colorado Plateau* centers roughly on the spot known as the Four Corners, where the states of Utah, Colorado, Arizona, and New Mexico meet at a common point. The rocks near the surface of the Colorado Plateau are mostly flat-lying beds of sedimentary rock thousands of feet above sea level. These rocks are well exposed at the Grand Canyon in Arizona.

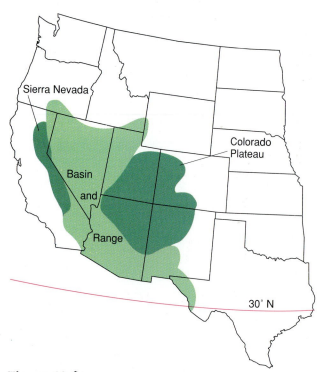

Figure 13.6
The Colorado Plateau and the Basin and Range province in the southwestern United States.

Because the rock layers are well above sea level, they are vulnerable to erosion by the little rain that does fall in the region. Flat-lying layers of resistant rock, such as sandstone, limestone, and lava flows, form **plateaus**—broad, flat-topped areas elevated above the surrounding land and bounded, at least in part, by cliffs. As erosion removes the rock at its base, the cliff is gradually eroded back into the plateau (figure 13.7*A*). Remnants of the resistant rock layer may be left behind, forming flat-topped mesas or narrow buttes (figure 13.7*B*). A **mesa** is a broad, flat-topped hill bounded by cliffs and capped with a resistant rock layer. A **butte** is a narrow pinnacle of resistant rock with a flat top and very steep sides. Most buttes form by continued erosion of mesas. (The term *butte* is also used in other parts of the country for any isolated hill.)

The Colorado Plateau is also marked by peculiar steplike folds called *monoclines*. Erosion of monoclines leaves resistant rock layers protruding above the surface as ridges (figure 13.8). A steeply tilted layer erodes to form a *hogback*, a ridge with equal slopes on its sides. A gently tilted layer forms a *cuesta*, with one steep side and one gently sloping side.

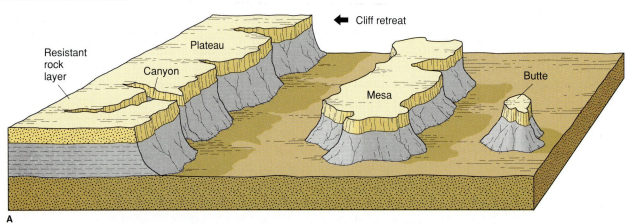

Figure 13.7
Characteristic landforms of the Colorado Plateau. (*A*) Erosional retreat of a cliff at the edge of a plateau can leave behind mesas and buttes as erosional remnants of the plateau. This is an example of parallel retreat of slopes, which was discussed in chapter 10. (*B*) Mesas and buttes in Monument Valley, Arizona, an area of eroded, horizontal, sedimentary rocks.

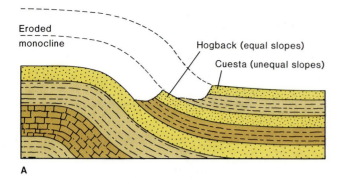

Eroded monocline

Hogback (equal slopes)

Cuesta (unequal slopes)

A

B

Figure 13.8
(A) Steplike monocline folds often erode so that resistant rock layers form hogbacks and cuestas. (B) Monocline near Mexican Hat, Utah. Rocks on far left and far right are horizontal; rocks in the center are steeply tilted downward to the left.
Photo by D. A. Rahm, courtesy Rahm Memorial Collection, Western Washington University.

Figure 13.9
Basin and Range topography in Death Valley, California. In the distance the fault-bounded Panamint Mountains rise more than 11,000 feet (3 kilometers) above Death Valley. Giant alluvial fans at the base of the mountains show a braided stream pattern. Fine-grained sediments and salt deposits form the playas in the foreground.

The *Basin and Range province* is characterized by rugged mountain ranges separated by flat valley floors (figure 13.9). The blocks of rock that form the mountain ranges and the valley floors are bounded by **faults,** cracks in the earth along which some rock movement has taken place. (The chapter titled "Geologic Structures" discusses faults in more detail.) In the Basin and Range province the rock movement on either side of the faults has lifted the mountains up and dropped the valleys down (figure 13.10). Fault-controlled topography is found throughout the Basin and Range province, which covers almost all of Nevada and portions of bordering states as well as southern Arizona and New Mexico.

Heavy rainfall in the mountain ranges causes rapid erosion of the steep mountain fronts and resulting deposition on the valley floors (figure 13.11). Rock debris from the mountains, picked up by flash floods and mudflows, is deposited in the form of alluvial fans at the base of the mountain ranges. Alluvial fans (described in the chapter on streams) build up where stream channels widen as they flow out of narrow canyons onto the open valley floors.

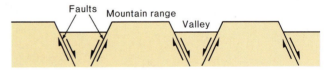

Faults   Mountain range   Valley

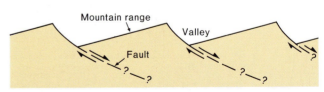

Mountain range   Valley

Fault

Figure 13.10
Two possible origins of the mountains and valleys of the Basin and Range province. In each case the mountains have been lifted up and the valleys dropped down. Movement along the faults is shown by arrows.

Deserts and Wind Action   289

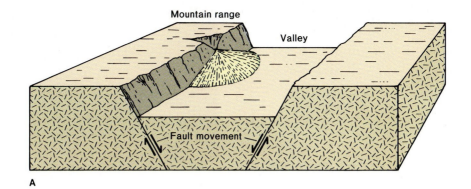

Mountain range

Valley

Fault movement

A

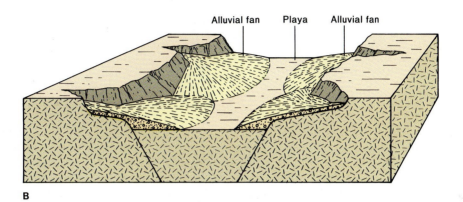

Alluvial fan   Playa   Alluvial fan

B

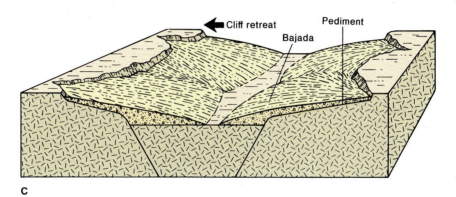

Cliff retreat   Pediment

Bajada

C

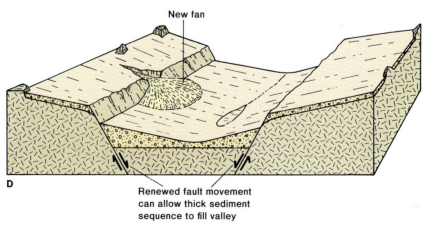

New fan

D

Renewed fault movement
can allow thick sediment
sequence to fill valley

Figure 13.11
Origin of some Basin and Range topography.

Figure 13.12
Alluvial fans at the base of mountain canyons, Death Valley, California. The white salt flat in the foreground is part of a playa.
Photo by Frank M. Hanna.

Figure 13.13
Mud-cracked playa surface.

Although most of the sediment carried by runoff is deposited in alluvial fans, some fine clay may be carried in suspension onto the flat valley floor. If no outlet drains the valley, runoff water may collect and form a **playa lake** on the valley floor. Playa lakes are usually very shallow and temporary, lasting for only a few days after a rainstorm. After the lake evaporates, a thin layer of fine mud may be left on the valley floor. The mud dries in the sun, forming a **playa,** a very flat, dry lake bed of hard, mud-cracked clay (figures 13.12 and 13.13). If the runoff contained a large amount of dissolved salt or if seeping ground water brings salt to the surface, a playa may be covered with a bright white layer of dried salt instead of mud.

Continued deposition near the base of the mountains may form a **bajada,** a broad, gently sloping depositional surface formed by the coalescing of individual alluvial fans (figure 13.11). A bajada is much more extensive than a single alluvial fan and may have a gently rolling surface resulting from the merging of the cone-shaped fans.

Erosion of the mountain can eventually form a **pediment,** which is a gently sloping surface, usually covered with a thin veneer of gravel, cut into the solid rock of the mountain (figure 13.11). A pediment develops uphill from a bajada as the mountain front retreats. It can be difficult to distinguish a pediment from the surface of the bajada downhill, since both have the same slope and gravel cover. The pediment, however, is an erosional surface, usually underlain by solid rock, while the bajada surface is depositional and may be underlain by hundreds or even thousands of feet of sediment.

An abrupt change in slope marks the upper limit of the pediment, where it meets the steep mountain front. Many geologists who have studied desert erosion believe that as this steep mountain front erodes, it retreats uphill, maintaining a relatively constant angle of slope. (This is *parallel retreat* of a slope—see figure 10.42.)

Notice that rock structure, not climate, largely controls the fact that plateaus and cliffs are found in the Colorado Plateau, while mountain ranges, broad valleys, alluvial fans, and pediments are found in the Basin and Range province. Features such as plateaus, mesas, and alluvial fans can also be found in humid climates wherever the rock structure is favorable to their development; they are not controlled by climate. Features such as steep canyons, playa lakes, thin soil, and sparse vegetation, however, *are* often controlled by climate.

Wind Action

Wind can be an important agent of erosion and deposition in any climate, as long as sediment particles are loose and dry. Wind differs from running water in two important ways. Because air is less dense than water, wind can erode only fine sediment—sand, silt, and clay. But wind is not confined to channels as running water is, so wind can have a widespread effect over vast areas.

In general, the faster the wind blows, the more sediment it can move. Wind velocity is determined by air temperature. As air warms and cools, it changes density, and these density changes create air pressure differences that cause wind. Wet climates and cloud cover help buffer changes in air temperature, but in dry climates daily temperature changes can be extreme. In a desert, the temperature may range from 50°F at night to more than 100°F in the daytime (10°C to 40°C). Because of these temperature fluctuations, wind is generally stronger in deserts than in humid regions. Desert winds often exceed 60 miles per hour (100 kilometers per hour).

A

B

C

Figure 13.14
(*A*) Wind erosion of soil from a cultivated field. (*B*) Approaching dust storm, Prowers County, Colorado, late afternoon (1930s). The storm lasted almost three hours, with wind speeds of approximately 30 miles (50 kilometers) per hour. (*C*) A dust storm rises above Death Valley, California. Strong desert winds remove fine-grained soil particles, leaving a stony, residual soil.

Photos *A* and *B* by U.S. Department of Agriculture, Soil Conservation Service.

Although strong winds are also associated with rainstorms and hurricanes, these winds seldom erode sediment because the rain wets the surface sediment. Wet sediment is heavy and cohesive and will not be blown away. Strong winds in the desert, however, blow over loose, dry sediment, so wind is an effective erosional agent in dry climates. (As we said earlier in the chapter, running water in the form of flash floods is a far more important agent than wind, even though the wind can be very strong.)

Wind Erosion and Transportation
Thick, choking *dust storms* are one example of wind action (figure 13.14). "Dust Bowl" conditions in the 1930s in the agricultural prairie states lasted for several years due to droughts and poor soil-conservation practices. Loose silt and clay are easily picked up from barren dry soil, such as in a cultivated field. Wind erosion is even greater if the soil is disturbed by animals or vehicles. Silt and clay can remain suspended in turbulent air for a long time, so a strong wind may carry a dust cloud thousands of feet upward and hundreds of miles horizontally. Dust storms

of the 1930s frequently blacked out the midday sun. Fertile soil was lost over vast regions, ruining many farms. Streets and rivers downwind were filled with thick dust deposits.

Wind-blown sediment is sometimes picked up on land and carried out to sea. Particles from the Sahara Desert in Africa have been collected from the air over the islands of the West Indies after having been carried across the entire Atlantic Ocean. A substantial amount of the fine-grained sediment that settles to the sea floor is land-derived sediment that the wind has deposited on the sea surface. Ships 500 miles (800 kilometers) offshore have reported dustfalls a few millimeters thick covering their decks.

Volcanic ash can be carried by wind for very great distances. An explosive volcanic eruption can blast ash more than 10 miles (15 kilometers) upward into the air. Such ash may be caught in the high-altitude *jet streams,* narrow belts of strong winds with velocities sometimes greater than 200 miles (300 kilometers) per hour. Following the 1980 eruption of Mount St. Helens in western Washington, a visible ash layer blanketed parts of Washington, Idaho, and Montana to the east. At high altitudes, St. Helens ash could be detected blowing over New York and out over the Atlantic Ocean, 3,000 miles (5,000 kilometers) from the volcano. But the St. Helens eruption was a relatively small one. Ash from the 1883 Krakatoa eruption in Indonesia circled the globe for two years, causing spectacular sunsets and a slight, but measurable, drop in temperature as the ash reflected sunlight back into space. The 1815 eruption of Tambora, also in Indonesia, put so much ash into the air that there were blizzards, crop failures, and famine in New England and northern Europe in 1816, "the year without a summer." Slightly lower temperatures and brilliant sunsets also marked the 1982 eruption of El Chichón in Mexico.

Because sand grains are heavier than silt and clay, sand moves close to the ground in the leaping pattern called *saltation* (as does some sediment in streams). High-speed winds can cause *sandstorms,* clouds of sand moving rapidly near the land surface. The high-speed sand in such a storm can sandblast smooth surfaces on hard rock and scour the windshields and paint of automobiles. Because of the weight of the sand grains, however, sand rarely rises more than 3 feet (1 meter) above a flat land surface, even under extremely strong winds. Therefore, most of the sandblasting action of wind occurs close to the ground (figure 13.15). Telephone poles in regions of wind-driven sand often are severely abraded near the ground. To prevent this abrasion, desert residents pile small stones or wrap sheet metal around the base of the poles.

Wind seldom moves particles larger than sand grains, but wind-blown sand may sculpture isolated pebbles, cobbles, or boulders into **ventifacts**—rocks with flat, wind-abraded surfaces (figure 13.16). If the wind direction shifts, or if the stone is turned, more than one flat face may develop on the ventifact.

Figure 13.15
Wind erosion near the ground has sandblasted the lower half of this 6-foot (2-meter) basalt outcrop, Death Valley, California.

Figure 13.16
Large granite ventifact, Sweetwater County, Wyoming.
Photo by W. H. Bradley, U.S. Geological Survey.

Deflation The removal of clay, silt, and sand particles from the land surface by wind is called **deflation.** If the sediment at the land surface is made up only of fine particles, the erosion of these particles by the wind can lower the land surface substantially. A **blowout** is a depression on the land surface caused by wind erosion (figure 13.17).

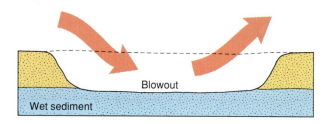

Figure 13.17
Deflation by wind erosion can form a blowout in loose, dry sediment. Deflation stops at the water table.

Figure 13.18
Large blowout near Harrison, Nebraska. Pillar top is the original level of land, before wind erosion lowered the land surface by more than 10 feet (3 meters).
Photo by N. H. Darton, U.S. Geological Survey.

Blowouts are common in the Great Plains states (figure 13.18). One in Wyoming measures 9 by 3 miles (15 by 5 kilometers) and is 150 feet (45 meters) deep. The enormous Qattara Depression in western Egypt, more than 150 miles (250 kilometers) long and more than 300 feet (100 meters) *below* sea level, has been attributed to wind deflation. Deflation can continue to deepen a blowout in fine-grained sediment until it reaches wet, cohesive sediment at the water table.

If the sediment near the land surface is made up of both fine and coarse particles, the wind removes the fine particles and leaves the coarse gravel behind (figure 13.19). As this gravel becomes concentrated by the removal of the fine particles, it may eventually form a thin layer of closely packed gravel called a **desert pavement** (or sometimes *pebble armor*). Other soil processes may help form desert pavement. The pebble layer protects the underlying sediment from further deflation (figures 13.19 and 13.20).

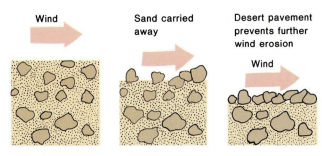

Figure 13.19
Deflation by wind erosion can develop a desert pavement in rocky sediment.

Figure 13.20
Desert pavement in Mojave Desert, California. Fine sand lies under the single layer of pebbles.

Figure 13.21
Vertical road cuts in loess, Vicksburg, Mississippi.
Photo by E. W. Shaw, U.S. Geological Survey.

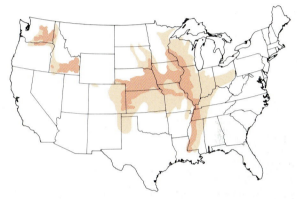

Figure 13.22
Loess deposits in the United States. Orange color shows major deposits. Tan color represents thin, discontinuous deposits.
From U.S. Bureau of Reclamation, 1960, and various other sources.

Wind Deposition

Loess Loess is a deposit of wind-blown silt and clay composed of unweathered, angular grains of quartz, feldspar, and other minerals and weakly consolidated by calcareous cement. Loess has a high porosity, often near 60%. Deposits of loess may blanket hills and valleys downwind of a source of fine sediment, such as a desert or a region of glacial outwash.

China has extensive loess deposits, more than 300 feet (100 meters) thick in places. Wind from the Gobi Desert carried the silt and clay that formed these deposits. Loess is easy to dig into and has the peculiar ability to stand as a vertical cliff without slumping (figure 13.21), perhaps because of its cement or perhaps because the fine, angular, sediment grains interlock with one another. The Chinese have for centuries dug cavelike homes in loess cliffs. When a large earthquake shook China in 1920, however, many of these cliffs collapsed, burying alive about 100,000 people.

During the glacial ages of the Pleistocene Epoch (chapter on glaciers), the rivers that drained what is now the midwestern United States transported and deposited vast amounts of glacial outwash. Later, winds eroded silt and clay (originally glacial rock flour) from the flood plains of these rivers and blanketed large areas of the Midwest with a cover of loess (figure 13.22). Soils that have developed from the loess are usually fertile and productive. The grain fields of much of the Midwest (and Pacific Northwest) are planted on these rich soils.

Desert Varnish

Many rocks on the surface of deserts are darkened by a chemical coating known as *desert varnish.* Although the interior of the rocks may be light colored, a hard, often shiny, coating of dark manganese oxide and clay minerals can build up on the rock surface over long periods of time (box figure 1). Although no one is quite certain how this coating develops, it seems to be added to the rocks from the outside, for even white quartzite pebbles with no internal source of manganese or clay minerals can develop desert varnish. One current hypothesis is that the clay is windblown, perhaps sticking to rocks dampened by dew. A film of clay on a rock may draw manganese-containing solutions upward from the soil by capillary action, and the presence of the clay minerals may help

Box 13.2 Figure 1
Petroglyphs carved on this rock cut through the dark desert varnish to show the lighter color of the interior of the rock, Valley of Fire, Nevada.
Photo by J. Freeberg, U.S. Geological Survey.

deposit the dark manganese oxide that cements the clay to the rock. Another hypothesis is that the oxide is deposited biologically by microbes. Regardless of how the varnish forms, the longer a rock is exposed on a desert land surface, the darker it becomes.

Figure 13.23
Coastal dunes formed from beach sand blowing inland, Pismo Beach, California.
Photo by Frank M. Hanna.

Sand dunes **Sand dunes** are mounds of loose sand grains heaped up by the wind. Dunes are most likely to develop in areas with strong winds that generally blow in the same direction. Patches of dunes are scattered throughout the southwestern United States desert. More extensive fields of dunes occur on some of the other deserts of the world, such as the Sahara Desert of Africa, which contains vast *sand seas.* Dunes are also commonly found just landward of beaches (figure 13.23), where sand is blown inland.

Beach dunes are common along the shores of the Great Lakes and along both coasts of the United States. Braided rivers (see the chapter on streams) can also be sources of sand for dune fields.

The mineral composition of the sand grains in sand dunes depends on both the character of the original sand source and the intensity of chemical weathering in the region. Many dunes, particularly those near beaches in humid regions, are composed largely of quartz grains because quartz is so resistant to chemical weathering. Inland dunes often contain unstable feldspar and rock fragments in addition to quartz. Some dunes are formed mostly of carbonate grains, particularly those near tropical beaches. At White Sands, New Mexico, dunes are made of gypsum grains, eroded by wind from playa lake beds.

Sand grains found in dunes are commonly well-sorted and well-rounded because wind is very selective as it moves sediment. Fine-grained silt and clay are carried much farther than sand, and grains coarser than sand are left behind when sand moves. The result is a dune made solely of sand grains, often all very nearly the same size. The prevalence of well-rounded grains in many dunes also may be due to selective sorting by the wind. Rounded grains roll more easily than angular grains, and so the wind may take only the rounded grains away from a source to form dunes. Wind will often selectively roll oolitic grains from a carbonate beach of mixed oolitic and skeletal grains.

Most sand dunes are asymmetric in cross section, with a gentle slope facing the wind and a steeper slope on the downwind side. The steep downwind slope of a dune is

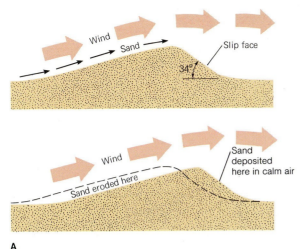

A

Figure 13.24
(*A*) A sand dune forms with a gentle upwind slope and a steeper slip face on the downwind side. Sand eroded from the upwind side of the dune is deposited on the slip face. Movement of sand causes the dune to move slowly downwind. (*B*) Strong desert winds (60 miles per hour) blowing to the left remove sand from the gently sloping upwind side of this dune. The sand settles onto the steep slip face on the left.

B

called the **slip face** (figure 13.24). It forms from loose, cascading sand that generally keeps the slope at the *angle of repose*, which is about 34° for loose, dry sand. Sand grains are blown up the gentle slope and over the top of the dune. Loose sand builds up near the top of the slip face until it becomes oversteepened and the sand grains tumble down the slip face. (The cross-bedding in sandstone formed from lithified sand dunes is evidence of similar sand movement in the past.)

In passing over a dune, the wind erodes sand from the gentle upwind slope and deposits the same sand on the slip face downwind. As a result, the entire dune moves slowly in a downwind direction. The rate of dune motion is much slower than the speed of the wind, of course, because only a thin layer of sand on the surface of the dune is moving at any one time. The dune may move only 30 to 50 feet (10 to 15 meters) per year. Over many years, however, the movement of dunes can be significant, a fact not always appreciated by people who build homes close to moving sand dunes.

If a dune becomes overgrown with grass or other vegetation, movement stops. The Sand Hills of north-central Nebraska are large dunes, formed during the Pleistocene or Holocene Epochs, that have become stabilized by vegetation. The migration of many beach dunes toward beach homes and roads has been stopped by planting a cover of beach grass over the dunes. Dune-buggy tires can trample and kill the grass, however, and start the dunes moving again.

Sand moving over a dune surface often forms *wind ripples*—small, low ridges of sand produced by saltation of the grains (figure 13.25). The ripples are similar to those formed in sediment by a water current (see the chapter on sedimentary rocks). Because sand moves perpendicularly to the long dimension of the ripples, a rippled sand surface indicates the direction of sand movement.

Figure 13.25
Wind ripples on a sand dune (ripples are a few inches apart). Prevailing wind blows from left to right.

Types of dunes As figure 13.26 shows, dunes tend to develop certain characteristic shapes, depending on (1) the wind's velocity and direction (that is, whether constant or shifting); (2) the sand supply available; and (3) how the vegetation cover, if any, is distributed.

Where the sand supply is limited, a type of dune called a **barchan** generally develops. The barchan is a crescent-shaped dune with the horns of the crescent pointing downwind (figure 13.27). Barchans are usually separated from one another and move across a bare rock surface. If more sand is available, the wind may develop a **transverse dune,** a relatively straight, elongate dune oriented perpendicular to the wind (figure 13.28). A **parabolic dune,** which commonly forms around a blowout, particularly near a beach, is deeply curved. The horns point upwind and are apt to be anchored by vegetation. It requires abundant sand. All three of these dune shapes develop under a steady wind direction, and all three have steep slip faces on the downwind sides.

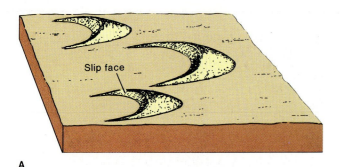

A

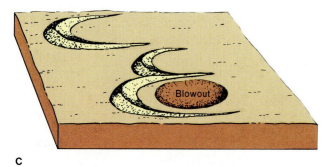

B

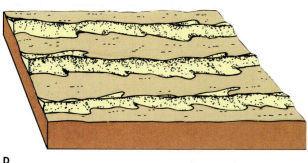

C

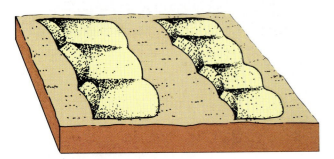

D

Figure 13.26
Types of sand dunes. (*A*) Barchan. (*B*) Transverse dunes.
(*C*) Parabolic dune. (*D*) Longitudinal dunes (seifs).

Figure 13.27
These barchan dunes are advancing as much as 50 feet (15 meters) a year over this barren valley floor in southern California.
Photo by John S. Shelton.

Figure 13.28
Transverse dunes, Oregon.
Photo by D. A. Rahm, courtesy Rahm Memorial Collection, Western Washington University. pg. 300

One of the largest types of dune is the **longitudinal dune** or *seif* (figure 13.29). It is a large, symmetrical ridge of sand parallel to the wind direction, which often varies slightly. Longitudinal dunes in the Sahara Desert are as high as 650 feet (200 meters) and more than 75 miles (120 kilometers) in length.

Not all dunes can be classified by an easily recognizable shape. Many of them are quite irregular.

Wind Action on Mars

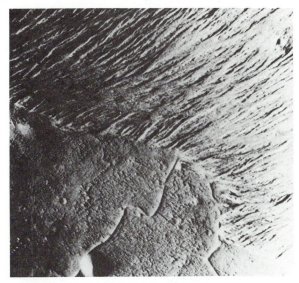

Astrogeology Box 13.1 Figure 1
Wind eroded terrain southwest of Olympus Mons.
Photo by Dr. Michael H. Carr/National Space Science Data Center.

Because Mars has an atmosphere, winds blow, eroding material, transporting it to other locations, and depositing it. Because the atmosphere is not very dense (only about 1/200th that of Earth's atmosphere), high winds are necessary to transport sand and dust. Wind velocities exceeding 200 kilometers per hour have been observed, and because some photographs show large portions of Martian shield volcanoes such as Olympus Mons obscured by dust storms, it is known that the wind carries some particles more than 15 kilometers high. The lower gravitational field of Mars permits the wind to carry particles four times as high as a wind of the same velocity would on Earth. Dust in the atmosphere scatters sunlight, making the Martian sky pink.

Some linear grooves and streamlined ridges on Mars (box figure 1) have been attributed to wind erosion, as has a cliff more than a kilometer high at the base of Olympus Mons. Several older meteorite craters show evidence of having been worn down or abraded. On the other hand, the preservation of many younger craters suggests that wind erosion has been very slow.

Localized dust storms are quite common on Mars. In addition, once each Martian year, a great dust storm covers a large portion of the planet. It may be started by the heating of dust particles in the Martian atmosphere. This large dust storm is responsible for seasonal changes in light and dark markings on the Martian surface. As the storm dies down, it deposits a light-colored blanket of particles over the Martian surface. As local winds remove the light particles from some areas, they appear darker.

Figure 13.29
Longitudinal dunes in the Sahara Desert, Algeria. Photo from Gemini spacecraft at an altitude of about 65 miles (100 kilometers).
Photo from NASA.

Dark- and light-colored streaks (box figure 2) are common on the Martian surface, particularly downwind from craters or other features that present obstacles to the flow of air across the surface. Changes in these markings are commonly observed after major dust storms, suggesting that these features are due to differential erosion and deposition of wind-blown particles.

Sand dunes (box figure 3) also have been observed on Mars. They are particularly common on the floors of some craters, where they form vast dune fields. The most common type is the crescent-shaped *barchan*, but there are other types, such as *longitudinal dunes* extending for hundreds of kilometers in parallel arrays.

Spacecraft landed on the surface of Mars have photographed dunelike features in several locations. Most of the photographed surface, however, has the appearance of *desert pavement* (box figure 4). Many Martian rocks look like *ventifacts* and are etched and pitted, most likely by wind erosion.

Astrogeology Box 13.1 Figure 2
200 sq. km. area of dark- and light-colored streaks on the planet Mars.
Photo by Dr. Michael H. Carr/National Space Science Data Center.

Astrogeology Box 13.1 Figure 3
Dune-like features on the planet Mars.
NASA.

Astrogeology Box 13.1 Figure 4
Probable desert pavement and ventifacts on the planet Mars.
NASA.

Summary

Deserts are located in regions where less than 10 inches (25 cm) of rain falls in a year. Such regions are found primarily in belts of descending air at 30° North and South latitude. Arid regions also may be due to the *rain shadow* of a mountain range, great distance from the sea, and proximity to a cold ocean current.

Desert landscapes differ from those in humid regions in lacking through-flowing streams and in having internal drainage and many local, rising base levels. *Flash floods* caused by desert thunderstorms are effective agents of erosion despite the low rainfall. Limestone is resistant in deserts. Thin soil and slow rates of creep may give desert topography an angular look.

Parts of the southwestern United States are a desert, its topography primarily determined by rock structure. Flat-lying sedimentary rocks of the Colorado Plateau are sculptured into cliffs, *plateaus, mesas,* and *buttes.* The fault-controlled topography of the Basin and Range province is marked by *alluvial fans, bajadas, playas,* and *pediments.*

Although wind erosion is most intense in regions of low moisture, streams are usually more effective than wind in sculpturing landscapes, regardless of climate.

Fine-grained sediment can be carried long distances by wind, even across entire continents and oceans.

Sand moves by *saltation* close to the ground, occasionally carving *ventifacts.*

Wind can *deflate* a region, creating a *blowout* in fine sediment or *desert pavement* in sediment that includes gravel.

Sand dunes move slowly downwind as sand is removed from the gentle upwind slope and deposited on the steeper *slip face* downwind.

Dunes are classified as *barchans, transverse dunes, parabolic dunes,* and *longitudinal dunes,* but many dunes do not resemble these types. Dune type depends on wind strength and direction, sand supply, and vegetation.

Terms to Remember

bajada	mesa
barchan	parabolic dune
blowout	pediment
butte	plateau
deflation	playa
desert	playa lake
desert pavement	rain shadow
fault	sand dune
flash flood	slip face
loess	transverse dune
longitudinal dune (seif)	ventifact

Questions for Review

1. What are two reasons why parts of the southwestern United States have an arid climate?
2. Sketch a cross section of an idealized dune, labeling the slip face and indicating the wind direction. Why does the dune move?
3. Describe the geologic structure and sketch the major landforms of:
   a. the Colorado Plateau;
   b. the Basin and Range province.
4. How does a flash flood in a dry region differ from most floods in a humid region?
5. Give two reasons why wind is a more effective agent of erosion in a desert than in a humid region.
6. Describe a desert pavement and discuss its origin.
7. Name four types of sand dunes and describe the conditions under which each forms.

Questions for Thought

1. How does a pediment differ from a bajada? Discuss the differences and similarities between the two features, particularly in regard to appearance and origin.
2. Study the photos of sand dunes in this chapter. Which way does the prevailing wind blow in each case?

Supplementary Readings

Bagnold, R. A. 1941. *The physics of blown sand and desert dunes.* New York: William Morrow, 1954 (reprinted).

Blackwelder, E. 1954. Geomorphic processes in the desert. *Bulletin of California Division of Mines* 170:11–20. Sacramento: Department of Natural Resources.

Cooke, R. U., and A. Warren. 1973. *Geomorphology in deserts.* Berkeley: University of California Press.

Glennie, K. W. 1970. *Desert sedimentary environments.* New York: Elsevier.

Greeley, R., and J. Iversen. 1985. *Wind as a geological process.* Cambridge: Cambridge University Press.

McGinnies, W. G., B. J. Goldman, and P. Paylore, eds. 1968. *Deserts of the world.* Tucson, Arizona: University of Arizona Press.

McKee, E. D., ed. 1979. *A study of global sand seas.* U.S. Geological Survey Professional Paper 1052.

Mabbutt, J. A. 1977. *Desert landforms.* Cambridge, Mass.: MIT Press.

Pewe, T. L., ed. 1981. *Desert dust: origin, characteristics, and effect on man.* Boulder, Colorado: Geological Society of America Special Paper 186.

Ritter, D. F. 1986. *Process geomorphology.* 2d ed. Dubuque, Iowa: Wm. C. Brown Publishers.

Thornbury, W. D. 1969. *Principles of geomorphology.* 2d ed. New York: John Wiley & Sons.

The previous five chapters have dealt with the geologic agents of mass wasting, streams, ground water, glaciers, and wind. Water waves are another agent of erosion, transportation, and deposition of sediment. Along the shores of oceans and lakes, waves break against the land, building it up in some places and tearing it down in others.

The energy of the waves comes from the wind. This energy is used to a large extent in eroding and transporting sediment along the shoreline. Understanding how waves travel and move sediment can help you see how easily the balance of supply, transportation, and deposition of beach sediment can be disturbed. Such disturbances can be natural or man-made, and the changes that result often destroy beachfront homes and block harbors with sand.

Beaches have been called "rivers of sand" because breaking waves, as they sort and transport sediment, tend to move sand parallel to the shoreline. In this chapter we look at how beaches are formed and also examine the influence of wave action on such coastal features as sea cliffs, barrier islands, and terraces.

14

Waves, Beaches, and Coasts

Wave-eroded stacks, Big Sur, California.
David Muench Photography.

Figure 14.1
These beach homes in Florida were built too close to the ocean, and were destroyed by storm waves in 1985.
Photo by Richard A. Davis, Jr.

If you spend a week at the shore during the summer, you may not notice any great change in the appearance of the beach while you are there. Even if you spend the whole summer at the seaside, nothing much seems to happen to the beach during those months. Tides rise and fall every day and waves strike the shore, but the sand that you walk on one day looks very much like the sand that you walk on the next day. The shape of the beach does not appear to change, nor does the sand seem to move very much.

On most beaches, however, the sand is moving, in some cases quite rapidly. The beach looks the same from day to day only because new sand is being supplied at about the same rate that old sand is being removed.

Where is the sand going? Some sand is carried out to deep water. Some is piled up and stored high on the beach. But on most shores much more of the sand moves along parallel to the beach in relatively shallow water. In this way, loose sand grains travel hundreds of feet per day along some coasts, especially those subject to strong waves.

On some beaches, sand is being removed faster than it is being replenished. When this happens, beaches become narrower and less attractive for swimming. Where erosion is severe, buildings close to the beach can be undermined and destroyed by waves as the beach disappears (figure 14.1). The sand moved from the beach may be redeposited in inconvenient places, such as across the mouth of a harbor, where it must be dredged out periodically. Because moving sand can create many problems for people in coastal towns and cities, it is important to understand something of how and why the sand moves.

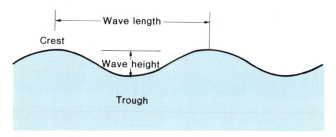

Figure 14.2
Wave height is the vertical distance between the wave crest and the wave trough. Wave length is the horizontal distance between two crests.

Water Waves

The energy that moves sand along a beach comes from the wind-driven water waves that break upon the shore. As wind blows over the surface of an ocean or a lake, some of the wind's energy is transferred to the water surface, forming the waves that move through the water. Wave shapes can vary. Short, choppy "seas" in and near a storm create a confused sea surface, often with considerable white foam as strong winds blow the tops off of waves. Long, rolling "swells" form a regular series of similar-sized waves on shores that may be thousands of miles from the storms that generated the waves. (Summer surfing waves in southern California are usually generated by large storms south of Australia in the southern hemisphere winter.) When waves break against the shore, a large portion of their energy is spent moving sand along the beach.

The height of waves is the key factor in determining wave energy. **Wave height** is the vertical distance between the **crest,** which is the high point of a wave, and the **trough,** which is the low point (figure 14.2). In the open ocean, normal waves have heights of 1 to 15 feet (about 0.3 to 5 meters), although during violent storms, including hurricanes, waves can be more than 50 feet (15 meters) high. The highest wind wave ever measured was 112 feet (34 meters) by the anxious crew of a ship in the north Pacific in 1933. (The highest "tidal wave" ever measured, caused by a submarine earthquake rather than wind, was 278 feet, or 85 meters, in the Ryukyu island chain south of Japan in 1971. See chapter on "Earthquakes.")

Wavelength is the horizontal distance between two wave crests (or two troughs). Most ocean wind waves are between 130 and 1,300 feet (40 to 400 meters) in length, and move at speeds of 15 to 55 miles per hour (25 to 90 kilometers per hour) in deep water.

The movement of water in a wave is like the movement of wheat in a field when wind blows across it. You can see the ripple caused by wind blowing across a wheat field, but the wheat does not pile up at the end of the field. Each stalk of wheat bends over when the wind strikes it and then returns to its original position. A particle of water moves in an *orbit,* a nearly circular path, as the wave passes; the particle, too, essentially returns to its original position after the wave has passed. In deep water, when a wave moves across the water surface, energy moves with the wave; but the water, like the wheat, does not move with the wave.

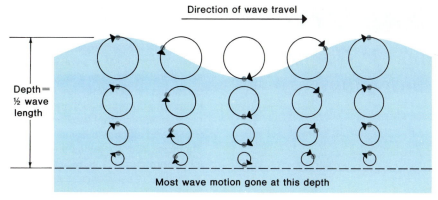

Direction of wave travel

Depth = ½ wave length

Most wave motion gone at this depth

Figure 14.3
Orbital motion of water in waves dies out with depth. At the surface the diameter of the orbits is equal to the wave height.

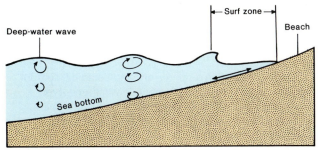

Surf zone

Beach

Deep-water wave

Sea bottom

Figure 14.4
As a deep-water wave approaches shore, it begins to "feel" the sea bottom and slow down. Circular water orbits flatten and the wave peaks and breaks. In the foamy surf zone, water moves back and forth rather than in orbits.

At the surface, the diameter of the orbital path of a water particle is equal to the height of the wave (figure 14.3). Below the surface, the orbits decrease in size until the motion is essentially gone at a depth equal to half the wavelength. This is why a submarine can cruise in deep, calm water beneath surface ships that are being tossed by the orbital motion of large waves.

Surf

As waves move from deep water to shallow water near shore, they begin to be affected by the ocean bottom. A wave first begins to "feel bottom" at the level of lowest orbital motion—that is, when the depth to the bottom equals half the wavelength. For example, a wave 500 feet (150 meters) long will begin to be influenced by the bottom at a water depth of 250 feet (75 meters).

In shallow water the presence of the bottom interferes with the circular orbits, which flatten into ovals (figure 14.4). The waves slow down and their length decreases. Meanwhile, the sloping bottom wedges the moving water upward, increasing the wave height. Because the height is increasing while the length is decreasing, the waves become steeper and steeper until they break. A **breaker** is a wave that has become so steep that the crest of the wave topples forward, moving faster than the main body of the wave. The breaker then advances as a turbulent, often foamy, mass. Breakers collectively are called **surf.** Water in the surf zone has lost its orbital motion and moves back and forth, alternating between onshore and offshore flow.

Nearshore Circulation

Wave Refraction

Most waves do not come straight into shore. A wave crest usually arrives at an angle to the shoreline (figure 14.5). One end of the wave breaks first, and then the rest breaks progressively along the shore.

This angled approach of a wave toward shore can change the direction of wave travel. One end of the wave reaches shallow water first. This end of the wave "feels bottom" and slows down while the rest of the wave continues at its deep-water speed (figure 14.6). As more and more of the wave comes into contact with the bottom, more of the wave slows down. As the wave slows progressively along its length, the wave crest changes direction and becomes more nearly parallel to the shoreline. This process is called **wave refraction** (figures 14.5 and 14.6).

Longshore Currents

Although most wave crests become nearly parallel to shore as they are refracted, waves do not generally strike *exactly* parallel to shore. Even after refraction, a small angle remains between the wave crest and the shoreline. As a result, the water in the wave is pushed both *up* the beach toward land and *along* the beach parallel to shore.

Each wave that arrives at an angle to the shore pushes more water parallel to the shoreline. Eventually a moving mass of water called a **longshore current** develops parallel to the shoreline (figure 14.6). The width of the longshore current is equal to the width of the surf zone. The seaward edge of the current is the outer edge of the surf zone, where waves are just beginning to break; the landward edge is the shoreline. A longshore current can be very strong, particularly when the waves are large. Such a current can carry swimmers hundreds of yards (meters) parallel to shore before they are aware that they are being swept along. It is these longshore currents that transport most of the beach sand parallel to shore.

BOX 14.1

Rip Currents—A Common Cause of Drowning

Rip currents are narrow currents that flow straight out to sea in the surf zone, returning water seaward that breaking waves have pushed ashore. Rip currents travel at the water surface and die out with depth. They pulsate in strength, flowing most rapidly just after a set of large waves has carried a large amount of water onto shore. Rip currents can be important transporters of sediment, as they carry fine-grained sediment out of the surf zone into deep water.

As a single wave comes toward shore, its height varies from place to place. Rip currents tend to develop locally where wave height is low. Rip currents that are fixed in position are apt to be found over channels or hollows on the sea floor, because depressions on the bottom reduce wave height. Complex wave interactions can also lower wave height, and rip currents that form because of wave interactions tend to shift position along the shore. Such shifting rip currents are usually spaced at regular intervals along the beach.

Rip currents are fed by water within the surf zone. They flow rapidly out through the surf zone and then die out quickly. Where waves are nearly parallel to a shoreline, longshore feeder currents of equal strength develop in the surf zone on either side of a rip current (box figure 1*A*). Where waves strike the shore at an angle and set up a strong, unidirectional, longshore current, a rip current is fed from one side by the longshore current, which increases in strength as it nears the rip current (box figure 1*B*). Rip currents are also found alongside points of land and man-made obstacles such as jetties and piers, which can deflect longshore currents seaward.

You can easily learn to spot rip currents at a beach. Look for discoloration in the water where sediment is being picked up in the surf zone and moved seaward (box figure 2). Another sign is incoming waves breaking early within a rip current as they meet the opposing flow. The diffuse heads of rip currents outside the surf zone may be marked at the edge with foam lines. Even on very calm days rips can often be identified by subtle changes in the water surface, such as a different pattern of water ripples or light reflection off the water.

Getting caught in a rip current and being carried out to sea can panic an inexperienced swimmer—even though the trip will stop some distance beyond the surf zone as the rip dies out. A swimmer frightened by being carried away from land and into breaking waves can grow exhausted fighting the current to get back to shore. The thing to remember is that rip currents are narrow. Therefore, you can get out of a rip easily by swimming *parallel* to the beach instead of struggling against the current.

Surfers, on the other hand, often look for rip currents and paddle intentionally into them to get a quick ride out into the high breakers.

Figure 14.5
These waves are arriving at an angle to the shoreline. They break progressively along the shore, from the top of the photo to the bottom. Note that waves outside the surf zone make a steeper angle with the shoreline than waves within the surf zone; refraction has bent the waves in shallow water.
© Frank M. Hanna.

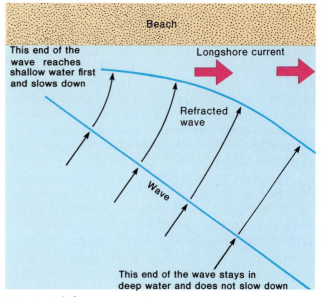

Figure 14.6
Wave refraction changes the wave direction, bending the wave so it becomes more parallel to shore. The angled approach of waves to shore sets up a longshore current parallel to the shoreline.

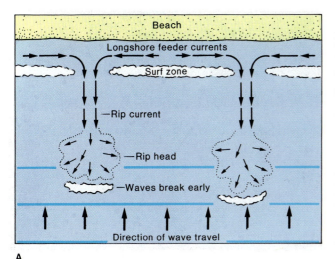
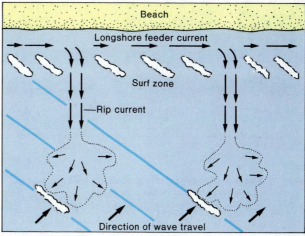

A

B

Box 14.1 Figure 1
Rip currents and their feeder currents can develop
regardless of the angle of approach of waves. (*A*) Waves
approach parallel to shore; feeder currents on both sides of
rip currents. (*B*) Waves approach at an angle to the shore;
feeder current on only one side of rip current.

Box 14.1 Figure 2
Rip currents marked by discolored patches of water
extending seaward of the surf zone.
Photo by D. A. Rahm, courtesy Rahm Memorial Collection, Western
Washington University.

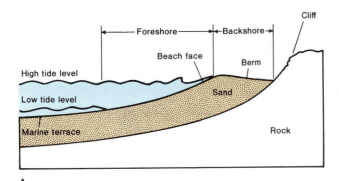

Figure 14.7
(*A*) Parts of a beach. (*B*) The beach face (on the left) and berm (on the right) on a northern California beach.
Photo by K. Lajoie, U.S. Geological Survey.

Beaches

A **beach** is defined as a strip of sediment (usually sand or gravel) that extends from the low-water line inland to a cliff or a zone of permanent vegetation. Waves break on beaches, and rising and falling tides may regularly change the amount of beach sediment that is exposed above water. A beach can be divided into the foreshore and the backshore (figure 14.7).

The *foreshore* of a beach is the zone that is regularly covered and uncovered by the rise and fall of tides. The steepest part of the foreshore is the **beach face,** which is the section exposed to wave action, particularly at high tide. Offshore from the beach face there is usually a **marine terrace,** a broad, gently sloping platform that may be exposed at low tide if the shore has significant tidal action. Marine terraces may be *wave-built* terraces constructed of sediment carried away from the shore by waves, or they may be *wave-cut* rock benches or platforms, perhaps thinly covered with a layer of sediment.

The upper part of the beach, landward of the high-water line, is the *backshore*. It is usually dry, being covered by waves only during severe storms. The backshore is made up of one or more **berms**—wave-deposited sediment platforms that are flat or slope slightly landward (figure 14.7*B*).

The sediment underlying both the beach face and the berm is usually sand, typically with a high percentage of quartz grains (because of quartz's resistance to chemical weathering). Other minerals may be present, particularly heavy minerals that lag behind as lighter minerals are carried away by waves, currents, or sea breezes. Heavy minerals are often dark in color ("black sands"). In some places waves have so concentrated heavy, metal-bearing minerals that they can be mined, as are the titanium-bearing beach sands in a few parts of Florida and Australia. Tropical beaches may be made largely of carbonate grains from offshore algae, coral, and other sources. Some Hawaiian beaches are made of abraded fragments of volcanic rock. Gravel beaches are found in regions subjected to the high energy of large waves. The beach face of a

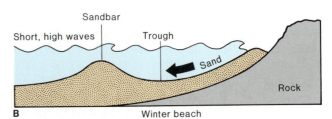

Figure 14.8
Seasonal cycle of a beach caused by differing wave types. (*A*) Summer beach. (*B*) Winter beach. Waves may break once on the winter sandbar, then re-form and break again on the beach face.

gravel shoreline has a much steeper slope than the face of a sand beach. *Shingle,* a regional name for beach gravel, sometimes is reserved for a distinctive gravel formed of coarse, flat pebbles.

In seasonal climates, beaches go through a summer-winter cycle (figure 14.8) because winter waves are usually higher and closer together than summer waves. During summer, long, low waves wash sand from deeper water onto the beach and build out a wide berm. In winter, the short, high storm waves erode sand from the beach and narrow the berm. Offshore, in less turbulent water, the sediment settles to the bottom, building an underwater sandbar (more or less parallel to the beach) that serves as a "storage facility" for the next summer's sand supply. Each season the beach changes in shape until it comes into equilibrium with the prevailing wave type.

Many winter beaches can be dangerous because of high waves and narrowed beaches. Several beaches along the Pacific coast of the United States are nearly free of accidents in the summer, when they are heavily used, but are regularly marked by drownings in the winter as beachwalkers are swept out to sea by large storm waves.

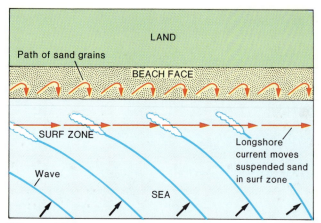

Figure 14.9
Longshore drift of sand on the beach face and by a longshore current within the surf zone.

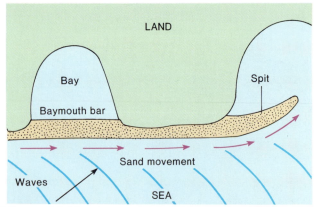

Figure 14.10
Longshore drift of sand can form spits and baymouth bars.

Longshore Drift of Sediment

Longshore drift is the movement of sediment parallel to shore when waves strike the shoreline at an angle. Figure 14.9 shows the two ways in which this movement of sediment (usually sand) occurs. Some longshore drift takes place directly on the beach face when waves wash up on land. A wave washing up on the beach at an angle tends to wash sand along at the same angle. After the wave has washed up as far as it can go, the water returns to the sea by running down the beach face by the shortest possible route, that is, straight downhill to the shoreline, not back along the oblique route it came up. (Wave run-up is known as *swash,* the return as *backwash.*) The net effect of this motion is to move the sand in a series of arcs along the beach face.

Much more sand is moved by longshore transport in the surf zone, where waves are breaking into foam. The turbulence of the breakers erodes sand from the sea bottom and keeps it suspended. Even a weak longshore current can move the suspended sand parallel to the shoreline. The sand in the longshore current moves in the same direction as the sand drift on the beach face (figure 14.9).

Eventually the sand that has moved along the shore by these processes is deposited. Sediment may build up off a point of land to form a **spit,** a fingerlike ridge of sediment that extends out into open water (figures 14.10 and 14.11). A **baymouth bar,** a ridge of sediment that cuts a bay off from the ocean, is formed by sediment migrating across what was earlier an open bay (figures 14.10 and 14.12). Off the western coast of the United States, a considerable amount of drifting sand is carried into the heads of underwater canyons, where the sediments slide down into deep, quiet water.

A striking, but rare, feature formed by longshore drift is a *tombolo,* a bar of sediment connecting a former island to the mainland. As shown in figure 14.13, waves are refracted around an island in such a way that they tend to converge behind the island. The waves sweep sand along the mainland (and from the island) and deposit it at this zone of convergence, forming a bar that grows outward from the mainland and eventually connects to the island.

Figure 14.11
This elongated spit has nearly sealed off the bay on the right, near Lincoln City, Oregon. Sand moves into the photo.
Photo by Paul D. Komar.

Figure 14.12
A baymouth bar has sealed off this bay from the ocean as sand migrated to the right across the mouth of the bay. Near Eureka, California.
Photo by D. A. Rahm, courtesy Rahm Memorial Collection, Western Washington University.

Waves, Beaches, and Coasts   **309**

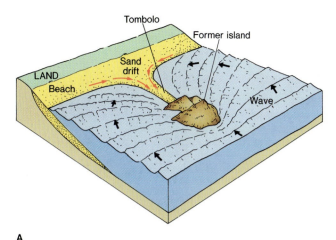

A

Figure 14.13

(*A*) A tombolo connects a former island to the mainland. It forms by sand drift resulting from wave refraction caused by the island. Some sand may also be supplied by erosion of the island. (*B*) A tombolo has connected this rock, once an island, to the shore. Note the waves bending around the two sides of the rock. Near Santa Cruz, California.

B

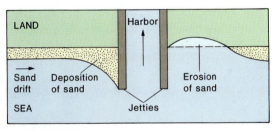

A

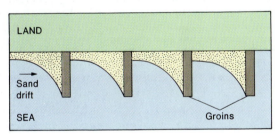

B

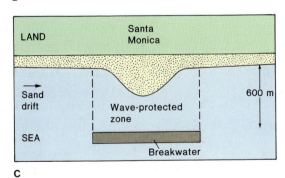

C

Figure 14.14

Sand piles up against obstructions and in areas deprived of wave energy.

Human Interference with Sand Drift

Several man-made features can interrupt the flow of sand along a beach (figure 14.14). *Jetties,* for example, are rock walls designed to protect the entrance of a harbor from sediment deposition and storm waves. Usually built in pairs, they protrude above the surface of the water. Figures 14.14 and 14.15 show how sand piles up against one jetty while the beach next to the other, deprived of a sand supply, erodes back into the shore.

Groins are sometimes built in an attempt to protect beaches that are losing sand from longshore drifting. These short walls are built perpendicular to shore to trap moving sand and widen a beach (figures 14.14 and 14.15).

Sand deposition also occurs when a stretch of shore is protected from wave action by a *breakwater,* an offshore structure built to absorb the force of large breaking waves and provide quiet water near shore. When the city of Santa Monica in California built a rock breakwater parallel to the shore to create a protected small-boat anchorage, the lessening of wave action on the shore behind the breakwater allowed sand to build up there (figure 14.14), threatening eventually to fill in the anchorage. The city had to buy a dredge to remove the sand from the protected area and redeposit it farther along the shore where the waves could resume moving sediment.

A beach attempts to come into equilibrium with the waves that strike it. The type and amount of sediment, the position of the sediment, and especially the movement of the sediment, adjust to the incoming wave energy. Whenever human activity interferes with sand drift or wave action, the beach responds by changing its configuration, usually through erosion or deposition in a nearby part of the beach.

A

B

Figure 14.15
(*A*) Jetties at Manasquan Inlet, New Jersey. Sand drift to the right has piled sand against the left jetty, and removed sand near the right jetty. (*B*) Groins at Ocean City, New Jersey. Sand drift is to the right.
Photos by S. Jeffress Williams.

Sources of Sand on Beaches

Some beach sand comes from the erosion of local rock, such as points of land or cliffs nearby. On a few beaches replenishment comes from sand stored outside the surf zone in the deeper water offshore. But the greater part of the sand on most beaches comes from river sediment brought down to the ocean. Waves pick up this sediment and move it along the beach by longshore drift.

What happens to a beach if all the rivers contributing sand to it are dammed? Although damming a river may be desirable for many reasons (flood control, power generation, water supply, recreation), when a river is dammed, its sediment load no longer reaches the sea. The sand that supplied the beach in the past now comes to rest in the quiet waters of the reservoir behind the dam. Longshore drift, however, continues even though little new sand is being supplied, and the result is a net loss of sand from beaches. Beaches without a sand supply eventually disappear. To prevent this, some coastal communities have set up expensive programs of draining reservoirs and trucking the trapped sand down to the beaches.

Coasts and Coastal Features

A beach is just a small part of the **coast,** which is all the land near the sea, including the beach and a strip of land inland from it. Coasts can be rocky, mountainous, and cliffed, as in northern New England and on the Pacific shore of North America; or they can be broad, gently sloping plains, as along much of the southeastern United States. Wave erosion and deposition can greatly modify coasts from their original shapes. Many coasts have been drowned during the past 10,000 years by the rise in sea level caused by the melting of the Pleistocene glaciers (see the chapter on glaciers). Other coasts have been lifted up by tectonic forces at a rate greater than the rise in sea level so that sea-floor features are now exposed on dry land.

Erosional Coasts

A great many steep, rocky coasts have been visibly changed by wave erosion. Soluble rocks such as limestone dissolve as waves wash against them, and more durable rocks such as granite are fractured by the enormous pressures caused by waves slamming into rock (wave impact pressures have been measured as high as 6 atmospheres—6 tons per square foot).

An irregular coast with bays separated by rocky **headlands** (points of land) can be gradually straightened by wave action. Because wave refraction bends waves approaching such a coast until they are nearly parallel to shore, most of the waves' energy is concentrated on the headlands, while the bays receive smaller, diverging waves (figure 14.16). Rocky cliffs form from wave erosion on the headlands. The eroded material is deposited in the quieter water of nearby bays, forming broad beaches. **Coastal straightening** of an irregular shore gradually takes place through wave erosion of headlands and wave deposition in bays (figure 14.17).

Wave erosion of headlands produces **sea cliffs,** steep slopes that retreat inland by mass wasting as wave erosion undercuts them (figure 14.18). At the base of sea cliffs are sometimes found *sea caves,* cavities eroded by wave action along zones of weakness in the cliff rock. As headlands on irregular coasts are eroded landward, sea cliffs enlarge until the entire coast is marked by a retreating cliff (figure 14.17). On some exposed coasts the rate of cliff retreat can be quite rapid, particularly if the rock is weakly consolidated. Some sea cliffs north of San Diego, California, and at Cape Cod National Seashore in Massachusetts, are retreating at an average rate of 3 feet (1 meter) per year. Because sea-cliff erosion in weak rock is often in the form of large, infrequent slumps (see chapter on "Mass Wasting"), some portions of these coasts may

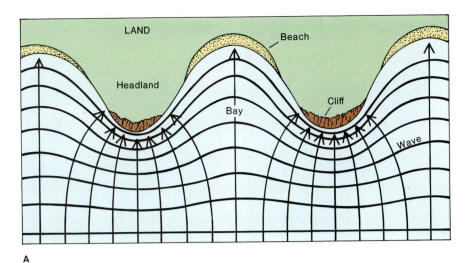

A

B

Figure 14.16
(A) Wave refraction on an irregular coast. Shallow water slows waves off headlands while the same waves move faster through the deep bays. Arrows show energy concentrated on headlands, spread out in bays. (B) Wave refraction around a point of land, Rincon Point, California. Note waves at right center have been bent 90 degrees from their original direction (parallel to the bottom of the photo).
Photo by Frank M. Hanna.

retreat 30 to 100 feet (10 to 30 meters) in a single storm. Some of these cliffs have "ocean-view" homes and hotels at their very edges. Sea cliffs in hard, durable rock such as granite and schist retreat much more slowly.

Seawalls may be constructed along the base of retreating cliffs to prevent wave erosion. Seawalls of giant pieces of broken stone (riprap) or concrete tetrahedrons are designed to absorb wave energy rather than allow it to erode cliff rock. Vertical or concave seawalls of concrete are designed to reflect wave energy seaward rather than allow it to impact the shore. Reflection of waves from a seawall will increase the amount of wave energy just offshore, often increasing the amount of sand erosion offshore. Thus a seawall designed to protect a sea cliff (and

the buildings at its edge) may totally destroy a sand beach at the base of the cliff. Seawalls are difficult and expensive to build and maintain, and they may destroy beaches, but political pressure to build more of them will increase as the sea level rises in the future (box 14.2).

Wave erosion produces other distinctive features in association with sea cliffs. A **wave-cut platform** (or *terrace*) is a horizontal bench of rock formed beneath the surf zone as a coast retreats by wave erosion (figure 14.19). The platform widens as the sea cliffs retreat. The depth of water above a wave-cut platform is generally 20 feet (6 meters) or less, coinciding with the depth at which turbulent breakers actively erode the sea bottom. **Stacks** are erosional remnants of headlands left behind as the coast

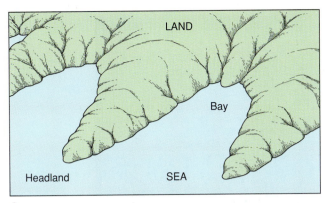

A

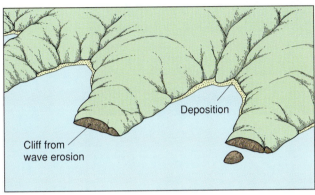

B

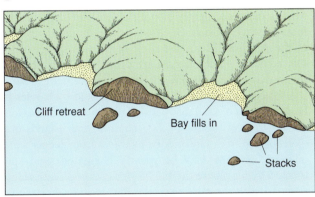

C

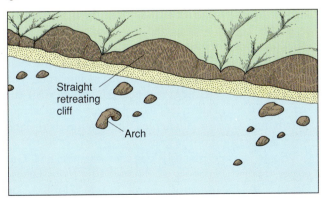

D

Figure 14.17
Coastal straightening of an irregular coastline by wave erosion of headlands and wave deposition of sediment in bays. Continued erosion produces a straight, retreating cliff.

Figure 14.18
Retreating wave-cut cliff, La Jolla, California. The boulder beach and the sea caves at the waterline are signs of an actively eroding cliff.

Figure 14.19
A wave-cut platform (the wide, horizontal bench of dark rock at the base of the cliffs) is exposed at low tide at La Jolla, California.

Figure 14.20
Several stacks show how this point is retreating due to wave erosion near Mendocino, California. One stack has been cut through by waves to form an arch.

Figure 14.21
Arch formed by wave erosion of a stack, Pacific City, Oregon.
Photo by D. A. Rahm, courtesy Rahm Memorial Collection, Western Washington University.

retreats inland (figure 14.20). They form small, rocky islands off retreating coasts, often directly off headlands (figure 14.17). **Arches** (or *sea arches*) are bridges of rock left above openings eroded in headlands or stacks by waves (figure 14.21). The openings are eroded in spots where the rock is weaker than normal, perhaps because of closely spaced fractures.

Depositional Coasts

Many coasts are gently sloping plains and show few effects of wave erosion. Such coasts are found along most of the Atlantic Ocean and Gulf of Mexico shores of the United States. These coasts are primarily shaped by sediment deposition, particularly by longshore drift of sand.

Coasts such as these are often marked by **barrier islands**—ridges of sand that parallel the shoreline and extend above sea level (figure 14.22). These barrier islands may have formed from sand eroded by waves from deeper water offshore, or they may be greatly elongated sand spits formed by longshore drift. The slowly rising sea level associated with the melting of the Pleistocene glaciers may have been a factor in their development. A protected lagoon separates barrier islands from the mainland. Because the lagoon is protected from waves, it provides a quiet waterway for boats. A series of such lagoons stretches almost continuously from New York to Florida, and many also exist along the Gulf Coast, forming an important route for barge traffic. As tides rise and fall, strong tidal currents may wash in and out of gaps between barrier islands, distributing sand in submerged *tidal deltas* both landward and seaward of the gaps.

Some barrier islands along the Atlantic and Gulf coasts are densely populated. Atlantic City (New Jersey), Ocean City (Maryland), Miami Beach (Florida), and Galveston (Texas) are examples of cities built largely on barrier islands. In some of these cities, houses, luxury hotels, and condominiums are clustered near the edge of the sea; many are built upon the loose sand of the island (figure 14.23). These developed areas are vulnerable to late-summer hurricanes that sooner or later bring huge storm waves onto these coasts, eroding the sand and undermining the building foundations at the water's edge.

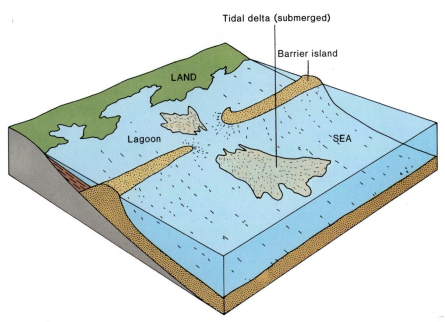

B

A

Figure 14.22
(*A*) A barrier island on a gently sloping coast. A lagoon separates the barrier island from the mainland, and tidal currents flowing in and out of gaps in the barrier island deposit sediment as submerged tidal deltas. (*B*) A barrier island near Pensacola, Florida. Open ocean to left, lagoon to right, mainland Florida on far right. Light-colored lobes of sand within lagoon were eroded from the barrier island by hurricane waves.
Photo by Frank M. Hanna.

Figure 14.23
Hotels built upon the loose sand of a barrier island, Miami Beach, Florida. The lagoon and mainland Florida are visible at the upper left.
Photo by Florida Division of Tourism.

Nonmarine deposition may also shape a coast. Rapid sedimentation in *deltas* by rivers can build a coast seaward (figure 10.31). *Glacial deposition* can form shoreline features. Several islands off the New England coast were glacially deposited; Long Island, New York, formed from two end moraines.

Drowned Coasts

Drowned (or *submergent*) coasts are common because sea level has been rising worldwide for the past 10,000 years. During the glacial ages of the Pleistocene (chapter titled "Glaciers and Glaciation"), sea level was 300 to 600 feet (100 to 200 meters) below its present level. The shallow sea floor near the continents was then dry land, and rivers flowed across it, cutting valleys. As the great ice sheets melted, sea level began to rise, drowning the river valleys. These drowned river mouths, called **estuaries,** mark many coasts today (figure 14.24). They extend inland as long arms of the sea. Fresh water from rivers mixes with the sea water to make most estuaries brackish. The quiet, protected environment of estuaries makes them very rich in marine life, particularly the larval forms of numerous species. Unfortunately, cities and factories built on many estuaries to take advantage of quiet harbors are severely polluting the water and the sediment of the estuaries. The poor circulation that characterizes most estuaries hinders the flushing away of this pollution, and estuary shellfish are sometimes inedible as a result.

Drowned coasts may be marked by fiords, glacially cut valleys flooded by rising sea level (figure 12.37). They form in the same way as estuaries, except they were cut by glacial ice rather than rivers during low sea-level stands.

A

B

Figure 14.24
(*A*) An estuary formed as rising sea level drowned a river valley, Malibu, California. Sand spits have migrated across the mouth of the estuary from two directions. Tidal currents and river flow keep the estuary open to the sea. (*B*) Landsat satellite photo of estuaries, Albemarle and Pamlico Sounds, North Carolina. Barrier islands are visible in upper right. Infrared image shows vegetation as red.
Photo *A* by Frank M. Hanna. Photo *B* by NASA.

BOX 14.2

The Dangers of High Sea Level

A slight rise in sea level can be devastating to coastal communities on a low-lying shore. In the United States many coasts have very low relief. From New Jersey south to Florida, and along the Gulf of Mexico to southern Texas, the coastal land is very flat. A small rise in sea level can send sea water many miles inland.

Hurricanes, which are common along these coasts, tend to cause short-term rises in sea level called *storm surges.* Hurricanes consist of strong winds rotating counterclockwise (in the northern hemisphere) around a region of very low air pressure. They may be 300 miles (500 kilometers) in diameter and contain winds up to 180 miles (300 kilometers) per hour. The low air pressure in the center of a hurricane literally sucks up the sea surface into a broad dome, and this rise in sea level is accentuated by strong onshore winds, which pile water against shore (box figure 1). Storm surges can easily raise sea level 15 feet (about 5 meters). A storm surge 26 feet (8 meters) high struck Galveston, Texas, in 1900. High storm waves on top of the high sea level destroyed countless buildings, and 6,000 people died, many by drowning. Hurricane-tracking programs cut the death tolls of storm surges today by providing advance warning to coastal communities. In September, 1989, hurricane Hugo hit Charleston, South Carolina, with 135 mile-per-hour winds and a 17-foot storm surge, causing more than $1 billion in damage. Because more than 500,000 people were evacuated along the low-lying coasts, however, the death toll was only about 25.

Long-term changes in sea level also occur. Sea level has been rising steadily for more than 10,000 years as the Pleistocene glaciers melted. The remaining continental glaciers on Greenland and Antarctica continue to recede, adding water to the oceans and raising sea level. A series of recent predictions by scientists indicate that global sea level may stand 4 to 7 feet (1 to 2 meters) higher by the year 2100. The burning of coal and oil may contribute to this

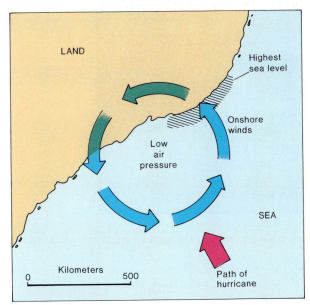

Box 14.2 Figure 1
Strong onshore winds in a hurricane pile water against the shore, forming a storm surge (high sea level) that may cause severe flooding on a low-lying coast.

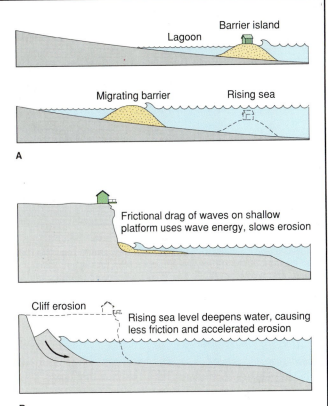

Box 14.2 Figure 2
Rising sea level can cause erosion of both gentle and steep coasts, leading to destruction of buildings.

rise by adding carbon dioxide to the atmosphere; this has the effect of trapping more of the sun's energy on earth, warming the air and accelerating the melting of glaciers, by the "greenhouse effect." As the air warms, so does the ocean, and thermal expansion of the water as it warms contributes to the rise of sea level.

Although a sea-level rise of 4 feet may seem slight, it is not. On low-lying coasts the effect of ocean waves (particularly storm waves) would extend much farther inland than before, and many parts of coastal communities would be destroyed. Vast areas now covered with beach-front homes and hotels would become uninhabitable due to flooding and to the landward migration of barrier islands (box figure 2). On steep coasts the erosion of coastal cliffs would accelerate, and choice ocean-view building sites would crumble into the sea. Careful planning for the use of coastal land in your lifetime will be necessary to lessen this anticipated destruction.

Developed land at the ocean's edge will likely face a grim choice in the future—abandonment of existing buildings or expensive "armoring" of the coast with structures such as seawalls in hopes of protecting the buildings from wave damage (box figure 3).

Box 14.2 Figure 3
Seawall at Cape May, New Jersey. Note the lack of a sand beach seaward of the seawall. What would this area look like after a 7-foot rise in sea level?
Photo by S. Jeffress Williams.

Figure 14.25
Uplifted marine terrace near Half Moon Bay, California. The flat land surface at the top of the sea cliffs was eroded by wave action, and elevated above sea level by tectonic uplift.
Photo by K. Lajoie, U.S. Geological Survey.

Uplifted Coasts

Uplifted (or *emergent*) coasts have been elevated by deep-seated tectonic forces. The land has risen faster than sea level, so parts of the old sea floor are now dry land.

Marine terraces form just offshore from the beach face, as described earlier in this chapter. These terraces can be wave-cut platforms caused by erosion of rock associated with cliff retreat, or they can be wave-built terraces caused by deposition of sediment. If the shore is elevated by tectonic uplift, these flat surfaces will become visible as *uplifted marine terraces* (figure 14.25). They formed below the ocean surface but are visible now because of uplift. The tectonically unstable Pacific coast of the United States and Canada has many areas marked by uplifted terraces, along with the erosional coast features described earlier.

Coasts Shaped by Organisms

The growth of coral and algal *reefs* offshore can shape the character of a coast. The reefs act as a barrier to strong waves, protecting the shoreline from most wave erosion. Carbonate sediments blanket the sea floor on both sides of a reef and usually form a carbonate sand beach on land (figure 6.18). Southernmost Florida has a coast of this type.

Branching *mangrove roots* dominate many parts of the southeastern United States coast. The roots dampen wave and current action, creating a quiet environment that provides a haven for the larval forms of many marine organisms, and may trap fine-grained sediment. Mangroves also deposit layers of organic peat on low-lying coasts.

Summary

Wind blowing over the sea surface forms waves, which transfer some of the wind's energy to shorelines. Orbital water motion extends to a depth equal to half the wave length.

As a wave moves into shallow water, the bottom flattens the orbital motion and causes the wave to slow and peak up, eventually forming a *breaker* whose crest topples forward. The turbulence of *surf* is an important agent of sediment erosion and transportation.

Wave refraction bends wave crests and makes them more parallel to shore. Few waves actually become parallel to the shore, and so *longshore currents* are set up in the surf zone.

A beach consists of a *berm, beach face,* and *marine terrace.* Summer beaches have a wide berm and a smooth offshore profile. Winter beaches are narrow, with offshore bars.

Longshore drift of sand is caused by the waves hitting the beach face at an angle and also by longshore currents.

Deposition of sand that is drifting along the shore can form *spits* and *baymouth bars.* Drifting sand may also be deposited against jetties or groins or inside breakwaters.

Rivers supply most sand to beaches, although local erosion may also contribute sediment. If the river supply of sand is cut off by dams, the beaches gradually disappear.

Coasts may be erosional or depositional, drowned or uplifted, or shaped by organisms such as corals and mangroves.

Coastal straightening by waves is caused by headland erosion and by deposition within bays.

A coast retreating under wave erosion can be marked by *sea cliffs*, a *wave-cut platform*, *stacks*, and *arches*.

Waves can form *barrier islands* off gently sloping coasts. River and glacial deposition can also shape coasts.

Drowned coasts are marked by *estuaries* and fiords. *Uplifted marine terraces* characterize coasts that have risen faster than the recent rise in sea level.

Terms to Remember

arch (sea arch)
barrier island
baymouth bar
beach
beach face
berm
breaker
coast
coastal straightening
crest (of wave)
estuary
headland
longshore current

longshore drift
marine terrace
rip current
sea cliff
spit
stack
surf
trough (of wave)
wave-cut platform
wave height
wavelength
wave refraction

Questions for Review

1. Show in a sketch how longshore drift of sand can form a baymouth bar.
2. In a sketch, show how and why sand moves along a beach face when waves approach a beach at an angle.
3. How are summer beaches different from winter beaches? Discuss the reasons for these differences.
4. What would happen to the beaches of most coasts if all the rivers flowing to the sea were dammed? Why?
5. What does the presence of an estuary imply about the recent geologic history of a region?
6. Describe how waves can straighten an irregular coastline.
7. Describe the transition of deep-water waves into surf.
8. Show in a sketch the refraction of waves approaching a straight coast at an angle. Explain why refraction occurs.
9. What is a longshore current? Why does it occur?

Questions for Thought

1. What might happen to a sandy beach located in a downdrift direction from a newly constructed seawall?
2. Waterfalls are uncommon where rivers enter the sea. Why?

Supplementary Readings

Bascom, W. 1980. *Waves and beaches*. Rev. ed. New York: Doubleday Anchor Books.

Bird, E. C. 1985. *Coastline changes: a global review*. New York: John Wiley & Sons.

Davis, R. A., Jr., and R. L. Ethington. 1976. *Beach and nearshore sedimentation*. Tulsa: Society of Economic Paleontologists and Mineralogists, Special Publication 24.

Dolan, R., B. Hayden, and H. Lins. 1980. Barrier islands. *American Scientist* 68: 16–25.

Inman, D. L., and B. M. Brush. 1973. The coastal challenge. *Science* 180 (4094): 20–32.

Kaufman, W., and O. H. Pilkey, Jr. 1983. *The beaches are moving: the drowning of America's shoreline*. Durham, N.C.: Duke University Press.

King, C. A. M. 1959. *Beaches and coasts*. London: Edward Arnold Ltd.

Komar, P. D. 1976. *Beach processes and sedimentation*. Englewood Cliffs, N.J.: Prentice-Hall.

Shepard, F. P. 1973. *Submarine geology*. 3d ed. New York: Harper & Row.

Shepard, F. P., and H. R. Wanless. 1971. *Our changing coastlines*. New York: McGraw-Hill.

Snead, R. E. 1982. *Coastal landforms and surface features: a photographic atlas and glossary*. Stroudsburg, Pennsylvania: Hutchinson Ross.

In previous chapters we have discussed how rock at the earth's surface is affected by erosional agents such as wind and water. We now shift our focus to changes in bed rock caused by powerful forces originating deep within the earth. In this chapter we explain how rocks respond to these tectonic forces.

The main purpose of this chapter is to help you recognize certain geologic structures, understand the forces that caused them, and thus determine the geologic history of an area.

Some principles dealt with in chapter 8 should help you interpret the way structures develop in an area and the sequence. The principles of original horizontality, of superposition, and of cross-cutting relationships are as important to structural geology as they are to determining relative time.

Subsequent chapters will require an understanding and knowledge of structural geology as presented in this chapter. To understand earthquakes, for instance, one must know about faults. Appreciating how major mountain belts and the continents have evolved (chapter 20) calls for a comprehension of faulting and folding. Understanding plate tectonic theory as a whole (chapter 19) also requires a knowledge of structural geology, for it was primarily to explain certain structural phenomena that plate tectonic theory was developed.

Finally, understanding structural geology can help us more fully appreciate the problem of finding more of the earth's dwindling natural resources. Chapter 21 discusses the association of certain geologic structures with petroleum deposits and other valuable resources.

15

Geologic Structures

Folded sedimentary rock on the West side of the Salton Sea, California.
Photo by Frank M. Hanna.

Figure 15.1
Folded and faulted sedimentary bed rock exposed in a road cut near Palmdale, California.
Photo by C. C. Plummer.

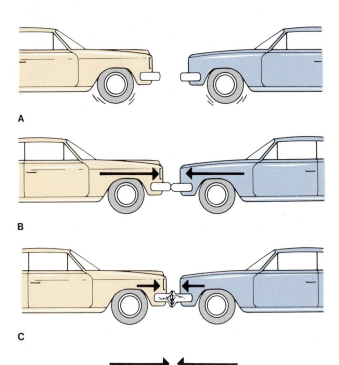

A

B

C

D

Figure 15.2
Colliding vehicles show the relationship between stress and strain. (*A*) Two cars moving toward each other. (*B*) Stress. (*C*) Strain. (*D*) Arrows used to represent compressive stress.

In a broad sense structural geology can be thought of as the study of the architecture of the earth's crust, its deformational features, and their mutual relations and origins. For our purposes, **structural geology** can be defined as the branch of geology concerned with the shapes, arrangement, and interrelationships of bedrock units and the forces that cause them.

Tectonic Forces at Work

Stress and Strain

Tectonic forces (as defined in chapter 1) move and deform parts of the earth's crust. The relationship between *movements* within the earth and the resulting *deformation* of the crustal rocks is as follows. The movement of a large or small part of the crust creates **stress,** a force that acts on a body (or rock unit) and tends to change its size or shape. The adjustment of the rock unit to stress is called **strain,** the change of the rock in size (volume) or shape in response to stress. Figure 15.1 is an example of strained, layered rock.

The relationship between stress and strain can be illustrated by what happens when a car's bumper is crumpled in a collision with another moving vehicle (figure 15.2). When the two cars meet, both bumpers are stressed. The heavier the cars and the faster they are moving toward each other, the greater the stress. The result of the stress is that the bumpers crumple; in other words, they become strained.

The forces represented by the two cars moving toward each other are indicated in the figure by arrows. In this particular case the forces are said to be **compressive;** they tend to *shorten* the body (or bodies) involved. Both cars are a bit shorter because of the collision. Compressive forces are represented by a pair of arrows pointing toward each other ($\longrightarrow$ $\longleftarrow$).

The opposite of compressive force is **tensional force,** which tends to *elongate,* or pull apart, a body. If you stretch a rubber band, you are applying tension to it. Keeping to our analogy of the crashed cars, a tow truck (figure 15.3) can bend or even pull off the rear bumper by exerting too much tensional stress. Tensional forces are represented by a pair of arrows pointing away from each other ($\longleftarrow$ $\longrightarrow$).

Shear stress, a third type, is due to forces parallel but in opposite directions to each other. It causes strain *parallel* to the direction of the forces. If our two cars had not crashed head-on but had slid past each other, the strain would have broken door handles and outside mirrors and scraped the doors (figure 15.4). Shear stresses are represented by a pair of parallel arrows pointing in opposite directions ($\rightleftharpoons$).

In solid materials, stress can cause three types of strain or deformation: plastic, elastic, and fracture. In **plastic strain** a body is molded or bent under stress and does not return to its original shape after the stress is released. The crumpled car bumper is an example of plastic strain; so is the behavior of rocks during regional metamorphism or ice in glacier flow. When rock undergoes plastic strain, it is said to be *ductile.*

If a deformed body recovers its original shape after the stress is released, the strain is **elastic.** A mattress is deformed when a person is on it but recovers its original shape when the person gets out of bed. Rock is elastic to a certain extent. Parts of northern Europe and North America are still in the process of "bouncing back" to their former elevations after having been depressed by the weight of the last continental Pleistocene ice sheet.

In strain by **fracturing,** the body under stress cracks or breaks, as the name suggests. *Brittle* objects tend to fracture rather than yield. Some rocks initially yield elastically or plastically, but if the intensity of stress is increased, they may fracture.

A sedimentary rock exposed at the earth's surface is brittle; it will fracture if you hit it with a hammer. How then do sedimentary rocks, such as those shown in figure 15.1, become bent (or deformed plastically)? The answer is that either stress was applied very slowly or that the rock was deformed under considerable confining pressure (buried under more rock).

Note, however, that there are some fractures (faults) disrupting the bent layers in figure 15.1. This tells us that although the rock was ductile initially, the stress became too intense and the rock fractured.

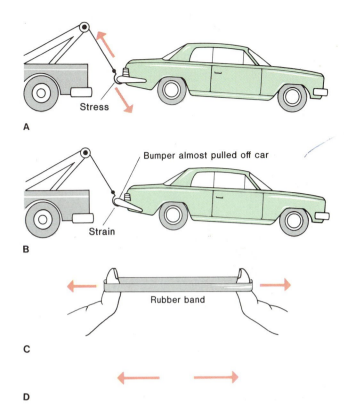

Figure 15.3
Tensional stress and resulting strain. (**A**) Tow truck with inexperienced operator attempts to pull car away by hooking tow cable onto rear bumper. (**B**) Bumper is pulled out of shape (if not pulled off) by tensional forces. (**C**) Rubber band being stretched. Tensional stress causes stretching (strain) of the rubber band. (**D**) Arrows used to represent tensional stress.

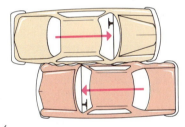

Figure 15.4
Shear stress as shown by two cars scraping past each other. The resulting strain is the deformation of the cars along their sides. Arrows indicate direction of forces.

Stress and Strain of Bed Rock

Some scientists have described the crust of the earth as "mobile" or "restless" because bed rock is moving and being deformed in many parts of the world. Bed rock may be displaced suddenly during earthquakes. By contrast, much of the movement of rock within the earth's crust is continuous and very slow, at rates of less than a millimeter per year.

Figure 15.5
Displaced ditch near Hollister, California.
Photo by B. Amundson.

Compared with most geologic processes, the present-day movement of the crust in much of California is very rapid. A large slice of the coastal portion of the state is moving relentlessly northward relative to the rest of North America and has apparently been doing so for millions of years. Some of the movement is jerky and associated with earthquakes (discussed in the following chapter), but elsewhere it is essentially continuous and smooth.

Around Hollister, California, homes and buildings are being slowly torn apart because they straddle an active fault (figure 15.5). A **fault** is a fracture in bed rock along which movement has taken place. The fault that goes through Hollister (part of the San Andreas fault system) is one of several major cracks extending deep into the earth's crust, separating the northward-moving coastal portion of California from the rest of the state. Hollister residents whose homes "ride" the fault report hearing almost constant creaking, evidently due to motion along the fault. Yet the movement—about one centimeter per year (roughly the same rate at which a fingernail grows)—is slow enough that walls can be patched with plaster as they crack.

Geologically rapid movement of the crust can be observed in young, developing mountain regions such as the California Coast Ranges, which have been forming throughout the Cenozoic Era (the last 65 million years). In other parts of the world, however, the continents and sea floors are shifting very slowly up and down as well as moving laterally. Some of these motions can be detected by precise, repeated surveying.

Structures as a Record of the Geologic Past

Some geologic structures that give us clues to the past have been described in earlier chapters. Batholiths, stocks, dikes, and sills, for example, are keys to past igneous activity (see chapter on intrusive activity). In this chapter

we are mainly concerned with types of structures that can provide a record of crustal deformation no longer active. Often we look for very old structures, once buried but now exposed by erosion.

The study of geologic structures is of more than academic interest. The petroleum and mining industries, for example, employ geologists to look for geologic structures associated with oil and metallic ore deposits. Understanding geologic structures is also important in evaluating problems related to engineering decisions and environmental planning, such as the siting of dams or nuclear reactors, and even the building of houses.

Implications of Horizontal and Inclined Layers of Rock

Layered rocks generally are easier to interpret than non-layered rocks. Therefore, our descriptions are mainly of deformed sedimentary or volcanic layered rocks.

To determine what type of structure is present in an area, and to decipher the sequence of events that caused that structure, we need to find out what has happened to layered rocks since they were deposited as sediments or lava flows. According to the principle of *original horizontality,* layers of sedimentary rock (or lava flows) began as horizontal beds or strata. Where we find essentially horizontal, layered, marine sedimentary rock, such as that exposed in the walls of the Grand Canyon, we conclude that the layers have been uplifted to their present position from beneath an ancient sea (figure 15.6). The uplifting has been uniform over a large area, so tilting is negligible. In this case the forces that caused the uplift can be shown with arrows of equal length, implying that the forces were equally strong throughout the area (figure 15.6*B*).

Where rock layers are observed to be inclined rather than horizontal, however, the forces responsible for the tilting must have been unequal—for example, as indicated by the arrows in figure 15.6*C*.

Geologic Maps and Field Methods

In an ideal situation, a geologist studying structures would be able to fly over an area and see the local and regional patterns of bed rock from above. Sometimes this is possible, but more often soil and vegetation conceal the bed rock. Therefore, geologists ordinarily use observations from a number of individual *outcrops* (exposures of bed rock at the surface) in determining the patterns of geologic structures. The characteristics of rock at each outcrop in an area are plotted on a map by means of appropriate symbols. With the data that can be collected, a geologist can make inferences about those parts of the area he or she cannot observe. The symbols on the field map, or rough draft, are thus converted into a **geologic map** of a given area. On such a map are plotted the distribution and nature of rock units, the occurrence of structural features (folds, faults, joints, etc.), ore deposits, and so forth. Sometimes surficial features, such as deposits by former glaciation, are included, but these may be shown separately on a different type of geologic map.

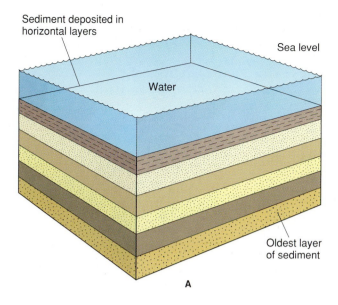

Sediment deposited in horizontal layers

Water

Sea level

Oldest layer of sediment

A

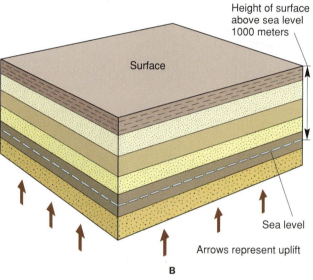

Height of surface above sea level 1000 meters

Surface

Sea level

Arrows represent uplift

B

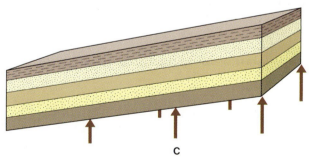

C

Figure 15.6
An area (*A*) during deposition of sediment below sea level and (*B*) after uplift has occurred. (*C*) Tilted layers of sedimentary rock, showing possible distributions of forces. The length of arrows indicates the relative strength of forces.

Anyone trained in the use of geologic maps can find on them considerable information about local geologic structures because standard symbols and terms are used on the maps and the accompanying reports. For example, the symbol ⊕ on a geologic map denotes horizontal bedding in an outcrop. Different colors or patterns on a geologic map may represent distinct rock units, such as *formations* (defined in the chapter on sedimentary rocks).

Figure 15.7
Sedimentary beds along the coast of South Australia. The geologist is walking in the direction of strike. The direction of dip is to the right of the photograph.
Photo by C. C. Plummer.

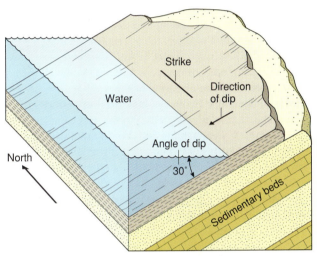

Strike

Direction of dip

Water

Angle of dip

North

30°

Sedimentary beds

Figure 15.8
Strike, angle of dip, and direction of dip.

Strike and dip   Where bedding has been tilted (figure 15.7), someone studying a geologic map of the area would want to know the extent and direction of tilting. By convention, this is determined by plotting the relationship between a surface of an inclined bed and an imaginary horizontal plane. You can understand the relationship by looking carefully at figure 15.8, which represents sedimentary beds cropping out alongside a lake (the lake surface provides a convenient horizontal plane for this discussion).

Strike is the compass direction of a line formed by the intersection of an inclined plane with a horizontal plane. In this example, the inclined plane is a bedding plane. You can see from figure 15.8 that the beds are striking from north to south. Customarily only the northerly direction (of the strike line) is given, so we simply say that these beds strike north.

Observe that the **angle of dip** is measured downward from the horizontal plane to the bedding plane (an inclined plane). Note that the angle of dip (30° in the

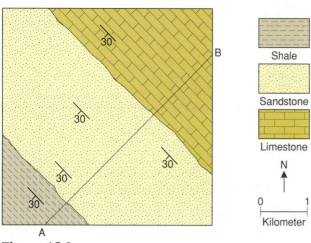

Figure 15.9
A geologic map of an area with three sedimentary formations. (Each formation may contain many individual sedimentary layers, as explained in chapter 6.) Beds strike northwest and dip 30° to the southwest.

figure) is measured within a vertical plane that is perpendicular to both the bedding and the horizontal planes.

The **direction of dip** is the compass direction in which the angle of dip is measured. If you could roll a ball down a bedding surface, the compass direction in which the ball rolled would be the direction of dip.

The dip angle is always measured at right angles to the strike, that is, perpendicular to the strike line as shown in figure 15.8. Because the beds could dip away from the strike line in either of two possible directions, the general direction of dip is also specified—in this example, west.

Besides recording strike and dip measurements in a field notebook, a geologist who is mapping an area draws strike and dip symbols on the field map, such as Y or ⅄ for each outcrop with dipping or tilted beds. The long line of the symbol is aligned with the compass direction of the strike; the small tick, which is always drawn perpendicular to the strike line, is put on one side or the other, depending on which way the beds actually dip. The angle of dip is given as a number next to the appropriate symbol on the map. Thus, ³⁰ Y indicates that the bed is dipping 30° from the horizontal. These symbols are used in figure 15.9, which is a geologic map that shows all the sedimentary layers striking northwest and dipping 30° to the southwest.

On a map the intersection of the two lines at the center of each strike and dip symbol represents the location of the outcrop where the strike and dip of the bed rock were measured. A specially designed compass called a Brunton (after the inventor) is used by geologists for this purpose (figure 15.10). A Brunton compass contains a level and a device for measuring angles of inclination.

Beds with vertical dip require a unique symbol because they dip neither to the left nor the right of the direction of dip. The symbol used is ✗, which (assuming that the top of the page is north) indicates that the beds are striking northeast and that they are vertical.

Figure 15.10
Determining the strike of a steeply dipping bed.

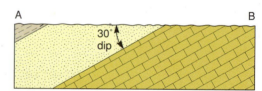

Figure 15.11
Geologic cross section for line *A–B* on the geologic map in figure 15.9.

Geologic cross sections A **geologic cross section** is a vertical representation of a portion of the earth. Geologic cross sections help people visualize geology in three dimensions. They are used extensively throughout this and many other geology books. Figure 15.11 shows a geologic cross section constructed between points *A* and *B* on the geologic map shown in figure 15.9.

Folds

Folds are bends in layered bed rock. Folded rock can be compared to several layers of rugs or blankets that have been pushed into a series of arches and troughs. Oftentimes folds in rock can be seen in roadcuts or other exposures (figure 15.12). When the arches and troughs of folds are concealed (or when they exist on a grand scale),

Figure 15.12
Folded rock, Calico Hills, California.
Photo by C. C. Plummer.

Figure 15.13
Marble slab that has sagged under its own weight.
Photo by W. T. Lee, U.S. Geological Survey.

geologists can still determine the presence of folds by noticing repeated reversals in the direction of dip taken on outcrops in the field or shown on a geologic map.

The fact that the rock is folded shows that it was strained plastically rather than elastically or by fracturing. Yet the rock exposed in outcrops is generally brittle and shatters when struck with a hammer. We can speculate that at the time of folding the rock layers were buried fairly deeply. Under such conditions of high pressure and high temperatures, the layered rock tended to yield plastically to stress and to bend rather than break. Rock also tends to bend rather than break if stress is applied very slowly. The marble slab shown in figure 15.13 sagged under its own weight during a period of over a hundred years, but rapidly applied stress, such as a hammer blow, would probably break it.

Geometry of Folds

The shapes and patterns of folds reveal much about the nature and extent of the forces of deformation that created them. Familiarity with the geometry of folds helps one understand how to interpret them. Figure 15.14 illustrates the relationship between the arches (anticlines) and the troughs (synclines) of folds.

An **anticline** is an arched fold. Usually the rock layers dip away from the **hinge line** (or axis) of the fold. It can be visualized as an upfold, that is, one that opens downward. The downfolded counterpart of an anticline is a **syncline,** a troughlike fold. The layered rock usually dips toward the syncline's hinge line. Think of it as a downfold opening upward. In the series of folds shown in figure 15.14, two anticlines are separated by a syncline. Each anticline and adjacent syncline share a **limb.** Note the hinge lines on the crests of the two anticlines and bottom of the syncline. Similar hinge lines would be underground

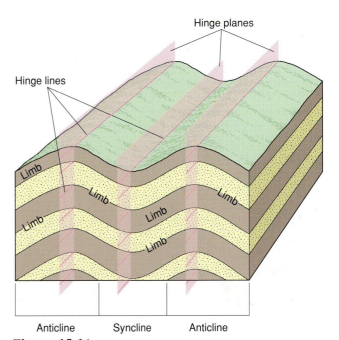

Figure 15.14
Two anticlines and a syncline.

at the contacts between any two adjacent folded layers. For each anticline and the syncline, the hinge lines are within the shaded vertical planes. Each of these planes is a **hinge plane,** a plane containing all of the hinge lines of a fold.

It is important to remember that anticlines are not necessarily related to ridges nor synclines to valleys, because valleys and ridges are nearly always erosional features. In an area that has been eroded to a plain, the presence of underlying anticlines and synclines is determined by the direction of dipping beds in exposed bed rock, as shown in figure 15.15. (In the field, of course, the cross sections are not exposed to view as they are in the diagram.)

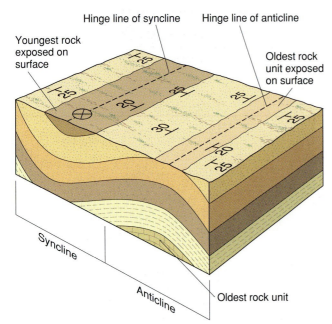

Figure 15.15 labels: Hinge line of syncline, Hinge line of anticline, Youngest rock exposed on surface, Oldest rock unit exposed on surface, Syncline, Anticline, Oldest rock unit

Figure 15.15
Folded rock. This view is a *block diagram*. Its top represents the land surface and its two visible sides are vertical cross sections. The surface has been eroded to a nearly horizontal plain. Side views are interpretations based on what the geologist notices on the surface.

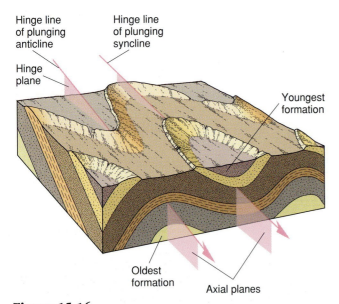

Figure 15.16 labels: Hinge line of plunging anticline, Hinge line of plunging syncline, Hinge plane, Youngest formation, Oldest formation, Axial planes

Figure 15.16
Plunging folds: anticline on left and right, syncline in center. The hinge lines are at an angle to the block diagram, penetrating the surface and emerging from the front cross section.

Figure 15.15 also illustrates how determining the relative ages of the rock layers, or beds, can tell us whether a structure is an anticline or a syncline. Observe that the oldest rocks are exposed along the hinge line of the anticline. This is because lower layers in the originally flat-lying sedimentary or volcanic rock were moved upward and are now in the core of the anticline. The youngest

A

B

Figure 15.17
Plunging folds. (*A*) Anticline in Utah plunging in the direction of the upper part of the photo. (*B*) The front portion, or "nose," of an anticline in Spain.
Photo *A* by Frank M. Hanna. Photo *B* by C. C. Plummer.

rocks, on the other hand, which were originally in the upper layers, were folded downward and now crop out along the synclinal hinge line.

Plunging fold The examples shown so far have been of folds with horizontal hinge lines. These are the easiest to visualize. In nature, however, anticlines and synclines are apt to be **plunging folds**—that is, folds in which the hinge lines are not horizontal. On a surface leveled by erosion, the patterns of exposed strata (beds) resemble V's or horseshoes (figures 15.16 and 15.17) rather than the striped patterns of nonplunging folds. However, plunging anticlines and synclines are distinguished from one another in the same way as are nonplunging folds—by directions of dip or by relative ages of beds.

Structural domes and structural basins A **structural dome** is a structure in which the beds dip away from a central point. In cross section, a dome resembles an anticline. In a **structural basin,** the beds dip toward a central point; in cross section, it is comparable to a syncline (figure 15.18).

Domes and basins tend to be features on a grand scale (some are more than a hundred kilometers across), formed

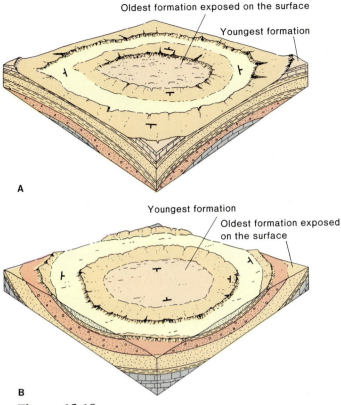

A

B

Figure 15.18
(*A*) Structural dome. (*B*) Structural basin.

Figure 15.19
Dome near Casper, Wyoming. The ridges are sedimentary layers that are resistant to erosion. Beds dip away from the center of the dome.
Photo by D. A. Rahm, courtesy of Rahm Memorial Collection, Western Washington University.

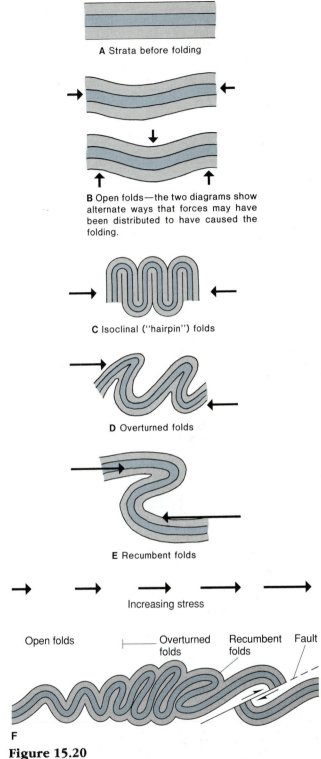

A Strata before folding

B Open folds—the two diagrams show alternate ways that forces may have been distributed to have caused the folding.

C Isoclinal ("hairpin") folds

D Overturned folds

E Recumbent folds

Increasing stress

Open folds — Overturned folds — Recumbent folds — Fault

F

Figure 15.20
Cross sections of some types of folds and possible directions of stress implied by each. The length of the arrows in A through E is proportional to the intensity of the forces involved.

Interpreting Folds

Folds occur in many varieties and in any size. Some are studied under the microscope, while others can have adjacent axes tens of kilometers apart. Some folds are a kilometer or more in height. Figure 15.20 shows several types of folds, each one revealing something about the stress that created the pattern. **Open folds** (figure 15.21*A*) have limbs that dip gently. All other factors being equal, the more

by uplift somewhat greater (for domes) or less (for basins) than that of the rest of a region. Most of Michigan and parts of adjoining states and Ontario are on a large structural basin. Domes of similar size are found in other parts of the Middle West. Smaller domes are found in the Rocky Mountains (figure 15.19).

Geologic Structures   **329**

A

B

C

Figure 15.21
Various types of folds. (A) Open folds in Spain. (B) Overturned syncline in Spain. (Notice that both limbs are dipping to the right of the photo.) (C) Recumbent folds in the Canadian Rockies.
Photos by C. C. Plummer.

open the fold, the less intense the forces that created it. By contrast, an **isoclinal fold,** one in which limbs are parallel to one another, implies intense compressive stress.

Overturned folds (figure 15.21*B*), in which limbs dip in the *same direction,* imply that compressive forces caused the upper part to override the lower part of the fold pattern. Looking at an outcrop in which only the overturned limb of a fold is exposed, you would probably conclude

that the youngest bed is at the top. However, the principles of *superposition* (see chapter titled "Time and Geology") cannot be applied to determine top and bottom for overturned beds. You must either see the rest of the fold or find features within the beds that indicate the original top or upward direction.

Recumbent folds (figure 15.21*C*) are overturned to such an extent that the limbs are essentially horizontal. Note that the terms *syncline* and *anticline* (as used here) are not applicable to recumbent folds.

Is There Oil Beneath My Property?—First Check the Geologic Structure

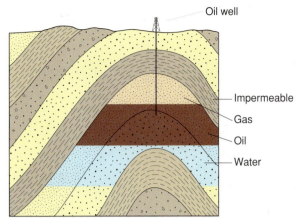

Box 15.1 Figure 1
Anticlinal trap. Gas, oil, and water saturate permeable rock.

An "oil pool" can exist only under certain conditions. Crude oil does not fill caves underground as the term *pool* suggests. Oil simply occupies the pore spaces of certain sedimentary rocks, such as poorly cemented sandstone, in which space exists between grains. Natural gas (being lighter) often occupies the pore spaces above the crude oil, while water (being heavier) saturates the rock below the oil pool (box figure 1).

A *source rock,* which is always a sedimentary rock, must be present for oil to form. The sediment of the source rock has to include organisms buried during sedimentation. This organic matter partially decomposes into petroleum and natural gas. Once formed, the droplets of petroleum tend to migrate, following fractures and interconnecting pore spaces. Being lighter than rock and water, the petroleum migrates upward.

If it is not blocked by impermeable rock, the oil may migrate all the way to the surface, where it dissipates and is permanently lost for human use. Natural seepage of petroleum exists both on land and offshore. Where impermeable rock blocks the oil droplets' path of migration, an oil pool may accumulate below the rock, much like helium-filled balloons might collect under a domed ceiling. For any significant amount of oil to collect, the rock below the impermeable rock must be porous as well as permeable. Such a rock, when it contains oil, is called a *reservoir rock.*

Another condition is that the geologic structure must be one that favors the accumulation and retention of petroleum. An "anticlinal trap" is one of the best structures for holding oil. As oil became a

major energy source and the demand for it increased, most of the newly discovered wells penetrated anticlinal traps. Geologists discovered these by looking for indications of anticlines cropping out on the earth's surface. As time went on, other types of structures were also found to be oil traps. Many of these were difficult to find because of the lack of telltale surface patterns indicating favorable underground structures. Box figure 2 illustrates some types of traps other than anticlinal ones that have a potential for oil production. (Unconformities and faults are described later in this chapter; stratigraphic traps are due to changes in rock type within a bed as described in the chapter on sedimentary rocks.)

At present, oil companies rely on detailed and sophisticated geologic studies of an area they hope has the potential for an "oil strike." The petroleum industry also depends heavily on geophysical techniques (see chapter titled "The Earth's Interior") for determining, by indirect means, the subsurface structural geology. (The geology of petroleum is described in more detail in the chapter on geologic resources.)

Even when everything indicates that conditions are excellent for oil to be present underground, there is no guarantee that oil will be found. Eventually an oil

Fractures in Rock

If a rock is brittle, or if the forces on it are exerted faster than it can bend to accommodate the strain, the rock fractures, commonly with some movement or displacement. If essentially no displacement occurs, a fracture or crack in bed rock is called a **joint.** If the rock on either side of a fracture moves, the fracture is a *fault* (as defined earlier). Most rock at or near the surface is brittle, so nearly all exposed bed rock is jointed to some extent.

Joints

In discussing volcanoes, we described *columnar jointing,* in which hexagonal columns form as the result of contraction of a cooling, solidified lava flow. *Exfoliation* (sometimes called sheet jointing), a type of jointing caused by expansion, has been discussed along with weathering.

Parallel joints in bed rock can be seen in many places (figure 15.22). Joints that are oriented in one direction, approximately parallel to one another, make up a **joint set.**

Box 15.1 *continued*

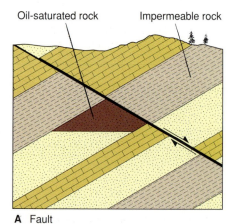

Oil-saturated rock   Impermeable rock

A Fault

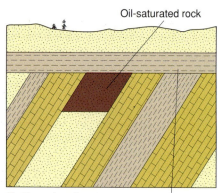

Oil-saturated rock

B Unconformity

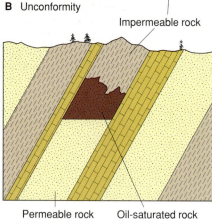

Impermeable rock

Permeable rock   Oil-saturated rock

C Stratigraphic

Box 15.1 Figure 2
Other types of oil traps.

company must commit perhaps a million dollars or more to drill a test well, or "wildcat" well. Statistics indicate that the chance of a test well yielding commercial quantities of oil is much less than one in ten. As more and more of the world's supply of petroleum is used up, what is left becomes increasingly harder—and costlier—to find.

Figure 15.22
Vertical joints in nearly horizontal sedimentary rock at Bryce Canyon, Utah. The pinnacles are due to erosion along the vertical joints.
Photo by C. C. Plummer.

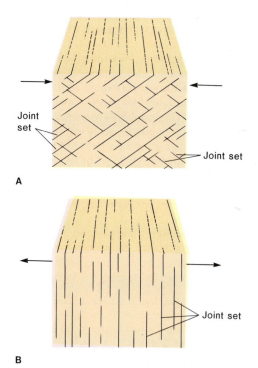

Joint set

Joint set

A

Joint set

B

Figure 15.23
Joint sets. Arrows indicate directions of principal stress. (A) Two joint sets created by compressional stress. (B) Joint set created by tensional stress.

More than one joint set may be present in the same outcrop. Figure 15.23 shows how compressive stress generally produces two joint sets that cut across each other and intersect the direction of stress. Tensional stress, by contrast, tends to produce a single joint set that is perpendicular to the direction of stress.

Faults and Nuclear Power Plants

In the 1960s nuclear power was expected to provide most of the additional electrical energy needed for the rest of the century. The West Coast of the United States seemed ideal for locating future nuclear power plants for the following reasons. (1) Much of the coast is sparsely populated, so if a radiation leak did occur, it would affect only the few people living near the power plant. (2) The urban centers requiring most of the electric power are close enough to the coast so that the costs of building power lines and transmitting electricity could be kept to a minimum. (3) The ocean water from the Pacific could efficiently cool a nuclear reactor—a major consideration since two-thirds of the heat generated by a nuclear power plant is wasted heat that must be dispersed. Because of the mountainous coastline, power plants could be constructed out of reach of potential floods.

In spite of these advantages, problems have arisen at specific sites selected for the plants. Geologic studies of each site are always necessary to ensure that there is no danger from landsliding and that the underlying material, usually bed rock, has the strength to support the structure.

Furthermore, a potential site must be carefully studied to be sure there are no active faults. The West Coast is one of the most tectonically active regions in the United States and includes many closely spaced faults, active and inactive. If a nuclear power plant were built on an active fault, even a minor amount of movement might rupture the nuclear reactor. To complicate matters further, determining whether a fault is active or inactive is seldom easy; evidence of fault motion in the recent geologic past must be found. Different government agencies define an active fault in different ways. For some, an active fault is one in which motion has occurred within the past 10,000 years; others extend the time limit to the past 100,000 years.

Most earthquakes are caused by fault movement (as explained in chapter titled "Earthquakes"). Federal regulations prohibit building a nuclear power plant within a quarter-mile of a known active fault. Moreover, before a nuclear power plant building permit will be issued, the entire earthquake history of the larger region must be carefully analyzed to ensure that a more distant but exceptionally strong earthquake could not affect the plant.

In 1976 a nuclear power plant at Eureka, California, was shut down because of concern about its safety in the event of an earthquake. The power company then spent $21 million strengthening it to comply with federal standards. In 1980 one of the strongest earthquakes in that region's history took place near the still shut down power plant. Although the plant was undamaged, the power company dropped its plans to reactivate the nuclear reactor.

The Diablo Canyon power plant in southern California was completed and ready to be activated in 1979 when new concerns about nearby active faults caused the government to withhold an operating permit. Not until 1984 was the power company allowed to begin testing the nuclear reactor. Eventually, the power plant received its operating permits and the utility company could begin recovering the half billion dollars of construction costs by selling electricity.

Because of public fears about the safety of nuclear power plants, it seems likely that many nuclear plants originally projected for the West Coast will never be built.

Geologists sometimes find valuable ore deposits by studying a joint system. For example, gold-bearing hydrothermal solutions may migrate upward through a set of joints and deposit quartz and gold in the cracks. Accurate information about joints also is important in the planning and construction of large engineering projects, particularly dams and reservoirs. If the bed rock at a proposed location is intensely jointed, the possibility of dam failure or reservoir leakage may make that site too hazardous.

Faults

Faults were defined earlier as fractures in bed rock along which movement has taken place. The displacement may be only a centimeter or may involve hundreds of kilometers. An active fault is generally regarded as one along which movement has taken place during recorded time (these are the misnamed "earthquake faults"). Most faults, however, are no longer active.

The nature of past movement ordinarily can be discerned where a fault is exposed in an outcrop. The geologist looks for dislocated beds or other features of the rock that might show how much displacement has occurred and the relative direction of movement. In some faults the contact between the two displaced sides is a crack. In others the rock has been broken or ground to a fractured or pulverized mass sandwiched between the displaced sides.

Geologists describe fault movement in terms of direction of slippage: dip-slip, strike-slip, or oblique-slip (figure 15.24). In a **dip-slip fault,** movement is parallel to the dip of the fault surface. A dip-slip fault implies *vertical* motion of the blocks of rock on either side of the fault.

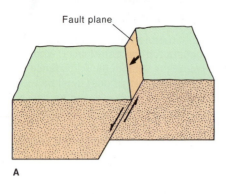

Fault plane

A

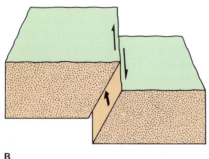

B

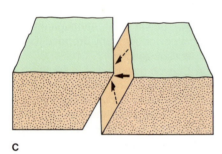

C

Figure 15.24
Three types of faults illustrated by displaced blocks. Heavier arrows show direction in which block to the left moved. (*A*) Dip-slip movement. (*B*) Strike-slip movement. (*C*) Oblique-slip movement. Dashed arrows show dip-slip and strike-slip components of movement.

A **strike-slip fault** indicates *horizontal* motion parallel to the strike of the fault surface. An **oblique-slip fault** has both strike-slip and dip-slip components.

Dip-slip faults   Normal and reverse faults, the most common types of dip-slip faults, are distinguished from each other on the basis of the relative movement of the *footwall block* and the *hanging-wall block*. The **footwall** is the underlying surface of an inclined fault plane, whereas the overlying surface is the **hanging wall.** These old mining terms brought into geology are illustrated in figure 15.25. If a miner is tunneling along the strike of a fault, his feet are on the footwall and his lantern is hung on the hanging wall.

In a **normal fault** (figures 15.26 and 15.27), the hanging-wall block has moved downward relative to the footwall block. The relative movement is represented on a geological cross section by a pair of arrows, because we cannot generally tell which block actually moved. As

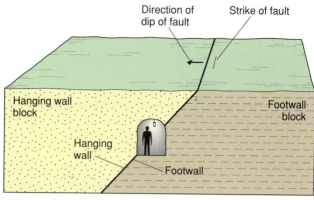

Figure 15.25
Relationship between the hanging wall block and footwall block of a fault.

shown in figure 15.26, a normal fault may be caused by either vertical compressive stresses (figure 15.26*A*) or horizontal tensional stress (figure 15.26*C*). Under the tensional stress, the downthrown hanging-wall block sags downward along the fault to compensate for the pulling apart of the rocks. Sometimes a block bounded by normal faults will drop down, creating a *graben,* as shown in figure 15.26*D*. (*Graben* is the German word for "grave.") *Rifts* are grabens associated with spreading centers, either along mid-oceanic ridges or on continents (see chapters titled "The Sea Floor" and "Plate Tectonics").

If a block bounded by normal faults is uplifted sufficiently, it becomes a fault-block mountain range. (This is also called a *horst,* the opposite of a graben.) The Basin and Range province (described in other chapters) of Nevada and portions of adjoining states is characterized by numerous mountain ranges separated from adjoining valleys by normal faults.

In a **reverse fault,** the hanging-wall block has moved upward relative to the footwall block. As shown in figure 15.28, horizontal compressive stress is the likely cause of a reverse fault.

A **thrust fault** is a reverse fault in which the dip of the fault plane is at a low angle to horizontal (figures 15.28*C* and 15.29). In some mountain regions it is not uncommon for the upper plate (or hanging-wall block) of a thrust fault to have overridden the lower plate (footwall block) for several tens of kilometers.

Strike-slip faults   The displacement of strike-slip faults is either left-lateral or right-lateral and can be determined by looking across the fault. For instance, if an active fault has displaced a trail (figure 15.30), a person walking along the trail would stop where it is truncated by the fault. If the person looks across the fault and sees the displaced trail to the left, it is a **left-lateral fault.** A **right-lateral fault** is one in which displacement is to the right as seen from across the fault. Again, we cannot tell which side actually moved, so pairs of arrows are used to indicate relative movement.

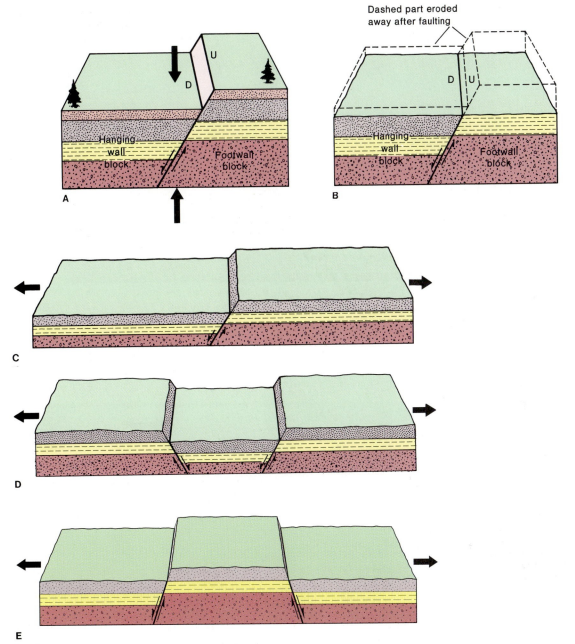

Figure 15.26
Normal faults. (*A*) Diagram shows the fault before erosion and the geometric relationships of the fault and how it might have been caused by vertical compressive stress. (*B*) The same area after erosion. (*C*) How tensional stress could cause a normal fault. (*D*) A graben. (*E*) A horst.

Figure 15.27
Normal fault. Sedimentary beds have been offset approximately one meter.
Photo by Frank M. Hanna.

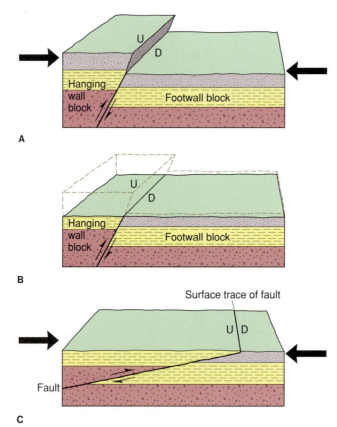

A

B

C

Figure 15.28
(*A*) A reverse fault. The fault is unaffected by erosion. Arrows indicate compressive stress. (*B*) Diagram shows area after erosion; dashed lines indicate portion eroded away. (*C*) Thrust fault. Note in the thrust fault that fault motion has placed older rock layers over younger rocks.

A

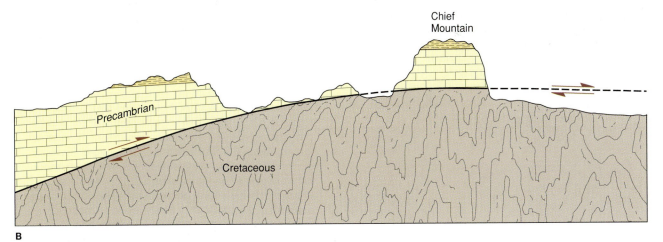

B

Figure 15.29
(*A*) Chief Mountain in Glacier National Park, Montana, is an erosional remnant of a major thrust fault. (*B*) Cross-section of the area. Older (Precambrian) rocks have been thrust over younger (Cretaceous) rocks. Dashed lines show where the fault has been eroded away.
Photo by Frank M. Hanna.

Chief
Mountain

Precambrian

Cretaceous

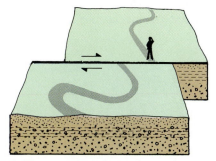

Figure 15.30
Strike-slip fault (right-lateral).

BOX 15.3

California's Greatest Fault—The San Andreas

The San Andreas fault in California (box figure 1) is probably the best known geologic structure in the United States, but the geologists, seismologists, and other scientists who monitor it or who have studied it admit readily that our knowledge of its activity and its history is far from complete. Actually, the San Andreas is only the longest of a number of subparallel faults that transect western California and make up an extensive fault system. Shorter faults continue southward into Mexico, ending in the Gulf of California.

The San Andreas extends for a thousand kilometers through western California, slowly moving Los Angeles toward San Francisco (it is a right-lateral fault). Given a rate of movement now averaging about 2 centimeters per year, Los Angeles could be a western suburb of San Francisco (or San Francisco an eastern suburb of Los Angeles) in some 25 million years. Sudden movement along parts of the fault system has caused major earthquakes. During the 1906 earthquake that destroyed much of San Francisco, bed rock along the San Andreas fault was displaced as much as 5 meters. The amount of horizontal movement was determined by measuring the displacement of fences and other features that straddled the fault.

The fault itself is a zone of broken and ground-up rock, usually a hundred meters or more wide, and its presence is easy to determine throughout most of its length. Along the fault trace are long, straight valleys (formed by erosion and subsidence) that show quite different terrain on either side. Stream channels follow much of the fault zone because the weak, ground-up material along the fault is easily eroded. Locally, elongate lakes (called sag ponds) are found where the ground-up material has settled more than the surface of adjacent parts of the fault zone. The fault was named after one of these ponds, San Andreas Lake, just south of San Francisco (box figure 2).

One can visually follow the fault northward from San Andreas Lake into the southwestern suburbs of San Francisco. There the fault zone is hidden by recently built housing tracts. Apparently the builders and residents have chosen to ignore the hazards of living on the nation's most famous fault.

Box 15.3 Figure 1
Map showing the San Andreas fault.

Box 15.3 Figure 2
Part of the San Andreas fault. View northward toward San Francisco. Lakes occupy the fault zone. Photo by B. Amundson.

Geologists have been unable to agree on the total displacement of the fault or on how long it has been active. Some believe movement began in the Mesozoic Era (over 65 million years ago); most feel that it began later, probably sometime in the Cenozoic Era. The difficulty in establishing an age for the inception of the faulting lies in finding clear evidence of displaced bed rock. What geologists would like to find, if it exists, is a rock unit that can be dated (by fossil fauna or other means) and that was formed in the fault zone (or its general area) just about the time faulting was beginning. Rock on both sides of the fault zone would have to be clearly identifiable as having been the same unit before displacement began.

Geologically young features that cross the fault, such as displaced stream channels (box figure 3), are not uncommon. Similarly, ancient rocks that undoubtedly were there before faulting began are recognized as having been displaced. Many California geologists believe that the belt of granitic rock just west of the fault was once the southern continuation of the granitic batholiths of the Sierra Nevada (box figure 4A), which are more than 80 million years old. But these extremes tell us only that the age of the San Andreas is somewhere between approximately 80 million years and a few thousand years, when the stream channel in box figure 3 carved its course across the fault.

The search for displaced features that would close the bracket between these two time extremes has yielded controversial results for two reasons. First, the age of displaced features may not be determinable. Second, there is a disagreement over which rock units were part of a single unit before faulting. A total displacement of about 500 kilometers seems reasonable to many geologists familiar with the geology of the fault. However, some geologists argue for much less movement.

One hypothesis, based on plate tectonics, places the beginning of strike-slip movement for the San Andreas fault at about 30 million years ago. According to this hypothesis, the Baja California peninsula has split away from Mexico; as the Gulf of California widens, the block of crust west of the San Andreas fault is pushed continuously northward (box figure 4B).

A

B

Box 15.3 Figure 4
(*A*) Reconstruction of California and Mexico as they may have been before faulting. (*B*) Continuous opening of the Gulf of California creates motion along the San Andreas fault.
After Tanya Atwater, 1970, *Geological Society of America Bulletin*.

Box 15.3 Figure 3
Stream channel displaced by the San Andreas fault. The arrows on either side of the fault trace indicate relative motion.
Photo by C. C. Plummer.

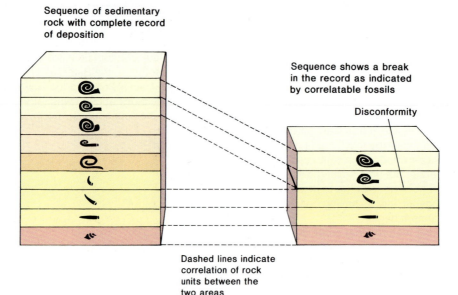

Sequence of sedimentary
rock with complete record
of deposition

Sequence shows a break
in the record as indicated
by correlatable fossils

Disconformity

Dashed lines indicate
correlation of rock
units between the
two areas

Figure 15.31
Schematic representation of a disconformity. The
disconformity is in the block on the right. Dashed lines indicate
correlation of rock between the two areas.

Unconformities

An **unconformity** is a surface (or contact) that represents a *break in the geologic record,* with the rock unit immediately above the contact being considerably younger than the rock beneath. Most unconformities are buried erosion surfaces. Unconformities are classified into three types—disconformities, angular unconformities, and nonconformities—with each type having important implications for the geologic history of the area in which it occurs. Unconformities were discussed in chapter 8 in illustrating how relative age can be determined.

Disconformities

In a **disconformity,** the contact representing missing rock strata is parallel to beds above and below it. Probably what has happened is that older rocks were eroded away parallel to the bedding plane; renewed deposition later buried the erosion surface (figure 15.31).

Because it often appears to be just another sedimentary contact (or bedding plane) in a sequence of sedimentary rock, a disconformity is the hardest type of unconformity to detect in the field. Rarely, a telltale weathered zone is preserved below a disconformity. Usually the disconformity can be detected only by studying fossils from the beds in a sequence of sedimentary rocks. If certain fossil beds are absent, indicating that a portion of geologic time is missing from the sedimentary record, it can be inferred that a disconformity is present in the sequence. Although it is most likely that some rock layers are missing because erosion followed deposition, in some instances possibly neither erosion nor deposition took place for a certain amount of geologic time.

Angular Unconformities

An **angular unconformity** is a contact in which younger strata overlie an erosion surface on tilted or folded layered rock. It implies the following sequence of events, from oldest to youngest: (1) deposition and lithification of sedimentary rock (or solidification of successive lava flows if the rock is volcanic); (2) uplift accompanied by folding or tilting of the layers; (3) erosion; (4) renewed deposition (usually preceded by subsidence) on top of the erosion surface (figures 15.32 and 15.33).

If fossils are abundant in the rocks above and below an angular unconformity, the laws of *faunal succession* and *cross-cutting relationships* (chapter 8) can be used to determine the relative time of folding and tilting. For example, if the youngest fossils identified in the folded sequence are late Paleozoic and the fossils in the horizontal layer just over the unconformity contain middle Mesozoic fossils, the folding must have occurred between late Paleozoic and middle Mesozoic times.

Nonconformities

A **nonconformity** is a contact in which an erosion surface on plutonic or metamorphic rock has been covered by younger sedimentary or volcanic rock (figure 15.34). A nonconformity generally indicates deep or long-continued erosion before subsequent burial, because metamorphic or plutonic rocks form at considerable depths in the earth's crust.

The geologic history implied by a nonconformity, shown in figure 15.35, is (1) crystallization of igneous or metamorphic rock at depth; (2) erosion of several kilometers or more of overlying rock (the great amount of erosion further implies considerable uplift of this portion of the earth's crust); (3) deposition of new sediment, which eventually becomes sedimentary rock, on the ancient erosion surface.

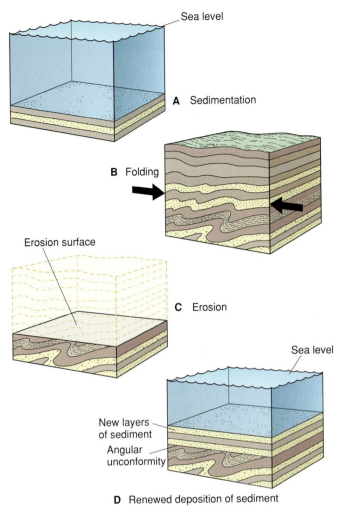

A Sedimentation

B Folding

Erosion surface

C Erosion

Sea level

New layers of sediment

Angular unconformity

D Renewed deposition of sediment

Figure 15.32
Development of an angular unconformity. (*A*) Deposition of sediment. (*B*) Folding. (*C*) Erosion. (*D*) Renewed deposition of sediment.

Figure 15.34
A nonconformity in Grand Canyon, Arizona. Horizontal Paleozoic sedimentary rocks overlie vertically foliated Precambrian metamorphic rocks.
Photo by Frank M. Hanna.

Figure 15.33
An angular unconformity in Grand Canyon, Arizona.

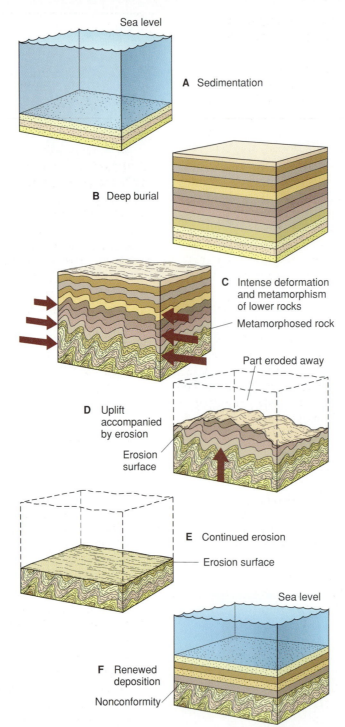

Figure 15.35
Sequence of events implied by a nonconformity underlain by metamorphic rock.

Summary

Present-day deformation of the earth's crust is studied by observing active faults. *Faults* and other geologic structures provide information about the nature and intensity of forces at work in the geologic past. Strained rock (changed in size or shape) records past stresses.

A geologic map shows the structural characteristics of a region. *Strike* and *dip* symbols on geologic maps indicate the attitudes of inclined surfaces such as bedding planes. The strike and dip of a bedding surface indicate the relationship between the inclined plane and a horizontal plane.

If rock layers bend rather than break, they become folded. Rock layers are folded into *anticlines* and *synclines* and more complex structures. If the hinge line of a fold is not horizontal, the fold is *plunging*. Older beds exposed in the core of a fold indicate an anticline, whereas younger beds in the center of the structure indicate a syncline. In places where folded rock has been eroded to a plain, an anticline can usually be distinguished from a syncline by whether the beds dip toward the center (syncline) or away from the center (anticline).

Fractures in rock are either *joints* or *faults*. A joint indicates that movement has not occurred on either side of the fracture; displaced rock along a fracture indicates a fault. In a *dip-slip* fault, bed rock is vertically displaced. In a *strike-slip* fault, which can be either left-lateral or right-lateral, horizontal movement has occurred. An *oblique-slip* fault involves both strike-slip and dip-slip motion.

Dip-slip faults are either *normal* or *reverse*, depending on the motion of the hanging-wall block relative to the footwall block. A reverse fault with a low angle of dip for the fault plane is a *thrust fault*. Reverse faults are thought to be the product of compressive horizontal forces, while normal faults are caused by vertical forces or horizontal tensional forces.

Unconformities are buried erosion surfaces that help geologists determine the relative sequence of events in the geologic past. Beds above and below a *disconformity* are parallel, generally indicating less intense activity in the earth's crust. An *angular unconformity* implies that folding or tilting of rocks took place before or around the time of erosion. A *nonconformity* implies deep erosion because metamorphic or plutonic rocks have been exposed and subsequently buried by younger rock.

Terms to Remember

angle of dip	nonconformity
angular unconformity	normal fault
anticline	oblique-slip fault
compressive force	open fold
dip-slip fault	overturned fold
direction of dip	plastic strain
disconformity	plunging fold
elastic strain	recumbent fold
fault	reverse fault
fold	right-lateral fault
footwall	shear stress
fracturing	strain
geologic cross section	stress
geologic map	strike
hanging wall	strike-slip fault
hinge line	structural basin
hinge plane	structural dome
isoclinal fold	structural geology
joint	syncline
joint set	tensional force
left-lateral fault	thrust fault
limb	unconformity

Questions for Review

1. On a geologic map, if no cross sections were available, how could you distinguish an anticline from a syncline?
2. If you locate a dip-slip fault while doing field work, what kind of evidence would you look for to determine whether the fault is normal or reverse?
3. What is the difference between stress and strain?
4. Name several geologic structures described in earlier chapters.
5. What is the difference between strike, direction of dip, and angle of dip?
6. Draw a simple geologic map, using strike and dip symbols for a syncline plunging to the west.
7. What sequence of events in the geologic past is implied by each of the three types of unconformities?
8. How does a structural dome differ from a plunging anticline?
9. Most anticlines have both limbs dipping away from their axes. For which kind of fold is this not the case?
10. Faults and unconformities are two types of contacts. Name the other types of contacts described in previous chapters.

Questions for Thought

1. What criteria would you use to distinguish a nonconformity from an intrusive contact of a stock?
2. Can a fault always be distinguished from an unconformity?
3. In what parts of North America would you expect to find the most intensely folded rock?
4. A subduction zone can be regarded as a very large example of what type of fault?
5. Why do some horizontal compressive forces cause thrust faults while others cause strike-slip faults?
6. What features in sedimentary or volcanic rock layers would you look for to tell you that the rock was part of the overturned limb of a fold?
7. Can you identify and name the various geologic structures shown in the figures in the chapter titled "Time and Geology"?

Supplementary Readings

Davis, G. H. 1984. *Structural geology of rocks and regions.* New York: John Wiley & Sons.

Dennis, J. G. 1987. *Structural geology.* Dubuque, Iowa: Wm. C. Brown Publishers.

Kligfield, R. and P. A. Geiser. 1989. *Geology and geometry of thrust belts.* London: Unwin Hyman.

Ragan, D. M. 1984. *Structural geology: An introduction to geometrical techniques.* 3d ed. New York: John Wiley & Sons.

This chapter will help you understand the nature and origin of earthquakes. We discuss the seismic waves created by earthquakes and how the quakes are measured and located by studying these waves. We also describe some effects of earthquakes, such as ground motion and displacement, damage to buildings, and quake-caused fires, landslides, and tsunamis ("tidal waves").

Earthquakes are largely confined to a few narrow belts on earth. This distribution was once puzzling to geologists, but here we show how the concept of plate tectonics neatly explains it.

As geologists learn more about earthquake behavior, there is the intriguing possibility that we will be able to predict—and even control—earthquakes. We conclude the chapter with a look at this developing branch of earth study.

CHAPTER

16

Earthquakes

The upper deck of Interstate 880 (the Nimitz Freeway) in Oakland, California, collapsed onto the lower deck, killing more than 40 people during the Loma Prieta earthquake, October 17, 1989.
Los Angeles Times photo.

On April 18, 1906, at 5:12 in the morning, part of California slid abruptly past the rest of the state during a great earthquake. A visible scar 280 miles (450 kilometers) long was left where the earth was torn along coastal northern California. The land was displaced horizontally as much as 16 feet (5 meters). The quake, located on the San Andreas fault near San Francisco, shook the ground for one full minute.

Buildings toppled in San Francisco, and broken gas mains fed fires that raged for three days. Broken water mains hampered fire fighting. The fires were finally extinguished when buildings were dynamited to create a firebreak. Terrified and homeless people moved to refugee camps set up in city parks. Looters were shot on sight. As the city gradually recovered from the shock of the devastation, it was found that at least 3,000 people had died and $400 million (in 1906 dollars) of damage had been done. Perhaps 90% of the destruction was caused by the fires.

At 5:30 P.M. on March 27, 1964, southern Alaska was rocked by an earthquake that lasted for three minutes. Although the force of this earthquake was twice as strong as the 1906 San Francisco earthquake, loss of life and property was relatively low because of Alaska's small population—15 people died as a direct result of the shaking, and damage amounted to slightly over $300 million (in 1964 dollars). The tremor was felt over an area of more than 350,000 square miles (1 million square kilometers). A section of the earth's surface 30 by 125 miles (50 by 200 kilometers) was raised as much as 40 feet (13 meters), and a similar block of land sank 3 to 6 feet (1 to 2 meters). Horizontal movement was slight. In Anchorage, 90 miles (150 kilometers) from the center of the earthquake, landslides wrecked parts of the city. The greatest loss of life was caused by large sea waves generated by land movement associated with the earthquake—almost 100 people drowned in Alaska, and a few people as far away as Oregon and northern California, as the waves spread over the Pacific Ocean.

At 5:04 P.M. on October 17, 1989, just 20 minutes before game 3 of the World Series, the San Francisco Bay area was severely shaken by an earthquake on the San Andreas fault. The quake, located midway between Santa Cruz and San Jose, did not tear the ground surface. Ground shaking lasted for 15 seconds and collapsed some buildings and freeway overpasses built upon the soft "bay fill" sediment near the water's edge in San Francisco and Oakland. A section of the Bay Bridge between these two cities collapsed. In the Marina district of San Francisco raging fires were fed by broken gas mains, just as in 1906. Once again, broken water pipes hampered fire fighting; the fires were extinguished by fireboats and by firehoses using water pumped from the bay. Very severe damage occurred in smaller towns near the center of the quake, such as Santa Cruz and Watsonville. The death toll was 67, and damage was estimated as high as $7 billion.

Causes of Earthquakes

What causes earthquakes? An **earthquake** is a trembling or shaking of the ground caused by the sudden release of energy stored in the rocks beneath the earth's surface. As described in chapter 15, great forces acting deep in the earth may put a *stress* on the rock, which may bend or change in volume (*strain*). If you bend a stick of wood, your hands put a stress (the force) on the stick; its bending (a change in shape) is the strain.

Like a bending stick, rock can deform only so far before it breaks. When a rock breaks, waves of energy are sent out through the earth. These are **seismic waves,** the waves of energy produced by an earthquake. It is the seismic waves that cause the ground to tremble and shake during an earthquake.

The sudden release of energy when rock breaks may cause one huge mass of rock to slide past another mass of rock into a different relative position. As you know from the chapter titled "Geologic Structures," the crack between the two rock masses is a *fault*. This explanation of why earthquakes take place is called the **elastic rebound theory** (figure 16.1). It involves the sudden release of progressively stored strain in rocks, causing movement along a fault. Deep-seated internal forces (*tectonic forces*) act on a mass of rock over many decades. Initially the rock bends, but does not break. More and more energy is stored in the rock as the bending becomes more severe. Eventually the energy stored in the rock exceeds the breaking strength of the rock, and the rock breaks suddenly, causing an earthquake. Two masses of rock move past one another along a fault. The movement may be vertical, horizontal, or both. The strain on the rock is released; the energy is expended by moving the rock into new positions and by creating seismic waves.

The brittle behavior of rock assumed in the elastic rebound theory is characteristic only of rocks near the earth's surface. Rocks at depth are subject to increased temperature and pressure, which tend to reduce brittleness. Deep rocks flow plastically (*ductile* behavior) instead of breaking (*brittle* behavior).

Most earthquakes are associated with movement on faults, but in some quakes the connection with faulting may be hard to establish. Most earthquakes in the eastern United States are not associated with known faults or surface displacement. Two recent California quakes, in Coalinga in 1983 and Whittier in 1987, occurred on deeply buried anticlines. Although thrust faults seem to be associated with the anticlines, neither quake caused a break at the surface. Earthquakes also occur during explosive volcanic eruptions and as magma forcibly fills underground magma chambers prior to many eruptions; these quakes may not be associated with fault movement at all.

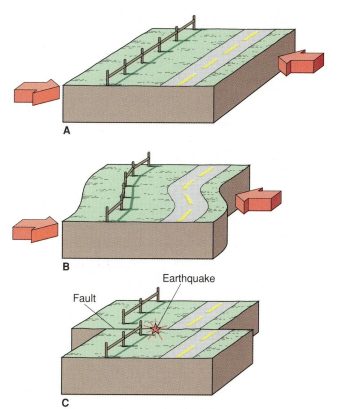

D

Figure 16.1
The elastic rebound theory of the cause of earthquakes.
(*A*) Rock with stress acting on it. (*B*) Stress has caused strain in the rock. Strain builds up over a long period of time.
(*C*) Rock breaks suddenly, releasing energy, with rock movement along a fault. Horizontal motion is shown; rocks can also move vertically. (*D*) Horizontal offset of rows in a lettuce field, 1979, El Centro, California.
Photo by Univ. of Colorado; courtesy National Geophysical Data Center, Boulder, Colorado.

Seismic Waves

The point within the earth where seismic waves originate is called the **focus** of the earthquakes (figure 16.2). This is the center of the earthquake, the point of initial movement on a fault. The point on the earth's surface directly above the focus is the **epicenter.**

Two types of seismic waves radiate outward from the earthquake focus. **Body waves** are seismic waves that travel through the earth's interior, spreading outward from the focus in all directions, like sound waves moving through air. **Surface waves** are seismic waves that travel on the earth's surface away from the epicenter, like water waves spreading out from a pebble thrown into a pond. Rock movement associated with seismic surface waves dies out with depth into the earth, just as water movement in ocean waves dies out with depth.

There are two main types of body waves, both shown in figure 16.3. A **P wave** is a compressional (or longitudinal) wave in which rock vibrates *parallel* to the direction of wave propagation, that is, in the same direction as the waves are moving. Because it is a very fast wave, traveling through surface rocks at speeds of 4 to 7 kilometers

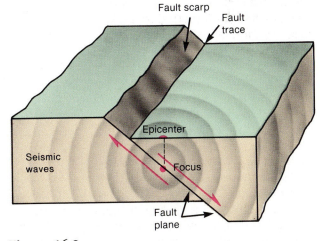

Figure 16.2
The focus of an earthquake is the point where rocks first break along a fault; seismic waves radiate from the focus. The epicenter is the point on the earth's surface directly above the focus.

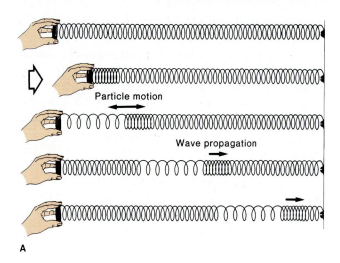

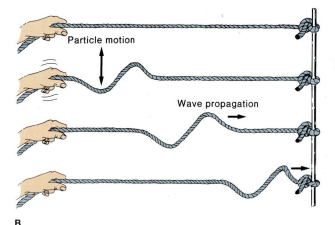

Figure 16.3
Particle motion in seismic waves. (**A**) A P wave is illustrated by a sudden push on the end of a stretched spring. The particles vibrate *parallel* to the direction of wave propagation. (**B**) An S wave is illustrated by shaking a loop along a stretched rope. The particles vibrate *perpendicular* to the direction of wave propagation.

per second (9,000 to more than 15,000 miles per hour), a P wave is the first (or *primary*) wave to arrive at a recording station following an earthquake. The second type of body wave is called an **S wave** (*secondary*) and is a slower, transverse wave that travels at 2 to 5 kilometers per second. An S wave is propagated by a shearing motion much like that in a stretched, shaken rope. The rock vibrates *perpendicular* to the direction of wave propagation, that is, crosswise to the direction the waves are moving. Both P waves and S waves pass easily through solid rock. A P wave can also pass through a fluid (gas or liquid), but an S wave cannot. We discuss the importance of this fact in the chapter titled "The Earth's Interior."

Surface waves are the slowest waves set off by quakes. In general, surface waves cause more property damage than body waves because surface waves produce more ground movement and travel more slowly, so they take longer to pass.

✳Locating and Measuring Earthquakes

The invention of instruments that could accurately record seismic waves was an important scientific advance. They measure the amount of ground motion and can be used to find the location of an earthquake.

The instrument used to detect seismic waves is a *seismometer*. The principle of the seismometer is to keep a heavy suspended mass as motionless as possible—suspending it by springs or hanging it as a pendulum from the frame of the instrument (figure 16.4). When the ground moves, the frame of the instrument moves with it. However, the inertia of the heavy mass suspended inside keeps the mass motionless to act as a point of reference in determining the amount of ground motion.

A seismometer by itself cannot record the motion that it measures. A **seismograph** is a seismometer with a recording device that produces a permanent record of earth

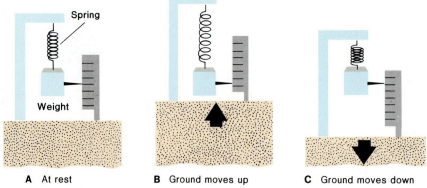

A At rest   **B** Ground moves up   **C** Ground moves down

Figure 16.4
A simple seismometer for detecting vertical rock motion. The indicator needle moves on the scale as ground motion stretches and compresses a spring. Frame and scale move with the ground. Inertia of the weight keeps it and the needle relatively motionless.

motion, usually in the form of a wiggly line drawn on a moving strip of paper (figure 16.5). The paper record of earth vibration is called a **seismogram.**

A network of seismograph stations is maintained all over the world to record and study earthquakes (and nuclear bomb explosions). Within minutes after an earthquake occurs, distant seismographs begin to pick up seismic waves. A large earthquake can be detected by seismographs all over the world.

Because the different types of seismic waves travel at different speeds, they arrive at a seismograph station in a definite order, first the P waves, then the S waves, and finally the surface waves. These three different waves can be distinguished on the paper seismograms. By analyzing these seismograms, geologists can learn a great deal about an earthquake, including its location and size.

Determining the Location of an Earthquake

P, S, and surface waves all start out from an earthquake at essentially the same time. As they travel away from the quake, the three types of waves gradually separate because they are traveling at different speeds. On a seismogram from a station close to the earthquake, the first arrival of the P wave is separated from the first arrival of the S wave by a short distance on the paper record (figure 16.6). At a recording station far from the earthquake, however, the first arrivals of these waves will be recorded much farther apart on the seismogram. The farther the seismic waves travel, the longer the time intervals between the arrivals of P and S waves and the more they are separated on the seismograms.

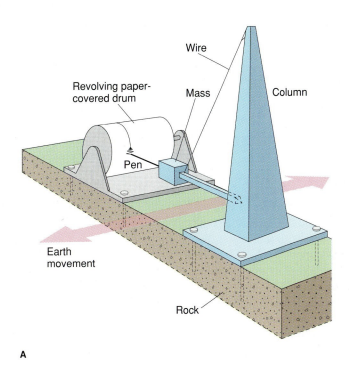

Wire

Column

Mass

Revolving paper-covered drum

Pen

Earth movement

Rock

A

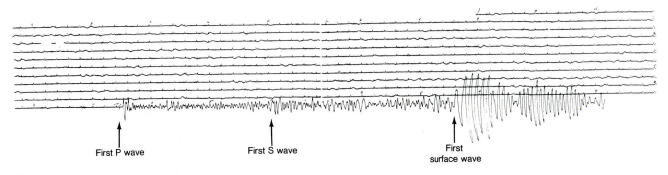

First P wave

First S wave

First surface wave

B

Figure 16.5

(A) A seismograph for horizontal motion. Modern seismographs record earth motion on moving strips of paper. The mass is suspended by a wire from the column and swings like a pendulum when the ground moves horizontally. A pen attached to the mass records the motion on a moving strip of paper.
(B) A seismogram of a 1967 earthquake in Taiwan, magnitude 6.2, recorded in Berkeley, California, 6,300 miles away. First arrivals of P, S, and surface waves are shown.
Courtesy University of California, Berkeley.

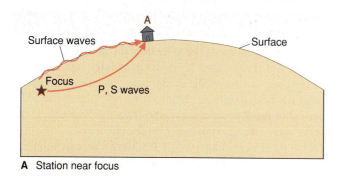

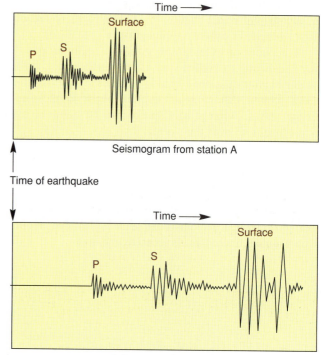

Seismogram from station A

Time of earthquake

Seismogram from station B

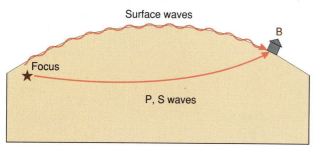

Figure 16.6
Intervals between P waves, S waves, and surface waves increase with distance from the focus.

Because the time interval between the first arrivals of P and S waves increases with distance from the focus of an earthquake, this interval can be used to determine the distance from the seismograph station to a quake. The increase in the P–S interval is regular with increasing distance for several thousand kilometers and so can be graphed in a **travel-time curve,** which plots seismic-wave arrival time against distance (figure 16.7).

In practice, a station records the P and S waves from a quake, then matches the interval between the waves to a standard travel-time curve. By reading directly from the graph, a station can determine, for example, that an earthquake has occurred 5,500 kilometers (3,400 miles) away. This determination can often be made very rapidly, even while the ground is still trembling from the quake.

A single station can determine only the distance to a quake, not the direction. A circle is drawn on a globe with the center of the circle being the station and its radius the distance to the quake (figure 16.8). The scientists at the station know that the quake occurred somewhere on that circle, but from the information recorded they are not able to tell exactly where. With information from other stations, however, they can pinpoint the location of the quake. If three or more stations have determined the distance to a single quake, a circle is drawn for each station. If this is done on a globe, the intersection of the circles locates the quake.

Analyses of seismograms can also indicate at what depth beneath the surface the quake occurred. Most earthquakes occur relatively close to the earth's surface, although a few occur much deeper. The maximum **depth**

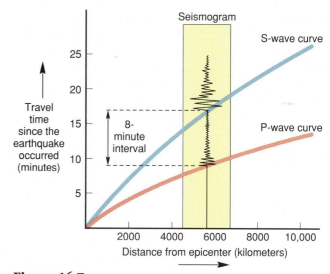

Figure 16.7
A travel-time curve is used to determine the distance to an earthquake. Note that the time interval between the first arrival of P and S waves increases with distance from the epicenter. This seismogram has an interval of 8 minutes, so the earthquake occurred 5,500 kilometers away.

of focus—the distance between focus and epicenter—for earthquakes seems to be about 670 kilometers. Quakes are classified into three groups according to their depth of focus, as shown in table 16.1. Note that intermediate- and deep-focus earthquakes combined account for only 15% of the total quake energy released. This is because deep rocks behave plastically rather than brittlely.

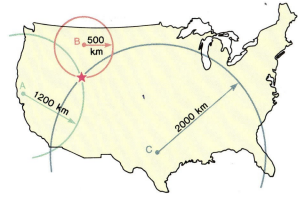

Figure 16.8
Locating an earthquake. The distance from each of three stations (*A, B, C*) is determined from seismograms and travel-time curves. Each distance is used for the radius of a circle about the station. The star shows the location of the earthquake, where the three circles intersect.

TABLE 16.1

Percentage of Total Quake Energy

Shallow focus	0–70 km deep	85%
Intermediate focus	70–350 km	12%
Deep focus	350–700 km	3%

Earthquake Strength

The strength (or size) of earthquakes is measured in two ways. One method is to find out how much and what kind of damage the quake has caused. This determines the **intensity,** which is a measure of an earthquake's effect on people and buildings. Intensities are expressed as Roman numerals ranging from I to XII on the **modified Mercalli scale** (table 16.2); higher numbers indicate greater damage.

Although intensities are widely reported at earthquake locations throughout the world, using intensity as a measure of earthquake strength has a number of drawbacks. Because damage generally lessens with distance from a quake's epicenter, different locations report different intensities for the same earthquake (figure 16.9). Moreover, damage to buildings and other structures depends greatly on the type of geologic material on which a structure was built as well as the type of construction. Houses built on solid rock normally are damaged far less than houses built upon loose sediment, such as delta mud or bay fill. Damage estimates are also subjective: people may exaggerate damage reports consciously or unconsciously. Intensity maps can be drawn for a single earthquake to show the approximate damage over a wide region (figure 16.9). But such maps cannot be drawn for uninhabited areas (the central ocean, for instance), and so not all quakes can be assigned intensities. The one big advantage of intensity ratings is that no instruments are required.

The second method of measuring the strength of a quake is to calculate the amount of energy released at the earthquake's focus. In practice this is done by measuring

TABLE 16.2

Modified Mercalli Intensity Scale of 1931 (Abridged)

I. Not felt except by a very few under especially favorable circumstances.

II. Felt only by a few persons at rest, especially on upper floors of buildings. Delicately suspended objects may swing.

III. Felt quite noticeably indoors, especially on upper floors of buildings, but many people do not recognize it as an earthquake. Standing motor cars may rock slightly. Vibration like passing of truck. Duration estimated.

IV. During the day felt indoors by many, outdoors by few. At night some awakened. Dishes, windows, doors disturbed; walls made cracking sound. Sensation like heavy truck striking building. Standing motor cars rocked noticeably.

V. Felt by nearly everyone; many awakened. Some dishes, windows, etc., broken; a few instances of cracked plaster; unstable objects overturned. Disturbance of trees, poles, and other tall objects sometimes noticed. Pendulum clocks may stop.

VI. Felt by all; many frightened and run outdoors. Some heavy furniture moved; a few instances of fallen plaster or damaged chimneys. Damage slight.

VII. Everybody runs outdoors. Damage *negligible* in buildings of good design and construction; *slight* to moderate in well-built ordinary structures; *considerable* in poorly built or badly designed structures; some chimneys broken. Noticed by persons driving motor cars.

VIII. Damage *slight* in specially designed structures; *considerable* in ordinary substantial buildings with partial collapse; *great* in poorly built structures. Panel walls thrown out of frame structures. Fall of chimneys, factory stacks, columns, monuments, walls. Heavy furniture overturned. Sand and mud ejected in small amounts. Changes in well water. Persons driving motor cars disturbed.

IX. Damage *considerable* in specially designed structures; well-designed frame structures thrown out of plumb; *great* in substantial buildings, with partial collapse. Buildings shifted off foundations. Ground cracked conspicuously. Underground pipes broken.

X. Some well-built wooden structures destroyed; most masonry and frame structures destroyed with foundations; ground badly cracked. Rails bent. Considerable landslides from river banks and steep slopes. Shifted sand and mud. Water splashed (slopped) over banks.

XI. Few, if any, (masonry) structures remain standing. Bridges destroyed. Broad fissures in ground. Underground pipe lines completely out of service. Earth slumps and land slips in soft ground. Rails bent greatly.

XII. Damage total. Waves seen on ground surface. Lines of sight and level distorted. Objects thrown upward into the air.

From Wood and Neumann, 1931, *Bulletin Seismological Society of America.*

the height (amplitude) of one of the wiggles on a seismogram. This method is commonly used in the United States. The larger an earthquake, the more the ground vibrates. After measuring a specific wave on a seismogram, and correcting for the type of seismograph and distance from the quake, scientists can assign a number called the **magnitude.** It is a measure of the energy released during the earthquake. The **Richter scale** (table 16.3) is a numerical scale of magnitudes from 0 to 8.6, the higher numbers indicating larger earthquakes. The largest quake measured so far is 8.6, but because larger quakes might be measured in the future, 8.6 is not the upper limit of the

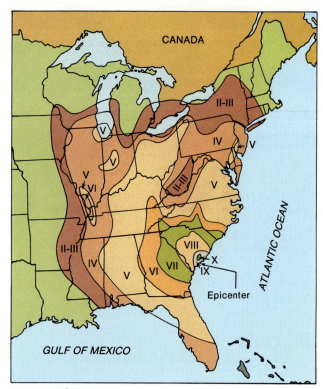

Figure 16.9
Zones of different intensity from the 1886 Charleston, South Carolina, earthquake.
U.S. Geological Survey.

TABLE 16.3

Richter Scale Magnitudes

Magnitude		Number Per Year (World)
2	Just felt	More than 100,000
4.5	Damage begins	A few thousand
7	"Major"	16 to 18
8	"Great"	1 or 2
8.6	Maximum recorded	

Some Representative Magnitudes

1906	San Francisco	8.25
1960	Chile	8.5
1964	Alaska	8.6
1970	Peru	7.75
1971	San Fernando Valley, Cal.	6.4
1976	China	7.6
1980	N. California	7.4
1983	Coalinga, Cal.	6.7
1983	Challis, Idaho	7.3
1983	Adirondack Mts., New York	5.2
1983	Hawaii	6.6
1985	Mexico	8.1, 7.5
1987	Lawrenceville, Ill.	5.0
1987	Whittier, Cal.	5.9
1988	Quebec	6.0
1988	Armenia, USSR	6.9, 5.8
1989	Tadzhikistan, USSR	5.4
1989	Loma Prieta, Cal.	7.1

Richter scale. It seems probable, however, that values near 9 represent a limit of elastic strength beyond which rocks will break. Therefore, earthquakes larger than magnitude 9 are unlikely.

Because the Richter scale is logarithmic, the difference between two consecutive whole numbers on the scale means an increase of 10 times in the amplitude of the earth's vibrations. It has been estimated that a tenfold increase in the size of the earth vibrations is caused by an increase of about 31.5 times in terms of energy. A quake of magnitude 5, for example, releases 31.5 times more energy than one of magnitude 4. A magnitude 6 quake is almost 1,000 times (31.5 × 31.5) more powerful in terms of energy released than a magnitude 4 quake.

Although a seismograph is required to measure magnitude, this measure has many advantages over intensity as an indicator of earthquake strength. A worldwide network of standard seismograph stations now makes determining magnitude a routine matter; and the press reports Richter-scale numbers for all earthquakes of interest to the United States. A single magnitude number can be assigned to a single earthquake, whereas intensity varies for a single earthquake, depending on local damage. Magnitudes can be reported for all quakes, even those in distant uninhabited areas where no property is harmed.

Earthquakes in the United States

Figure 16.10*A* shows the locations of all moderate-to-large earthquakes that occurred in the United States through 1970 (small quakes, which are very common, have been omitted). Note that only a few localities are relatively earthquake-free. Nearly all parts of the United States are subject to earthquakes, although the quakes in some regions are seldom destructive.

Figure 16.10*B* leaves out moderate earthquakes and shows only large, destructive quakes in the United States through 1970. Note that most of the large quakes occur in the western states, generally along well-known faults.

Geologists have mapped regions of seismic risk in the United States (figure 16.11) primarily on the assumption that large earthquakes will occur in the future in places where they have occurred in the past.

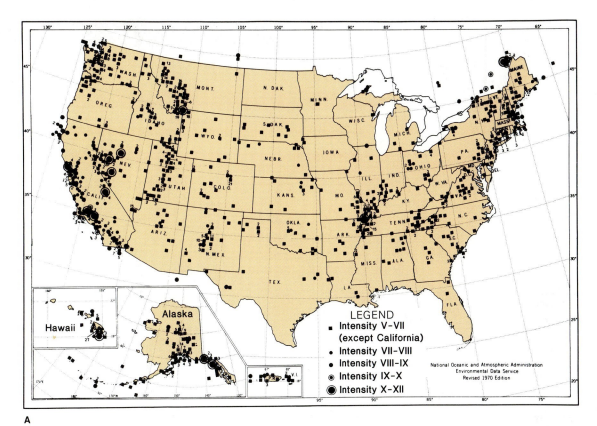

A

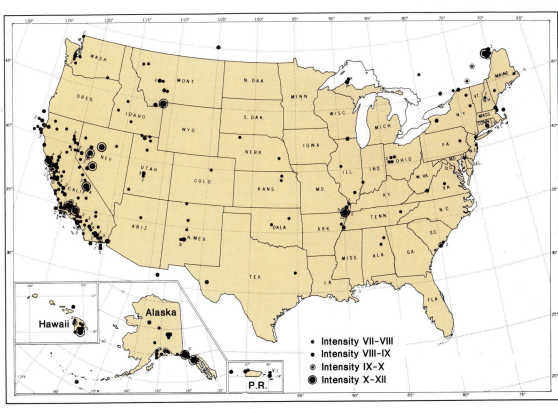

B

Figure 16.10

(**A**) Locations of earthquakes (intensity V and above) in the
United States and southern Canada through 1970.
(**B**) Locations of destructive earthquakes (intensity VII and
above) in the United States and southern Canada through 1970.

National Oceanic and Atmospheric Administration.

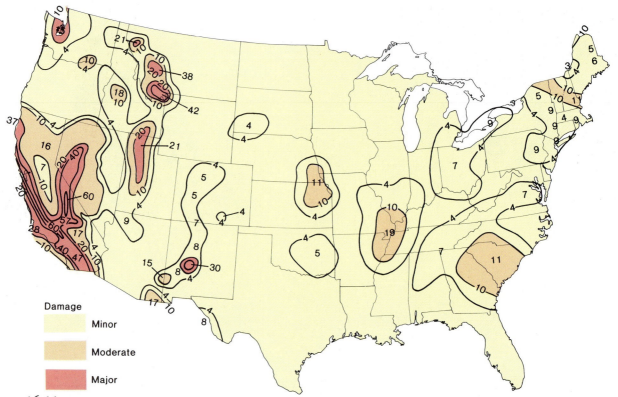

Figure 16.11

A 1976 map of seismic risk. The risk is shown as the expected maximum horizontal acceleration of the ground during an earthquake, shown as a percent of gravity. Damage usually begins at 10% of gravity; accelerations of 250% of gravity have been measured in some locations. Acceleration shown has only a 1-in-10 chance of being exceeded in a 50-year period. Maximum acceleration (in California) is 80% of gravity.

U.S. Geological Survey.

Damage
- Minor
- Moderate
- Major

Earthquakes in the Eastern United States—Rare but Occasionally Strong

Although most earthquakes in the western United States are clearly related to known faults, earthquakes that occur east of the Rocky Mountains are more difficult to relate to faulting. Earthquakes in the eastern United States are uncommon, generally small, and deeper than western United States quakes; they are not associated with surface displacement of the ground or known faults. (They may lie along the landward extensions of oceanic fracture zones—see chapter titled "The Sea Floor.")

Although large quakes are extremely rare in the central and eastern United States (figure 16.10*B*), when they do occur they can be very destructive and widely felt, because the earth's crust is older, cooler, and more brittle in the east than in the west. The Saint Lawrence River Valley along the Canadian border has had several intensity IX and X earthquakes, most recently in 1944. Plymouth, Massachusetts, had an intensity IX quake in 1638, and a quake of intensity VIII occurred in 1775 near Cambridge, Massachusetts. In 1929 in Attica, New York, an earthquake of intensity IX knocked over 250 chimneys. A series of quakes (intensity XI) that occurred near New Madrid, Missouri, in the winter of 1811–1812 may have been the largest recorded earthquakes ever to occur in North America. The quakes knocked over chimneys as far away as Richmond, Virginia. The 1886 quake in Charleston, South Carolina, (intensity X) was felt throughout almost half the United States (figure 16.9) and killed sixty people. Moderate quakes hit Arkansas and New Hampshire in 1982, and in 1983 a quake of 5.2 magnitude rocked New York's Adirondack Mountains. A 5.0-magnitude quake near Lawrenceville, Illinois, was felt from Kansas to South Carolina to Ontario in 1987. In 1988 a 6.0-magnitude quake north of Quebec City was felt as far away as Indiana and Washington, D.C.

Because the origin of these quakes is not clear, and because they are not associated with surface faults that can be monitored, the prediction of quakes in the eastern United States will be much more difficult than in the western states.

A

B

C

Figure 16.12
Earthquake damage to man-made structures. (**A**) Fallen statue at Stanford University, San Francisco earthquake, 1906. (**B**) Wreckage of San Francisco City Hall, 1906. (**C**) Collapsed freeway overpasses, San Fernando Valley, California, 1971.
Photos *A* and *B* by W. C. Mendenhall, U.S. Geological Survey. Photo *C* by R. Kachadoorian, U.S. Geological Survey.

Effects of Earthquakes

Ground motion is the trembling and shaking of the land that can cause buildings to vibrate. During small quakes, windows and walls may crack from such vibration. In a very large quake the ground motion may be visible. It can be strong enough to topple large structures such as bridges and office buildings (figure 16.12). Most people injured or killed in an earthquake are hit by falling debris from buildings. Because proper building construction can greatly reduce the dangers, building codes need to be strict in earthquake-prone areas. As we have seen, the location of buildings also needs to be controlled; buildings built on soft sediment are damaged more than buildings on hard rock.

Fire is a particularly serious problem just after an earthquake because of broken gas and water mains and fallen electrical wires. Although fire was the cause of most of the damage to San Francisco in 1906, changes in building construction and improved fire-fighting methods have reduced (but not eliminated) the fire danger to modern cities. The stubborn Marina district fires in San Francisco in 1989 attest to modern dangers of broken gas and water mains (figure 16.13).

Landslides can be triggered by the shaking of the ground (figure 16.14). The 1959 Madison Canyon landslide in Montana was triggered by a nearby quake of magnitude 7.2. Landslides and subsidence caused extensive

Figure 16.13
Earthquake-triggered fire, Marina district, San Francisco, 1989.
Photo: Gamma-Liaison.

A

B

Figure 16.14
(A) Landslide near Anchorage caused by the Alaskan earthquake in 1964. (B) A type of ground failure called liquefaction occurred in 1964 in Niigata, Japan. Earthquake shaking caused the ground to behave like quicksand, and

earthquake-resistant apartment buildings toppled over intact.
Photo A by U.S. Geological Survey. Photo B by National Geophysical Data Center.

damage in downtown and suburban Anchorage during the 1964 Alaskan quake (magnitude 8.6). The 1970 Peruvian earthquake (magnitude 7.75) set off thousands of landslides in the steep Andes Mountains, burying more than 17,000 people (box 9.1). In 1920 in China over 100,000 people living in hollowed-out caves in cliffs of loess (chapter on "Deserts and Wind Action") were killed when a quake collapsed the cliffs.

Permanent displacement of the land surface may be the result of rock movement along a fault. Rocks can move vertically, those on one side of a fault rising while those on the other side drop. Rocks can also move horizontally, those on one side of a fault sliding past those on the other

side. Both vertical and horizontal movement can occur during a single quake. Such movement can affect huge areas, although the displacement in a single earthquake seldom exceeds 25 feet (8 meters) in any direction and almost never exceeds 50 feet (15 meters). The trace of a fault on the earth's surface may appear as a low cliff, called a *scarp,* or as a closed tear in the ground (figure 16.15). In rare instances small cracks open during a quake (but not to the extent that Hollywood films often portray). Ground displacement during quakes can tear apart buildings, roads, and pipelines that cross faults. Sudden subsidence of land near the sea can cause flooding and drownings.

A

B

D

C

E

Figure 16.15
Varieties of ground displacement caused by earthquakes.
(*A*) Sixteen-foot scarp (cliff) formed by vertical ground motion, Alaska, 1964. (*B*) Tearing of the ground near Olema, California, 1906. (*C*) Fence offset nearly 10 feet near Woodville, California, 1906. (*D*) Fence compressed by ground movement, Gallatin County, Montana, 1959. (*E*) Compression of concrete freeway, San Fernando Valley, California, 1971.

Photo *A* by U.S. Geological Survey. Photos *B* and *C* by G. K. Gilbert, U.S. Geological Survey. Photo *D* by I. J. Witkind, U.S. Geological Survey.

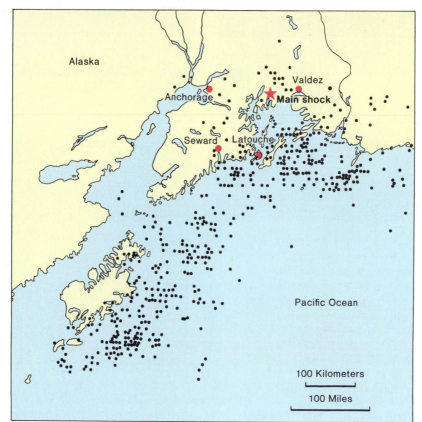

Figure 16.16
Aftershocks associated with the 1964 Alaskan earthquake,
from March 27 to December 31, 1964. The aftershocks range
from 4.4 to 6.7 magnitude (numerous smaller aftershocks are
not mapped).
From U.S. Geological Survey.

Aftershocks are small earthquakes that follow the
main shock (figure 16.16). In the months after the Alaskan
earthquake of 1964 there were several thousand aftershocks, tapering off in frequency. Although aftershocks
are smaller than the main quake, they can cause considerable damage, particularly to structures previously
weakened by the powerful main shock. A long period of
aftershocks can be extremely unsettling to people who have
lived through the main shock.

Tsunamis

The sudden movement of the sea floor upward or downward during a submarine earthquake can generate very
large sea waves, popularly called "tidal waves." Because
the ocean tides have nothing to do with generating these
huge waves, the Japanese term **tsunami** is preferred by geologists. Tsunamis are also called **seismic sea waves.** They
usually are caused by great earthquakes (magnitude 8[+])
that disturb the sea floor, but they also result from submarine landslides or volcanic explosions. When a large
section of sea floor suddenly rises or falls during a quake
(figure 16.17), all the water over the moving area is lifted
or dropped for an instant. As the water returns to sea level,
it sets up a long, low wave that spreads very rapidly over
the ocean (figure 16.18).

Tsunamis are unlike ordinary water waves on the sea
surface. A large wind-generated wave may have a wavelength of 1,300 feet (400 meters) and be moving in deep

Figure 16.17
Uplift of the sea floor caused by the Alaskan earthquake, 1964,
near the town of Latouche. The trace of the Hanning Bay fault
runs diagonally through the center of the photograph. The
block of rock on the right has been displaced upward as much
as 16 feet (5 meters). The large, light-colored area to the right
of the fault is bleached reef rock that was once sea floor but is
now above sea level.
Photo by G. Plafker, U.S. Geological Survey.

water at a speed of 55 miles per hour (90 kilometers per
hour). The wave height when it breaks on shore may be
only 2 to 10 feet (0.6 to 3 meters), although in the middle
of hurricanes the waves can be more than 50 feet (15
meters) high. A tsunami, however, may have a wavelength of 100 *miles* (160 kilometers), and may be moving

The Mexico City Earthquake of 1985

A

B

Box 16.2 Figure 1
Mexico City earthquake, 1985. (*A*) This 15-story building collapsed completely, crushing all its occupants as its reinforced-concrete floors "pancaked" together. (*B*) The upper floors of the left portion of this hotel collapsed. The lower floors and the entire right portion of the hotel remained intact. Note similar destruction in the left rear, and undamaged buildings nearby.
Photos by M. Celebi, U.S. Geological Survey.

On September 19, 1985, central Mexico was rocked by an earthquake measuring 8.1 on the Richter scale. Thirty-six hours later, a magnitude-7.5 aftershock occurred nearby. The first quake was almost as large as the 1906 San Francisco quake. The second was thirty times bigger than the 1971 San Fernando Valley earthquake north of Los Angeles. This one-two punch in Mexico caused some severe damage in Mexico City, 200 miles (350 kilometers) from the epicenter. At least 5,000 people died and 30,000 were injured in a population of 18 million. Damage totaled $5 billion.

The ground shaking was relatively mild in most parts of Mexico, but it was amplified by some buildings and by the lake-bed sediments beneath Mexico City. Most buildings are not rigidly stiff but slightly flexible; they sway gently like large pendulums when struck by wind or seismic waves. The time necessary for a single back-and-forth oscillation is called a period, and it varies with a building's height and mass. Natural bodies of rock and sediment vibrate the same way, with periods that vary with the body's size and density.

When earthquake waves struck Mexico City, many were vibrating with a two-second period. The body of lake sediment beneath the city has a natural period of two seconds also, so the wave motion was amplified by the sediments. This type of amplification occurs when you push a child on a swing—if you push gently in time with the swing's natural period, the swing goes higher and higher. The water-saturated sediment began to move like a sloshing waterbed. The ground moved back and forth by 16 inches (40 centimeters) every two seconds, and it did this fifteen or twenty times.

This shaking was devastating to buildings with a natural two-second period (generally those five to twenty stories high), and several hundred buildings collapsed as they further amplified the shaking. Many structures weakened by the main shock collapsed during the large aftershock. Most shorter and taller buildings had other periods of vibration and rode out the shaking with little damage. Although 800 buildings collapsed or were seriously damaged, most of the city's 600,000 structures survived with little or no damage. The Mexico City quake vividly illustrates the fact that proper building design can greatly lessen earthquake damage.

at 450 miles per hour (725 kilometers per hour). In deep water the wave height may be only 2 to 6 feet (0.6 to 2 meters), but near shore the tsunami may peak up to heights of 50 to 100 feet (15 to 30 meters). This great increase in wave height near shore is caused by bottom topography; only a few localities have the combination of gently sloping offshore shelf and funnel-shaped bay that force tsunamis to awesome heights (the record height was 278 feet, or 85 meters, in 1971 in the Ryukyu islands south of Japan). Along most coastlines tsunami height is very small.

Although the speed of the wave slows drastically as it moves through shallow water, a tsunami can still hit some shores as a very large, very fast wave. Because of its extremely long wavelength, a tsunami does not withdraw quickly as normal waves do. The water keeps on rising for five to ten minutes, causing great flooding before the wave withdraws. A tsunami's long duration and great height can bring widespread destruction to the entire shore zone (figure 16.19).

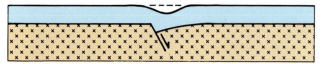

A Before earthquake

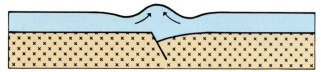

B Sudden displacement of sea floor causes sea level to drop momentarily

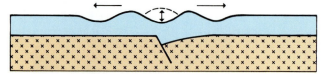

C Water rushes into depression and overcorrects, raising sea level slightly

D Sea level oscillates before coming to rest; long, low waves (tsunami) are sent out over sea surface

Figure 16.18
Generation of a tsunami by a submarine earthquake (rock and water displacements are not drawn to scale).

A tsunami that formed after the 1960 Chilean quake crossed the Pacific Ocean and did extensive damage in Japan. One of the most destructive tsunamis of modern times was generated April 1, 1946, by an earthquake offshore from Alaska. It devastated the city of Hilo, Hawaii, causing 159 deaths. The tsunami following the 1964 Alaskan quake drowned twelve people in Crescent City, California, a small coastal town near the Oregon border. Wave damage near an epicenter can be awesome. The 1946 Alaska tsunami destroyed the Scotch Cap lighthouse on nearby Unimak Island, sweeping it off its concrete base, which was 35 feet (10 meters) above sea level, and killing its five occupants. The wave also swept away a radio tower, whose base was 103 feet (31 meters) above sea level.

Distribution of Earthquakes

Most earthquakes are concentrated in narrow geographic belts (figure 16.20A), although *some* earthquakes have occurred in most regions on earth. The boundaries of plates in the plate-tectonic theory are defined by these earthquake belts (figure 16.20B). The most important concentration of earthquakes by far is in the **circum-Pacific belt,** which encircles the rim of the Pacific Ocean. Within this belt occur approximately 80% of the world's shallow-focus quakes, 90% of the intermediate-focus quakes, and essentially 100% of the deep-focus quakes (figure 16.21).

A

B

Figure 16.19
Tsunami damage in Alaskan earthquake, 1964. (**A**) Fishing boat carried inland in Resurrection Bay at Seward. (**B**) Spruce trees 2 feet (0.7m) thick broken by a wave at an elevation of 100 feet (30m) above sea level near Valdez.
Photo *A* by National Geophysical Data Center. Photo *B* by U.S. Geological Survey.

Another major concentration of earthquakes is in the **Mediterranean-Himalayan belt,** which runs through the Mediterranean Sea, crosses the Mideast and the Himalayas, and passes through the East Indies to meet the circum-Pacific belt north of Australia.

A number of shallow-focus earthquakes occur in two other significant locations on earth. One is along the summit or crest of the *mid-oceanic ridge,* a huge underwater mountain range that runs through all the world's oceans (figure 1.13 and the chapter on the sea floor). A few earthquakes have also been recorded in isolated spots usually associated with basaltic volcanoes, such as those of Hawaii.

In most parts of the circum-Pacific belt, earthquakes, andesitic volcanoes, and *oceanic trenches* (chapter titled "The Sea Floor") appear to be closely associated. Careful determination of the locations and depths of focus of earthquakes has revealed the existence of distinct *earthquake zones* that begin at oceanic trenches and slope landward and downward into the earth at an angle of about 30° to 60° (figures 16.21 and 16.22). Such zones are called **Benioff zones** after the man who first recognized them.

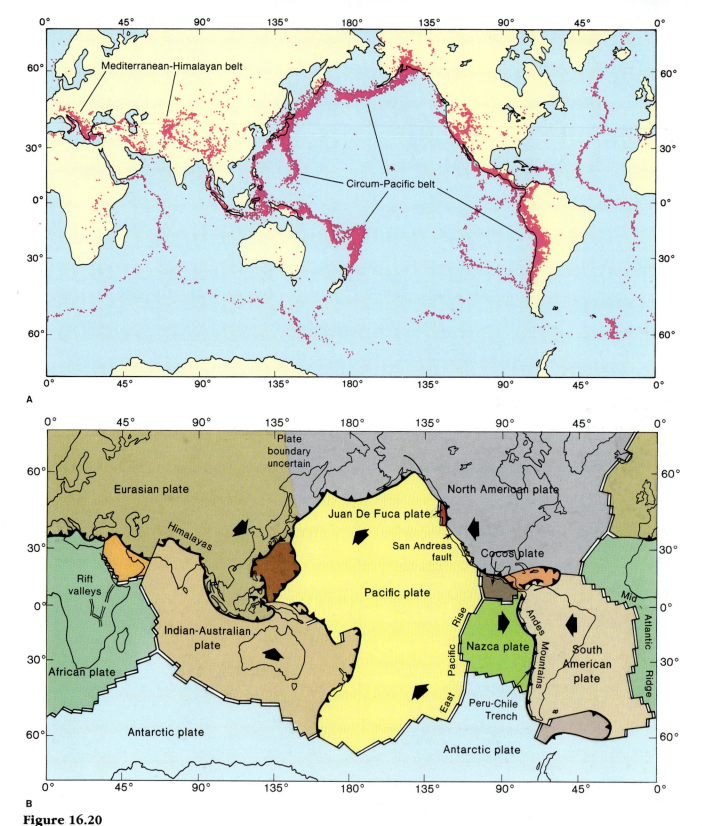

A

B

Figure 16.20

(A) World distribution of earthquakes from 1961 to 1967; epicenters of quakes with depth of focus between 0 and 700 kilometers. A map of quakes during the 1970s or 1980s would show the same pattern.

From Barazangi and Dorman, *Bulletin of Seismological Society of America*, 1969.

(B) The major plates of the world in the theory of plate tectonics. Compare the locations of plate boundaries with earthquake locations shown in figure 16.20A. Double lines show diverging plate boundaries, single lines show transform boundaries. Heavy lines with triangles show converging boundaries; triangles point down subduction zone.

Modified from Hamilton, U.S. Geological Survey.

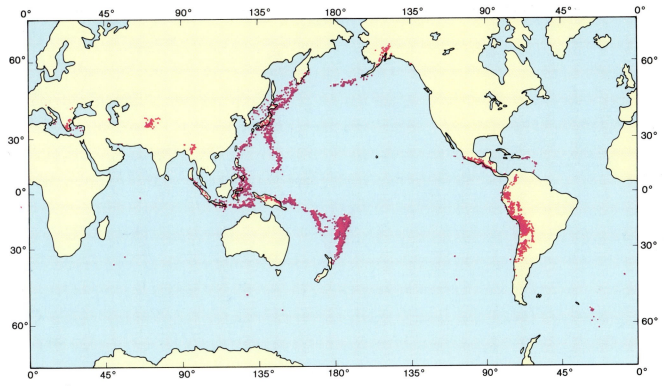

Figure 16.21
Distribution of quakes with depth of focus between 100 and 700 kilometers, 1961–1967. These quakes define Benioff zones.
From Barazangi and Dorman, *Bulletin of Seismological Society of America*, 1969.

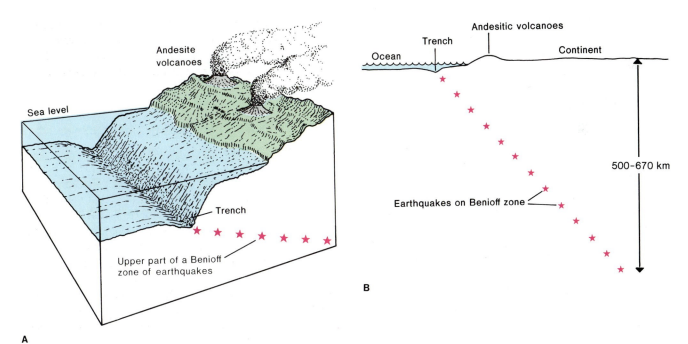

A

B

Figure 16.22
A Benioff zone of earthquakes begins at an oceanic trench and dips under a continent (such as South America). Andesitic volcanoes form near the edge of the continent. (*A*) Upper part of Benioff zone. (*B*) A Benioff zone may extend to a depth of 670 kilometers.

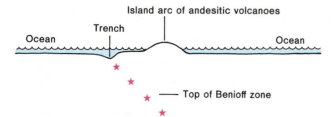

Figure 16.23
A Benioff zone can begin at a trench and dip under an island arc (such as the Aleutian islands). Only the upper part of the Benioff zone is shown.

All Benioff zones seem to slope under a continent or a curved line of islands called an **island arc** (figure 16.23). Andesitic volcanoes may form the islands of the island arc, or they may be found near the edge of a continent that overlies a Benioff zone.

Most of the circum-Pacific belt is made up of Benioff zones associated in this manner with oceanic trenches and andesitic volcanoes. Parts of the Mediterranean-Himalayan belt represent Benioff zones, too, notably in the eastern Mediterranean Sea and in the East Indies. Essentially all the world's intermediate- and deep-focus earthquakes occur in Benioff zones.

First-Motion Studies of Earthquakes

By studying seismograms of an earthquake on a distant fault, geologists can tell which way rocks moved along that fault. Rock motion is determined by examining seismograms from many locations surrounding a quake. Each seismograph station can tell whether the first rock motion recorded there was a push or a pull (figure 16.24). If the rock moved toward the station (a push), then the pen drawing the seismogram is deflected up. If the first motion is away from the station (a pull), then the pen is deflected down.

If an earthquake occurs on a fault as shown in figure 16.25A, large areas around the fault will receive a push as first motion, while different areas will receive a pull. Any station within the shaded area marked A will receive a push, for the rock is moving from the epicenter toward those stations, as shown by the arrows in the figure. All stations in area C will also receive a push, but areas B and D will record a pull as the first motion.

In figure 16.25B you can see the same pattern of pushes and pulls, but in this case the pattern is caused by a fault with a different orientation. Either fault can cause the same pattern, if the rock moves in the direction shown by the arrows. In other words, there are two possible solutions to any pattern of first motions.

If the orientation of the fault trace on the earth's surface is known, as it is for most faults on land (and for a few on the ocean floor), the correct choice of the two solutions can be made. But if the orientation of the fault is not known—as is the usual case for earthquakes at sea or at great depth—the choice between the two possible solutions can be difficult. One solution may be more *likely*,

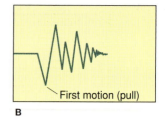
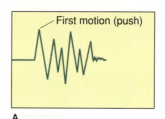

Figure 16.24
Seismograms showing how first horizontal motions of rocks along a fault are determined. (*A*) If the first motion is a push (from the epicenter to the seismograph station), the seismogram trace is deflected upward. (*B*) If it is a pull (away from the station), the deflection is downward.

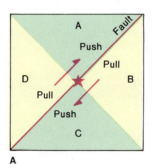
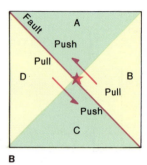

Figure 16.25
Map view of two possible solutions for the same pattern of first motion. Each solution has a different fault orientation. If the fault orientation is known, the correct solution can be chosen. The star marks the epicenter, and rock motion is shown by the arrows.

based on the study of topography, other faults, or similar factors. But such a solution is by no means *proven* if the fault orientation is not known. As you read about first-motion studies, keep in mind that there are always two solutions to any pattern.

Earthquakes and Plate Tectonics

One of the great attractions of the concept of plate tectonics is its ability to explain the distribution of earthquakes and the rock motion associated with them.

As described briefly in chapter 1, the concept of plate tectonics is that the earth's surface is divided into a few giant *plates*. Plates are rigid slabs of rock, thousands of miles wide and 100 kilometers (60 miles) thick, that move across the earth's surface. Because the plates include continents and sea floors on their upper surfaces, the plate tectonics concept means that the continents and sea floors are moving. The plates change not only position but size and shape (as we discuss in more detail in the chapter titled "Plate Tectonics").

Earthquakes occur commonly at the edges of plates but only occasionally in the middle of a plate. The close correspondence between plate edges and earthquake belts can be seen by comparing the map of earthquake distribution in figure 16.20A with the plate map in figure 16.20B.

This correspondence is hardly surprising—plate boundaries are identified and *defined* by earthquakes. According to plate tectonics, earthquakes are caused by the

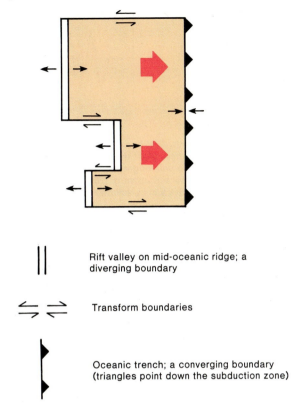

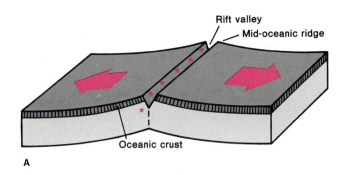

A

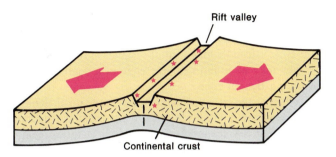

B

Figure 16.27
Diverging plate boundaries. (*A*) On the ocean floor. (*B*) On a continent. Each is marked by a rift valley and shallow-focus earthquakes (shown as stars). Depth of rift valleys is exaggerated.

|| Rift valley on mid-oceanic ridge; a diverging boundary

⇌ ⇌ Transform boundaries

▶ Oceanic trench; a converging boundary (triangles point down the subduction zone)

Figure 16.26
Relation of three types of plate boundaries to one another. Plate moves to right (as shown by the large arrows), away from rift valley at diverging boundary and toward oceanic trench at converging boundary.

interactions of plates along plate boundaries. Therefore, narrow bands of earthquakes are used to outline plates on plate maps. This can be clearly seen in the east Pacific Ocean off South America, where the Nazca plate (figure 16.20*B*) is almost completely outlined by earthquake epicenters (figure 16.20*A*).

The earthquakes on the western border of the Nazca plate are shallow-focus quakes, and they occur in a narrow belt along the crest of the mid-oceanic ridge here, locally called the East Pacific Rise. The quakes along the eastern boundary occupy a broader belt that lies mostly within South America. This belt includes shallow-, intermediate-, and deep-focus earthquakes in a Benioff zone that begins at the Peru-Chile Trench just offshore and slopes steeply down under South America to the east. The Nazca plate moves eastward, away from the spreading center on the crest of the mid-oceanic ridge and toward the subduction zone at the trench, where the plate plunges down into the mantle. The plate's western boundary is located at the crest of the East Pacific Rise, and its eastern boundary is at the bottom of the Peru-Chile Trench.

Earthquakes at Plate Boundaries
As you have learned, there are three types of plate boundaries, *diverging boundaries* where plates move away from each other, *transform boundaries* where plates move horizontally past each other, and *converging boundaries* where plates move toward each other. Each type of boundary has a characteristic pattern of earthquake distribution and rock motion.

One example of the relation of the three types of boundaries to each other is shown in figure 16.26. Note how the motion at each boundary (shown by the pairs of small arrows) agrees with the motion of the plate (shown by the large arrows).

Diverging boundaries   At a spreading center, where plates move away from each other, earthquakes are shallow and restricted to a narrow band. A diverging boundary on the sea floor is marked by the crest of the mid-oceanic ridge and the *rift valley* that is often (but not always) found on the ridge crest (figure 16.27*A*). The earthquakes are located along the sides of the rift valley and beneath its floor. The rock motion that is deduced from first-motion studies shows that the faults here are normal faults, parallel to the rift valley. (This is the most likely solution of the first-motion studies. Keep in mind as you read this section that there are always two solutions for first motion; we give only one—the more likely solution.) If this interpretation of the first motions is correct, it implies that the ridge crest is being subjected to horizontal tension, which is apparently tearing the sea floor open here, creating the rift valley and causing the earthquakes.

A diverging boundary within a *continent* is usually also marked by a rift valley, shallow-focus quakes, and normal faults (figure 16.27*B*). The African Rift Valleys in eastern Africa (figure 16.20*B*) seem to be such a boundary. Horizontal tension here may be tearing eastern Africa slowly apart, creating the rift valleys, some of which contain lakes.

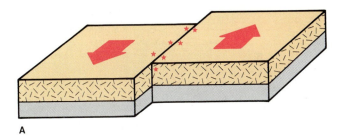

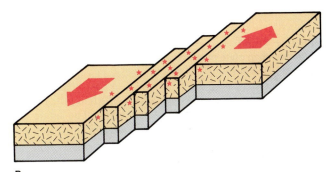

Figure 16.28
Transform boundaries. (*A*) Narrow band of shallow-focus earthquakes shown as stars along single fault. (*B*) Broad band of earthquakes along a system of parallel faults.

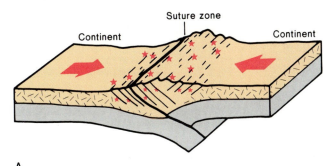

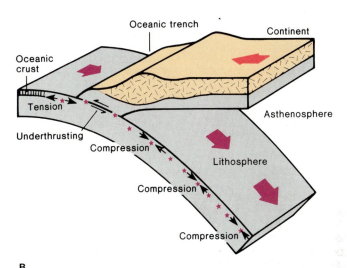

Figure 16.29
(*A*) A converging boundary marked by the collision of two continents. A very broad zone of shallow-focus earthquakes occurs along a complex system of faults. (*B*) A converging boundary with ocean floor subducting under a continent. Earthquakes occur near the top of the subducting plate due to tension, underthrusting, and compression.

Transform boundaries Where two plates move past each other along a transform boundary, the earthquakes are shallow. First-motion studies indicate strike-slip motion on faults parallel to the boundary. The earthquakes may be aligned in a narrow band along one fault (figure 16.28*A*), or they may form a broader zone if plate motion is taken up by movement along a system of parallel faults (figure 16.28*B*). The San Andreas fault in California (figure 16.20*B*) may be an example of a single fault forming a plate boundary, but many geologists believe that the broad zone of seismicity along the Basin and Range faults in the western United States (figure 16.10*A*) shows that a system of faults both parallel and at an angle to the San Andreas forms the plate boundary here.

Converging boundaries Converging boundaries are of two general types, one marked by the *collision* of two continents (figure 16.29*A*), the other marked by *subduction* of the ocean floor under a continent (figure 16.29*B*) or another piece of sea floor. Each type has a characteristic pattern of earthquakes.

Collision boundaries are characterized by broad zones of shallow earthquakes on a complex system of faults (figure 16.29*A*). Some of the faults are parallel to the dip of the suture zone that marks the line of collision; some are not. One continent usually overrides the other slightly (continents are not dense enough to be subducted), creating thick crust and a mountain range. The Himalaya Mountains are thought to represent such a boundary (figure 16.20*B*). The seismic zone is so broad and complex at such boundaries that other criteria, such as detailed geologic maps, must be used to identify the position of the suture zone at the plate boundary.

During *subduction,* earthquakes occur for several different reasons (figure 16.29*B*). As a dense oceanic plate bends to go down at a trench, it stretches slightly and normal faults occur as a result of *tension*. This gives a block-faulted character to the outer (seaward) wall of a trench. For some distance below the trench, the subducting plate is in contact with the overlying plate. First-motion studies of earthquakes at these shallow depths show that the quakes are caused by shallow-angle thrust-faulting. This is the motion expected as one plate slides beneath another, a process commonly called *underthrusting.*

At greater depths, where the descending plate is out of contact with the overlying plate, earthquakes are common, but the reasons for them are not obvious. The quakes are confined to a thin zone, only 20 to 30 kilometers thick, within the lithosphere of the descending plate, which is about 100 kilometers thick (figure 16.29*B*). This zone is thought to be near the top of the lithosphere, where the rock is colder and more brittle, but there is little real evidence for this. It is clear, however, that these deep earthquakes cannot be located at the upper *surface* of the plate, for there is very little friction between

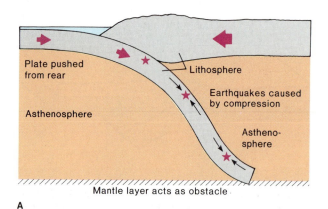

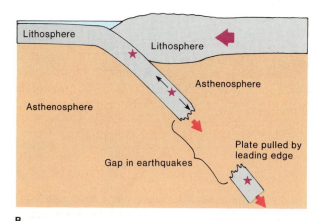

Figure 16.30
Cross-section views of subducting plates. (*A*) Quakes caused by compression, perhaps resulting from the plate being pushed from the rear (or hitting an obstacle). (*B*) Quakes caused by tension, perhaps resulting from the plate being pulled downward.

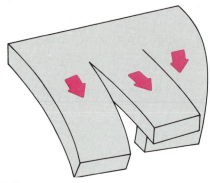

Figure 16.31
A descending plate may tear into segments, some with steeper subduction angles than others.

the cold, descending lithosphere and the hot plastic asthenosphere above it. The deep quakes must be located completely *within* the descending lithosphere.

First motions of these quakes imply that along many subduction zones the quakes are caused by *compression* parallel to the dip of the plate (figure 16.29*B*). This could mean that the plates are being pushed from the rear, or that they are running into some obstacle at depth, or both (figure 16.30*A*).

Earthquake first motion on other subduction zones, however, implies *tension* parallel to the dip of the descending plates (figure 16.30*B*). This might mean that the plates are being pulled along by their leading edges. Descending lithosphere is slightly denser than the surrounding asthenosphere (because it is colder and because low-density minerals within a descending plate may either "distill" off the plate by partial melting or be converted to denser minerals by pressure). The inference of tension is supported by evidence that some plates break apart as they sink, resulting in gaps in the earthquake pattern along some subduction zones (figure 16.30*B*).

Subduction Angle
The distribution of earthquakes can be used to determine the angle of subduction of a downgoing plate. Subduction angles vary considerably from trench to trench, and in some

cases they become steeper or gentler with depth. The reasons for these variations are not clear. The angle may depend on a complex interaction between spreading rate, cooling rate, plate density, duration of subduction, and other factors.

Descending plates may tear or break into segments or they may become folded, usually below a depth of 200 kilometers. Individual segments may descend at different angles (figure 16.31), which can be determined from the distribution of earthquakes. The segmentation or folding may be required geometrically to allow subduction of curved rigid plates to successively greater depths. The plate forms with a radius of curvature equal to that of the earth's surface, and during subduction it sinks to regions of a progressively smaller radius.

In summary, earthquakes are very closely related to plate tectonics. Most plate boundaries are defined by the distribution of earthquakes, and plate motion can be deduced by the first motions of the quakes. Analysis of first motions can also help determine the type and orientation of forces that act on plates, such as tension and compression. Quake distribution with depth indicates the angle of subduction and has shown that some plates change subduction angle and even break up as they descend. A few quakes, such as those that occur in the center of plates, cannot easily be related to plate motion.

Earthquake Prediction

People who live in earthquake-prone regions are plagued by unscientific predictions of impending earthquakes by popular writers and self-proclaimed prophets. Several techniques are being developed for *scientifically* forecasting a coming earthquake. Most of these methods involve monitoring slight changes that occur in rock next to a fault before the rock breaks and moves.

Just as a bent stick may crackle and pop before it breaks with a loud snap, a rock gives warning signals that it is about to break. Before a large quake, small cracks may open within the rock, causing small tremors, or *microseisms,* to increase. The *properties of the rock* next to the fault may be changed by the opening of such cracks. If these changes are monitored, they may give some warning of an imminent quake, especially if the magnetism or electrical resistivity of the rock or the speed of seismic waves is observed to be changing.

The opening of tiny cracks changes the rock's porosity, so *water levels in wells* often rise or fall before quakes. The cracks provide pathways for the release of radioactive radon gas from rocks (radon is a product of radioactive decay of uranium and other elements). An *increase in radon emission from wells* may be a prelude to an earthquake. A very local method of predicting quakes is to time the *interval between eruptions of Old Faithful geyser* in Yellowstone National Park. Long-term records of the time between eruptions have shown that this interval changes in a regular way before a large local earthquake, probably because of porosity changes within the surrounding rock.

In some areas the *surface of the earth tilts and changes elevation* slightly before an earthquake. Through the use of highly sensitive instruments that measure increasing strain in rocks, scientists can sometimes predict quakes with considerable accuracy.

Chinese scientists claim successful predictions by watching *animal behavior*—horses become skittish and snakes leave their holes shortly before a quake. United States scientists are conducting a few pilot programs along these lines, but many are skeptical of the Chinese claims.

Japanese and Russian geologists were the first to predict earthquakes successfully, and Chinese geologists have recently made some very accurate predictions. In 1975 a 7.3-magnitude earthquake near Haicheng in northeastern China was predicted five hours before it happened. Alerted by a series of *foreshocks,* authorities evacuated millions of people from their homes; many watched outdoor movies in the open town square. Half the buildings in Haicheng were destroyed, along with many entire villages, but only a few hundred lives were lost. In grim contrast, however, the Chinese program failed to predict the 1976 Tangshan earthquake (magnitude 7.6), which struck with no warning and killed an estimated 500,000 people.

Patterns of earthquakes in space and time are being recognized, particularly on the faults along some plate boundaries. If one plate is sliding diagonally under another plate, a series of earthquakes may occur in a regular way along the plate boundary (figure 16.32). If such a pattern is known to repeat itself in the same area over several centuries, it becomes possible to predict where the next large quake might occur. Along some long-active faults, there are short inactive segments called *seismic gaps* where earthquakes have not occurred for a long time. These gaps are widely thought to be the most likely sites for future earthquakes, so many of them are being carefully studied.

Geologists in the United States are becoming increasingly confident of their ability to predict earthquakes along some segments of some faults. In 1988 the U.S. Geological Survey estimated a 50 percent chance of a magnitude-7 quake along the San Andreas fault, with the section of the fault near Santa Cruz identified as the most likely location of such a quake. In 1989 a 7.1 occurred on this very section. Since the techniques are new and in some

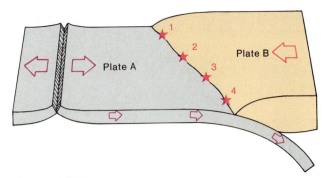

Figure 16.32
As plate *A* slides diagonally under plate *B*, earthquakes occur in a regular sequence from one to four. The entire sequence may repeat.

cases only partly understood, some errors will undoubtedly be made. Many faults are not monitored at all because of lack of money and personnel, so we will never have a warning of impending quakes in some regions. For large urban areas near active faults, however, earthquake prediction techniques are rapidly lessening the danger of an earthquake's occurring without warning.

Earthquake Control

Controlling the timing of earthquakes is an intriguing idea for the distant future. The potential for death and destruction in some cities near faults is so great that the planned triggering of earthquakes might be desirable. If people knew the exact time at which an earthquake was to occur, they could try to protect their belongings and temporarily evacuate the area, much as people do when a hurricane warning is given.

One method of earthquake control was discovered by accident when water under high pressure was pumped down into the ground. The water may release the strain that has accumulated in the rocks near a fault. The pressure of the water may force the rock walls of the fault apart and lubricate them so that the rocks move, causing small quakes. Such earthquakes have occurred around a disposal well near Denver and in an oil field in Colorado where water was forced underground. Without water, rock strain would build up until a large quake occurred. With the water reducing the friction, strain can be released in smaller, timed quakes.

Could such a method of quake release be used on the dangerous portions of the San Andreas fault? Perhaps, but only after years of testing, and then only as a trial in a relatively uninhabited section of the fault. The cost of such a program might prove to be prohibitive, and the liability problems would be enormous. Still, it is exciting to consider that in the future the danger of quakes along some faults might be greatly lessened by controlled release of strain.

It is important to realize that such experiments release strain that is already there. The advantage is in timing the release of the strain. The water does not *cause* the quake; it allows a quake that would occur sometime anyway to be *controlled.*

Waiting for the Big One

The San Andreas fault, running north-south for 800 miles (1300 kilometers) through California, is a right-lateral fault capable of generating great earthquakes of magnitude 8 or more. The 1906 earthquake (estimated magnitude 8.25) near San Francisco caused a 280-mile (450-kilometer) scar in northern California (box figure 1). The portion of the fault nearest Los Angeles last broke in 1857 in a quake that was probably of comparable size. The ground has not broken in either of these regions since these quakes. Each old break is now a seismic gap, where rock strain is being stored prior to the next giant quake.

Despite the extensive press coverage given the 1989 Loma Prieta ("San Francisco") quake, that quake was not the "Big One" long predicted by geologists. The 1989 quake had a magnitude of 7.1; the 1906 quake has been estimated at magnitude 8.25. The ground shaking in an 8.25 quake is 14 times as great as the shaking in a 7.1 quake; the energy released in the bigger quake was 53 times that released in 1989. In other words, it would take 53 quakes the size of the 1989 quake to equal the energy released in 1906, or in

Box 16.3 Figure 1
The two major breaks on the San Andreas fault in California. Each break occurred during a giant earthquake. Each old break is now a seismic gap—the future site for another giant earthquake.
From California Division of Mines and Geology.

the expected Big One. Although Bay area residents refer to the 1989 quake as the "Pretty Big One," it did not reduce the strain building up on the San Andreas fault to harmless levels. A giant quake exceeding magnitude 8 could strike either the northern section or the southern section of the San Andreas tomorrow.

Summary

Earthquakes usually occur when rocks break and move along a fault to release strain that has gradually built up in the rock. Volcanic activity can also cause earthquakes.

Seismic waves move out from the earthquake's *focus.* *Body waves* (P waves and S waves) move through the earth's interior, and *surface waves* move on the earth's surface.

Seismographs record seismic waves on *seismograms,* which can be used to determine an earthquake's strength, location, and depth of focus. Most earthquakes are shallow-focus quakes, but some occur as deep as 670 kilometers below the earth's surface.

The time interval between first arrivals of P and S waves is used to determine the distance between the seismograph and the *epicenter.* Three or more stations are needed to determine the location of earthquakes.

Earthquake *intensity* is determined by assessing damage and is measured on the *modified Mercalli scale.*

Earthquake *magnitude* is determined by the amplitude of seismic waves on a seismogram and is measured on the *Richter scale.*

The most noticeable effects of earthquakes are ground motion and displacement (which destroy buildings and thereby injure or kill people), fire, landslides, and *tsunamis. Aftershocks* can continue to cause damage months after the main shock.

Earthquakes are generally distributed in belts. The *circum-Pacific belt* contains most of the world's earthquakes. Earthquakes also occur on the Mediterranean-Himalayan belt, the crest of the mid-oceanic ridge, and in association with basaltic volcanoes.

Which section will break first? Because the southern section has been inactive longer, it may be the likelier candidate. A magnitude-8 earthquake here would be about 500 times as powerful as the 1971 San Fernando Valley earthquake, which killed sixty-five people and caused a half billion dollars' worth of damage. Damage could be in the 100s of billions of dollars; death tolls could be in the thousands if a quake of magnitude 8 struck during weekday business hours, when city buildings and streets are crowded with people.

Spotty evidence suggests that great earthquakes have a recurrence interval of about 140 years on this portion of the San Andreas fault. Historic records in California do not go very far back in time, and much of the evidence involves radiometric dating of broken beds of carbon-rich lake muds. This number is only an average; 2 or 3 quakes spaced about 60 years apart occur in clusters, with the interval between clusters being 200 to 300 years. Prior to 1857 this portion last broke in about 1745. Adding this 112-year difference to 1857 gives 1969, and adding the average of 140 years to 1857 gives 1997, so the present danger to Los Angeles is clear. If the 1857 quake ended a cluster, however, the next big quake may not occur till 2057 or later.

The southern section of the fault is very heavily instrumented to detect changes in ground tilt, water-well levels, and other earthquake precursors. In the mid-1970s most geologists thought that the Big One was imminent when it was discovered that most of southern California had risen as much as 16 inches (0.4 meters) from 1959 to 1974, with the greatest uplift occurring very close to the fault (the "Palmdale bulge"). Rapid subsidence from 1974 to 1976, however, eliminated at least half of this uplift, and the episode of rising and falling land passed without a large quake.

But the northern portion of the San Andreas fault is dangerous, too. Prior to 1906 this section of the fault broke in another giant quake in 1838. These quakes were only 68 years apart, and 1906 plus 68 equals 1974, so the northern section may actually be overdue for a big quake.

Another way of estimating the recurrence interval is by rock displacement. In 1906 rocks were displaced about 15 feet (4.5 meters) at the epicenter, and we know that plate motion across the San Andreas fault is about 2 inches (5 centimeters) per year. It should therefore take about 90 years to store enough strain to move rocks 15 feet, so the next quake might be expected in 1996.

The U.S. Geological Survey recently estimated that the southern portion of the fault has about a 60 percent chance of an earthquake of magnitude 7.5 to 8.3 within the next 30 years, while the northern portion has only a 20 percent chance during the same period for the Big One. Although a quake near Los Angeles seems more probable, few geologists would be more than mildly surprised if the northern section of the San Andreas fault broke first.

Benioff zones of shallow-, intermediate-, and deep-focus earthquakes are associated with andesitic volcanoes, oceanic trenches, and the edges of continents or island arcs.

The concept of plate tectonics explains most earthquakes as being caused by interactions between two plates at their boundaries. Plate boundaries are generally defined by bands of earthquakes.

Diverging plate boundaries are marked by a narrow zone of shallow earthquakes along normal faults. Transform boundaries are marked by shallow quakes caused by strike-slip motion along one or more faults.

Converging boundaries where continents collide are marked by a very broad zone of shallow quakes. Converging boundaries involving deep subduction are marked by Benioff zones of quakes caused by tension, under-thrusting, and compression.

The distribution of quakes indicates subduction angles and shows that some plates break up as they descend.

The potential for successful earthquake prediction is rapidly increasing, largely through the measurement of rock properties near faults and the recognition of patterns of quakes at plate boundaries.

The timing of the release of accumulating rock strain may be controlled in the future by pumping water underground to trigger limited earthquakes.

Terms to Remember

aftershock	Mediterranean-Himalayan
Benioff zone	belt
body wave	modified Mercalli scale
circum-Pacific belt	P wave
depth of focus	Richter scale
earthquake	seismic wave
elastic rebound theory	seismogram
epicenter	seismograph
focus	surface wave
intensity	S wave
island arc	travel-time curve
magnitude	tsunami (seismic sea wave)

Questions for Review

1. Describe in detail how earthquake epicenters are located by seismograph stations.
2. What causes earthquakes?
3. Compare and contrast the concepts of intensity and magnitude of earthquakes.
4. Name and describe the various types of seismic waves.
5. Discuss the distribution of earthquakes with regard to location and depth of focus.
6. Show with a sketch how the concept of plate tectonics can explain the distribution of earthquakes in a Benioff zone and on the crest of the mid-oceanic ridge.
7. Describe several techniques that help scientists predict earthquakes.
8. How may the timing of earthquakes someday be controlled?
9. Describe several ways that earthquakes cause damage.
10. How do earthquakes cause tsunamis?
11. What are aftershocks?

Questions for Thought

1. What are some arguments in favor of and against predicting earthquakes? What would happen in your community if a prediction were made today that within a month a large earthquake would occur nearby?
2. Stone, brick, and adobe buildings often collapse completely during an earthquake, while wood-frame houses sustain little damage. Two-story buildings are damaged more than single-story buildings. Why?

Supplementary Readings

Anderson, D. L. 1971. The San Andreas fault. *Scientific American* (November). New York: W. H. Freeman.

Bolt, B. A. 1987. *Earthquakes,* 2d ed. New York: W. H. Freeman.

Gere, J. M., and H. C. Shah. 1984. *Terra non firma— Understanding and preparing for earthquakes.* New York: W. H. Freeman.

Hodgson, J. H. 1964. *Earthquakes and earth structure.* Englewood Cliffs, N.J.: Prentice-Hall.

Iacopi, R. 1964. *Earthquake country.* Menlo Park, Calif.: Lane Press.

Penick, J. L., Jr. 1981. *The New Madrid earthquakes.* 2d ed. Columbia, Mo.: University of Missouri Press.

Rayleigh, C. B., K. Sieh, L. R. Sykes, and D. L. Anderson. 1982. Forecasting southern California earthquakes. *Science* 217: 1097–1104.

Richter, C. F. 1958. *Elementary seismology.* New York: W. H. Freeman.

U.S. Geological Survey. *Earthquake information bulletin.*

U.S. Geological Survey, 1965–70. *The Alaska earthquake.* Professional Papers 541–546.

U.S. Geological Survey, 1971. *The San Fernando, California, earthquake of February 9, 1971.* Professional Paper 733.

Wesson, R. L., and R. E. Wallace. 1985. Predicting the next earthquake in California. *Scientific American* 252 (February): 35–43.

Yanev, P. 1974. *Peace of mind in earthquake country.* San Francisco: Chronicle Books.

The only rocks that geologists can study directly in place are those of the crust; and the earth's crust is but a thin skin of rock, making up less than 1% of the earth's total volume. To learn about the inside of the earth, geologists must study it *indirectly*, largely by using the tools of geophysics—that is, seismic waves and the measurement of gravity, heat flow, and earth magnetism.

The evidence from geophysics suggests that the earth is divided into three major parts—the crust on the earth's surface, the rocky mantle beneath the crust, and the metallic core at the center of the earth. The study of plate tectonics has shown that the crust and uppermost mantle can be conveniently divided into the brittle lithosphere and the plastic asthenosphere.

You will learn in this chapter how gravity measurements can indicate where certain regions of the crust and upper mantle are being held up or held down out of their natural position of equilibrium. We will also discuss the earth's magnetic field and its history of reversals. We will show how magnetic anomalies can indicate hidden ore and structures. We close with a discussion of the distribution and loss of the earth's heat.

17

The Earth's Interior

Serpentinized peridotite may be mantle rock brought to the surface during mountain building. Sierra Nevada, California.

Geologists who have considerable exposure to the public (such as in government offices and universities) occasionally encounter people who are convinced that geologists are misled about the interior of the earth. These people hold strong convictions about what is deep inside the earth. Many believe, for example, that the earth is hollow and is a storage place for water or even flying saucers.

What *do* geologists know about the earth's interior? How do they obtain information about the parts of the earth beneath the surface? Is there even a slight chance that geologists are wrong, that the earth really is hollow?

Geologists, in fact, are not able to sample rocks very far below the earth's surface. Some deep mines penetrate 3 kilometers (2 miles) into the earth, and a deep oil well may go as far as 8 kilometers beneath the surface; the deepest scientific well has reached 12 kilometers in the Soviet Union. Rock samples can be brought up from a mine or a well for geologists to study. But the earth has a radius of about 6,370 kilometers, so it is obvious that geologists can only scratch the surface when they try to study *directly* the rocks beneath their feet.

Deep parts of the earth are studied *indirectly,* however, largely through the branch of geology called **geophysics,** which is the application of physical laws and principles to a study of the earth. Geophysics includes the study of seismic waves and the earth's magnetic field, gravity, and heat. All these things tell us something about the nature of the deep part of the earth. Together they create a convincing picture of what makes up the earth's interior.

Evidence from Seismic Waves

Seismic waves from a large earthquake may pass entirely through the earth. A nuclear bomb explosion also generates seismic waves. Geologists obtain new information about the earth's interior after every large earthquake and bomb test.

One important way of learning about the earth's interior is the study of **seismic reflection,** the return of some of the energy of seismic waves to the earth's surface after the waves bounce off a rock boundary. If two rock layers of differing densities are separated by a fairly sharp boundary, seismic waves reflect off that boundary just as light reflects off a mirror. These reflected waves are recorded on a seismogram, which shows the amount of time the waves took to travel down to the boundary, reflect off it, and return to the surface. From the amount of time necessary for the round trip, geologists calculate the depth of the boundary (figure 17.1).

Another method used to locate rock boundaries is the study of **seismic refraction,** the bending of seismic waves as they pass from one material to another, which is similar to the way that light waves bend when they pass through the lenses of eyeglasses. As a seismic wave strikes a rock boundary, much of the energy of the wave passes across the boundary. As the wave crosses from one rock layer to another, it changes direction (figure 17.2). This change of

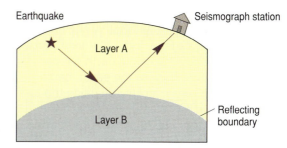

Figure 17.1
Seismic reflection. Seismic waves reflect from a rock boundary deep within the earth and return to a seismograph station on the surface.

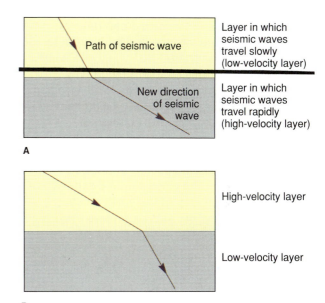

Figure 17.2
Seismic refraction occurs when seismic waves bend as they cross rock boundaries. (A) Low-velocity layer above high-velocity layer. (B) High-velocity layer above low-velocity layer.

direction, or refraction, occurs only if the velocity of seismic waves is different in each layer (which is generally true if the rock layers differ in density or strength).

The boundaries between such rock layers are usually distinct enough to be located by seismic refraction techniques, as shown in figure 17.3. Seismograph station 1 is receiving seismic waves that pass directly through the upper layer *A*. Stations farther from the epicenter, such as station 2, receive seismic waves from two pathways: (1) a direct path straight through layer *A* and (2) a refracted path through layer *A* to a higher-velocity layer *B* and back to layer *A*. Station 2 therefore receives the same wave twice.

Seismograph stations close to station 1 receive only the direct wave or possibly two waves, the direct (upper) wave arriving before the refracted (lower) wave. Stations near station 2 receive both the direct and the refracted waves. At some point between station 1 and station 2 there is a transformation from receiving the direct wave first to receiving the *refracted* wave first. Even though the refracted wave travels farther, it can arrive at a station first because most of its path is in the high-velocity layer *B*.

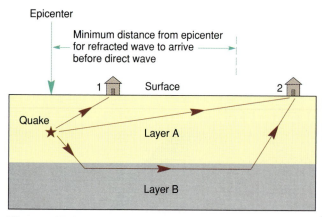

Figure 17.3
Seismic refraction can be used to detect boundaries between rock layers. See text for explanation.

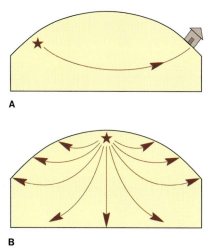

Figure 17.4
Curved paths of seismic waves caused by uniform rock with increasing seismic velocity with depth. (*A*) Path between earthquake and recording station. (*B*) Waves spreading out in all directions from earthquake focus.

The distance between this point of transformation and the epicenter of the earthquake is a function of the depth to the rock boundary between layers *A* and *B*. A series of portable seismographs can be set up in a line away from an explosion (a *seismic shot*) to find this distance, and the depth to the boundary can then be calculated. The velocities of seismic waves within the layers can also be found.

Figure 17.2 shows how waves bend as they travel downward into higher velocity layers. But why do waves return to the surface, as shown in figure 17.3? The answer is that advancing waves give off energy in all directions. Much of this energy continues to travel horizontally within layer *B* (figure 17.3). This energy passes beneath station 2 and out of the figure toward the right. A small part of the energy "leaks" upward into layer *A,* and it is this pathway that is shown in the figure. There are many other pathways for this wave's energy that are not shown here.

A sharp rock boundary is not necessary for the refraction of seismic waves. Even in a thick layer of uniform rock, the increasing pressure with depth tends to increase the velocity of the waves. The waves follow curved paths through such a layer, as shown in figure 17.4. To understand the reason for the curving path, visualize the thick rock layer as a stack of very thin layers, each with a slightly higher velocity than the one above. The curved path results from many small changes in direction as the wave passes through the many layers.

The Earth's Internal Structure

It was the study of seismic refraction and seismic reflection that enabled scientists to plot the three main zones of the earth's interior (figure 17.5). The **crust** is the outer layer of rock, which forms a thin skin on the earth's surface. Below the crust lies the **mantle,** a thick shell of rock that separates the crust above from the core below. The **core** is the central zone of the earth. It is probably metallic and the source of the earth's magnetic field.

The Earth's Crust

Studies of seismic waves have shown (1) that the earth's crust is thinner beneath the oceans than beneath the continents (figure 17.6) and (2) that seismic waves travel faster in oceanic crust than in continental crust. Because of this velocity difference, it is assumed that the two types of crust are made up of different kinds of rock.

Seismic P waves travel through oceanic crust at about 7 kilometers per second, which is also the speed at which they travel through basalt. Thus the seismic evidence suggests that oceanic crust is made of basalt. Samples of rocks taken from the sea floor by oceanographic ships verify this. Most of the sea-floor samples are basalt, although other rock types also are found (table 17.1).

Seismic P waves travel more slowly through continental crust—about 6 kilometers per second, the same speed at which they travel through granite. Continental crust is often referred to as being "granite." The term is put in quotation marks because most of the rocks exposed on the surface of continents are not granite. Since a single rock term cannot accurately describe crust that varies in composition, some geologists use the term *sial* (rocks high in *si*licon and *al*uminum) for continental crust and *sima* (rocks high in *si*licon and *ma*gnesium) for oceanic crust.

Seismic waves passing beneath mountain ranges indicate that the crust is thickest under mountains. A protruding *mountain root* of crust bulges down into the mantle.

The boundary that separates the crust from the mantle beneath it is called the **Mohorovičić discontinuity** (Moho for short). Note from figure 17.6 that the mantle lies closer to the earth's surface beneath the ocean than it does beneath continents. The idea behind an ambitious program called Project Mohole (begun during the early 1960s) was to use specially equipped ships to drill through the oceanic crust and obtain samples from the mantle. Although the project was abandoned because of high costs, ocean-floor drilling has become routine since then, but not to the great depth necessary to sample the mantle. Perhaps in the

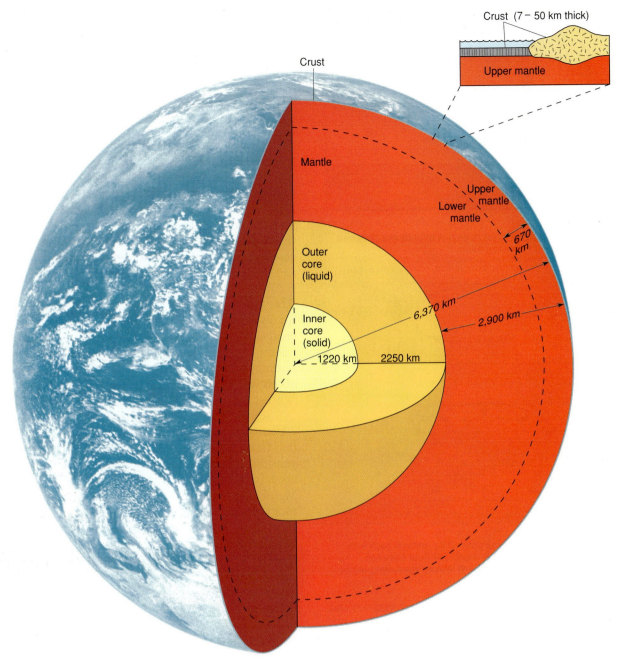

Crust (7 – 50 km thick)

Upper mantle

Crust

Mantle

Upper mantle

Lower mantle

670 km

Outer core (liquid)

6,370 km

2,900 km

Inner core (solid)

1220 km

2250 km

Figure 17.5
The earth's interior. Seismic waves show that the earth is divided into three main divisions: the crust, the mantle, and the core.

future the original concept of drilling to the mantle through oceanic crust will be revived. (Ocean drilling is discussed in more detail in the following two chapters.)

The Mantle

Because of the way seismic waves pass through the mantle, geologists believe that it, like the crust, is made of solid rock. Localized magma chambers of melted rock may occur as isolated pockets of liquid in both the crust and the upper mantle, but most of the mantle seems to be solid. Because P waves travel at about 8 kilometers per second in the upper mantle, it appears that the mantle is a different type of rock from either oceanic crust or continental crust. The best hypothesis that geologists can make

BOX 17.1

Deep Drilling on Continents

The structure and composition of most of the continental crust is unknown. Surface mapping and seismic reflection and refraction suggest that the continents are largely igneous and metamorphic rock, such as granite and gneiss, overlain by a veneer of sedimentary rocks. This sedimentary cover is generally thin, like icing on a cake, but it may thicken to 10 kilometers or more in giant sedimentary basins where the underlying "basement rock" has subsided. Although oil companies have drilled as deep as 8 kilometers on land, they drill in the sedimentary basins. The igneous and metamorphic basement, which averages 40 kilometers thick and makes up most of the continental crust, has rarely been sampled deeper than 2 or 3 kilometers.

Since 1970 the Soviet Union has been drilling a superdeep hole on the Kola Peninsula near Murmansk north of the Arctic Circle. The hole penetrates ancient Precambrian basement rocks, and is currently 12 kilometers deep, with a planned depth of 15 kilometers. The Kola hole is the deepest of 11 holes currently being drilled in the Soviet Union to depths of 10 to 15 kilometers. Deep drilling is as technically complex as space exploration. High pressures and 300°C temperatures require special equipment and techniques. The Soviets use a turbodrill that rotates under the pressure of circulating drilling mud. Unlike normal drilling operations, the lightweight aluminum drillpipe does not turn.

The drilling at Kola shows that seismic models for this area are wrong. The Soviets expected 4.7 kilometers of metamorphosed sedimentary and volcanic rock, then a granitic layer to a depth of 7 kilometers, and a "basaltic" layer below that. The granite, however, appeared at 6.8 kilometers and extends to more than 12 kilometers; the "basalt" has not yet been found. These results may mean that seismic surveys of continental crust are being systematically misinterpreted.

The Soviets also found open fractures and circulating fluids throughout the borehole. The fluids include hydrogen, helium, and methane (natural gas), as well as mineralized waters forming ore bodies. Copper-nickel ore was found deeper than theory predicted, and gold mineralization was present from 9.5 to 11 kilometers down. These results will change geologists' models of ore formation and fluid circulation underground.

Deep drilling is also underway in Sweden, France, and West Germany, and is planned in Canada and Japan. There are no firm plans for deep drilling in the United States.

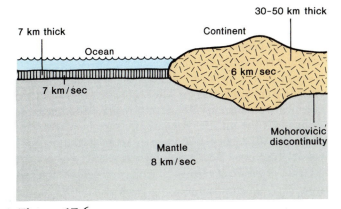

Figure 17.6
Thin crust with a P-wave velocity of 7 kilometers per second underlies the ocean. Thick crust with a lower velocity makes up continents. Mantle velocities are about 8 kilometers per second.

TABLE 17.1

Characteristics of Oceanic Crust and Continental Crust

	Oceanic Crust	Continental Crust
Thickness	7 km	30 to 50 km (thickest under mountains)
Seismic P-wave velocity	7 km/second	6 km/second (higher in lower crust)
Density	3.0 gm/cm³	2.7 gm/cm³
Probable composition	Basalt (gabbro in lower crust)	Granite, other plutonic rocks, schist, gneiss (with sedimentary rock cover)

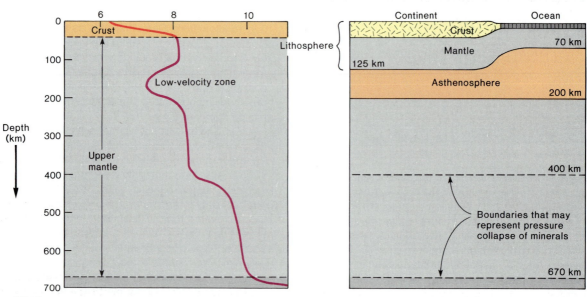

Figure 17.7
The concentric shell structure of the upper mantle. The velocity of seismic P waves generally increases with depth except in the low-velocity zone. The lithosphere contains the crust and the uppermost mantle. The plastic asthenosphere slows down seismic waves. Velocity increases at 400 and 670 kilometers may be caused by mineral collapse.

about the composition of the mantle is that it consists of ultramafic rock such as peridotite. *Ultramafic rock* is heavy igneous rock made up chiefly of ferromagnesian minerals such as olivine and pyroxene. Some ultramafic rocks contain garnet, and all of them lack feldspar.

The earth's crust and uppermost mantle together form the **lithosphere,** the outer shell of the earth that is relatively strong and brittle. The lithosphere makes up the plates of plate tectonic theory. The lithosphere is about 70 kilometers thick beneath oceans and 125 or more kilometers thick beneath continents. Its lower boundary is marked by a curious mantle layer in which seismic waves slow down (figure 17.7).

Generally, seismic waves increase in velocity with depth as increasing pressure alters the properties of the rock. Beginning at a depth of 70 to 125 kilometers, however, seismic waves travel more slowly than they do in shallower layers, and so this zone has been called the *low-velocity zone* (figure 17.7). This zone, extending to a depth of perhaps 200 kilometers, is also called, in plate tectonic theory, the **asthenosphere.** The rocks in this zone may be closer to their melting point than the rocks above or below the zone. (The rocks are probably not *hotter* than the rocks below—melting points are controlled by pressure as well as temperature.) These rocks may actually be partially melted, forming a crystal-and-liquid slush.

If the rocks of the asthenosphere are close to their melting point, this zone may be important for two reasons: (1) it may represent a zone where magma is likely to be generated; and (2) the rocks here may have relatively little strength and therefore are likely to flow. If mantle rocks in the asthenosphere are weaker than they are in the overlying lithosphere, then the asthenosphere can deform easily by plastic flow. Plates of brittle lithosphere probably move easily over the asthenosphere, which may act as a lubricating layer below.

Data from seismic reflection and refraction indicate several concentric layers in the mantle (figure 17.7), with prominent boundaries at 400 and 670 kilometers (670 kilometers is also the depth of the deepest earthquakes). It is doubtful that the layering is due to the presence of several different kinds of rock. Most geologists think that the chemical composition of the mantle rock is about the same throughout the mantle. Because pressure increases with depth into the earth, the boundaries between mantle layers possibly represent depths at which pressure collapses the internal structure of certain minerals into denser minerals (figure 17.7). For example, at a pressure equivalent to a depth of about 400 kilometers, the mineral *olivine* should collapse into a denser mineral called *spinel*. If the boundary at 670 kilometers also represents pressure-caused transformations of minerals, the entire mantle may have the same *chemical* composition throughout, although not the same *mineral* composition. However, some geologists think that the 670-kilometer boundary represents a chemical change as well as a physical change and separates the *upper mantle* from the chemically different *lower mantle* below.

Mantle Xenoliths—A Peek at the Deep

Box 17.2 Figure 1
Dunite xenoliths in basalt. The xenoliths are thought by many geologists to be pieces of mantle rock torn off and brought to the surface by ascending basaltic magma.

A *xenolith* is an inclusion of foreign rock within an igneous rock. Plutons sometimes contain xenoliths of nearby country rock, but xenoliths within some basalts may have come from deep in the earth's mantle.

Numerous basalts contain xenoliths of the relatively rare rocks *peridotite* and *dunite*. Peridotite is an igneous rock composed mostly of olivine and pyroxene; it lacks feldspar, which is common in most igneous rocks. Dunite is a related rock that is almost entirely olivine (box figure 1).

Because basalts appear to be generated in the mantle's asthenosphere, and because peridotite and dunite are uncommon at the earth's surface, most geologists suspect that these xenoliths may be solid pieces of the earth's mantle brought to the surface by

the erupting basalt. The density, seismic velocity, and chemical composition of peridotite and dunite make them reasonable candidates for mantle rock. Peridotites (usually serpentinized by metamorphism) also occur at the base of *ophiolite* sequences on land (see next chapter), which are widely interpreted to be slivers of oceanic crust and mantle rock attached to continents during mountain building.

The Core

Seismic-wave data provide the primary evidence for the existence of the core of the earth. (See the chapter on earthquakes for a discussion of seismic P and S waves.) Seismic waves do not reach certain areas on the opposite side of the earth from a large earthquake. Figure 17.8 shows how seismic P waves spread out from a quake until, at 103° of arc (11,500 km) from the epicenter, they suddenly disappear from seismograms. At more than 142° (15,500 km) from the epicenter, P waves reappear on seismograms. The region between 103° and 142°, which lacks P waves, is called the **P-wave shadow zone.**

The P-wave shadow zone can be explained by the refraction of P waves when they encounter the core boundary deep within the earth's interior. Because the paths of P waves can be accurately calculated, the size and shape of the core can be determined also. In figure 17.8, notice that the earth's core deflects the P waves and, in effect, "casts a shadow" where their energy does not reach the surface. In other words, P waves are missing within the shadow zone because they have been bent (refracted) by the core.

The chapter on "Earthquakes" explains that while P waves can travel through solids and fluids, S waves can travel only through solids. As figure 17.9 shows, an **S-wave shadow zone** also exists and is larger than the P-wave shadow zone. S waves are not recorded in the entire region more than 103° away from the epicenter. The S-wave shadow zone seems to indicate that S waves do not travel

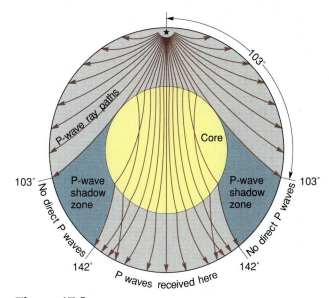

Figure 17.8
The P-wave shadow zone, caused by refraction of P waves within the earth's core.

through the core at all. If this is true, it implies that the core of the earth is a liquid, or at least acts like a liquid.

The way in which P waves are refracted within the earth's core (as shown by careful analysis of seismograms) suggests that the core has two parts, a *liquid outer core* and a *solid inner core* (figure 17.5).

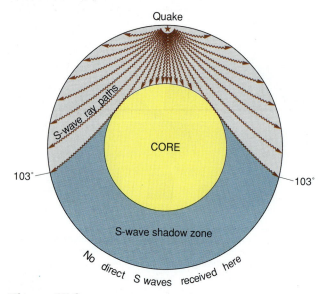

Figure 17.9
The S-wave shadow zone. Because no S waves pass through the core, the core is apparently a liquid (or acts like a liquid).

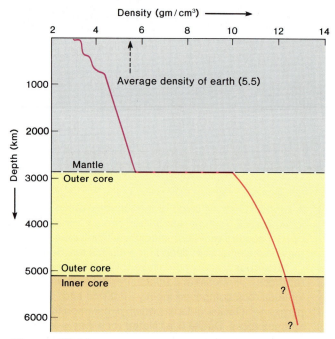

Figure 17.10
Density variations with depth into the earth. Note the great increase in density at the core-mantle boundary.

Composition of the core When evidence from astronomy and seismic-wave studies is combined with what we know about the properties of materials, it appears that the earth's core is made of metal—not rock—and that this metal is probably iron (along with a minor amount of silicon, sulfur, or nickel). How did geologists arrive at this conclusion?

The overall density of the earth is 5.5 gm/cm³, according to astronomers who calculate the speed of the earth's revolution about the sun and the speed of its rotation on its own axis. The crustal rocks are relatively light in weight, from 2.7 gm/cm³ for granite to 3.0 gm/cm³ for basalt. The ultramafic rock thought to make up the mantle probably has a density of 3.3 gm/cm³ in the upper mantle, although rock pressure should raise this value to about 5.5 gm/cm³ at the base of the mantle (figure 17.10).

If the crust and the mantle, which have approximately 85% of the earth's volume, are at or below the average density of the earth, then the core must be very heavy to bring the average up to 5.5 gm/cm³.

Calculations show that the core has to have a density of about 10 gm/cm³ at the core-mantle boundary, increasing to 12 or 13 gm/cm³ at the center of the earth (figure 17.10). This great density would be enough to give the earth an average density of 5.5 gm/cm³.

Under the great pressures existing in the core, iron would have a density slightly greater than that required in the core. Iron mixed with a small amount of a lighter element, such as sulfur or silicon, would have the required density. Therefore, many geologists think that such a mixture makes up the core.

But a study of density by itself is hardly convincing evidence that the core is mostly iron, for many other heavy substances could be there instead. The choice of iron as the major component of the core comes from looking at meteorites that have fallen to earth. Meteorites are thought by some scientists to be remnants of the basic material that created our own solar system. An estimated 10% of meteorites are composed of iron mixed with small amounts of nickel. Material similar to these meteorites may have helped create the earth, perhaps settling to the center of the earth because of its high density. The composition of these meteorites, then, may tell us what is in the earth's core. Nickel is heavier than iron, however, so a mixture of just iron and nickel would have a density greater than that required in the core. (The other 90% of meteorites are mostly ultramafic rock and perhaps represent material that formed the earth's mantle.)

Seismic and density data, together with assumptions based on meteorite composition, point to a core that is largely iron, with at least the outer part being liquid. The existence of the earth's magnetic field, which is discussed later in this chapter, also suggests a metallic core. Of course, no geologist has seen the core, nor is anyone likely to in the foreseeable future. But since so many lines of indirect evidence point to a liquid metal core, most scientists accept this theory as the best conclusion that can be made about the core's composition. Geologists could be wrong about the earth's interior, but the current model of a solid rock mantle and a liquid metallic core with a solid inner core is widely accepted because it is consistent with all available knowledge. A hollow earth is not.

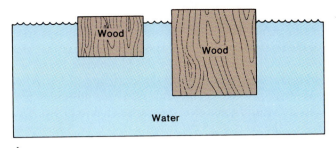

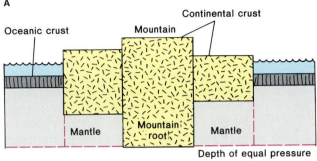

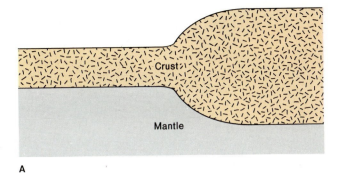

A

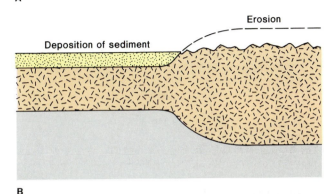

B

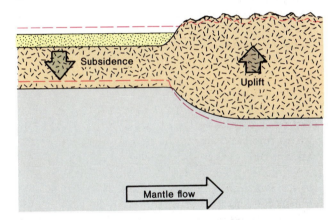

C

Figure 17.11
Isostatic balance. (*A*) Wood blocks float in water with most of their bulk submerged. (*B*) Crustal blocks "float" on mantle in approximately the same way. The thicker the block, the deeper it extends into the mantle.

Isostasy

Isostasy is a balance or *equilibrium* between adjacent blocks of brittle crust "floating" on the plastic upper mantle. Since crustal rocks weigh less than mantle rocks, the crust can be thought of as floating on the denser mantle much as wood floats on water (figure 17.11).

Blocks of wood floating on water rise or sink until they displace an amount of water equal to their own weight. The weight of the displaced water buoys up the wood blocks, allowing them to float. The higher a wood block appears above the water surface, the deeper the block extends under water. Thus a tall block has a deep "root."

In a greatly simplified way, crustal rocks can be thought of as tending to rise or sink gradually until they are balanced by the weight of displaced mantle rocks. This concept of vertical movement to reach equilibrium is called **isostatic adjustment.** Just as with the blocks of wood, once crustal blocks have come into isostatic balance, a tall block (a mountain range) extends deep into the mantle (a *mountain root,* as shown in figure 17.11).

Figure 17.11 shows both the blocks of wood and the blocks of crustal rock in isostatic balance. The weight of the wood is equal to the weight of the displaced water. Similarly, the weight of the crustal blocks is equal to the weight of the displaced mantle. As a result, the rocks (and overlying sea water) in figure 17.11 can be thought of as separated into vertical columns, each with the same pressure at its base. At some *depth of equal pressure* each column is in balance with the other columns, for each column has the same weight. A column of thick continental crust (a mountain and its root) has the same weight as a column containing thin continental crust and some of

Figure 17.12
Isostatic adjustment due to erosion and deposition of sediment. Rock within the mantle must flow to accommodate vertical motion of crustal blocks. Mantle flow occurs in the asthenosphere, deeper than shown in *C*.

the upper mantle. A column containing sea water, thin oceanic crust, and a thick section of heavy mantle weighs the same as the other two columns.

Visualizing the crust as isolated blocks moving vertically past each other does not really give a good picture of crustal structure. Generally, vertical faults do not separate the crust into discrete blocks. It is more accurate to think of the crust as bending in broad warps and flexures without vertical faults.

Let us look at some examples of isostatic balance (equilibrium) in crustal rocks. Suppose that two sections of crust of unequal thickness are next to each other as in figure 17.12. Sediment from the higher part, which is subject to more rapid erosion, is deposited on the lower part.

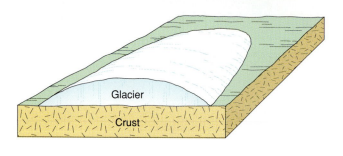

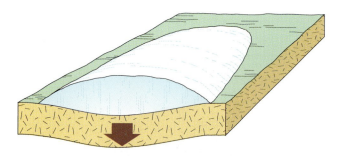

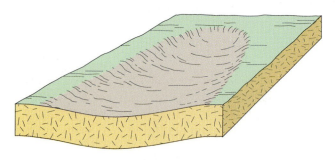

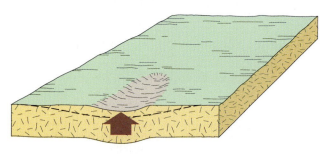

Figure 17.13
The weight of glaciers depresses the crust, and the crust rebounds when the ice melts.

The decrease in weight from the high part causes it to rise, while the increasing weight on the low part causes it to sink. These vertical movements (isostatic adjustment) do in fact take place whenever large volumes of material are eroded from or deposited on parts of the crust.

Rising or sinking of the crust, of course, requires plastic flow of the mantle to accommodate the motion. By measuring the rate of rising or sinking, the viscosity of the mantle can be calculated. The plastic flow of the mantle probably takes place within the asthenosphere in the upper mantle.

Another example of isostatic adjustment, caused by plastic mantle flow, is the upward movement of large areas of the crust since the glacial ages. The weight of the thick continental ice sheets during the Pleistocene Epoch depressed the crust underneath the ice (figure 17.13). After the melting of the ice, the crust rose back upward, a process that is still going on in some areas (figure 17.14). This rise of the crust after the removal of the ice is known as **crustal rebound.**

Isostatic adjustment may occur at subduction zones. As explained in chapters on igneous rocks and volcanism, a subducting plate may generate molten magma, which sometimes rises all the way to the surface to erupt as lava. However, some geologists believe that in certain cases the magma stops at the base of an overlying continent (or perhaps within the continent near its base). This accumulation of magma can locally thicken the continent when the magma cools (figure 17.15*A*). The thickening of continents from below (a theory not accepted by all geologists) causes the crust to be out of isostatic equilibrium. So it rises, reestablishing equilibrium and forming a mountain range (figure 17.15*B*).

Gravity Measurements

The force of gravity between two objects varies with the masses of the objects and the distance between them (figure 17.16):

$$\text{Force of gravity between } A \text{ and } B = \text{constant}\left(\frac{\text{mass}_A \times \text{mass}_B}{\text{distance}^2}\right).$$

The force increases with an increase in either mass. The gravitational attraction between the earth and the moon, for example, is vastly greater than the extremely small attraction that exists between two bowling balls. The equation also shows that force decreases with the square of the distance between the two objects. The farther two objects are apart, the less gravitational attraction there is between them.

A useful tool for studying the crust and upper mantle is the **gravity meter,** which measures the gravitational attraction between the earth and a mass within the instrument. One use of the gravity meter is to explore for local variations in rock density (mass = density × volume). Dense rock such as metal ores and ultramafic rock pulls strongly on the mass inside the meter (figure 17.17). The strong pull stretches a spring, and the amount of stretching can be very precisely determined. So a gravity meter can be used to explore for metallic ore deposits. A cavity or a body of low-density material such as sediment causes a much weaker pull on the meter's mass (figure 17.17). The use of a gravity meter to explore for salt domes and their associated traps for oil and gas is shown in Box 21.1.

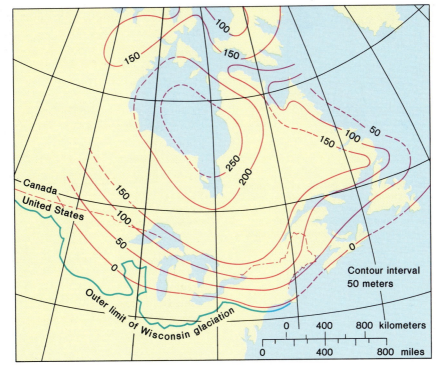

Figure 17.14
Uplift of land surface in Canada and the northern United States caused by crustal rebound after glaciers melted. Colored lines show the amount of uplift in meters since the ice disappeared.

From Phillip B. King, ''Tectonics of Quaternary Time in Middle North America,'' in *The Quaternary of the United States,* H. C. Wright, Jr. and David G. Frey, eds., fig. 4*A*, p. 836. Reprinted by permission of Princeton University Press.

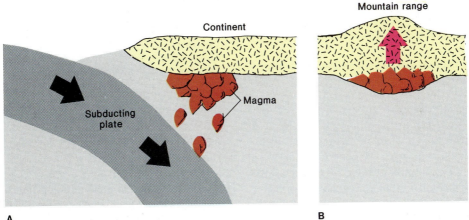

Figure 17.15
Isostatic uplift as a result of crustal thickening from below. (*A*) Rising blobs of magma accumulate at the base of a continent, thickening the crust (base of lithosphere below continent not shown). (*B*) Uplift produces a mountain range.

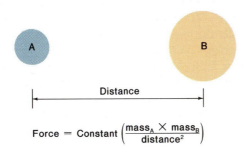

$$\text{Force} = \text{Constant} \left(\frac{\text{mass}_A \times \text{mass}_B}{\text{distance}^2} \right)$$

Figure 17.16
The force of gravitational attraction between two objects is a function of the masses of the objects and the distance between the centers of the objects.

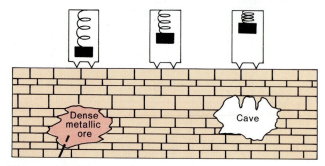

Figure 17.17
A gravity meter reading is affected by the density of the rocks beneath it. Dense rock pulls strongly on the mass within the meter, stretching a spring; a cavity exerts a weak pull on the mass. A gravity meter can be used to explore for hidden ore bodies, caves, and other features that have density contrasts with the surrounding rock.

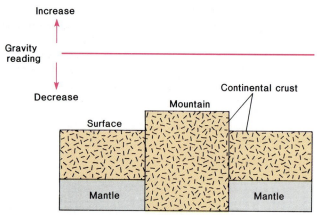

Figure 17.18
A region in isostatic balance gives a uniform gravity reading (no gravity anomalies).

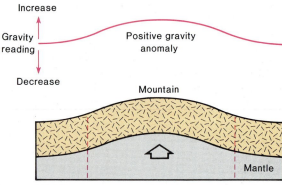

Figure 17.19
A region being held up out of isostatic equilibrium gives a positive gravity anomaly.

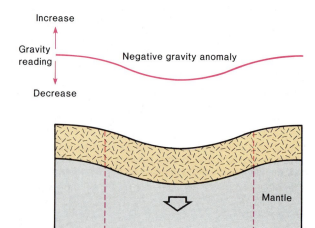

Figure 17.20
A region being held down out of isostatic equilibrium gives a negative gravity anomaly.

Another important use of a gravity meter is to discover whether regions are in isostatic equilibrium. If a region is in isostatic balance, as in figure 17.18, each column of rock has the same mass. If a gravity meter were carried across the rock columns, it would register the same amount of gravitational attraction for each column (after correcting for differences in elevation—gravitational attraction is less on a mountaintop than at sea level because the mountaintop is farther from the center of the earth).

Some regions, however, are held up out of isostatic equilibrium by deep tectonic forces. Figure 17.19 shows a region with uniformly thick crust. Tectonic forces are holding the center of the region up. This uplift creates a mountain range without a mountain root. There is a thicker section of heavy mantle rock under the mountain range than there is on either side of the mountain range. Therefore, the central "column" has more mass than the neighboring columns, and a gravity meter shows that the gravitational attraction is correspondingly greater over the central than over the side columns.

A gravity reading higher than the normal regional gravity is called a **positive gravity anomaly** (figure 17.19). It can indicate that tectonic forces are holding a region up out of isostatic equilibrium, as shown in figure 17.19. When

the forces stop acting, the land surface sinks until it reestablishes isostatic balance. The gravity anomaly then disappears. For the region shown in figure 17.19, equilibrium will be established when the land surface becomes level.

Positive gravity anomalies, particularly small ones, are also caused by local concentrations of dense rock such as metal ore. The gravity meter in figure 17.17 is registering a positive gravity anomaly over ore (the spring inside the meter is stretched). Since there can be more than one cause of a positive gravity anomaly, geologists may disagree about the interpretation of anomalies. Drilling into a region with a gravity anomaly usually discloses the reason for the anomaly.

A region can also be held down out of isostatic equilibrium, as shown in figure 17.20. The mass deficiency in such a region produces a **negative gravity anomaly**—a gravity reading lower than the normal regional gravity. Negative gravity anomalies indicate either that a region is being held down (figure 17.20) or that local mass deficiencies exist for other reasons (figure 17.17).

The greatest negative gravity anomalies in the world are found over oceanic trenches (see next two chapters). These negative anomalies are interpreted to mean that trenches are actively being held down and are out of isostatic balance.

The Earth's Magnetic Field

A region of magnetic force—a **magnetic field**—surrounds the earth. The invisible lines of magnetic force surrounding the earth deflect magnetized objects, such as compass needles, that are free to move. The field has north and south **magnetic poles,** one near the geographic North Pole, the other near the geographic South Pole. (Because it has two poles, the earth's field is called *dipolar.*) The strength of the magnetic field is greatest at the magnetic poles where magnetic lines of force appear to leave and enter the earth vertically (figure 17.21).

Because the compass is important in navigation, the earth's magnetism has been observed for centuries. It has long been known that the magnetic poles are displaced about 11½° from the geographic poles (about which the earth rotates). Furthermore, changes in the position of the magnetic poles have been well documented, especially since the time of the great explorations of the globe. The magnetic poles appear to be moving slowly around the geographic poles. The separation between the two types of poles has probably never been much greater than it is today.

More recently, geophysical studies have been directed toward the *source* of the earth's magnetism. The rate of the poles' changes in position, together with the strength of the magnetic field, strongly suggests that the magnetic field is generated within the liquid metal of the outer core rather than within the solid rock of the crust or the mantle.

How is the earth's magnetic field generated? A number of hypotheses have been put forth. One widely accepted hypothesis suggests that the magnetic field is created by electric currents within the slowly circulating liquid part of the core. This hypothesis requires the core to be an electrical conductor. Metals are good conductors of electricity, whereas rock is generally a poor electrical conductor. Indirectly, this is evidence that the core is metallic.

Magnetic Reversals

In the 1950s evidence began to accumulate that the earth's magnetic field has periodically reversed its polarity in the past. Such a change in the polarity of the earth's magnetic field is a **magnetic reversal.** During a time of *normal polarity,* magnetic lines of force leave the earth near the geographic South Pole and reenter the earth near the geographic North Pole (figure 17.21). During a time of *reversed polarity,* the magnetic lines of force run the other way, leaving the earth near the North Pole and entering the earth near the South Pole (figure 17.21). In other words, during a magnetic reversal, the north magnetic pole and the south magnetic pole exchange positions.

Many rocks contain a record of the strength and direction of the earth's magnetic field *at the time the rocks formed.* When the mineral magnetite, for example, is crystallizing in a cooling lava flow, the atoms within the crystals respond to the earth's magnetic field and form magnetic alignments that "point" toward the north magnetic pole. As the rocks solidify, this magnetic record is

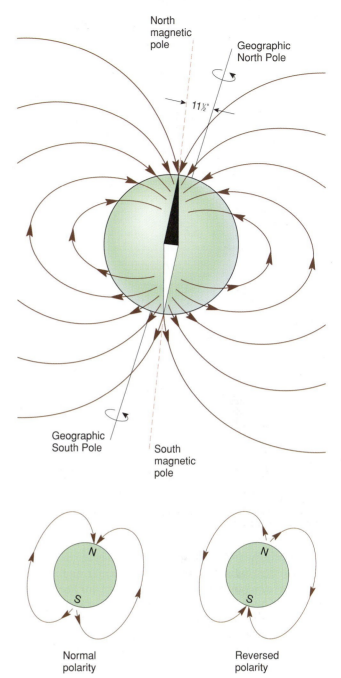

Figure 17.21
The earth's magnetic field.

permanently trapped in the rock (figure 17.22). Unless the rock is heated again this magnetic record is retained, and when studied reveals the direction of the earth's magnetic field at the time the lava cooled. Other rock types, including sedimentary rocks stained red by iron compounds, also record former magnetic field directions. The study of ancient magnetic fields is called **paleomagnetism.**

Most of the evidence for magnetic reversals comes from lava flows on the continents. Paleomagnetic studies of a series of stacked lava flows often show that some of the lava flows have a magnetic orientation directly opposite to the earth's present orientation (figure 17.23). That is, at the time these lava flows cooled, the magnetic

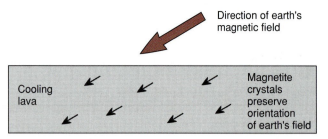

Figure 17.22
Some rocks preserve a record of the earth's magnetic field.

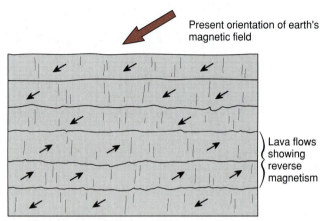

Figure 17.23
Cross section of stacked lava flows showing evidence of magnetic reversals.

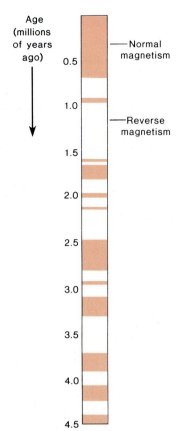

Figure 17.24
Magnetic reversals during the past 4.5 million years. Color represents normal magnetism; white represents reverse magnetism.
From A. Cox, 1969, *Science,* vol. 163, p. 240, copyright 1969 by the American Association for the Advancement of Science.

poles had exchanged positions. During this time of magnetic reversal, a compass needle would have pointed south rather than north. Many periods of normal and reverse magnetization are recorded in continental lava flows. They are worldwide events. Since lava flows can be dated radiometrically, the time of these reversals in the earth's past can be determined. The times of normal and reversed magnetization are shown for the past 4.5 million years in figure 17.24. Records for tens of millions of years suggest that the earth's field reverses about every 500,000 years. The present normal orientation has lasted for the past 700,000 years.

A magnetic reversal can have some profound effects on earth. The strength of the earth's magnetic field probably declines to near zero before the orientation reverses; then the field strength increases to its usual values, but in the opposite orientation. This collapse of the earth's magnetic field means that deadly ultraviolet radiation from the sun would be much more intense at the earth's surface. When the magnetic field is at its usual strength, it shields the earth from these rays, but when the field collapses, this shielding is lost. Ultraviolet radiation affects organisms; the extinction of some species and the appearance of new species by mutation have been correlated with some magnetic reversals.

Magnetic Anomalies

A **magnetometer** is an instrument used to measure the strength of the earth's magnetic field. A magnetometer can be carried over the land surface or flown over land or sea. At sea, magnetometers can also be towed behind ships.

The strength of the earth's magnetic field varies from place to place. As with gravity, a deviation from average readings is called an *anomaly.* Very broad regional magnetic anomalies may be due to *circulation patterns in the liquid outer core* or to other deep-seated causes. Smaller anomalies generally reflect *variations in rock type,* for the magnetism of near-surface rocks adds to the main magnetic field generated in the core. Rocks differ in their magnetism, depending upon their content of iron-containing minerals.

A **positive magnetic anomaly** is a reading of magnetic field strength that is higher than the regional average. Figure 17.25 shows three geologic situations that can cause positive magnetic anomalies. In figure 17.25*A,* a body of magnetite ore (a highly magnetic ore of the metal iron) has been emplaced in a bed of limestone by hot solutions rising along a fracture. The magnetism of the iron ore adds to the magnetic field of the earth, giving a stronger magnetic field measurement at the surface (a positive

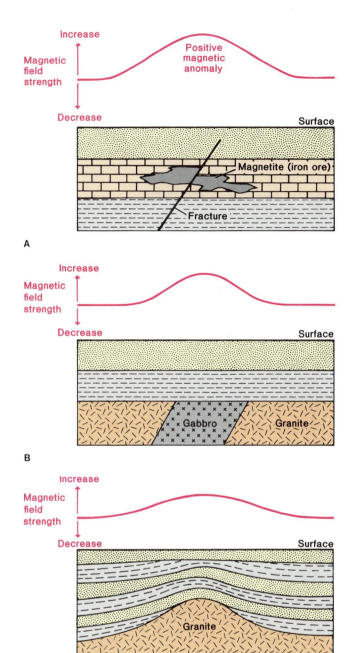

A

B

C

Figure 17.25
Positive magnetic anomalies can indicate hidden ore and geologic structures.

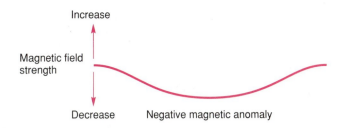

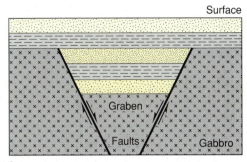

Figure 17.26
A graben filled with sediment can give a negative magnetic anomaly if the sediment contains fewer magnetic minerals than the rock beneath it.

each example shows horizontal sedimentary rocks at the surface, with no surface hint of the subsurface geology. The magnetometer helps find hidden ores and geologic structures.

A **negative magnetic anomaly** is a reading of magnetic field strength that is lower than the regional average. Figure 17.26 shows how a negative anomaly can be produced by a downdropped fault block (a *graben*) in igneous rock. The thick sedimentary fill above the graben is less magnetic than is the igneous rock, so a weaker field (a negative magnetic anomaly) develops over the thick sediment.

Not all local magnetic anomalies are caused by variations in rock type. The linear magnetic anomalies found at sea are apparently caused by a *variation in the direction of magnetism,* as you will see in the chapter on "Plate Tectonics."

Heat Within the Earth

Geothermal Gradient

The temperature increase with depth into the earth is called the **geothermal gradient.** The geothermal gradient can be measured on land in abandoned wells or on the sea floor by dropping specially designed probes into the mud. The average temperature increase is 25°C per kilometer (about 75°F per mile) of depth. Some regions have a much higher gradient, indicating concentrations of heat at shallow depths. Such regions have a potential for generating *geothermal energy* (chapter titled "Ground Water").

The temperature increase creates a problem in deep mines, such as in a 3-kilometer-deep gold mine in South Africa, where the temperature is close to the boiling point

anomaly). In figure 17.25*B* a large dike of gabbro has intruded into granitic basement rock. Because gabbro contains more ferromagnesian minerals than granite, gabbro is more magnetic and causes a positive magnetic anomaly. Figure 17.25*C* shows a granitic basement high (perhaps originally a hill) that has influenced later sediment deposits, causing a draping of the layers as the sediments on the hilltop compacted less than the thicker sediments to the sides. Such a structure can form an *oil trap* (see chapter 21). The granite in the hill contains more iron in its ferromagnesian minerals than the surrounding sedimentary rocks, so a small positive magnetic anomaly occurs where the granite is closer to the surface. Note how

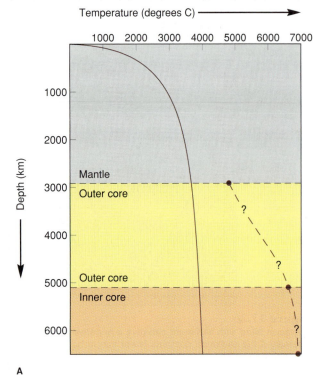

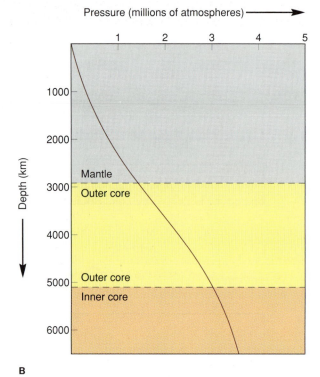

Figure 17.27
Distribution of (**A**) temperature and (**B**) pressure with depth into the earth. Part (**A**) shows the conventional temperature curve as well as three new points from recent experiments.

A

B

of water. Deep mines must be cooled by air-conditioning for the miners to survive. High temperatures at depth also complicate the drilling of deep oil wells. A well drilled to a depth of 7 or 8 kilometers must pass through rock with a temperature of 200°C. At such high temperatures, a tough steel drilling pipe will become soft and flexible unless it is cooled with a special mud solution pumped down the hole.

Geologists believe that the geothermal gradient must taper off sharply a short distance into the earth. The high values of 25°C/kilometer recorded near the earth's surface could not continue very far into the earth. If they did, the temperature would be 2500°C at the shallow depth of 100 kilometers. This temperature is above the melting point of all rocks at that depth—even though the increased pressure with depth into the earth increases the melting point of rocks. Seismic evidence seems to indicate a solid, not molten, mantle, so the geothermal gradient must drop to values as low as 1°C/kilometer within the mantle (figure 17.27*A*).

At the boundary between the inner core and the outer core, there would be some constraints on possible temperatures if the core is molten metal above the boundary and solid metal below. The weight of the thick rock layer of the mantle and the liquid metal of the outer core raises the pressure at this boundary (figure 17.27*B*) to about 3 million atmospheres. (An *atmosphere* of pressure is the force per unit area caused by the weight of the air in the atmosphere. It is about 1 kilogram per square centimeter, or 14.7 pounds per square inch.)

For many years geologists were reasonably certain that the temperature at the boundary between the inner core and the outer core was 3700°C ± 500°C. Calculations extrapolating from the surface downward showed that this temperature would melt the outer metal core at the 3-million-atmosphere pressure found there, while greater pressure below this boundary would keep the metal of the inner core solid. The temperature at the center of the earth was assumed to be about 4000°C.

Recent laboratory experiments with pressure anvils and giant guns have created (for a millionth of a second) the enormous pressures found at the center of the earth. The measured temperature at this pressure was far higher than expected. New estimates of the earth's internal temperatures have resulted: 4800°C at the core-mantle boundary, 6600°C at the inner-core/outer-core boundary, and 6900°C at the earth's center. Both the old and new estimates of temperature are shown in figure 17.27.

Heat Flow

A small but measurable amount of heat from the earth's interior is being lost gradually through the earth's surface. This gradual loss of heat through the earth's surface is called the **heat flow.** What is the origin of the heat? It could be "original" heat from the time that the earth formed, that is, *if* the earth formed as a hot mass that is now cooling down. Or the heat could be a by-product of the decay of radioactive isotopes inside the earth. Radioactive decay *may* actually be warming up the earth. Geologists are not sure whether the earth formed as a hot or cold mass, or whether the earth is now cooling off or

Planetary Interiors

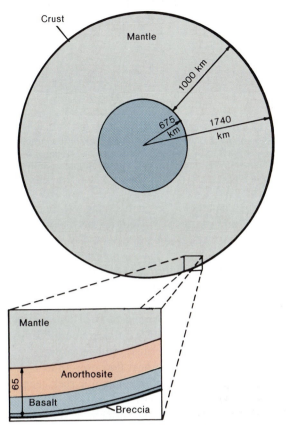

Astrogeology Box 17.1 Figure 1
Cross section of the Moon.

We have attempted to learn about the interiors of the Moon and other planets by analyzing seismic waves, studying variations in densities, measuring magnetic fields, and using other techniques that have also been used to study the Earth's interior.

Very little is known about the Moon's interior. Some information has been obtained by studying changes in the velocities of seismic waves as they travel through the Moon's interior during a moonquake. There are very few moonquakes, however, and they are very small. Most moonquakes are produced by tidal stresses caused by Earth's gravitational pull; a few may be produced by meteorite impact or landsliding.

In the dark maria regions, the crust of the Moon consists of a thin layer (about 1.5 kilometers thick) of highly shattered rocks overlying, first, a layer of basalt (about 25 kilometers thick) and below that a layer of anorthosite (about 40 kilometers thick). Below the lunar crust is the lunar mantle (box figure 1). The composition of the lunar mantle is not known, but it is probably composed of rocks more mafic than those in the crust, perhaps rocks rich in olivine and pyroxene. The mantle is at least 1,000 kilometers thick. The Moon's core is small and may be composed of a nickel-iron alloy. It may be partially molten because seismic S waves appear not to pass through it.

Some lunar rocks are slightly magnetic. The Moon does not have a magnetic field at the present time, but the magnetism in these lunar rocks may be "left over" from a time when the Moon did possess a magnetic field.

The density of Mercury is unusually high, probably indicating the presence of a large nickel-iron core extending outward from the center for as much as three-quarters of the planet's radius. Common on Mercury are cliffs apparently caused by faulting. It has been suggested that this faulting may have occurred as a result of the shrinking of Mercury's core as it cooled. Unlike the Moon, Mercury has a weak magnetic field, about 1% as strong as that of the Earth. This may indicate that Mercury's core is partially molten.

Mercury, nearest the Sun, has the largest core relative to planet size, followed by Venus, and then the Earth, which are progressively farther from the Sun. All these cores are apparently iron-nickel alloys. Still farther from the Sun, Mars' core may be mostly iron sulfide. Apparently core sizes and their compositions are related to distance from the Sun. The crusts of these four planets closest to the Sun are made up almost entirely of silicates.

Marsquakes occur and, along with studies of the gravitational field, have provided some information about the Martian interior. The Martian crust varies in thickness but is generally between 15 and 80 kilometers thick, being thickest under uplifted areas such as the Tharsis Ridge. The crust contains large amounts of water ice and other substances that evaporate at relatively low temperatures. The low density of Mars suggests that any Martian core must be small. Mars lacks a magnetic field, but 5% to 10% of the material at the surface is magnetic.

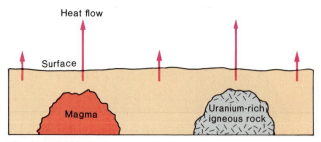

Figure 17.28
Some regions have higher heat flow than others; the amount of heat flow is indicated by the length of the arrow. Regions of high heat flow may be underlain by cooling magma or uranium-rich igneous rock.

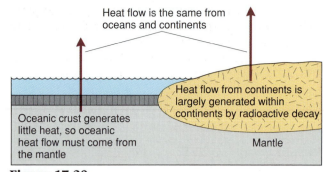

Figure 17.29
The average heat flow from oceans and continents is the same, but the origin of the heat differs from the ocean to continents.

warming up. Changes in the earth's temperature are extremely slow, and trying to work out its thermal history is a slow, often frustrating, job.

Some regions on earth have a high heat flow. More heat is being lost through the earth's surface in these regions than is normal. High heat flow is usually caused by the presence of a magma body or still-cooling pluton near the surface (figure 17.28). An old body of igneous rock that is rich in uranium and other radioactive isotopes can cause a high heat flow, too, because radioactive decay produces heat as it occurs. High heat flow over an extensive area may be due to the rise of warm mantle rock beneath abnormally thin crust.

The average heat flow from continents is the same as the average heat flow from the sea floor, a surprising fact if you consider the greater concentration of radioactive material in continental rock (figure 17.29). The unexpectedly high average heat flow under the ocean may be due to hot mantle rock rising slowly by convection under parts of the ocean (chapter 1). Regional patterns of high heat flow and low heat flow on the sea floor may also be explained by convection of mantle rock (see the chapter on "Plate Tectonics").

Summary

The interior of the earth is studied indirectly by *geophysics*—a study of seismic waves, gravity, earth magnetism, and earth heat.

Seismic reflection and *seismic refraction* can indicate the presence of boundaries between rock layers.

The earth is divided into three major units—the *crust,* the *mantle,* and the *core.*

The earth's crust beneath oceans is 7 kilometers thick and made of basalt. Continental crust is 30 to 50 kilometers thick and made of rock that has the same seismic velocity as granite.

The *Mohorovičić discontinuity* separates the crust from the mantle.

The mantle is a layer of solid rock 2,900 kilometers thick and is probably composed of an ultramafic rock such as peridotite. Seismic waves show the mantle has a structure of concentric shells, perhaps caused by pressure transformations of minerals.

The *lithosphere,* which forms plates, is made up of brittle crust and upper mantle. It is 70 to 125 kilometers thick and moves over the plastic asthenosphere.

The *asthenosphere* extends from the base of the lithosphere to the depth of 200 kilometers and may represent rock close to its melting point (seismic waves slow down here). It is probably the region of most magma generation and isostatic adjustment.

Seismic-wave shadow zones show the core has a radius of 3,450 kilometers and is divided into a liquid outer core and a solid inner core. A core composition of iron (perhaps mixed with sulfur, silicon, or nickel) is suggested by the earth's density, the composition of meteorites, and the existence of the earth's magnetic field.

Isostasy is an equilibrium between crustal columns floating on plastic mantle. *Isostatic adjustment* occurs when weight is added to or subtracted from a column of rock. *Crustal rebound* is isostatic adjustment that occurs after the melting of glacial ice.

A *gravity meter* can be used to study variations in rock density or to find regions that are out of isostatic equilibrium.

A *positive gravity anomaly* forms over dense rock or over regions being held up out of isostatic balance. A *negative gravity anomaly* indicates low-density rock or a region being held down.

The earth's *magnetic field* has two *magnetic poles,* probably generated by slow circulation and electric currents in the earth's outer core.

Some rocks record the earth's magnetism at the time they form. *Paleomagnetism* is the study of ancient magnetic fields.

Magnetic reversals of polarity occurred in the past, with the north magnetic pole and south magnetic pole exchanging positions. Radiometric dating of rocks can show the age of the reversals.

A *magnetometer* measures the strength of the earth's magnetic field.

A *positive magnetic anomaly* develops over rock that is more magnetic than neighboring rock. A *negative magnetic anomaly* indicates rock with low magnetism.

Magnetic anomalies can also be caused by circulation patterns in the earth's core and by variations in the direction of rock magnetism.

The *geothermal gradient* is about 25°C/kilometer near the earth's surface but decreases rapidly with depth. The temperature at the center of the earth may be 6900°C. *Heat flow* measurements show that heat loss per unit area from continents and oceans is about the same, perhaps because of convection of hot mantle rock beneath the oceans.

Terms to Remember

asthenosphere	magnetic reversal
core	magnetometer
crust	mantle
crustal rebound	Mohorovičić discontinuity
geophysics	negative gravity anomaly
geothermal gradient	negative magnetic anomaly
gravity meter	paleomagnetism
heat flow	positive gravity anomaly
isostasy	positive magnetic anomaly
isostatic adjustment	P-wave shadow zone
lithosphere	seismic reflection
magnetic field	seismic refraction
magnetic pole	S-wave shadow zone

Questions for Review

1. Describe how seismic reflection and seismic refraction show the presence of layers within the earth.
2. Sketch a cross section of the entire earth showing the main subdivisions of the earth's interior and giving the name, thickness, and probable composition of each.
3. What facts make it probable that the earth's core is composed of iron?
4. Describe the differences between continental crust and oceanic crust.
5. What is a gravity anomaly, and what does it generally indicate about the rocks in the region where it is found?
6. Discuss seismic-wave shadow zones and what they indicate about the earth's interior.
7. Describe the earth's magnetic field. Where is it generated?
8. What is the temperature distribution with depth into the earth?
9. Heat flow has been found to be about equal through continents and the sea floor. Why was this unexpected? What might cause this equality?
10. What is the Mohorovičić discontinuity?
11. What is the asthenosphere? Why is it important?
12. How does the lithosphere differ from the asthenosphere?
13. What is a magnetic reversal? What is the evidence for magnetic reversals?
14. What is a magnetic anomaly? How are magnetic anomalies measured at sea?

Questions for Thought

1. If the earth were hollow, what evidence would we have from seismic-wave shadow zones? Could a hollow earth have an average density of 5.5 gm/cm³? Could a hollow earth have a magnetic field?
2. What isostatic adjustment of the sea floor takes place as huge amounts of sediment are deposited on it? Does this adjustment have any effect on sea level? Does the addition of the sediment to the sea have any other effect on sea level? (What effect does a boulder have if it is dropped into a bathtub filled to the brim?)
3. Subsidence of the earth's surface sometimes occurs as reservoirs fill behind newly built dams. Why?

Supplementary Readings

Bolt, B. A. 1973. The fine structure of the earth's interior. *Scientific American* (March). New York: W. H. Freeman.

Bolt, B. A. 1982. *Inside the earth.* New York: W. H. Freeman.

Burchfiel, B. C. 1983. The continental crust. *Scientific American* (September) 249: 86–98. New York: W. H. Freeman.

Carrigan, C. R., and D. Gubbins. 1979. The source of the earth's magnetic field. *Scientific American* (February). New York: W. H. Freeman.

Clark, S. P. 1971. *Structure of the earth.* Englewood Cliffs, N.J.: Prentice-Hall.

Jeanloz, R. 1983. The earth's core. *Scientific American* (September) 249: 40–49. New York: W. H. Freeman.

McKenzie, D. P. 1983. The earth's mantle. *Scientific American* (September) 249: 50–62. New York: W. H. Freeman.

Strangway, D. W. 1970. *History of the earth's magnetic field.* New York: McGraw-Hill.

Sumner, J. S. 1969. *Geophysics, geologic structures, and tectonics.* Dubuque, Iowa: Wm. C. Brown Publishers.

Wyllie, P. J. 1975. The earth's mantle. *Scientific American* (March). New York: W. H. Freeman.

The rocks and topography of the sea floor are different from those on land. To understand the evidence for plate tectonics in the next chapter, you need to understand the nature of major sea-floor features such as the mid-oceanic ridge, oceanic trenches, and fracture zones, as well as the surprisingly young age of the sea-floor rocks.

This chapter and the next are an excellent example of how the scientific method works. This chapter is concerned with the physical *description* of most sea-floor features—the data-gathering part of the scientific method. The next chapter shows how the theory of plate tectonics explains the *origin* of many of these features. Geologists generally agree upon the descriptions of features but often disagree on their interpretations. As you read, keep a clear distinction in your mind between *data* and the *hypotheses* used to explain the data.

18

The Sea Floor

Oceanographers prepare to take a core sample of the deep ocean floor.

A

B

C

Figure 18.1
Bottom sampling devices. (*A*) Rock dredge for sampling hard rock. An unusually large rock is caught in the mouth of the dredge. Additional rocks are visible in the chain bag. (*B*) Corer for sampling sea-floor sediment. Sediment is caught inside the pipe when the corer is dropped to the sea floor. The bomb-shaped part of the corer adds weight to the core pipe (mostly under water). (*C*) A core of sea-floor sediment. The sediment has been forced out of the core barrel and sawed in half lengthwise, so that the sediment layers are visible.

Photo *A* from Scripps Institution of Oceanography, University of California, San Diego. Photo *C* by D. Hopkins, U.S. Geological Survey.

About 4.5 billion years ago the earth began to form by the accretion of small, cold chunks of rock and metal that surrounded the sun. As the earth grew it began to heat up because of the heat of collisional impact, gravitational compaction, and radioactive decay of elements such as uranium. The temperature of the earth rose until its iron melted and "fell" to the earth's center to form its core. Violent volcanic activity occurred at this time, releasing great quantities of water vapor and other gases from the earth's interior and perhaps even covering the earth's surface with a thick, red-hot sea of lava. The earth began to cool as its growth and internal reorganization slowed down and as the amount of radioactive material was reduced by decay. Eventually the earth's surface became solid rock, cool enough to permit the condensation of billowing clouds of volcanic water vapor to form liquid water. Thus the modern oceans were born, perhaps 4 billion years ago. The oceans grew in size as volcanic *degassing* of the earth continued, and became salty as the water picked up chlorine from other volcanic gases, and sodium (and calcium and magnesium) from the chemical weathering of minerals on the earth's surface.

Oceans cover more than 70% of the earth's surface. Even though the rocks of the sea floor are widespread, they are difficult to study. Geologists have to rely on small samples of rock taken from the sea floor and brought to the surface, or they must study the rocks indirectly by means

A

B

Figure 18.2
(*A*) A drilling ship for sampling both sediment and rock from the deep ocean floor. (*B*) The small research submersible ALVIN of Woods Hole Oceanographic Institution in Massachusetts; it is capable of taking three oceanographers to a depth of 12,000 feet (about 4,000 meters).
Photo *A* by U.S. Geological Survey; Photo *B* by Woods Hole Oceanographic Institution.

of instruments on board ships. Despite the difficulties, however, a great deal has been learned about the sea floor in the past few decades. Much of the information that led to the concept of plate tectonics was obtained by geologists working at sea.

Methods of Studying the Sea Floor

Samples of rock and sediments can be taken from the sea floor in several ways (figure 18.1). Rocks can be broken from the sea floor by a *rock dredge,* which is an open steel container dragged over the ocean bottom at the end of a cable. Sediments can be sampled with a *corer,* a weighted steel pipe dropped vertically into the mud and sand of the ocean floor.

Both rocks and sediments can be sampled by means of *sea-floor drilling.* Offshore oil platforms drill holes in the relatively shallow sea floor near shore. A ship with a drilling derrick on its deck can drill a hole in the deep sea floor far from land (figure 18.2*A*). The drill cuts long, rod-like rock cores from the ocean floor. Hundreds of such holes have been drilled in the sea floor, primarily by cooperative government-funded programs headed by the United States, and the rock and sediment cores recovered from these holes have revolutionized the field of marine geology. In the 1950s more was known about the moon's surface than about the floor of the sea. Sea-floor drilling has been instrumental in expanding our knowledge of sea-floor features and history. Small research submarines, more correctly called *submersibles,* can take geologists to many parts of the sea floor to observe, photograph, and sample rock and sediment (figure 18.2*B*).

A basic tool for indirectly studying the sea floor is the *echo sounder,* which draws profiles of submarine topography (figure 18.3). A sound sent downward from a ship bounces off the sea floor and returns to the ship. The water depth is determined from the time it takes the sound to

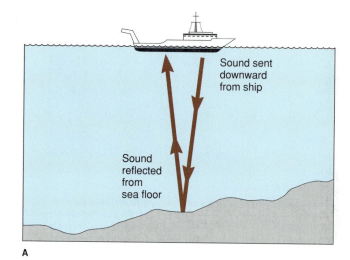

Sound sent downward from ship

Sound reflected from sea floor

A

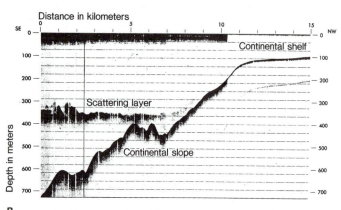

B

Figure 18.3
Echo-sounding. (*A*) A sound bounces off the sea floor and returns to a ship. (*B*) An echo-sounder record of the continental shelf and continental slope off Virginia. The scattering layer is caused by a dense concentration of small marine organisms.
Echo-sounder record courtesy of Peter Rona.

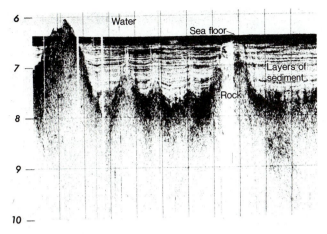

Figure 18.4
Seismic profiler record of an abyssal plain, showing sediment layers that have buried an irregular rock surface in the Atlantic Ocean.

From Vogt, et al, in Hart, *The Earth's Crust and Upper Mantle*, p. 574, 1969, copyrighted by the American Geophysical Union.

Figure 18.5
Photograph of the sea floor on the side of a South Pacific island at a depth of 1000 meters. The rocks are basalt. Rippled sediment between the rocks indicates that strong currents move sea-floor sediment here. Field of view is about 5 by 7 feet.

From Lamont-Doherty Geological Observatory.

make the round trip. A *seismic profiler* works on essentially the same principles as the echo sounder but uses a louder noise at lower frequency. This sound penetrates the bottom of the sea and reflects from layers within the rock and sediment. The seismic profiler gives more information than the echo sounder. It records water depth and reveals the internal structure of the rocks and sediments of the sea floor, such as bedding planes, folds and faults, and unconformities (figure 18.4). *Magnetic, gravity,* and *seismic refraction* surveys (see the previous chapter) also can be made at sea. *Deep-sea cameras* can be lowered to the bottom to photograph the rock and sediment (figure 18.5).

Features of the Sea Floor

Figure 18.6, a simplified profile of the sea floor, shows that continents have two types of margins. A *passive continental margin* includes a continental shelf, continental slope, and continental rise. An *abyssal plain* usually forms a remarkably flat ocean floor beyond the continental rise. An *active continental margin,* associated with earthquakes and volcanoes, has a continental shelf and slope, but the slope extends much deeper to form one wall of an *oceanic trench.* Abyssal plains are seldom found off active margins. The deep ocean floor seaward of trenches is hilly and irregular, lacking the extreme flatness of abyssal plains. Encircling the globe is a *mid-oceanic ridge,* usually (but not always) near the center of an ocean. Conical *seamounts* stick up from the sea floor in some regions. Some important submarine features do not show in this profile view—in particular, fracture zones, submarine canyons, and aseismic ridges.

Continental Shelves and Continental Slopes

Almost all continental edges are marked by a shallow continental shelf and a steeper continental slope that leads down to the deep ocean floor (figure 18.7). Figures such as 18.6 and 18.7 are usually drawn with great *vertical exaggeration,* which makes submarine slopes appear much steeper than they really are.

A **continental shelf,** a shallow submarine platform at the edge of a continent, inclines very gently seaward, generally at an angle of 0.1°. Continental shelves vary in width. On the Pacific coast of North America the shelf is only a few kilometers wide, but off Newfoundland in the Atlantic Ocean it is about 500 kilometers (300 miles) wide. Portions of the shelves in the Arctic Ocean off Siberia and northern Europe are even wider. Water depth over a continental shelf tends to increase regularly away from land, with the outer edge of the shelf being about 100 to 200 meters below sea level.

Continental shelves are *topographic features,* defined by their depth, flatness, and gentle seaward tilt. Their *geologic origin* varies from place to place and is related to plate tectonics, so it will be discussed in the next chapter. Some generalities about shelves, however, are worth noting here.

The continental shelves of the world are usually covered with relatively young sediment, in most cases derived from land. The sediment is usually sand near shore, where the bottom is shallow and influenced by wave action. Fine-grained mud is deposited farther offshore in deeper, quieter water.

The outer part of a wide shelf is often covered with coarse sediment that was deposited near shore during a time of lower sea level. The advance and retreat of continental glaciers during the Pleistocene Epoch caused sea level to rise and fall several times by 100 to 200 meters. This resulted in a complex history of sedimentation for continental shelves as they were alternately covered with sea water and exposed as dry land.

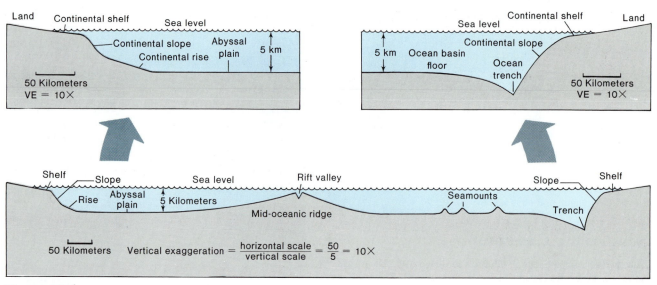

Figure 18.6
Profiles of sea-floor topography. The vertical scales differ from the horizontal scales, causing vertical exaggeration, which makes slopes appear steeper than they really are. The bars for the horizontal scale are 50 kilometers long, while the same distance vertically represents only 5 kilometers, so the drawings have a vertical exaggeration of 10 times.

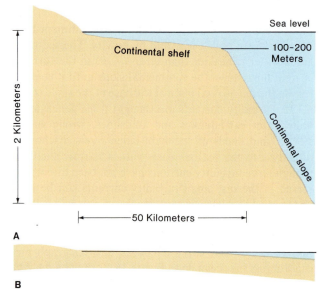

Figure 18.7
Continental shelf and continental slope. (**A**) Vertical exaggeration 25 ×. The continental slope has an actual slope of only 4 or 5 degrees, but the great vertical exaggeration of this drawing makes it appear to be sloping about 60 degrees. (**B**) Same profile with no vertical exaggeration.

Marine seismic surveys and drilling at sea have shown that the young sediments on many continental shelves are underlain by thick sequences of sandstone, shale, and sometimes limestone. These sedimentary rocks were deposited during the Tertiary Period (or sometimes before) and appear to have been deposited in much the same way as the modern shelf sediments. Beneath these rocks is the thick continental crust (figure 18.8); the continental shelves are truly part of the continents, even though the shelves are covered by sea water.

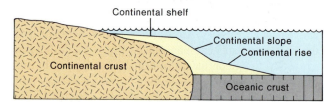

Figure 18.8
The continental shelf lies upon continental crust, and the continental rise lies upon oceanic crust. The transition from continental crust to oceanic crust lies under the continental slope.

A **continental slope** is a relatively steep slope that extends from a depth of 100 to 200 meters at the edge of the continental shelf down to oceanic depths. The average angle of slope for a continental slope is 4° to 5°, although locally some parts are much steeper.

Because the continental slopes are more difficult to study than the continental shelves, less is known about them. The greater depth of water and the locally steep inclines on the continental slopes hinder rock dredging and drilling and make the results of seismic refraction and reflection harder to interpret. This is unfortunate, for the rocks that underlie the slopes are of particular interest to marine geologists, who believe that in this area the thick continental crust (beneath the land and the continental shelves) grades into thin oceanic crust (underneath the deep ocean floor), as shown in figure 18.8.

Although relatively little is known about continental slopes, it is clear that their character and origin vary greatly from place to place. Because these variations can be best described within the context of plate tectonics, we will postpone discussing them until the next chapter.

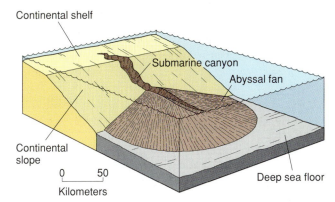

Figure 18.9
Submarine canyon and abyssal fan.

Submarine Canyons

Submarine canyons are V-shaped valleys that run across continental shelves and down continental slopes (figure 18.9). On narrow continental shelves, such as those off the Pacific coast of the United States, the heads of submarine canyons may be so close to shore that they lie within the surf zone. On wide shelves, such as those off the Atlantic coast of the United States, canyon heads usually begin near the outer edge of the continental shelf tens of kilometers from shore. Great fan-shaped deposits of sediment called **abyssal fans** are found at the base of many submarine canyons (figure 18.9). Abyssal fans are made up of land-derived sediment that has moved down the submarine canyons. Along continental margins that are cut by submarine canyons, many coalescing abyssal fans may build up at the base of the continental slope.

Submarine canyons are erosional features, but how rock and sediment are removed from the steep-walled canyons is a controversial question. Erosional agents probably vary in relative importance from canyon to canyon. Divers have filmed *down-canyon movement of sand* in slow, glacierlike flow and in more rapid sand falls (figure 18.10). This sand movement, which has been observed to cause erosion of rock, is particularly common in Pacific coast canyons, which collect great quantities of sand from longshore drift. *Bottom currents* have been measured moving up and down the canyons in a pattern of regularly alternating flow, in some cases apparently caused by ocean tides. The origin of these currents is not well understood, but they often move fast enough to erode and transport sediment. *River erosion* may have helped to cut the upper part of canyons when the drop in sea level during the Pleistocene glaciations left canyon heads above the water. Many (but not all) submarine canyons are found off land canyons or rivers, which tends to support the view that river erosion helped shape them. It is unlikely, however, that the deeper parts of submarine canyons were ever exposed as dry land.

Turbidity Currents

In addition to the canyon-cutting processes just described, turbidity currents probably play the major role in canyon erosion.

Figure 18.10
A 10-meter-high sand fall in a submarine canyon near the southern tip of Baja California, Mexico. The sand is beach sand, fed into the nearshore canyon head by longshore currents.
From Scripps Institution of Oceanography, University of California, San Diego.

Turbidity currents are great masses of sediment-laden water that are pulled downhill by gravity. The sediment-laden water is heavier than clear water, so the turbidity current flows down the continental slope until it comes to rest on the flat abyssal plain at the base of the slope (figure 6.13). Turbidity currents are thought to be generated by underwater earthquakes and landslides, strong surface storms, and floods of sediment-laden rivers discharging directly into the sea on coasts with a narrow shelf. Although large turbidity currents have not been directly observed in the sea, small turbidity currents can be made and studied in the laboratory. Much indirect evidence also indicates that turbidity currents occur in the sea.

The best evidence comes from the breaking of submarine cables that carry telephone and telegraph messages across the ocean floor. Figure 18.11 shows a downhill sequence of cable breaks that followed a 1929 earthquake in the Grand Banks region of the northwest Atlantic. This sequence of cable breaks has been interpreted to be the result of an earthquake-caused turbidity current flowing rapidly down the continental slope.

If cable breaks are caused by turbidity currents, they give good evidence of the currents' dramatic size, speed, and energy. Breaks in Grand Banks cables continued for more than thirteen hours after the 1929 earthquake, the last of the series occurring more than 700 kilometers from the epicenter. The velocity of the flow that caused the breaks has been calculated to be from 15 to 60 kilometers per hour. Sections of cable more than 100 kilometers long were broken off and carried away, both ends of a missing section being broken simultaneously. Attempts to find broken cable sections were fruitless, and it is assumed that they were buried by sediment.

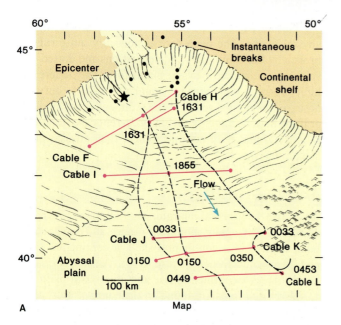

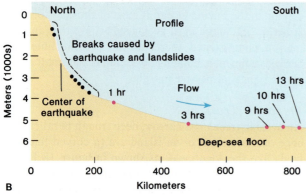

Figure 18.11
Submarine cable breaks following the Grand Banks earthquake of 1929. (**A**) Map view of the cable breaks. Black dots near the epicenter show locations of cable breaks that were simultaneous with the earthquake (cables not shown for these breaks). Colored dots show cable breaks that followed the earthquake, with the time of each break shown (on the 24-hour clock). Segments of cable more than 100 kilometers long were broken simultaneously at both ends and then carried away. Dashes show sea-floor channels that probably concentrated the flow of a turbidity current, increasing its velocity. (**B**) Profile showing the time elapsed between the quake and each cable break.
A from H. W. Menard, 1964, *Marine Geology of the Pacific*, copyright McGraw-Hill, Inc. *B* from B. C. Heezen and M. Ewing, 1952, *American Journal of Science*.

The Grand Banks 1929 cable breaks were not unique. Cables crossing submarine canyons are broken frequently, particularly after river floods and earthquakes. In the submarine canyons off the Congo River of Africa and off the Magdalena River in Colombia, for example, cables break every few years.

Additional indirect evidence for the existence of turbidity currents comes from the graded bedding and shallow-water fossils in the sediments that make up the continental rise and the abyssal plains, as described in the next section.

Passive Continental Margins

A **passive continental margin** (figure 18.6) includes a continental shelf, continental slope, and continental rise and generally extends down to an abyssal plain at a depth of about 5 kilometers (15,000 feet). It is called a passive margin because it usually develops on geologically quiet coasts, that lack earthquakes, volcanoes, and young mountain ranges.

Passive margins are found on the edges of most land masses bordering the Atlantic Ocean. They also border most parts of the Arctic and Indian oceans and a few parts of the Pacific Ocean.

The Continental Rise

Along the base of many parts of the continental slope lies the **continental rise,** a wedge of sediment that extends from the lower part of the continental slope to the deep sea floor. The continental rise, which slopes at about 0.5°, more gently than the continental slope, typically ends in a flat abyssal plain at a depth of about 5 kilometers. The rise rests upon oceanic crust (figure 18.8).

Types of deposition Sediments appear to be deposited on the continental rise in two ways—by turbidity currents flowing *down* the continental slope and by *contour currents* flowing *along* the continental slope.

Cores of sediment recovered from most parts of the continental rise show layers of fine sand or coarse silt interbedded with layers of fine-grained mud. The mineral grains and fossils of the coarser layers indicate that the sand and silt came from the shallow continental shelf. Some transporting agent must have carried these sediments from shallow water to deep water. The coarse layers also exhibit graded bedding, which indicates that they settled out of suspension; therefore, the transporting agent for these sediments was most likely turbidity currents. The continental rise in these locations probably formed as turbidity currents deposited abyssal fans at the base of a continental slope.

Sediments in some other parts of the continental rise, however, are uniformly fine-grained and show no graded bedding. This sediment appears to have been deposited by the regular ocean currents that flow along the sea bottom rather than by the intermittent turbidity currents that occasionally flow downslope.

A **contour current** is a bottom current that flows parallel to the slopes of the continental margin—*along* the contour rather than *down* the slope (figure 18.12). Such a current runs south along the continental margin of North America in the Atlantic Ocean. Flowing at the relatively slow speed of a few centimeters per second, this current carries a small amount of fine sediment from north to south. The current is thickest along the continental slope and gets progressively thinner seaward. The thick landward part of the current carries and deposits the most sediment. The thinner seaward edge of the current deposits less sediment. As a result, the deposit of sediment beneath

the current is wedge shaped, becoming thinner away from land. Similar contour currents apparently shape parts of the continental rise off other continents as well.

Abyssal Plains

Abyssal plains are very flat regions usually found at the base of the continental rise. Seismic profiling has shown that abyssal plains are formed of horizontal layers of sediment. The gradual deposition of sediment buried an older, more rugged topography that can be seen on seismic profiler records as a rock basement beneath the sediment layers (figure 18.4). Samples of abyssal plain sediment show that it is derived from land. Graded bedding within sediment layers suggests deposition by turbidity currents.

Abyssal plains are the flattest features on earth. They generally have slopes less than 1:1,000 (less than one meter of vertical drop for every 1,000 meters of horizontal distance) and some have slopes of only 1:10,000.

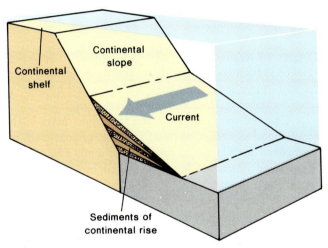

Figure 18.12
A contour current flowing along the continental margin shapes the continental rise.

Not all parts of the deep ocean basin floor consist of abyssal plains. The deep floor is normally quite rugged, broken by faults into hills and depressions and dotted with volcanic seamounts. Abyssal plains form only where turbidity currents can carry in enough sediment to bury and obscure this rugged relief. If the sediment is not available or if the bottom-hugging turbidity currents are stopped by a barrier such as an oceanic trench, then abyssal plains cannot develop.

Active Continental Margins

An **active continental margin,** characterized by earthquakes and by young mountain ranges and volcanoes on land, consists of a continental shelf, a continental slope, and an oceanic trench (figure 18.6). An active margin usually lacks a continental rise and an abyssal plain.

Active margins are found on the edges of most of the land masses bordering the Pacific Ocean and a few other places in the Atlantic and Indian oceans. A notable exception in the Pacific Ocean is most of the coast of North America. Although most geologists believe that an active margin once existed there, at the present time land-derived sediment is building abyssal fans and small abyssal plains typical of passive margins.

Oceanic Trenches

An **oceanic trench** is a narrow, deep trough parallel to the edge of a continent or an island arc (a curved line of islands like the Aleutians), as shown in figure 18.13. The continental slope on an active margin forms the landward wall of the trench, its steepness often increasing with depth. The slope is typically 4° to 5° on the upper part, steepening to 10° to 15° or even more near the bottom of the trench. The elongate oceanic trenches, often 8 to 10 kilometers deep, far exceed the average depth of abyssal plains on passive margins. The deepest spots on earth, almost 12 kilometers below sea level, are in oceanic trenches.

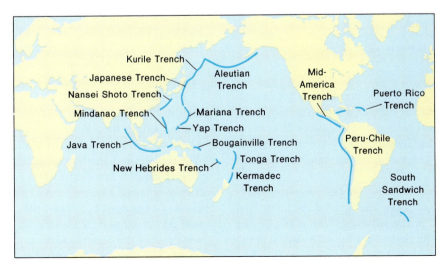

Figure 18.13
The distribution of oceanic trenches.

Associated with oceanic trenches are the *earthquakes of the Benioff seismic zones* (see chapter on "Earthquakes"), which begin at a trench and dip landward under continents or island arcs (figure 18.14). *Volcanoes* are found above the upper part of the Benioff zones and typically are arranged in long belts parallel to oceanic trenches. These belts of volcanoes form island arcs or erupt within young mountain ranges on the edges of continents. The rock produced by these volcanoes is usually andesite, a type of extrusive rock intermediate in composition between basaltic oceanic crust and "granitic" continental crust.

Oceanic trenches are marked by abnormally *low heat flow* compared to normal ocean crust. This implies that the crust in trenches may be colder than normal crust.

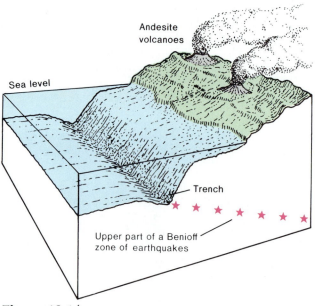

Figure 18.14
Active continental margin with an oceanic trench, a Benioff zone of earthquakes (only the upper part is shown), and a chain of andesitic volcanoes on land.

As you learned in the previous chapter, oceanic trenches are also characterized by very large *negative gravity anomalies,* the largest in the world. This implies that trenches are being held down, out of isostatic equilibrium.

The Mid-Oceanic Ridge

The **mid-oceanic ridge** is a giant undersea mountain range that extends around the world like the seams on a baseball (figures 18.6 and 18.15). The ridge, which is made up mostly of basalt, is more than 80,000 kilometers long and 1,500 to 2,500 kilometers wide. It rises 2 to 3 kilometers above the ocean floor.

Except in the Pacific Ocean, a **rift valley**—a large crack, apparently of tensional origin—runs down the crest of the ridge (figures 18.15 and 18.6). The rift valley is 1 to 2 kilometers deep and several kilometers wide—about the dimensions of the Grand Canyon in Arizona. The rift valley on the crest of the mid-oceanic ridge is a unique feature—no mountain range on land has such a valley running along its crest.

Geologic Activity on the Ridge

Associated with the rift valley at the crest of the mid-oceanic ridge and also with the riftless crest of the ridge in the Pacific Ocean, are *shallow-focus earthquakes* (chapter on "Earthquakes").

Careful measurements of the heat loss from the earth's interior through the crust have shown a very *high heat flow* on the crest of the mid-oceanic ridge. The heat loss at the ridge crest is many times the normal value found elsewhere in the ocean.

Basalt eruptions occur in and near the rift valley on the ridge crest. Sometimes these eruptions build up volcanoes that protrude above sea level as oceanic islands.

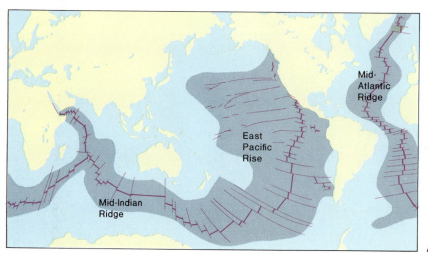

Figure 18.15
(**A**) The mid-oceanic ridge, offset by fracture zones. Darker lines indicate the ridge crest; lighter lines show the fracture zones. Figure continues on next page.

B

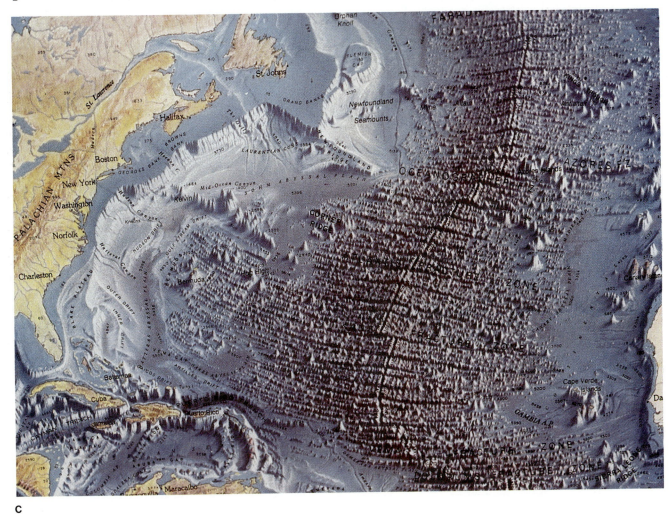

C

Figure 18.15 continued

(*B*) The world ocean floor; the mid-oceanic ridge is the dark ridge encircling the globe. (*C*) The mid-oceanic ridge in the North Atlantic Ocean, where it has the regional name the Mid-Atlantic Ridge. It covers one-third to one-half of the ocean floor, and its rift valley is equidistant from the continents on each side.

Figure 18.16
Underwater photograph of fresh pillow basalt on the floor of the rift valley of the mid-oceanic ridge in the North Atlantic Ocean.
Photo by Woods Hole Oceanographic Institution.

The large island of Iceland, which is mostly basaltic, appears to be a section of the mid-oceanic ridge elevated above sea level. Many geologists have studied the active volcanoes, high heat flow, and central rift valley of Iceland to learn about the mid-oceanic ridge. As Iceland is above sea level, however, it may not be a typical portion of the ridge.

In the summer of 1974 geologists were able to get a firsthand view of part of the submerged ridge and rift valley. A series of more than forty dives by submersibles, including ALVIN, carried French and American marine geologists directly into the rift valley in the North Atlantic Ocean. The project (called FAMOUS for French-American Mid-Ocean Undersea Study) allowed the ridge rock to be seen, photographed, and sampled directly, rather than indirectly from surface ships (figure 18.16).

The geologists on the FAMOUS project saw clear evidence of tensional cracks within the rift valley. These run parallel to the axis of the rift valley and range in width from hairline cracks to gaping fissures that ALVIN dived into. Fresh pillow basalts occur in a narrow band along the bottom of the rift valley, suggesting very recent volcanic activity there, although no active eruptions were observed. It appeared to the geologists that tensional cracking of the rift was continuous, and that sporadic volcanic activity occurred as a result of the rifting.

Fracture Zones

Fracture zones are major lines of weakness in the earth's crust that cross the mid-oceanic ridge at approximately right angles (figure 18.15). The rift valley of the mid-oceanic ridge is offset in many places across fracture zones (figure 18.17), and the sea floor on one side of a fracture zone is often at a different elevation than the sea floor on the other side. Shallow-focus earthquakes occur on fracture zones but are confined to those portions of the fracture zones between segments of the rift valley (figure 18.18). Fracture zones extend for thousands of kilometers across the ocean floor, generally heading straight for continental margins. Although fracture zones are difficult to trace where they are buried by the sediments of the abyssal plain and the continental rise, a number of geologists think that they can trace the extensions of fracture zones onto continents. Some earthquake epicenters and some major structural trends on continents appear to lie along the hypothetical extension of fracture zones onto the continents.

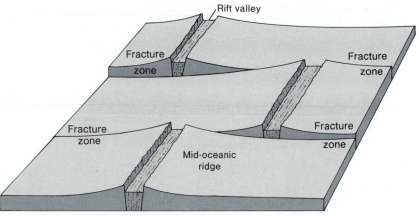

Figure 18.17
Fracture zones, which run perpendicular to the ridge crest and separate the ridge into segments, are often marked by steep cliffs up to 3 or 4 kilometers high.

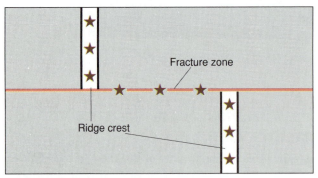

Figure 18.18
Shallow-focus earthquakes (shown by stars) occur on the ridge crest and on the portion of the fracture zone between the two segments of the ridge crest (map view).

Seamounts, Guyots, and Aseismic Ridges

Conical undersea mountains that rise 1,000 meters or more above the sea floor are called **seamounts** (figures 18.15 and 18.19). They sometimes rise above sea level to form islands. They are scattered on the flanks of the mid-oceanic ridge and on other parts of the sea floor, including abyssal plains. One area of the sea floor with a particularly high concentration of seamounts—one estimate is 10,000—is the southwestern Pacific. Rocks dredged from seamounts are nearly always basalt, so it is thought that most seamounts are extinct volcanoes. Of the thousands of seamounts on the sea floor, only a few are active volcanoes. Most of these are on the crest of the mid-oceanic ridge. A few others, such as the two active volcanoes on the island of Hawaii, are found at locations not associated with the ridge.

Guyots are flat-topped seamounts (figure 18.20) found mostly in the western Pacific Ocean. Most geologists think that the flat summits of guyots were cut by wave action. These flat tops are now many hundreds of meters below sea level, well below the level of wave erosion. If the guyot tops were cut by waves, the guyots must have subsided

after erosion took place. Evidence of such subsidence comes from the dredging of dead reef corals from guyot tops. Since such corals grow only in shallow water, they must have been carried to their present depths as the guyots sank.

Many of the guyots and seamounts on the sea floor are aligned in chains. Such volcanic chains, together with some other ridges on the sea floor, are given the name **aseismic ridges** (figure 18.21); that is, they are submarine ridges that are not associated with earthquakes. The name *aseismic* is used to distinguish these features from the much larger mid-oceanic ridge, where earthquakes occur along the rift valley.

Reefs

Reefs are wave-resistant ridges of coral, algae, and other calcareous organisms. They form in warm, shallow, sunlit water that is low in suspended sediment. Reefs stand above the surrounding sea floor, which is often covered with sediment derived from the reef (figure 6.18). Three important types are *fringing reefs, barrier reefs,* and *atolls* (figure 18.22).

Fringing reefs are flat, tablelike reefs attached directly to shore. The seaward edge is marked by a steep slope leading down into deeper water. Many of the reefs bordering the Hawaiian islands are of this type.

Barrier reefs parallel the shore but are separated from it by wide, deep lagoons. This type of reef is shown in figure 6.18. The lagoon has relatively quiet water because the reef shelters it by absorbing the energy of large, breaking waves. A barrier reef lies about 5 miles (8 kilometers) offshore of the Florida Keys, a string of islands south of Miami. On a much grander scale is the Great Barrier Reef off northeastern Australia. It extends for about 1,200 miles (2,000 kilometers) along the coast, and its seaward edge lies up to 150 miles (250 kilometers) from shore. Another long barrier reef lies along the eastern coast of the Yucatan Peninsula in Central America, and others surround many islands in the South Pacific.

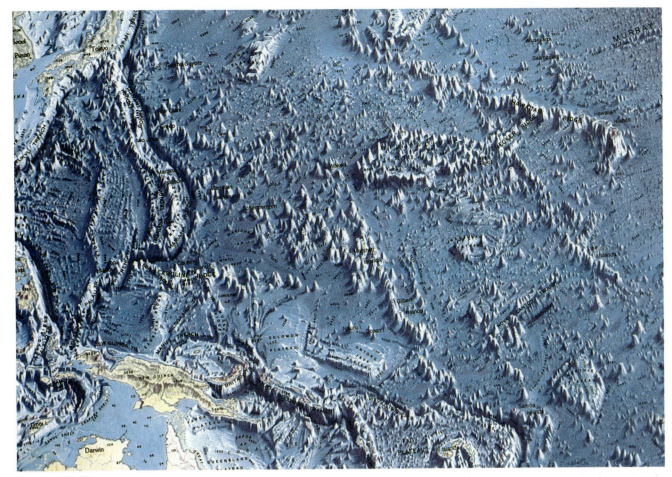

Figure 18.19
Seamounts in the western Pacific. Each conical mountain is a basaltic volcano. Some seamounts are aligned in chains as aseismic ridges.
A portion of the ''World Ocean Floor Panorama'' by Bruce C. Heezen and Marie Tharp. Copyright 1977 Marie Tharp.

Atolls are circular reefs that rim lagoons. They are surrounded by deep water. Small islands of calcareous sand may be built by waves at places along the reef ring. The diameter of atolls varies from 1 to more than 60 miles (100 kilometers). Numerous atolls dot the South Pacific. Bikini and Eniwetok atolls were used through 1958 for the testing of nuclear weapons by the United States.

Following the four-year cruise of the HMS *Beagle* in the 1830s, Charles Darwin proposed that these three types of reefs are related to one another by subsidence of a central volcanic island, as shown in figure 18.22. A fringing reef initially becomes established near the island's shore. As the volcano subsides because of tectonic lowering of the sea floor, the reef becomes a barrier reef. Less and less of the island remains above sea level, but the reef grows rapidly upward into shallow, sunlit water, maintaining its original size and shape. Finally the volcano disappears completely below sea level, and the reef becomes a circular atoll. Drilling through atolls in the 1950s showed that these reefs were built on deeply buried volcanic cores, thus confirming Darwin's hypothesis of 120 years before.

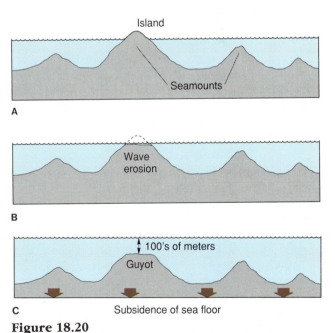

Figure 18.20
(A) Seamounts are conical mountains on the sea floor, occasionally rising above sea level to form islands. (B) The flat summit of a guyot was probably eroded by waves when the top of a seamount was above sea level. (C) The present depth of a guyot is due to subsidence.

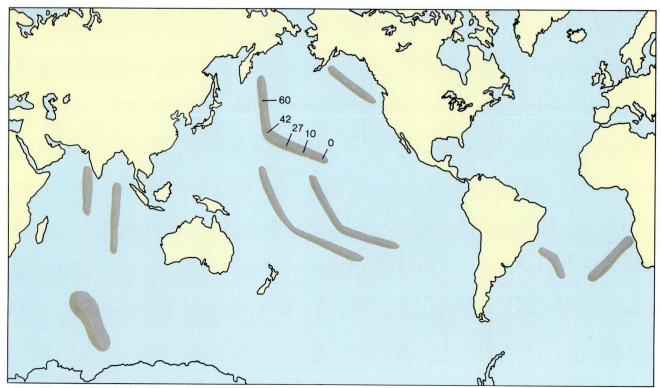

Figure 18.21
The distribution of the major aseismic ridges on the sea floor.
The numbers on the Hawaiian-Emperor chain are ages in
millions of years (see the end of chapter 19).
From W. Jason Morgan, 1972, *Geological Society of America Memoir 132,*
and other sources.

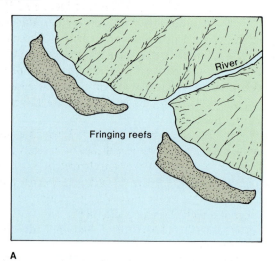

A

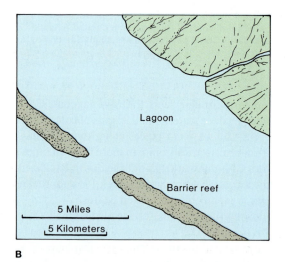

B

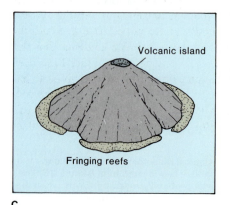

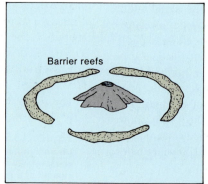

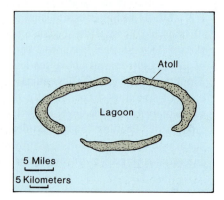

C

Figure 18.22
Types of coral-algal reefs. (*A*) Fringing reefs are attached
directly to shore. (*B*) Barrier reefs are separated from shore by
a lagoon. (*C*) Atolls are circular reefs with central lagoons.
Charles Darwin proposed that atolls form by the subsidence of
a central volcano.

Figure 18.23
Photograph (taken through a scanning-electron microscope) of pelagic sediment from the floor of the Pacific Ocean. The sediment is made up of microscopic skeletons of single-celled marine organisms (large objects are foraminifera; smaller, sievelike ones are radiolaria about 0.05mm in diameter).
Scripps Institution of Oceanography, University of California, San Diego.

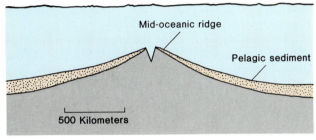

Figure 18.24
Pelagic sediment is thin or absent on the crest of the mid-oceanic ridge and becomes progressively thicker away from the ridge crest (distribution of sediment highly simplified).

Sediments of the Sea Floor

The basaltic crust of the sea floor is covered in many places with layers of sediment. This sediment is either *terrigenous,* derived from land, or *pelagic,* settling slowly through sea water.

Terrigenous sediment is land-derived sediment that has found its way to the sea floor. The sediment that makes up the continental rise and the abyssal plains is mostly terrigenous and apparently has been deposited by turbidity currents or similar processes. Once terrigenous sediment has found its way down the continental slope, contour currents may distribute it along the continental rise. On active continental margins, oceanic trenches may act as traps for terrigenous sediment and prevent it from spreading out onto the deep sea floor beyond the trenches.

Pelagic sediment is sediment that settles slowly through the ocean water. It is made up of fine-grained clay and the skeletons of microscopic organisms (figure 18.23). Fine-grained pelagic clay is found almost everywhere on the sea floor, although in some places it is masked by other types of sediments that accumulate rapidly. The clay is mostly derived from land; part of it may be volcanic ash. This sediment is carried out to sea primarily by wind, although rivers and ocean currents also help to distribute it.

Microscopic shells and skeletons of plants and animals also settle slowly to the sea floor when marine organisms of the surface waters die. In some parts of the sea, such as the polar and equatorial regions, great concentrations of these shells have built unusually thick pelagic deposits.

The constant slow rain of pelagic clay and shells occurs in all parts of the sea. Although the rate of accumulation varies from place to place, pelagic sediment should be expected on all parts of the sea floor.

Surprisingly, however, pelagic sediment is almost completely absent on the crest of the mid-oceanic ridge. Pelagic sediment is found on the flanks of the mid-oceanic ridge, often thickening away from the ridge crest (figure 18.24). But its absence on the ridge crest was an unexpected discovery about sea-floor sediment distribution.

Oceanic Crust and Ophiolites

As you have seen in the previous chapter, oceanic crust differs significantly from continental crust; it is both thinner and of a different composition than continental crust. Seismic reflection and seismic refraction surveys at sea have shown the oceanic crust to be about 7 kilometers thick and divided into three layers (figure 18.25).

The top layer (Layer 1), of variable thickness and character, is marine sediment. In an abyssal fan or on the continental rise, Layer 1 may consist of several kilometers of terrigenous sediment. On the upper flanks of the mid-oceanic ridge, there may be less than 100 meters of pelagic sediment. An average thickness for Layer 1 might be 0.5 kilometer.

Beneath the sediment is Layer 2, which is about 1.5 kilometers thick and has been drilled extensively by drilling ships. The layer is mostly basalt and highly fractured, containing pillows that form when hot lava erupts into cold water. Widely sampled by dredging, Layer 2 has also been observed directly by the geologists in the FAMOUS expedition to the rift valley of the mid-oceanic ridge in the Atlantic Ocean.

The lowest layer in the crust is Layer 3. It is 5 kilometers thick and is thought to consist of parallel vertical dikes ("sheeted dikes") in its upper part and sill-like gabbro bodies in its lower part. The evidence for this interpretation is scanty. A few gabbro samples have been dredged from steep submarine scarps where this layer has been exposed by faulting, and seismic velocities are consistent with these rock choices. The interpretation of this layer, however, is primarily based on the study of ophiolites on land.

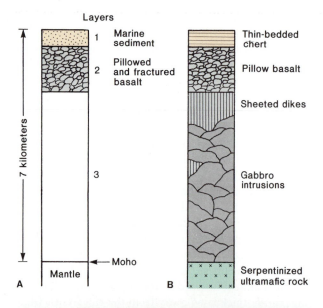

Layers

1 Marine sediment

2 Pillowed and fractured basalt

3

← Moho

Mantle

A

Thin-bedded chert

Pillow basalt

Sheeted dikes

Gabbro intrusions

Serpentinized ultramafic rock

B

7 kilometers

C

Figure 18.25
A comparison of oceanic crust and an ophiolite sequence.
(A) Structure of oceanic crust, determined from seismic studies
and drilling. Layer 3 is blank because it has not yet been
drilled. (B) Typical ophiolite sequence found in mountain
ranges on land. Thickness approximate—sequence is usually
highly faulted. (C) Pillow basalt from the upper part of an
ophiolite, northern California. These rocks formed as part of
the sea floor, when hot lava cooled quickly in cold seawater.

Ophiolites are distinctive rock sequences found in many mountain ranges on continents (figure 18.25). The top thin layer in an ophiolite sequence consists of marine sedimentary rock, often including thin-bedded chert. Below the sedimentary rock lies a zone of pillow basalt. The similarity between the upper two layers of an ophiolite and Layers 1 and 2 in oceanic crust is obvious.

Beneath the pillow basalt in a continental ophiolite sequence lies a zone of closely spaced dikes that are interpreted to be the feeder dikes for the pillowed lava flows above. Below this "sheeted-dike" complex is a zone of podlike intrusive bodies of gabbro, perhaps fat sills. The dikes and gabbro together may represent oceanic Layer 3, for their total thickness in a complete ophiolite is about 5 kilometers, and their seismic velocities are about 7 kilometers per second, which is the required velocity for oceanic crust (see previous chapter).

Beneath the gabbro in an ophiolite lies ultramafic rock such as peridotite (see previous chapter). The ultramafic rock has usually been converted to serpentine by metamorphism, so its original character is not always clear. If the upper part of an ophiolite represents Layers 1, 2, and 3 of oceanic crust, then the serpentinized ultramafic rock may be part of the upper mantle. The contact between the ultramafic rock and the overlying gabbro would then be the Mohorovičić discontinuity.

How could oceanic crust and upper mantle rocks find their way into mountain ranges on land? It is possible that slivers of sea floor were caught between converging plates and wedged upward onto continents. The next chapter deals with plate motions and describes how plate collisions occur.

The Age of the Sea Floor

As marine geologists began to determine the age of sea-floor rocks (by radiometric dating) and sediments (by fossils), an astonishing fact was discovered. All the rocks and sediments of the sea floor proved to be younger than 200 million years old. This was true only for rocks and sediments from the *deep* sea floor, not those from the continental margins. The rocks and sediments presently found on the deep sea floor formed during the Mesozoic and Cenozoic eras, but not earlier.

By contrast, as discussed in the chapter on Geologic Time, the earth is estimated to be 4.5 billion years old. Every continent contains some rocks formed during the Paleozoic Era and the Precambrian. Some of the Precambrian rocks on continents are more than 3 billion years old, and a few are almost 4 billion years old. Continents, therefore, preserve rocks from most of the earth's history. In sharp contrast to the continents is the deep sea floor, which covers more than half of the earth's surface but preserves less than one-twentieth of earth's history in its rocks and sediment.

BOX 18.1

Geologic Riches in the Sea

Many resources are currently being extracted from the sea floor and from sea water, and in some instances there is a great potential for increased extraction.

Offshore oil and gas are the most valuable resources now being taken from the sea. Approximately one-sixth of the United States' oil production (and more than one-quarter of world production) comes from drilling platforms set up on the continental shelf (box figure 1). Oil and gas have been found recently within deeper parts of the sea floor, such as the continental slope and continental rise. Producing oil from these deeper regions will be much more costly than present production; oil spills from wells in deep water would be especially hard to control.

Other important resources are dredged from the sea floor. *Phosphorite* can be recovered from shallow shelves and banks and used for fertilizers. *Gold, diamonds,* and heavy *black sands* (which are black because they contain metal-bearing minerals) are being separated from the surface sands and gravels of some continental shelves by specially designed ships.

Manganese nodules (box figure 2) cover many parts of the deep sea floor, notably in the central Pacific. These black, potato-sized lumps contain approximately 25% manganese, 15% iron, and up to 2% nickel and 2% copper, along with smaller amounts of cobalt. Although there are international legal problems concerning who owns them, larger industrial countries such as the United States may mine them, particularly for copper, nickel, and cobalt. The United States is also interested in manganese; it imports 95% of its manganese, which is critical to producing steel.

Metallic brines and sediments, first discovered in the Red Sea, are the result of hydrothermal processes active at the rift valley of the mid-oceanic ridge crest. The Red Sea sediments contain more than 1% copper and more than 3% zinc, together with impressive amounts of silver, gold, and lead (worth $25 billion according to a 1983 estimate). Because of their great value, the sediments will probably be mined even though they are at great depth. Deposits similar to those of the Red Sea—although not of such great economic potential—have been found on several other parts of the ridge.

Box 18.1 Figure 1
Offshore oil drilling platform. As many as 50 different wells can be drilled from a single platform.
Photo by U.S. Geological Survey.

Box 18.1 Figure 2
Dense concentration of manganese nodules on an abyssal plain in the South Pacific, depth 5,300 meters. Field of view about 5 by 7 feet.
From Lamont-Doherty Geological Observatory.

A few substances can be extracted from the salts dissolved in sea water. Approximately two-thirds of the world's production of *magnesium* and *bromine* is obtained from sea water, and in many regions *sodium chloride* (table salt) is obtained by solar evaporation of sea water.

The Sea Floor and Plate Tectonics

As we mentioned at the beginning, this chapter is concerned with the *description* of the sea floor. The *origin* of most sea-floor features is related to plate tectonics. The next chapter shows you how the theory of plate tectonics explains the existence and character of continental shelves and slopes, trenches, the mid-oceanic ridge, and fracture zones, as well as the very young age of the sea floor itself. In the workings of the scientific method, this chapter largely concerns *data*. The next chapter shows how *hypotheses* and *theories* account for these data.

Summary

The *continental shelf* and the steeper *continental slope* lie under water along the edges of continents. They are separated by a change in slope angle at a depth of about 100 meters.

Submarine canyons are cut into the continental slope and outer continental shelf by a combination of *turbidity currents,* sand flow and fall, bottom currents, and river erosion during times of lower sea level. Graded bedding and cable breaks suggest the existence of turbidity currents in the ocean.

Abyssal fans form as sediment collects at the base of submarine canyons.

A *passive continental margin* occurs off geologically quiet coasts and is marked by a continental rise and abyssal plains at the base of the continental slope.

The *continental rise* and *abyssal plains* may form from sediment deposited by turbidity currents.

The continental rise may also form from sediment deposited by *contour currents* at the base of the continental slope.

An *active continental margin* is marked by an *oceanic trench* at the base of the continental slope; the continental rise and abyssal plains are absent.

Oceanic trenches are twice as deep as abyssal plains, which generally lie at a depth of 5 kilometers. Associated with trenches are *Benioff zones* of earthquakes and *andesitic volcanism,* forming either an island arc or a chain of volcanoes near the edge of a continent. Trenches have low heat flow and negative gravity anomalies.

The *mid-oceanic ridge* is a globe-circling mountain range of basalt, located mainly in the middle of ocean basins. The crest of the ridge is marked by a *rift valley,* shallow-focus earthquakes, high heat flow, and active *basaltic volcanism.*

Fracture zones are lines of weakness that apparently offset the mid-oceanic ridge.

Seamounts are conical, submarine volcanoes, now mostly extinct. *Guyots* are flat-topped seamounts, probably leveled by wave erosion before subsiding.

Chains of seamounts and guyots form *aseismic ridges*.

Corals and algae living in warm, shallow water construct *fringing reefs, barrier reefs,* and *atolls.*

Terrigenous sediment is composed of land-derived sediment deposited near land by turbidity currents and other processes. *Pelagic sediment* is made up of wind-blown dust and microscopic skeletons that settle slowly to the sea floor.

The crest of the mid-oceanic ridge lacks pelagic sediment.

Ophiolites in continental mountain ranges probably represent slivers of oceanic crust somehow emplaced on land.

The oldest rocks on the deep sea floor are 200 million years old. The continents, in contrast, contain some rock that is 3 to 4 billion years old.

Terms to Remember

abyssal fan	guyot
abyssal plain	mid-oceanic ridge
active continental margin	oceanic trench
aseismic ridge	ophiolite
atoll	passive continental margin
barrier reef	pelagic sediment
continental rise	reef
continental shelf	rift valley
continental slope	seamount
contour current	submarine canyon
fracture zone	terrigenous sediment
fringing reef	turbidity current

Questions for Review

1. What is a submarine canyon? How do submarine canyons form?
2. Discuss the appearance, structure, and origin of abyssal plains.
3. Sketch a cross profile of the mid-oceanic ridge, showing the rift valley. Label your horizontal and vertical scales.
4. Sketch an active continental margin and a passive continental margin, labeling all their parts. Show approximate depths.
5. What is a fracture zone? Sketch the relation between fracture zones and the mid-oceanic ridge.
6. In a sketch, show the association between an oceanic trench, a Benioff zone of earthquakes, and volcanoes on the edge of a continent.
7. Describe two different origins for the continental rise.
8. What is a turbidity current? What is the evidence that turbidity currents occur on the sea floor?
9. Describe the appearance and origin of seamounts and guyots.
10. Describe the two main types of sea-floor sediment.
11. How does the age of sea-floor rocks compare with the age of continental rocks? Be specific.
12. Sketch a cross section of a fringing reef, a barrier reef, and an atoll.

Questions for Thought

1. How many possible origins can you think of for the rift valley on the mid-oceanic ridge?
2. What geologic dangers exist for structures such as offshore drilling platforms on the continental shelf? For deeper structures on the continental rise or on an abyssal plain?

Supplementary Readings

Anderson, R. N. 1986. *Marine geology—A planet earth perspective.* New York: John Wiley & Sons.

Dietz, R. S. 1964. Origin of continental slopes. *American Scientist* 52:50–69.

Heezen, B. C., and C. D. Hollister. 1971. *The face of the deep.* New York: Oxford University Press.

Keen, M. J. 1968. *An introduction to marine geology.* New York: Pergamon Press.

Kennett, J. 1982. *Marine geology.* Englewood Cliffs, N.J.: Prentice-Hall.

Menard, H. W. 1964. *Marine geology of the Pacific.* New York: McGraw-Hill.

Menard, H. W., ed. 1977. *Ocean science—Readings from Scientific American.* New York: W. H. Freeman.

Moore, J. R., ed. 1971. *Oceanography—Readings from Scientific American.* New York: W. H. Freeman.

Scientific American Editors. 1983. *The ocean.* New York: W. H. Freeman.

Scrutton, R. A., and M. Talwani, eds. 1982. *The ocean floor.* New York: Wiley-Interscience.

Seibold, E. and W. H. Berger. 1982. *The sea floor.* New York: Springer-Verlag.

Shepard, F. P. 1973. *Submarine geology.* 3d ed. New York: Harper & Row.

Turekian, K. K. 1976. *Oceans.* 2d ed. Englewood Cliffs, N.J.: Prentice-Hall.

(See also the list of readings for the next chapter.)

As you studied volcanoes, igneous and metamorphic rocks, and earthquakes, you learned how these topics are related to plate tectonics. In this chapter we take a closer look at plates and plate motion. We will pay particular attention to plate boundaries and the possible driving mechanisms for plate motion.

The history of the concept of plate tectonics is a good example of how scientists think and work and how a hypothesis can be proposed, discarded, modified, and then reborn. In the first part of this chapter we trace the evolution of an idea—how the earlier hypotheses of moving continents (continental drift) and a moving sea floor (sea-floor spreading) were combined to form the theory of plate tectonics.

Plate Tectonics

Plate motion has torn the Arabian peninsula (left) away from Africa (right), forming the Red Sea, the Gulf of Suez (bottom) and the Gulf of Aqaba (lower left).
Photo by NASA.

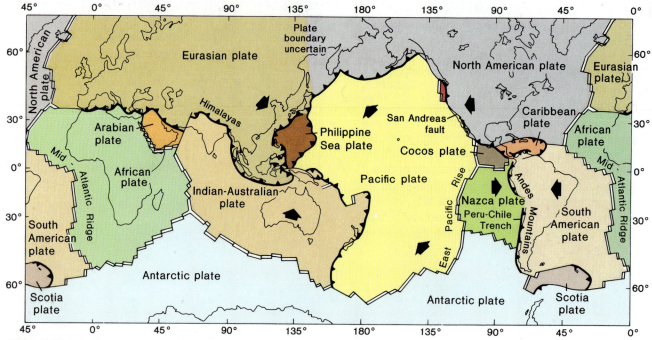

Figure 19.1
The major plates of the world. The western edge of the map repeats the eastern edge so that all plates can be shown unbroken. Double lines indicate spreading centers on diverging plate boundaries. Single lines show transform boundaries. Heavy lines with triangles show converging boundaries, with triangles pointing down subduction zones.
Modified from W. Hamilton, U.S. Geological Survey.

Plate tectonics has come to dominate geologic thought today because it can explain in a general way so many features found on earth. The basic idea of **plate tectonics** is that the earth's surface is divided into a few large, thick plates that move slowly and change in size. Intense geologic activity occurs at *plate boundaries* where plates move away from one another, past one another, or toward one another. The eight large plates shown in figure 19.1, plus a few dozen smaller plates, make up the outer shell of the earth (the crust and upper part of the mantle).

The concept of plate tectonics was born in the late 1960s by combining two preexisting ideas—continental drift and sea-floor spreading. **Continental drift** is the idea that continents move freely over the earth's surface, changing their positions relative to one another. **Sea-floor spreading** is a hypothesis that the sea floor forms at the crest of the mid-oceanic ridge, then moves horizontally away from the ridge crest toward an oceanic trench. The two sides of the ridge are moving in opposite directions like slow conveyor belts.

Before we take a close look at plates, we will examine the earlier ideas of moving continents and a moving sea floor.

The Early Case for Continental Drift

The idea of moving continents is not new. The similarity between the shape of the Atlantic coastlines of South America and Africa was noticed by Francis Bacon in 1620. In 1858 Antonio Snider showed on maps how these continents might once have been joined but then split apart and moved away from each other to form the Atlantic Ocean. The coastlines of other continents also can be made to fit together in "jigsaw puzzle" fashion. For example, the Atlantic coast of the United States fits fairly well against the Atlantic coast of northwestern Africa. Similarities in shoreline shape, of course, do not *prove* that the continents were ever together.

The Ideas of Alfred Wegener
In the early 1900s Alfred Wegener, a German meteorologist, carefully studied the fit of continents and assembled other evidence to make the strongest case he could for continental drift. He showed that the continents could fit together to form a giant supercontinent, which he called *Pangaea* (figure 19.2).

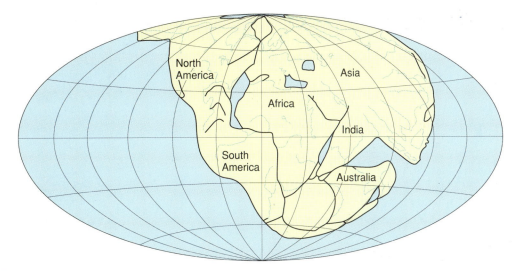

Figure 19.2
The supercontinent Pangaea.
From A. Wegener, 1928, *The Origin of Continents and Oceans*, reprinted and copyrighted, 1968, Dover Publications.

Fossils and paleoclimates Wegener showed that plant fossils of late Paleozoic age found on several different continents were quite similar. In particular, the plant fossil *Glossopteris* was found in rocks in South America, Africa, India, and Australia (figure 19.3). Although these localities are now widely separated from one another, in Wegener's reconstruction of Pangaea they fit closely together. If *Glossopteris* had developed while the continents were joined, the similarity of the fossils would be explained.

In addition to fossils, similar rocks also are found in India, Africa, South America, Australia, and Antarctica. All five localities contain rock sequences in which late Paleozoic tillites (lithified glacial till) are overlain by thick continental sedimentary rocks containing coal beds and *Glossopteris* fossils. Early Mesozoic lava flows overlie the sedimentary layers (figure 19.4). The five sequences of rocks are strikingly similar, even though today the localities are widely separated. Moreover, the younger rocks in the five localities are very dissimilar. Wegener interpreted this to mean that the rock sequences formed together as a single unit while the continents were joined as Pangaea. Beginning in early Mesozoic time, the continents split apart and began to migrate away from one another, splitting the rocks apart. The present-day continents are the original pieces that broke away from Pangaea.

Wegener also believed that the evidence for late Paleozoic glaciation on the continents of the southern hemisphere supported his idea of Pangaea (figure 19.5). If South America, Africa, India, and Australia were spread over the earth in Paleozoic time as they are today, a climate cold enough to produce extensive glaciation would have had to prevail over almost the whole world. Yet no evidence has been found of widespread Paleozoic glaciation in the northern hemisphere. In fact, the late Paleozoic

Figure 19.3
A leaf of *Glossopteris*.

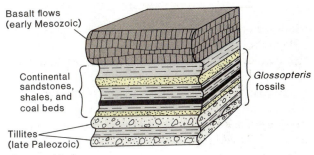

Figure 19.4
Rock sequences similar to this are found in India, Africa, South America, Australia, and Antarctica. The rocks in each of these localities contain the fossil plant *Glossopteris*.

coal beds of North America and Europe were being laid down at that time in swampy, probably warm environments. If the continents are arranged according to Wegener's Pangaea reconstruction, then glaciation in the southern hemisphere is confined to a much smaller area (figure 19.5), and the absence of widespread glaciation in the northern hemisphere becomes easier to explain.

Wegener also worked with evidence from sedimentary rocks in an attempt to reconstruct old climate zones. (The study of ancient climates is called *paleoclimatology*.) The earth today has distinct climate zones such

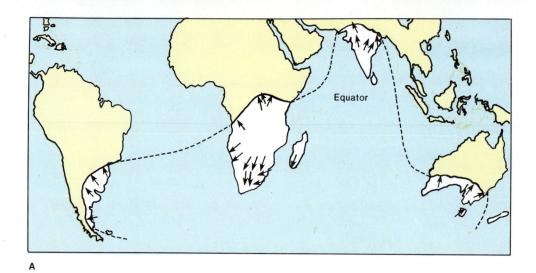

A

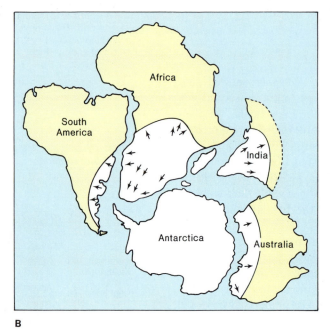

B

Figure 19.5
**Distribution of late Paleozoic glaciations; arrows show
direction of ice flow. (A) Continents in present positions show
wide distribution of glaciation (white land areas with flow
arrows). (B) Continents reassembled into Pangaea. Glaciated
region becomes much smaller.**
A from Arthur Holmes, 1965, *Principles of Physical Geology*, 2d ed.,
Ronald Press.

as cold polar regions and warmer tropical regions. Deserts
are found largely in two belts, one at 30° North latitude
and the other at 30° South latitude (chapter on "Deserts
and Wind Action"). Coral reefs are confined to warm
water near the equator (figure 19.6).

Wegener cataloged paleoclimatic evidence from the
sedimentary rocks of each geologic period to see if he could
find evidence of old climate zones. For example, glacial
till and striations would indicate a past cold climate and
perhaps proximity to the position of the geographic North
or South Pole in the past. Fossil coral reefs might indicate
an earlier position near the equator. Cross-bedded sandstones from old sand dunes might locate past desert belts.

Apparent polar wandering From such paleoclimatic evidence, Wegener found that many ancient climate belts were in different positions from the present belts
(figure 19.7). One way of explaining this shift in climate
belts through geologic time is the apparent movement of
the earth's geographic North Pole and South Pole, a process that is called **polar wandering** (figure 19.7*A*). Polar
wandering, however, is a deceptive term. The evidence can
actually be explained in two different ways:

1. The continents remained motionless and the poles
actually *did* move—polar wandering (figure 19.7*A*).

2. The poles stood still and the continents moved—
continental drift (figure 19.7*B*).

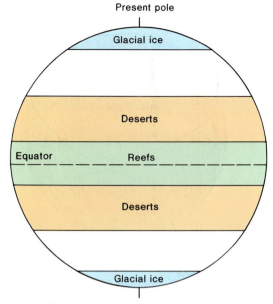

Figure 19.6
Generalized climate zones today produce belts of geologic features such as glaciers, deserts, and reefs.

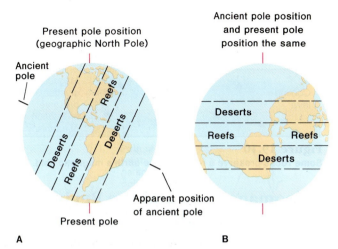

A   B

Figure 19.7
Two ways of interpreting the distribution of ancient climate belts. (*A*) Continents fixed, poles wander. (*B*) Poles fixed, continents drift. For simplicity, the continents in *B* are shown as having moved as a unit, without changing positions relative to one another. If continents move, they should change relative positions, complicating the pattern shown.

Wegener plotted curves of apparent polar wandering (figure 19.8). Since one interpretation of polar wandering data was that the continents moved, Wegener felt that this supported his concept of continental drift. (Notice that in only one interpretation of polar wandering do the poles actually move. You should keep in mind that when geologists use the term *polar wandering* they are referring to an *apparent* motion of the poles, which may or may not have actually occurred.)

Skepticism about Continental Drift
Although Wegener presented the best case possible in the early 1900s for continental drift, much of his evidence was not clear-cut. Fossil plants, for example, could have been spread from one continent to another by winds or ocean currents. The distribution over more than one continent does not *require* that the continents were all joined in the supercontinent, Pangaea. In addition, polar wandering might have been caused by moving poles rather than by moving continents. Because his evidence was not conclusive, Wegener's ideas were not widely accepted. This was particularly true in the United States, largely because of the mechanism Wegener proposed for continental drift.

Wegener proposed that continents plowed through the oceanic crust (figure 19.9), perhaps crumpling up mountain ranges on the leading edges of the continents where they pushed against the sea floor. Most geologists in the United States felt that this idea violated what was known about the strength of rocks at the time. In addition, it was difficult for geologists to conceive of a driving force that could move continents. So Wegener's ideas received little support in the United States or much of the northern hemisphere in the first half of the twentieth century. In the southern hemisphere, however, where Wegener's matches of fossils and rocks between continents were more evident, geologists were more impressed with the concept of continental drift.

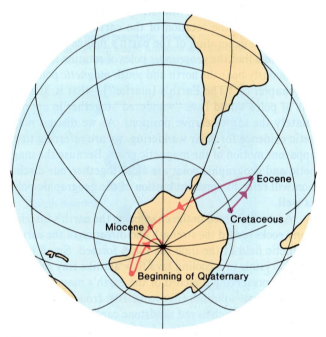

Figure 19.8
Apparent wandering of the South Pole since the Cretaceous Period as determined by Wegener from paleoclimate evidence. Wegener, of course, believed that *continents* rather than poles moved.
From A. Wegener, 1928, *The Origins of Continents and Oceans*, reprinted and copyrighted, 1968, Dover Publications.

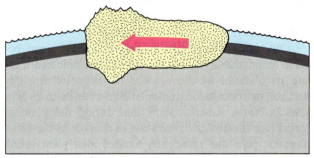

Figure 19.9
Wegener's concept of continental drift implied that continents drifted *through* oceanic crust, crumpling up mountain ranges on their leading edges as they pushed against oceanic crust.

Mountain belts evolve from marine-deposited rocks to towering peaks during periods of hundreds of millions of years. Ultimately the peaks are eroded to plains and become part of the stable interior of a continent. To appreciate the long and complex process of mountain building, you need to know much of the material covered in previous chapters. For instance, you must understand structural geology to appreciate what a particular pattern of folds and faults can tell us about the history of mountain building in a particular region. To understand how the rocks formed during the various stages of a mountain belt's history, you must know about volcanism, plutonism, sedimentation, and metamorphism. Your earlier study of weathering and erosion will help you understand how mountains are worn away. Plate tectonic theory has been strikingly effective in helping geologists make sense of all the complex aspects of mountain belts and the continental crust. For this reason, you need to thoroughly understand the material in chapter 19 before you can appreciate how continents evolve.

In this chapter, we first point out what geologists have observed of mountain belts. Next, we describe how these observations are interpreted, particularly in light of plate tectonic theory. Finally, we describe and discuss observations and interpretations that have arisen since the widespread acceptance of plate tectonic theory and the current conception of how continental crust forms and changes over time.

20

Mountain Belts and the Continental Crust

Satellite photo of part of the Himalaya Mountains (snow covered). Mount Everest, the highest mountain on Earth, is in the center. The brown area to the north is the Tibetan Plateau. In the foreground is the Ganges Plain.

Figure 20.1
View of glaciated peaks in one of the mountain ranges in the Andes mountain belt. A parallel but much lower range is visible at the extreme right skyline of the picture.
Photo by C. C. Plummer.

Mountains, as everyone knows, are large terrain features that rise more or less abruptly from surrounding levels. Several kinds of mountains are of geologic interest but are not related to the topic of this chapter—for example, erosional remnants of plateaus (mesas). We are concerned here with the earth's **major mountain belts,** chains thousands of kilometers long composed of numerous mountain ranges. A **mountain range** is a group of closely spaced mountains or parallel ridges (figure 20.1). Such ranges usually have been produced by intense deformation of part of the earth's crust or by extensive igneous activity.

The map in figure 20.2 shows that most of the world's mountains are in long chains that extend for thousands of kilometers. The Himalaya, the Andes, the Alps, and the Appalachians are examples of major mountain belts, each comprising numerous mountain ranges. Although it is natural to compare mountains in terms of their height and shape, these aspects reflect mainly the extent of uplift and erosion. They may be of little value in helping us understand the complex earlier history of a mountain range.

Our knowledge of the evolution of mountain belts has come from many decades of studying the geologic structures and the rocks exposed in the mountains themselves.

There the rocks that were once below the earth's surface can be seen and studied. Our knowledge of plutonism and metamorphism derives largely from the contributions of geologists working in mountainous regions. The general picture of how mountain belts formed is the result of putting together the work of many persons who did field work in many parts of the world.

Geologists find working in mountains difficult but challenging. Sometimes mountain ranges exhibit very complex interrelationships between rock units. Geologic structures that are necessary for understanding the events that took place may have been eroded away or may be buried underneath soil or glaciers. Mountaineering techniques may sometimes be the only way to reach rock exposures (figure 20.3).

Despite the complexities of individual ranges, the major characteristics of large mountain belts are surprisingly similar. We shall point out these general characteristics before presenting a sequence of events for the evolution of a "typical" mountain belt. As you read through each of the characteristics described next, think of what you have learned from previous chapters (especially the chapter on plate tectonics) that might help explain an observed feature of mountain belts.

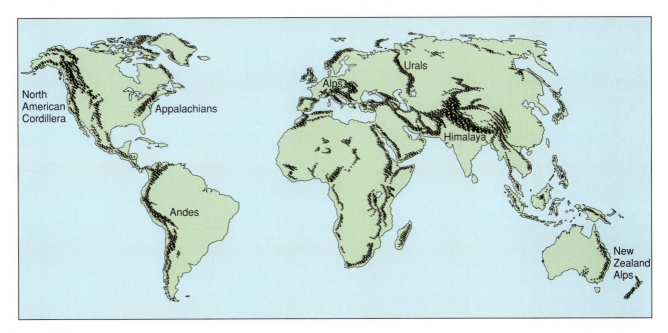

Figure 20.2
Map of the world showing major mountain belts.

Characteristics of Major Mountain Belts

Size and Alignment

Major mountain belts are very long compared to their width. For instance, the mountain belt that forms the western part of North America (the *North American Cordillera*) starts as the Aleutian Island arc in southwestern Alaska. The trend of the Aleutians is continuous with the trend of the mountain ranges that make up much of the Alaskan mainland (figure 20.4). Through Alaska the ranges trend more or less east-west, but in northern Canada they curve into a more north-south trend. The north-south ranges include the many individual ranges of western Canada and western United States. The belt is widest in the United States where it extends from the Coast Ranges of California and Oregon eastward to the Rocky Mountains of Montana, Wyoming, and Colorado. The belt narrows as the north-south trend continues through Mexico.

Ages of Mountain Belts and Continents

The major mountain belts of the world with higher mountain ranges tend to be geologically younger than those where mountains are lower. However, this is not always true. Most mountain regions show evidence that they were once high above sea level, were eroded to hills or low plains, and then rose again in a later episode of uplift. Such episodes of uplift and erosion may occur a number of times during the long history of a mountain range. Thus a newer range that is in a temporary erosional state may be low, or, by contrast, an old range in a temporary state of uplift may be high. Ultimately mountain ranges seem to stabilize and be eroded to plains.

Figure 20.3
Technical climbing is sometimes required for a geologist to work in mountains. Teton Range, Wyoming.
Photo by C. C. Plummer.

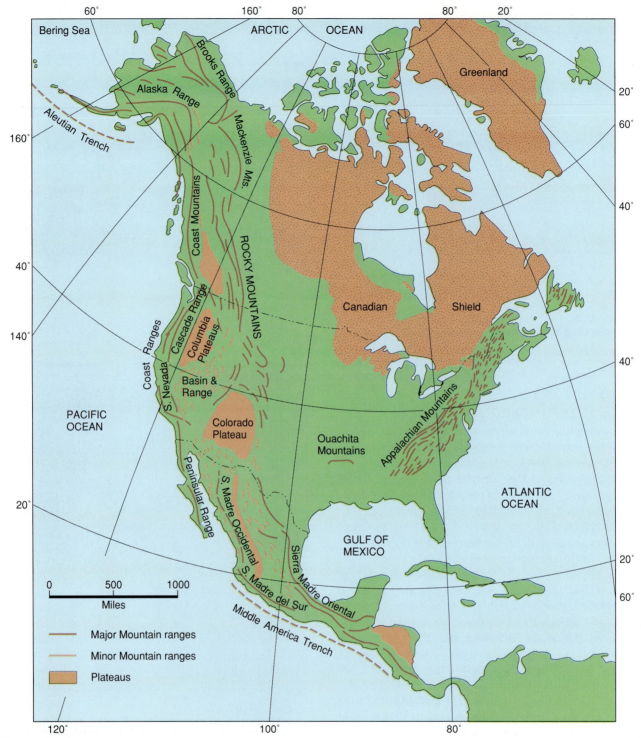

Figure 20.4
The mountain belts of North America. Some of the major ranges in the Cordillera are shown.

On the North American continent, the Appalachian Mountains extend from eastern Canada southward through the eastern United States into Alabama. In the Appalachians, fossils and rocks whose ages have been radioactively determined indicate that these mountains began to evolve earlier than the mountain belt along the western coast of North America. The interior plains between the Appalachians and the Rockies are considered to have evolved from mountain belts in the very distant geologic past (early Precambrian). The once deep-seated roots of the former Precambrian mountain belts are the *basement* rock for the now stable central part of the continent. Layers of Paleozoic and younger sedimentary rock cover most of that basement. The great age of the mountain-building episodes that preceded the Paleozoic sedimentation is confirmed by radioactively determined dates of over one billion years obtained from rocks in the few scattered locations where the basement is exposed. (These

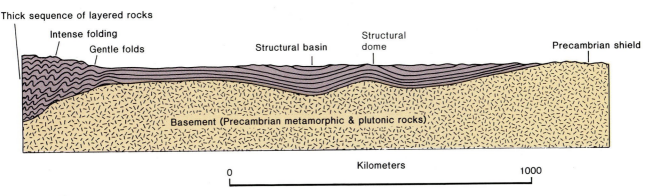

Thick sequence of layered rocks

Intense folding

Gentle folds

Structural basin

Structural dome

Precambrian shield

Basement (Precambrian metamorphic & plutonic rocks)

Kilometers

0   1000

Figure 20.5
Schematic cross section through part of a mountain belt (*left*)
and part of a continental interior (craton). Vertical scale is
exaggerated.

are the Grand Canyon in Arizona, the Ozark dome in
Missouri, the Black Hills of South Dakota, and the Adi-
rondacks in New York.) Continental interiors that have
been structurally stable for a prolonged period of time are
called **cratons** (figures 20.5 and 20.6).

Most of the central United States has a very thin
layer—only 1,000 to 2,000 meters—of sedimentary rocks
overlying its basement. However, in much of eastern
Canada no sedimentary rocks whatsoever cover the eroded
remnants of old mountain ranges. This region is a **Precambrian shield**—that is, a complex of Precambrian
metamorphic and plutonic rocks exposed over a large area.
Such shields and basement complexes of cratons (figure
20.6) represent the roots of mountain ranges that completed the deformation process more than a billion years
ago.

Thickness of Rock Layers

The relatively thin cover of sedimentary rock overlying
the basement in the craton contrasts sharply with the thick
sedimentary sequence of a mountain belt. In mountain
belts, layered sedimentary rock commonly is more than
10 kilometers thick. The sedimentary rock in cratons may
show no deformation, or it may have been gently warped
into basins and domes above the basement (figure 20.5).
By contrast, mountain belts are characterized by tightly
folded and faulted structures indicating intense deformation.

Most of the sedimentary rock in mountains is of
marine origin, indicating that today's highlands must have
been part of the sea floor in the geologic past. The part of
a mountain belt closest to a present-day ocean commonly
contains much volcanically derived rock—volcanic ash,
sediments from eroded volcanic rock, and great thicknesses of lava flows. This contrasts with the part of the
belt commonly adjacent to the craton, which tends to consist mostly of rocks similar to those on the craton—limestones, shales, and sandstones—with only minor amounts
(if any) of volcanic rock.

Figure 20.6
Satellite image of part of a craton in Western Australia.
Metamorphic rock surrounds 3.5 to 3 billion-year-old oval-
shaped domes of granite and gneiss. Gently dipping
sedimentary and volcanic rocks unconformably overlie the
granite-metamorphic basement complex. The area is 400 km
across.
Landsat mosaic produced by the Remote Sensing Applications Centre
Department of Land Administration, Western Australia.

Patterns of Folding and Faulting

Reconstructing the original position and determining the
thickness of layers of sedimentary and volcanic rock in
mountain belts is complicated because in most instances
the layered rocks have been folded and faulted at some
time after they were deposited. Folding and faulting are
generally more intense as one progresses from the stable

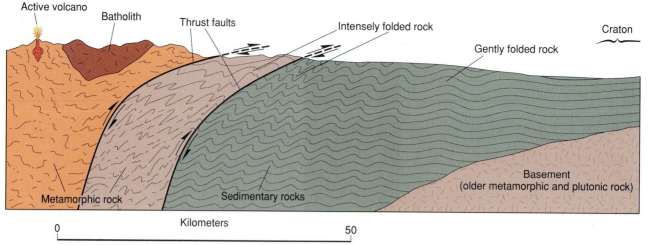

Figure 20.7
Schematic cross section of part of a mountain belt showing progressively more intense deformation from right to left of diagram. Vertical scale is expanded.

A

B

Figure 20.8
(*A*) Part of that portion of the Appalachian Mountains characterized by large, open folds. The ridges are sedimentary beds resistant to erosion and are part of the Cove Mountain syncline. The river is the Susquehanna. The area shown is near Harrisburg, Pennsylvania.
A Photo by John S. Shelton.

(*B*) Satellite image of that part of Pennsylvania.
B Photo by NASA.

interior of the continent further into the mountain belt (figure 20.7). Folds near the craton are generally open (figure 20.8), becoming progressively tighter toward the center of the mountain belt. Large overturned and recumbent folds (figure 20.9) may be exposed in the interior portion of a mountain belt. Reverse faults and large thrust faults, some indicating several tens of kilometers of movement, are commonly found along with the tight folds. These folds and faults indicate tremendous compressive stress. The sedimentary rocks of the Alps, for instance, are estimated to have covered an area of ocean floor about 500 kilometers wide when they were deposited. They were later compressed into the present width of the Alps, which is less than 200 kilometers.

Figure 20.9
Recumbent folds exposed on a mountainside in the Andes.
Dashed line is the trace of a bedding plane.
Photo by C. C. Plummer.

Metamorphism and Plutonism

A complex of regional metamorphic and plutonic rock is generally found in the mountain ranges of the most intensely deformed portions of major mountain belts (figure 20.7). Most of the metamorphic rocks were originally sedimentary and volcanic rocks that had been deeply buried and subjected to intense stress and high temperature. Where found, *migmatites* (interlayered granitic and metamorphic rocks) may represent those parts of the mountain belts that were once at even deeper levels in the crust, where higher temperatures caused partial melting of the rocks (as described in the chapter titled "Intrusive Activity and the Origin of Igneous Rocks"). Granite batholiths indicate that large volumes of magma may have accumulated from partial melting of the lower crust (or upper mantle) and then welled upward, eventually solidifying.

Episode of Normal Faulting

Older portions of some major mountain belts have undergone normal faulting (figure 20.10). Cross-cutting relationships show that the normal faulting occurred after the intense deformation that resulted in tight folding, thrust faulting, and metamorphism and after most batholiths had formed. The later stage of normal faulting is a result of vertical uplift or tensional stress. Either of these forces

contrasts sharply with the compressive stress that caused the folding, thrust faulting, and metamorphism of the original marine rocks.

Thickness and Density of Rocks

Geophysical investigations yield additional information about mountains and the continental crust. As discussed in the chapter titled "The Earth's Interior," gravity measurements indicate that the rocks of the continental crust and mountains are lighter (less dense) than those of the oceanic crust, and seismic velocities indicate a composition approximating that of granite. Furthermore, evidence from seismic studies supports the view that this lighter crust is much thicker beneath mountain belts than under the craton and that the crust is thicker under younger mountain belts than under older ones.

Features of Active Mountain Ranges

Frequent earthquakes are characteristic of portions of mountain belts that are geologically young and considered still "active." Also, deep ocean trenches are found parallel to many young mountain belts (the Andes, for example). Similar trenches lie off the coasts of island arcs, which are in many respects similar to very young mountain ranges. Isolated active volcanoes perched on top of older rock in a mountain range suggest that melting is still taking place at depth.

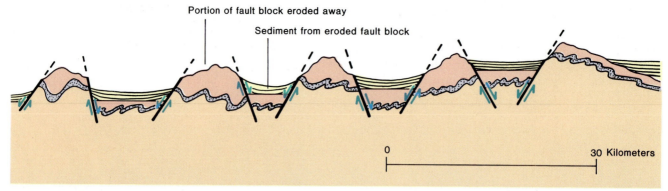

Portion of fault block eroded away

Sediment from eroded fault block

0   30 Kilometers

Figure 20.10
**Fault block mountains with uplift along normal faults. Only one
layer of previously folded rock is shown in order to highlight
the pattern of offset due to normal faulting.**

BOX 20.1

Ultramafic Rocks in Mountain Belts—From the Mantle to Talcum Powder

Ultramafic rocks (described in the chapter on intrusive
rocks) occur commonly in the portions of mountain
belts occupied by metamorphic and plutonic rocks.
Ultramafic rocks tend to crop out along long, narrow
zones that parallel the trend of a mountain belt. Most
geologists believe the bodies of ultramafic rocks
represent mantle material that was faulted into the
crust during the mountain-building process. Some of
the ultramafic bodies are found associated with
marine-deposited volcanic and sedimentary rocks in a
so-called *ophiolite sequence* (as described in the
chapter on the sea floor). This may represent a
segment of a former oceanic floor together with its
underlying mantle.

Ultramafic rocks in mountain belts commonly show
the effects of the metamorphism that has altered
adjacent rock units. Two of the foliated metamorphic
products of ultramafic rocks are of special interest.
One is *serpentinite,* a rock composed of the mineral
serpentine. Another is a rock composed mostly of the
mineral talc, commonly known as *soapstone.*

Serpentinite is a shiny, mottled, dark green and
black rock that looks rather like a snake's skin. It splits
apart easily along irregular, slippery, foliation surfaces.

Hillsides or slopes with serpentinite as bed rock are
sparsely vegetated because constant sliding prevents
soil and vegetation from building up. Houses built out
of ignorance on serpentinite hillsides also slide
downslope. Serpentinite is the official state rock of
California—a state in which a large number of homes
have been destroyed because they were built on
sliding hillsides. (Serpentinite, however, is not always
to blame.)

Most asbestos is a fibrous variety of serpentine
(see Box 2.2). Because it does not ignite or melt in fire,
asbestos has a number of valuable industrial
applications. Woven into cloth, it is used to make suits
for firefighters. It is also used as a fireproof insulation
for homes and other buildings and has commonly been
used in plaster for ceilings. It has now been virtually
banned from use in the United States.

Soapstone, which is less common than serpentinite,
is valuable mainly because of talc's softness (number
1 on Mohs' scale). Many sculptures (most notably
Eskimo carvings) are made from soapstone because
of the ease with which it can be cut. The best-known
product of talc, however, is talcum powder.

Talc's perfect cleavage has led to its unique
application in reflecting highway signs and center road
stripes. Finely crushed talc is mixed with paint. The
many fine, parallel cleavage flakes reflect light in a
specific direction depending on how the mixture is
brushed on the sign. For the signs showing different
daytime and nighttime speed limits, the daytime speed
limit is painted with ordinary black paint, and the
nighttime limit is painted with talc mixed into the white
paint. At night the light from a car's headlights reflects
back to the car from the cleavage surfaces of many
tiny flakes of talc.

The Evolution of a Mountain Belt

Although each mountain belt differs in details, plate tectonics has given us a general picture of how mountain belts evolve. The evolution begins when sediment is deposited; it ends hundreds of millions of years later when the former mountain belt has become part of a craton. Three stages in the history of a mountain belt will be described: (1) the accumulation, (2) the orogenic, and (3) the uplift and block-faulting stages.

The following paragraphs present an account of the first two stages from the perspective of plate tectonic theory. Specifically, they describe how a variety of mountain-building processes are controlled by either plate divergence or plate convergence.

The Accumulation Stage

As we noted earlier, mountain belts typically contain thick sequences of sedimentary and volcanic rocks. The accumulation of these great thicknesses (several kilometers) of sedimentary or volcanic rocks takes place in the **accumulation stage** of mountain building. Most of the sedimentary rocks and much of the volcanic material accumulate in a marine environment. The source for the sediment deposited in the water must be a nearby landmass, belonging to either an adjoining continent or part of a volcanic island arc.

Accumulation in an opening ocean basin At present, sediment is eroding off the eastern portion of North America and is accumulating in the Atlantic Ocean basin adjoining the continent (as described in detail in preceding chapters). Sedimentation has been going on here for tens of millions of years, and a great thickness of layered sedimentary rock has accumulated offshore. According to plate tectonics, accumulation of sediment off the Atlantic coast began when North America and Europe split apart (during the Jurassic Period, around the middle part of the Mesozoic Era). The two continents moved progressively farther apart from a spreading center, the mid-Atlantic ridge. As the Atlantic Ocean basin opened, a progressively bigger mass of land-derived sediment built outward from the land at either side of the ocean (figure 20.11).

Sedimentary rock layers that accumulate under conditions such as those just described are shales, sandstones, and limestones. Volcanic rocks are rare or absent. Where we presently find thick sequences of shale, sandstone, and limestone cropping out in a mountain range, it is reasonable to assume that these rocks were originally deposited under conditions similar to those along the present continental margin of eastern North America.

Accumulation along a converging boundary A large variety of rock types are created near a converging plate boundary. Volcanic rocks—most characteristically andesites—accumulate near the boundary either as pyroclastic layers or as flows. Sedimentary rocks, moreover,

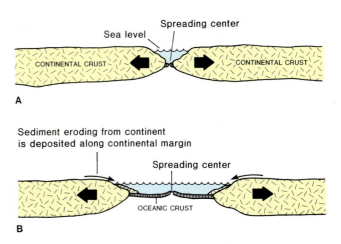

Figure 20.11
Sedimentation in an opening ocean basin. Not drawn to scale. (*A*) Beginning of continental breakup. (*B*) As spreading progresses, sediment is deposited on ocean floor along the continental margins. (*C*) The longer spreading continues, the greater the buildup of sediment.

may accumulate in equal or greater amounts than volcanic rocks. Limestone, however, is usually absent or present only in minor proportions. Shales and sandstones are the predominant sedimentary rock types. Typically, the sandstone contains grains of incompletely weathered igneous rock, indicating that it was derived from a nearby landmass containing volcanic or plutonic rocks. (This contrasts with the sandstone deposited off the east coast of North America, which tends to be composed mostly of grains of quartz.)

The source of this sedimentary as well as volcanic material is a **magmatic arc,** which, as described in the chapter on plate tectonics, is a chain of volcanoes or batholiths along a line (usually curved as seen from above). The magmatic arc has accumulated on and intruded the older igneous and metamorphic rocks of the continental crust (figure 20.12). The source of the magma is the underlying subduction zone.

Sediment eroded from the magmatic arc, as well as newly erupted pyroclastic debris, is transported to and deposited in the basins on either side of the arc. Usually the basin to the seaward side will be underwater, at least initially, and in time will fill with sediment. Once the basin has filled, sediment may be transported beyond it and accumulate outward onto the deep ocean floor. However, as layers accumulate on the edge of the oceanic crust, they are pulled at least partially down the subduction zone, along with the underlying oceanic crust.

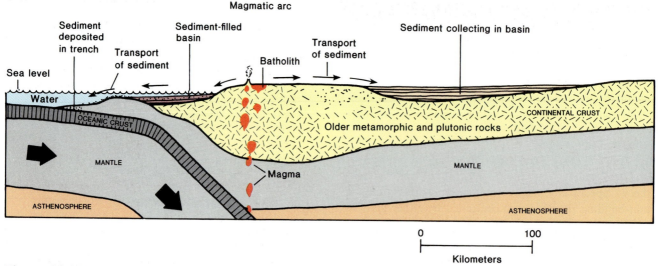

Figure 20.12
Igneous activity and sedimentation associated with a converging plate boundary.
Modified from W. R. Dickinson, 1977, in *Island Arcs, Deep Sea Trenches and Back-Arc Basins* (pp. 33–40), copyrighted by American Geophysical Union.

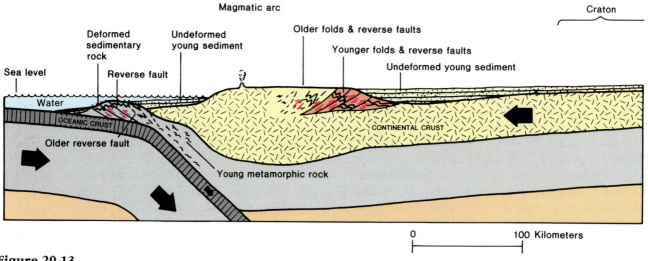

Figure 20.13
Deformation of layered rocks associated with a convergent boundary.
Modified from W. R. Dickinson, 1977, in *Island Arcs, Deep Sea Trenches and Back-Arc Basins* (pp. 33–40), copyrighted by American Geophysical Union.

The Orogenic Stage

Intense deformation follows or is contemporaneous with the accumulation stage. An **orogeny** is an episode of intense deformation of the rocks in a region; the deformation is usually accompanied by metamorphism and igneous activity. During an orogeny, layered sedimentary and volcanic rocks are compressed into folds and broken by reverse and thrust faults. The more deeply buried rocks, subjected to regional metamorphism, are converted to schists and gneisses. Magma generated from very deep crustal rocks (or perhaps even from the upper mantle) may work its way upward to erupt in volcanoes or form large batholiths.

Orogenies and ocean–continent convergence

The relationships among magmatism, metamorphism, and subduction are described in other chapters. Plate convergence also accounts for the folded and reverse-faulted layered rocks found in mountain belts. Newly formed layers of sedimentary and volcanic rock are caught between the oceanic plate and the craton and are compressed into folds and faults (for instance, the "younger folds and reverse faults" on the continental side of the magmatic arc in figure 20.13). As would be expected, rock that is pulled down the subduction zone is especially intensely deformed (for instance, the "deformed sedimentary rock" overlying oceanic crust in figure 20.13).

Geosynclines and Geoclines

In 1859 James Hall, an early American geologist with the New York Geological Survey, introduced the concept of a geosyncline (the name, however, came later). A **geosyncline,** as conceived by Hall, is a great, elongate downwarp of the earth's crust, about the length and width of a mountain belt, which accumulates thicknesses of thousands of meters of sedimentary and volcanic rock. Geosynclines were postulated to account for the great thicknesses of sedimentary and volcanic rock layers observed in mountain belts. For approximately a hundred years geologists accepted the idea that a geosyncline was a necessary predecessor of a mountain belt.

Geosynclines were thought to border continents along one edge of the subsiding trough and be bounded by a line of volcanic islands at the seaward side of the trough. In 1940 the volcanic portion of the geosyncline was named the *eugeosyncline,* and the nonvolcanic portion nearer the continent was called the *miogeosyncline.*

With the advent of plate tectonic theory, geologists realized that thick submarine accumulation need not be in a subsiding trough but can take place under a variety of conditions, as explained in this chapter. The term *geocline* was proposed to replace the older term (since the *syn* in *geosyncline* implied a trough shape). Similarly, *miogeocline* and *eugeocline* were coined to replace *miogeosyncline* and *eugeosyncline,* respectively. Many geologists now use these new terms for the rocks that build up during the accretion stage of mountain building. For instance, the miogeocline would be the sedimentary rocks that accumulate on the continental margin of an opening oceanic basin, and the eugeocline would be the sequence of rocks that form during oceanic-continental convergence.

Some geologists prefer to completely discard the geosynclinal–geoclinal terminology, while others feel that the original geosynclinal terms should be retained even though what they represent has been modified by plate tectonic theory.

The rise and fall of geosynclinal theory is a good example of how science works. For about a hundred years the geosynclinal concept was accepted by geologists because the available data tended to support it. However, when new data led to the formulation of plate tectonic theory, geologists recognized that the new theory explained the data better than the old theory.

It is important to note that where there are converging boundaries, accumulation and deformation are occurring simultaneously. In other words, the accumulation stage and the orogenic stage are taking place at the same time. This was not evident to early geologists, who did not have the perspective of plate tectonic theory for understanding mountain building (see Box 20.2). They generally assumed that the accretion stage ended before the orogenic stage began.

Arc–continent convergence Sometimes an island arc collides with a continent (figure 20.14). If an intervening ocean is destroyed by subduction (the subduction also causes the arc), the arc will approach the continent. When collision occurs, the arc, like a continent, is too buoyant to be subducted. Continued convergence of the two plates may cause the remaining sea floor to break away from the arc and create a new site of subduction and a new trench seaward of the arc (figure 20.14*C*). Note that the direction of the new subduction is opposite to the direction of the original subduction, but it still may supply the arc with magma. (Some geologists refer to these as "flipping subduction zones.") However, the arc has now become welded to the continent, increasing the size of the continent.

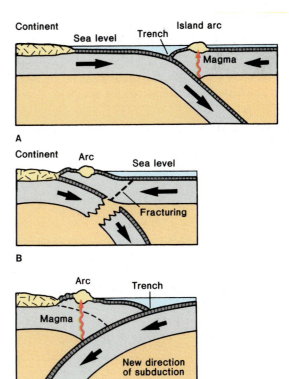

Figure 20.14
Arc–continent convergence can weld an island arc onto a continent. The direction of subduction changes after impact.

This type of collision apparently occurred in northern New Guinea (north of Australia). A similar collision may have added an island arc to the Sierra Nevada complex in California during Mesozoic time, when a subduction zone may have existed in what now is central California. Some geologists feel that much of westernmost North America along the Pacific shore has been built up by a series of arcs colliding with the craton (particularly north of central California).

Orogenies and continent–continent convergence The North Atlantic Ocean continues to widen due to spreading along the mid-Atlantic ridge. North America as well as the western Atlantic Ocean floor are propelled westward from the spreading center while the eastern Atlantic Ocean floor and Europe move eastward. Accretion continues on both sides of the Atlantic. An orogeny is not taking place because there is no subduction. But suppose spreading ceased along the present diverging boundary and the Atlantic Ocean began to close (figure 20.15). A new

Several years ago, a group of geologists began an ambitious project to learn more about the continental crust of North America. The project, which continues today, is called COCORP, an acronym for Consortium for Continental Reflection Profiling. Because it would be too costly and impractical to drill a series of holes into the deep crust, seismic reflection techniques are applied (see chapter titled "The Earth's Interior"). A group of 18-ton trucks with vibrating devices called "thumpers" simulate seismic waves. The wave reflections are monitored by an array of sensitive equipment, and the process is repeated a hundred meters or so further along a road. Eventually a mountain belt or region is traversed and the pattern of reflections compiled by computers can be studied.

The first profile was across the southern Appalachians. The findings dramatically altered how

geologists regard that and similar mountain belts. What they discovered was a major break several kilometers beneath the surface. The break is nearly horizontal in most places and represents a master fault—a thrust fault of more than a hundred kilometers (box figure 1). Below the master fault, the rocks are relatively intact. Above, sedimentary rocks have been deformed into folds and lesser faults in the frontal (western) portion (the Valley and Ridge geologic province); the eastern part is more rigid metamorphic rock that has moved as a block over the basement.

This has led to the concept of *thin-skinned tectonics* (also called thin-skinned thrusting), in which a large, relatively thin slab of the earth's crust has been shoved onto and over another continental mass.

When continental collision occurs, apparently the colliding rocks may not simply crumple in place, but one mass of crust might be shoved over the craton of the other. Why this should happen is still being debated. Box figure 1 is one interpretation of what may have taken place.

COCORP has since worked in other regions of the United States and the results have sometimes been used to modify previous geological interpretations. In the Rocky Mountains, oil companies have used COCORP data to identify thin-skinned thrusts beneath which are sedimentary rocks—and potential oil fields.

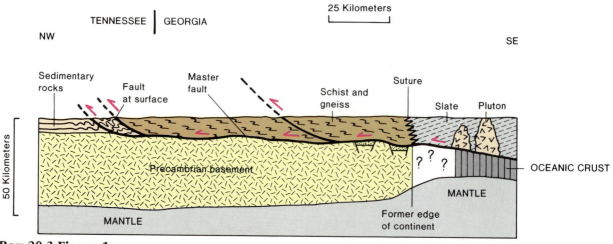

Box 20.3 Figure 1
Cross section across part of the southern Appalachians. According to this interpretation, based on COCORP data, the upper part of the crust has moved a long distance northwestward above the master fault. The area to the right of the suture may have been an island arc at one time.
Modified from F. A. Cook, 1983, *Geology*, vol. 11, p. 89.

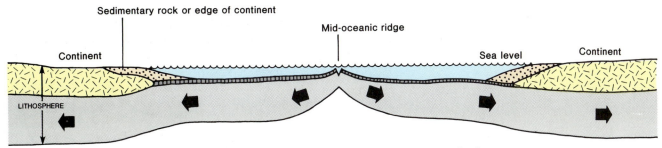

A Accumulation of sediment along the margins of two continents that are moving away from each other.

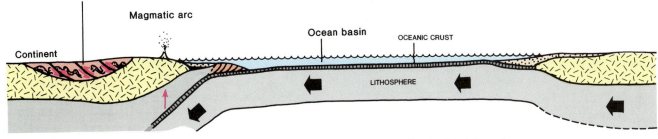

B Spreading ceases and ocean basin begins closing. A subduction zone develops at the continental margin. Sedimentary rock that had formed at the edge of the continent becomes folded and faulted.

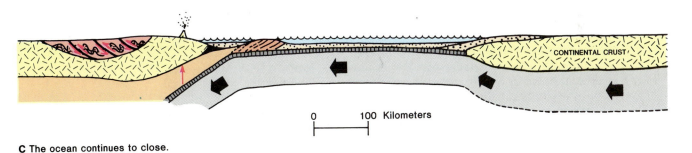

0   100   Kilometers

C The ocean continues to close.

D As ocean basin becomes completely closed, colliding sedimentary rocks are folded and faulted. Subduction stops when the more buoyant continental crust is partially subducted.

Figure 20.15
A mountain belt caused by continental-continental collision.
Modified from W. R. Dickinson, 1977, in *Island Arcs, Deep Sea Trenches and Back-Arc Basins* (pp. 33–40), copyrighted by American Geophysical Union.

subduction zone would develop at one or the other continental margin. As the subduction progressed, a magmatic arc would develop, probably near the seaward edge of the thick sequence of sediment that had accumulated while the ocean basin was opening. If the subduction zone were to develop beneath the east coast of North America, we would expect a chain of volcanoes to form near the eastern Atlantic coastline.

In time, the thick accumulations of sediment from both sides of the closing ocean would collide. During the continental collision the thick layers of sedimentary rocks would be compressed into folds and faults. Probably the continental crust would begin to follow the oceanic crust down the subduction zone. However, since the continental crust is not as dense as oceanic crust, it would only reach the depth where its own buoyancy prevented further downward movement. At this point subduction would stop and the two continents would be *sutured* together into a single continent.

A mountain belt that is entirely within a continent (rather than on the edge of a continent) is considered a product of a continent–continent convergence. The Himalayan belt is believed to have formed in just this way. In this case the collision was between Asia and India (India moved northward from the southern hemisphere).

Could the present Atlantic Ocean close and a new mountain belt develop between the two former continents of Europe and North America? Apparently that is just what happened in the past. During early Paleozoic time a thick sequence of sediment built outward from what was then the eastern edge of North America. Later in the Paleozoic the predecessor of the North Atlantic Ocean basin closed (the closing took millions of years). Volcanic and other rock types associated with a converging boundary were added to the eastern boundary adjacent to the previously accumulated sediments. The ocean closed completely, and Europe was sutured onto North America. As the Paleozoic Era was ending, the supercontinent once again split, roughly parallel to the old suture zone. The two present continents moved (and continue to move) farther and farther away from their present diverging boundary, the mid-Atlantic ridge.

What happened to the Appalachians would seem too implausible even for a science fiction plot. Yet, if one accepts the principles of plate tectonic theory and examines the rocks and structures in the Appalachians (and their counterparts in Europe), the argument for this sequence of events is not only plausible but convincing.

The Uplift and Block-faulting Stage
After the compressive stress of the orogeny ceases, there is a long period of uplift accompanied by erosion. During this stage, which lasts many millions of years, large regions in the mountain belt move vertically upward. Erosion may keep pace with uplift and the area remain low. Alternatively, uplift may temporarily outpace erosion, resulting in plateaus or mountain ranges. Eventually erosion triumphs, and a mountain belt slowly evolves into a part of the craton.

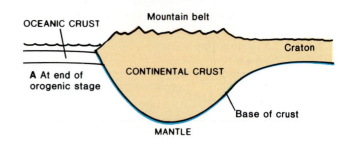

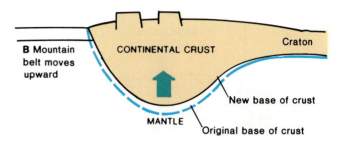

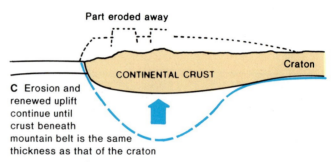

Figure 20.16
Isostasy in a mountain belt. The thickness of the continental crust is exaggerated.

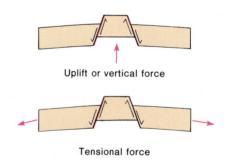

Figure 20.17
Normal faults bounding fault-block mountain ranges. Two ways in which forces might be distributed.

The most likely explanation for this stage is that after an orogeny the newly thickened continental crust adjusts isostatically (as explained in the chapter on the earth's interior and as shown in figure 20.16).

Along with the general uplift, greater uplift may occur along normal faults (or, less commonly, vertical faults) and create **fault-block mountain ranges.** Normal faults imply either vertical forces or horizontal tensional forces (figure 20.17). The fault-block mountain ranges are therefore caused either by tensional forces pulling apart the uplifting crust or by unequal vertical movement of parts of the crust.

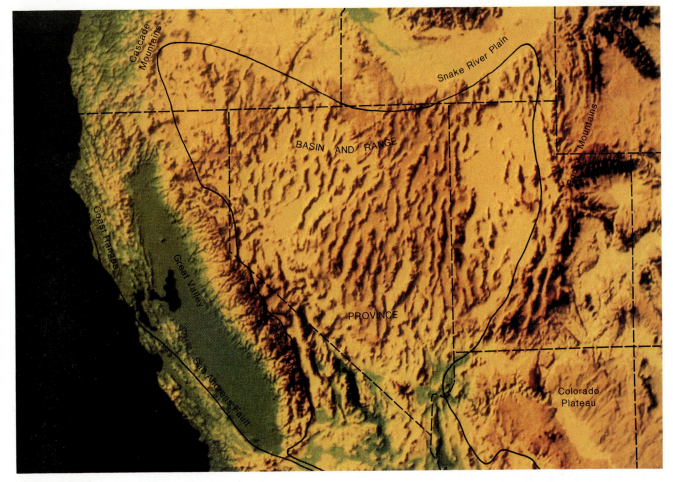

Figure 20.18
The Basin and Range and adjoining geological provinces.
Satellite and computer generated image prepared for Scientific American
by Thomas H. Jordan and J. Bernard Minster.

Block faulting is taking place in much of the western United States—the Basin and Range province (also called the Great Basin) of Nevada and parts of Utah, Arizona, New Mexico, Idaho, and California (figure 20.18). Hundreds of small, block-faulted mountain ranges are in evidence. They are separated by valleys that are filling with debris eroded from the mountains.

Although most fault-block mountain ranges are bounded by normal faults on either side of the range, some are tilted fault blocks in which the uplift has been much greater along one side of the range while the other side of the range has pivoted as if hinged (figure 20.19). The Sierra Nevada (California) and Teton (Wyoming) Range are tilted fault-block mountains (figure 20.20).

Isolated volcanic activity may be associated with this stage of a mountain belt's evolution. Eruptions occur along faults extending deep into the crust or the upper mantle.

Uplift is neither rapid nor continuous. A mountain range may suddenly move upward a few centimeters (or, more rarely, a few meters) and then not move again for hundreds of years. Erosion works relentlessly on newly uplifted mountains, carving the block into peaks during the long, spasmodic rise. Over the long time period, the later episodes of renewed faulting and uplift involve successively less and less vertical movement.

According to the concept of isostasy (chapter called "The Earth's Interior"), lighter, less dense continental crust "floats" higher on the mantle than the denser oceanic crust. The stable interior of the continent has achieved an equilibrium and is floating at the proper level for its thickness. Mountains, being thicker continental crust, "float" higher than the stable continent. As material is removed from mountains by erosion, the range floats upward to regain its isostatic balance.

Forces other than isostasy may be at work as well. This seems to be true for the Basin and Range province, where a hot mantle diapir beneath the crust seems likely (see Box 19.1) and the pattern of normal faults may be caused by regional tensional stress, along with uplift.

The evolution of a mountain belt can be considered complete when the belt becomes part of the craton, the stable continental interior. The forces of erosion and of uplift have been brought into balance. The surface is a low-lying plain. The crust shows little tendency toward uplift and may be considered isostatically balanced with respect to the underlying mantle (that is, it is "floating" at its proper level). The rocks exposed at the surface (or buried under a cover of sedimentary rock) are plutonic rocks and highly metamorphosed rocks—the same rocks that were very deeply buried during the accumulation and orogenic stages.

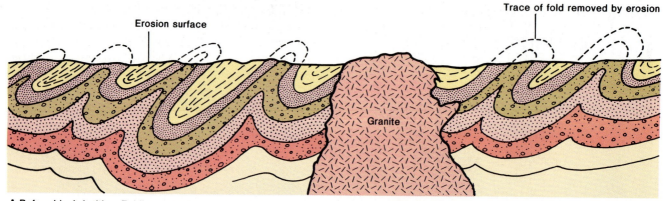

Erosion surface

Trace of fold removed by erosion

Granite

A Before block faulting. Folding and intrusion of a pluton during an orogeny has been followed by a period of erosion.

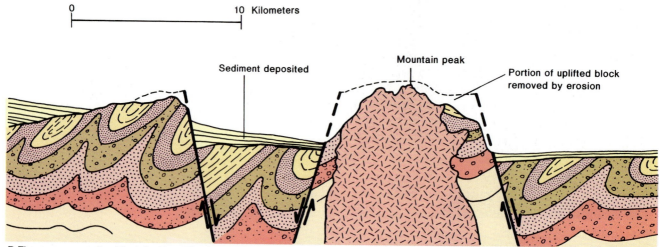

0   10   Kilometers

Sediment deposited

Mountain peak

Portion of uplifted block removed by erosion

B The same area after block-faulting. Tilted fault-block mountain range on left. Range to right is bounded by normal faults.

Figure 20.19
Development of fault-block mountain ranges.

Figure 20.20
The Teton Range, Wyoming, a fault-block range. The rocks exposed are Precambrian metamorphic and igneous rocks that were faulted upward. Extensive past glaciation is largely responsible for their rugged nature.
Photo by D. A. Rahm, courtesy Rahm Memorial Collection, Western Washington University.

The Growth of Continents

From what we have seen, continents grow as mountain belts evolving along continental margins. Accumulation and subduction add new continental crust beyond former coastlines. In the Paleozoic Era the Appalachians were added to the North American craton, and during the Mesozoic and Cenozoic eras the continent grew westward because of accumulation and orogenic processes in many parts of what is now the Cordillera. Therefore, if we radiometrically dated rocks that had been through an orogeny, starting in the Canadian Shield and working toward the east and west coasts, we should find the rocks to be progressively younger. In a very general way, this seems to be the case. However, there are some rather glaring exceptions.

Suspect and Exotic Terranes

In many parts of mountain belts are regions where the age and characteristics of the bed rock appear unrelated to that of adjacent regions. To help understand the geology of mountain belts, geologists have in recent years begun dividing major mountain belts into **tectonostratigraphic terranes** (or, more simply, **terranes**), regions within which there is geologic continuity. The geology in one terrane is markedly different from a neighboring terrane. Typically, a terrane covers thousands of square kilometers, but some terranes are considerably smaller. Alaska and western Canada have been subdivided into over fifty terranes (see figure 20.21). Terranes are named after major geographic features; for instance, Wrangellia, which is composed of

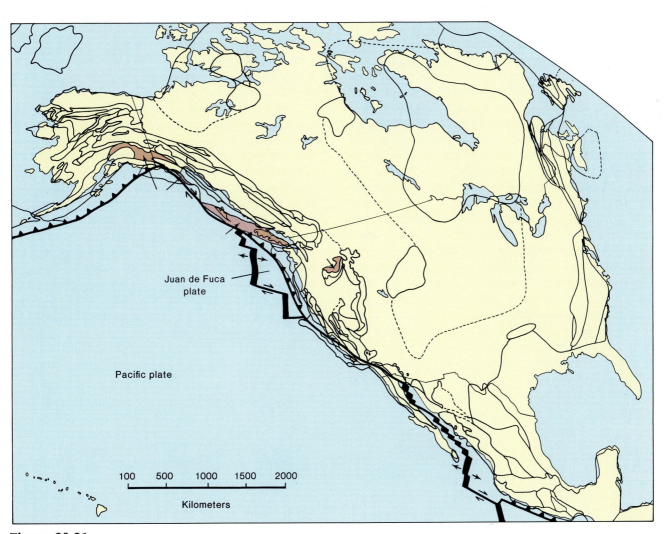

Figure 20.21
Terranes in North America. Lines are terrane boundaries.
Wrangellia, parts of which are in Alaska, western Canada and
Idaho, is shown in color.
After U.S. Geological Survey Open File Map 83–716.

parts of Alaska and Canada (with fragments of Washington and Idaho, according to some geologists) was named after the Wrangell Mountains of Alaska.

Many terranes appear to have formed essentially in place as a result of accretion and orogeny along the continent's margin. These are called **accreted terranes.** Other terranes have rock types and ages that do not seem related to the rest of the geology of the mountain belt and have been called **suspect terranes,** that is, terranes that may not have formed at their present site. If evidence indicates that a terrane did not form at or near its present site on a continent, the term **exotic terrane** is used.

A suspect terrane will have rock types and ages different from adjoining terranes, but to prove that it came from elsewhere in the world (and therefore is an exotic terrane), geologists will study the paleomagnetic poles (see chapter on plate tectonics) of the terrane's rocks. For an exotic terrane, the paleomagnetic poles for the rocks in the terrane will plot at totally different parts of the world from those of adjoining terranes that formed in place. This indicates that a particular terrane formed in a different part of the earth and drifted into the continent of which it is now a part. Some exotic terranes might have been *microcontinents* (such as present-day New Zealand) that moved considerable distances before crashing into other landmasses. Others may have been fragments of distant continents that split off and moved a long distance because of transform faulting. Imagine what might happen if the San Andreas fault remains active for another 100 million years or so. Not only would Los Angeles continue northward toward San Francisco and bypass it in about 10 million years (see Box 15.3), but the block of coastal California west of the fault would continue moving out to sea, becoming a large island with continental crust that drifts northward across the Pacific. Ultimately it would crash into and suture onto Alaska.

Figure 20.22 shows a tentative reconstruction of how parts of Alaska might have migrated in time. This is based on paleomagnetic data that indicate that parts of Alaska originally formed south of the equator and moved many thousands of kilometers to become part of the Cordillera.

Note from the diagram that the path of migration was not simple. Plates split, plates joined, and the direction of movement changed from time to time.

The Appalachians as well as mountain belts in other continents have also been divided into terranes. Even the Canadian Shield has been subdivided into terranes. Some geologists think they can determine, despite the great age and complexity of the shield's rocks, the extent to which some terranes traveled before crashing together (see Box 20.4).

What all of this suggests is that the earth is like a giant Rubic's cube (or rather, Rubic's sphere). Perhaps the new direction in which geology seems to be moving should be called Rubic's tectonics.

We should caution the reader that the terrane concept is controversial among geologists. While most would probably agree that some terranes are exotic, many geologists think the subdividing of Alaska and western Canada into fifty terranes is overdoing it and not supported by sufficient evidence. Only time and more painstaking gathering of evidence will allow geologists to determine the history of each alleged terrane.

Only a few years ago, many geologists thought that through the application of plate tectonic theory, we could easily determine the processes at work in each mountain belt and work back in time to understand the history of each of the earth's continents. Some suggested that there would hardly be major problems for earth scientists to solve in the future. The recognition of thin-skinned tectonics (see Box 20.3) and the recognition of diverse terranes indicate otherwise. Plate tectonics was a breakthrough, and a great many problems were solved; but with this great leap forward in the science have come new problems. New generations of geologists will have no shortage of challenges and no less excitement from solving newly discovered problems than did their predecessors who engineered the plate tectonics breakthrough.

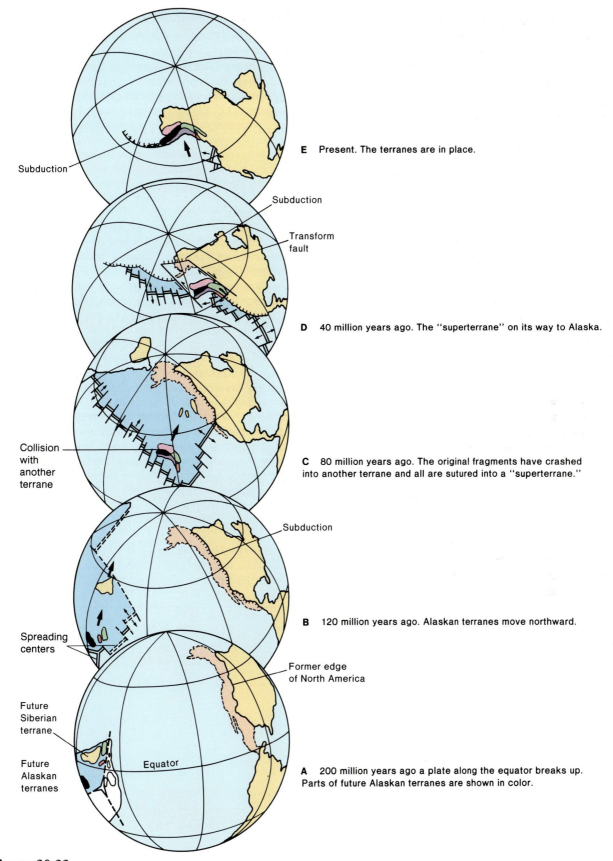

E Present. The terranes are in place.

Subduction

Subduction

Transform
fault

D 40 million years ago. The "superterrane" on its way to Alaska.

Collision
with
another
terrane

C 80 million years ago. The original fragments have crashed
into another terrane and all are sutured into a "superterrane."

Subduction

B 120 million years ago. Alaskan terranes move northward.

Spreading
centers

Former edge
of North America

Future
Siberian
terrane

Future
Alaskan
terranes

Equator

A 200 million years ago a plate along the equator breaks up.
Parts of future Alaskan terranes are shown in color.

Figure 20.22
**How fragments of the Southern Pacific Crust may have become
part of Alaska.**
Modified from D. B. Stone, B. C. Panuska, and D. R. Packer, 1982,
"Paleolatitudes versus time for southern Alaska," *Journal of Geophysical
Research,* vol. 87 (pp. 3697–3707), copyrighted by American Geophysical
Union.

BOX 20.4

Dance of the Continents

Collisions and breakup of continents have been going on for at least two billion years. Canadian geologist Paul Hoffman has described a cycle of formation and breakup of supercontinents, which he likens to a dance. The dancers are continental fragments and newly created crust that come together and later separate. Each dance cycle takes about a billion years and involves the building and dispersal of a supercontinent. Based on his studies of the Canadian Shield, Hoffman has described how the North American craton evolved through one "dance" cycle. The cycle is set to a symphony in four movements. It began two billion years ago (b.y.) and ended a billion years ago.

In the first movement (2.0–1.8 b.y.), seven microcontinents came together to form the beginnings of the North American (and part of what is now Europe's) craton. (Hoffman calls this the theory of the united plates of America.) New crust formed along some of the margins of the new continent through plate tectonic accretionary processes. Hoffman speculates that Precambrian shields now in other continents (e.g., Australia) were part of this supercontinent.

The second movement took place between 1.8 and 1.6 b.y. This was characterized by tectonic accretion of new crust in a belt extending from southern California to Sweden. It also featured felsic magmatism in the interior of the continent, away from converging boundaries. The magmatism is unusual in that it was anorogenic, not associated with subduction. The third movement (1.6–1.3 b.y.) featured more widespread, anorogenic magmatism in the continent's interior. In the fourth movement, we see orogenic activity once again becoming dominant and more new crust accreted to the continent. Basalt floods and rifting heralded the breakup of the continent. The symphony, or cycle, ended when the supercontinent fragmented— the "dancers" broke away.

The cycle would be repeated as the fragments came together during the Paleozoic to form the supercontinent Pangaea. Pangaea, of course, split apart into the present continents.

Hoffman also cites evidence that a similar cycle took place in the earlier Precambrian, prior to 2 b.y.

How is the cycle explained? The gathering of continental fragments is attributed to downwelling of cold mantle. The continental fragments are carried to the area of sinking mantle where they collect, much like leaves washed along a gutter collect over a drain. Once the supercontinent has formed, downwelling mantle (and subduction) moves to the edges of the continent. The mantle beneath the interior of the supercontinent now heats up, due to the insulation provided by the continental crust and fact that the downwelling cold mantle is now at the periphery of the large continent. Once the mantle beneath the interior of the continent heats sufficiently, hot spots develop resulting in anorogenic magmatism and eventual rifting and breakup of the supercontinent.

Further reading   Hoffman, P. F. Speculations on Laurentia's first gigayear (2.0 to 1.0 Ga). *Geology* 17 (1989): 135–38.

Summary

Major mountain belts are made up of a number of *mountain ranges*. Mountain belts are generally several thousand kilometers long but only a few hundred kilometers wide.

Mountain belts generally evolve as follows: (1) A thick sequence of sedimentary and volcanic rock accumulates (the *accumulation stage*). (2) The accumulation stage is either accompanied or followed by an *orogenic stage,* which involves intense compression of the layered rocks into folds and reverse (including thrust) faults, along with metamorphism and igneous activity. (3) The area is then subjected to a long period of uplift, perhaps block faulting, and erosion. Eventually the mountain belt is eroded down to a plain and incorporated into the *craton,* or stable interior of the continent.

According to the theory of plate tectonics, mountains on the edge of continents are formed by continent-oceanic convergence, and mountains in the interior of continents are formed by continent–continent collisions.

The uplift of a region following termination of an orogeny is generally attributed to isostatic adjustment of continental crust.

Continents grow larger by new mountain belts evolving along continental margins. They may also grow by the addition of terranes that may have traveled great distances before colliding with a continent.

Terms to Remember

accreted terrane	major mountain belt
accumulation stage	mountain range
craton	orogeny
exotic terrane	Precambrian shield
fault-block mountain	suspect terrane
range	terrane
geosyncline	(tectonostratigraphic
magmatic arc	terrane)

Questions for Review

1. Describe what takes place during an orogeny.
2. How are orogenies detected and dated?
3. What is the difference between the forces that could explain fault-block mountains and the forces that could account for an orogenic stage?
4. Explain how erosion and isostasy eventually produce stable, relatively thin, continental crust.
5. How do the sequences of sedimentary rocks in cratons differ from those in mountain belts?
6. What sequence of events accounts for a mountain belt that is bounded on either side by cratons?

Questions for Thought

1. What has the seismic study of the earth's interior contributed to our concept of how mountain belts form?
2. How do basalt and ultramafic rocks from the oceanic lithosphere become part of mountain belts?
3. Why is a craton locally warped into basins and domes?
4. Where in a mountain belt would you expect to find contact metamorphic rocks?
5. What causes an orogeny to end?
6. How could fossils in a terrane's rocks be used to indicate that it is an exotic terrane?

Supplementary Readings

Condie, K. C. 1989. *Plate tectonics and crustal evolution.* 3d ed. New York: Pergamon Press.

Dickinson, W. R. 1981. Plate tectonics and the continental margin of California. In *The geotectonic development of California,* ed. W. G. Ernst. Englewood Cliffs, N.J.: Prentice-Hall.

King, P. B. 1977. *The evolution of North America.* Princeton, N.J.: Princeton University Press.

McAlester, A. L., D. L. Eicher, and M. L. Rottman. 1984. *The history of the earth's crust.* Englewood Cliffs, N.J.: Prentice-Hall.

McPhee, J. A. 1981. *Basin and range.* New York: Farrar, Straus & Giroux.

McPhee, J. A. 1983. *In suspect terrain.* New York: Farrar, Straus & Giroux.

McPhee, J. A. 1986. *Rising from the plains.* New York: Farrar, Straus & Giroux.

Meissner, R. 1986. *The continental crust.* Orlando, Florida: Academic Press.

Miyashiro, A., K. Aki, and A. M. C. Sengor. 1982. *Orogeny.* New York: John Wiley & Sons.

Video:

The Appalachian story. VHS, Beta, 3/4 inch. Atlantic Geoscience Society, Atlantic Independent Media: Halifax, Nova Scotia.

Continental crust: Ancient and modern. VHS. The Media Guild: San Diego, California.

Throughout this book, we have mentioned human use of earth materials, most of which are nonrenewable. Our purpose in this chapter is to survey briefly some important geologic resources of economic value.

We first look at energy resources to see which ones might help replace our disappearing supplies of oil. Then we discuss metals and their relation to igneous rocks and plate tectonics and conclude with nonmetallic resources such as sand and gravel.

21

Geologic Resources

Bingham Canyon copper mine near Salt Lake City, Utah.
Photo courtesy of Kennecott Corporation/© Don Green

Nearly every man-made object depends on some geologic resources for its manufacture. An automobile contains substantial amounts of iron, chromium, manganese, nickel, platinum, tin, copper, lead, and aluminum in its body and engine and quartz sand in its window glass. It consumes petroleum in several forms—as fuel and lubricants, as synthetic rubber for tires, and as plastic for electrical parts, upholstery, and steering wheels. Dozens of other geologic resources go into automobiles, from tungsten in light bulb filaments to sulfur in battery acid.

People are beginning to realize how heavily dependent on geologic resources they are. Some have tried to limit their consumption of resources, or at least the rate at which their consumption *increases*. But it is impossible to stop consuming geologic resources. Think about some of the objects near you as you are reading this chapter. Your shirt may be partly polyester, which is made from petroleum, as are the plastic buttons. If you're wearing jeans, they may be made of fabric that is 100% cotton, but many jeans are made of shrinkproof fabrics that blend cotton with petroleum-based synthetic fibers. The brass zipper is made of copper and zinc. Some brands of jeans have pocket rivets made of copper. To make the leather tags on the back of some jeans, either aluminum or chromium was used during tanning. The fabric dye almost certainly came from petroleum. A pencil uses many geologic resources (figure 21.1) and a transistor radio many more. Eyeglasses are made of quartz sand and petroleum. Dental fillings are made of silver and other metals. Cassettes and videotapes are made of vinyl, a petroleum derivative, and iron or chromium particles from metal mines. Every one of us uses geologic resources, and therefore is directly responsible for the existence of mines and oil wells.

Types of Resources

Geologic resources are valuable materials of geologic origin that can be extracted from the earth. There are three main categories of geologic resources:

1. Energy resources—petroleum (oil and natural gas), coal, uranium, and a few others, such as geothermal resources.
2. Metals—iron, copper, aluminum, lead, zinc, gold, silver, and many more.
3. Nonmetallic resources—sand and gravel, limestone (for cement), sulfur, gems, gypsum, fertilizers, and many more. Ground water should also be regarded as an important geologic resource.

Geologic resources are sometimes called *mineral resources,* but the term, though widely used, is not accurate. Many geologic resources are not true minerals—for instance, most energy resources and many nonmetallic resources.

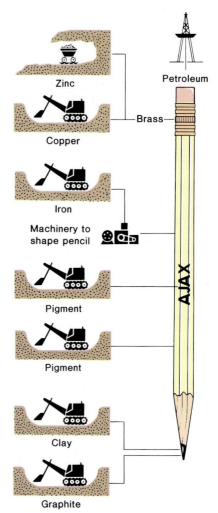

Figure 21.1
Mineral resources necessary to make a wooden pencil.

A very important point about geologic resources is that they are **nonrenewable resources,** except for ground water. They form so slowly that, at the present rapid rates of consumption, they can easily become exhausted. Some earth resources, such as food or timber, usually can be produced as fast as they are consumed. These are *renewable resources*. But petroleum, iron, lead, uranium, sulfur, sand and gravel, and essentially all other geologic resources are being used at rates far greater than the rates at which new deposits form. In some areas even ground water is being consumed faster than it is being replenished. This means that eventually these resources are going to run out or be priced so high because of their scarcity that their use will drop to insignificant levels. Recycling, new discoveries, and substitutes can help prolong the life of some resources, but it is likely that within your lifetime some of the resources in common use today will be scarce. This need not mean that civilization will come to a standstill, but it does mean that we must find inexpensive substitutes for these resources. Otherwise, skyrocketing costs will cause a drop in nearly everyone's material standard of living.

Energy Use

The United States consumes huge amounts of energy. By 1970 the United States was consuming eight times the amount of energy that it used in 1900. In 1973 and 1978 came steep price rises for petroleum. The need for conservation and fuel-efficient buildings and transportation became more and more apparent. From 1979 to 1985 total United States energy use declined and so did its dependence on petroleum (particularly imported oil). A precipitous drop in oil prices at the beginning of 1986, however, caused a sharp increase in both United States energy use and its reliance on imported oil.

The sources of United States energy have been changing in recent years in response to changing prices.

	1976	1980	1985	1988
Oil	47%	46%	42%	43%
Natural gas	27	26	24	23
Coal	19	20	24	24
Nuclear	3	4	5.5	7
Hydroelectric	4	4	4.5	3

Note that oil and natural gas account for almost two-thirds of the nation's energy supply; the fossil fuels (coal, oil, and gas) provide 90% of our energy.

Oil and Natural Gas

Within the petroleum industry, **petroleum** is a broad term that includes both crude oil and natural gas. That will be our usage. (In common usage *petroleum* is synonymous with crude oil; many geologists also use it this way.)

Crude oil is a liquid mixture of naturally occurring hydrocarbons (compounds containing hydrogen and carbon), which can be distilled to yield a great variety of products.

Natural gas is a *gaseous* mixture of naturally occurring hydrocarbons. Its origin and occurrence closely parallel that of oil. Many wells that recover oil also recover natural gas, although either may exist alone.

The Origin of Oil and Gas

Oil and natural gas seem to originate from organic matter in marine sediment. Microscopic organisms, such as diatoms and other single-celled algae, settle to the sea floor and accumulate in marine mud. The most likely environments for this are the continental shelves and the continental rise. The organic matter may partially decompose, using up the dissolved oxygen in the sediment. As soon as the oxygen is gone, decay stops and the remaining organic matter is preserved.

Continued sedimentation buries the organic matter and subjects it to higher temperatures and pressures, which cause physical and chemical changes in the organic compounds. In this way liquid and gaseous hydrocarbons form. As muddy sediments compact, the gas and small droplets of oil may be squeezed out of the mud and may move into more porous and permeable sandy layers nearby. Over long periods of time large accumulations of gas and oil can collect in the sandy layers. Both oil and gas are less dense than water, so they generally tend to rise upward through water-saturated rock and sediment.

The Occurrence of Oil and Gas

Economically important concentrations of petroleum are found underground where four specific conditions occur together: (1) a **source rock** (such as shale) containing organic matter that is converted to petroleum by burial and other post-depositional changes; (2) a **reservoir rock** (usually sandstone or limestone) that is sufficiently porous and permeable to store and transmit the petroleum; (3) a **trap,** a set of conditions to hold petroleum in a reservoir rock and prevent its escape by migration; and (4) deep enough burial (or *thermal maturity*) to "cook" the oil and gas out of the organic matter. All four conditions must be met; if one of the four is missing, the rock will hold no oil or gas.

Natural gas requires the same conditions as oil for accumulation. Gas can exist at greater depth than oil, however, and variations in source rock, depth of burial, and thermal history of the organic matter probably control whether gas, oil, or both accumulate in a region.

Figure 21.2 depicts several types of *structural traps* for oil and gas (some of these were also described in the chapter on "Geologic Structures"). *Anticlines* are the most common oil traps (figure 21.2*A*). Where oil and water occur together in folded sandstone beds, the oil droplets, being less dense than water, rise within the permeable sandstones toward the top of the fold. There the oil may be trapped by impermeable shale overlying the sandstone reservoir rocks. Since natural gas is less dense than oil, the gas collects in a pocket, under fairly high pressure, on the top of the oil.

Faults may create oil traps when permeable reservoir rocks break and slide next to impermeable rocks (figure 21.2*B*). Thrust faults are often associated with folds (figure 21.2*C*) because both are caused by compression. The backarc thrust belts (chapter on "Plate Tectonics") of both the Cordilleran and Appalachian mountain belts are currently being intensively explored for oil and gas.

A *stratigraphic trap* is a result of natural sedimentation rather than folding or faulting. It may be a lens of sandstone within a larger bed of shale (figure 21.2*D*). Another such trap is the narrow edge of a sandstone layer where it pinches out within a shale sequence (figure 21.2*E*). Oil can collect under *unconformities* if shale seals off a reservoir rock (figure 21.2*F*). Limestone *reefs* can form a variety of traps. The core of a reef is usually full of large openings formed by the irregular growth of coral and algae. Oil can collect both in a reef core and in the dipping beds of wave-broken debris on the reef flanks (figure 21.2*G*). *Salt domes* create a variety of traps described in Box 21.1.

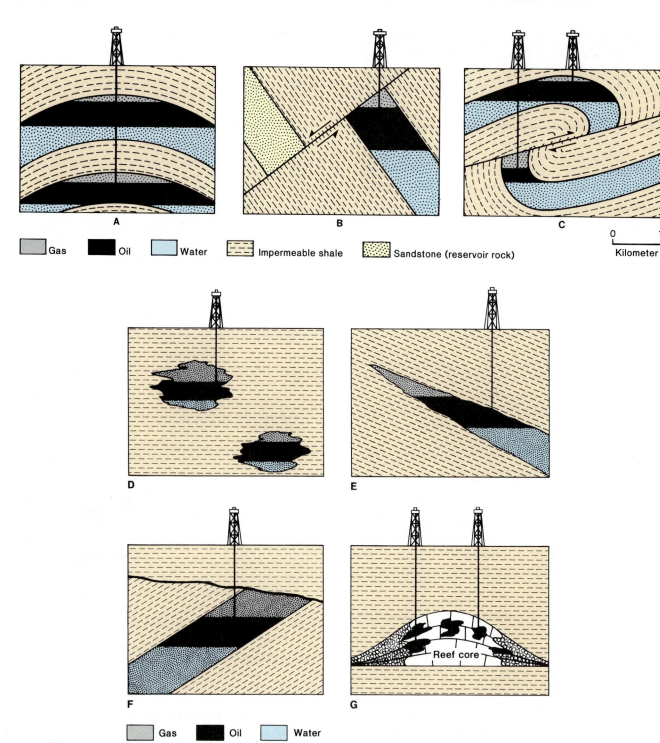

Figure 21.2
Structural traps for oil and gas: (*A*) Anticline. (*B*) Normal fault.
(*C*) Thrust fault. (Oil wells not drawn to scale.) Stratigraphic
traps for oil and gas: (*D*) Sandstone lenses in shale. (*E*) Narrow
edge or "pinch-out" of sandstone bed enclosed within dipping
shale. (*F*) Unconformity. (*G*) Small "patch" reef flanked by beds
of reef debris.

Legend (top): Gas · Oil · Water · Impermeable shale · Sandstone (reservoir rock)
Scale: 0 — 1 Kilometer
Legend (bottom): Gas · Oil · Water

BOX 21.1

Salt Domes

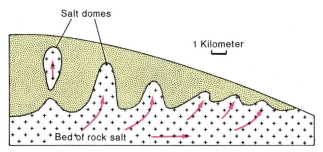

Box 21.1 Figure 1
Salt domes form as a bed of rock salt is loaded unevenly by a thick wedge of sediment. The salt flows toward the thin part of the sediment wedge and also flows upward as salt domes through the denser overlying sediment (and sedimentary rock).

Salt domes are vertical columns of rock salt that have risen upward through denser rock. They form from the thick layers of rock salt that mark some diverging continental margins. Subsidence of these continental margins carries the rock salt to great depth and allows it to be covered with a wedge of sedimentary rock (figures 19.25 and 19.29) perhaps 5 to 10 kilometers thick. The sedimentary rock on top of the salt layer is denser than the salt, and under the great pressure caused by the weight of the overlying rock the salt begins to deform plastically and flow upward. Large columns of salt rise from the top of the salt bed and then break off (box figure 1), perhaps assuming a streamlined teardrop shape one or more kilometers in diameter. As the salt continues to rise, it deforms the surrounding sedimentary rock, forming folds and faults that create oil traps, some of which may be under an overhanging mushroom-shaped top that is characteristic of some salt domes (box figure 2).

When the salt rises into the zone of active ground-water circulation near the surface, it dissolves. Therefore, many salt domes never reach the surface. If the rock salt contains gypsum, it may develop a *cap rock* on its upper surface as the halite dissolves but the less soluble gypsum does not. Gypsum, a calcium sulfate, contains sulfur that may form native elemental sulfur when it contacts bacteria and petroleum. Therefore, salt dome cap rocks often contain thick beds of bright yellow sulfur.

Salt domes, then, supply a variety of valuable resources. Oil and gas are found around and above the domes. The salt itself is mined underground. Cap-rock sulfur can be mined by drilling into it and dissolving it with hot water. In the United States there is a broad belt of salt domes along the Texas and Louisiana coasts. The belt extends offshore, onto the continental shelf and beyond, forming a major source of oil, gas, salt, and sulfur.

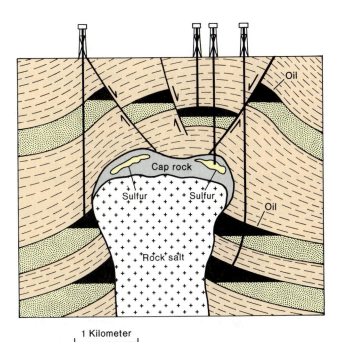

Box 21.1 Figure 2
A salt dome. Oil and gas are trapped in folds and along faults above the dome and within upturned sandstones along the flanks of the dome. Insoluble cap rock may contain recoverable sulfur.

How do we find salt domes that do not make it to the earth's surface? Some salt domes deform the surface. Some of the low, round islands in the coastal swamps of Louisiana were pushed up by salt domes. If solution of the salt by ground water is very rapid, the

---

The combination of circumstances that creates **oil pools,** or underground accumulations of oil, is not common. **Oil fields** are regions underlain by one or more oil pools. Figure 21.3 shows the location of most of the major North American oil fields. The two largest oil fields within the United States are in eastern Texas and on Alaska's North Slope. Most of the world's remaining oil lies in giant fields in the Middle East (especially in Saudi Arabia), the Soviet Union, Venezuela, and Mexico.

Recovering the Oil

When an oil pool or field has been discovered, wells are drilled into the ground. Permanent derricks used to be built to handle the long sections of drilling pipe. Now portable drilling rigs are set up to drill and are then dismantled and moved. When the well reaches a pool, oil usually rises up the well because of its density difference with water or because of the pressure of the expanding gas cap above

Box 21.1   *continued*

overlying rock may subside, so some salt domes are marked by surface depressions. Most salt domes, however, are found by geophysical surveys.

The density of rock salt is 2.2 grams per cubic centimeter, a value that does not change even if the rock salt is compressed. Sedimentary rocks within the Gulf Coast belt of domes have densities of 2.0 gm/cm³ at the surface, increasing to values of 2.5 or greater at depth due to compaction. The salt dome rises because it is out of isostatic equilibrium with the heavier rock around it.

The contrast in density between the salt and the surrounding rocks produces a negative gravity anomaly over the dome (box figure 3). Gravity surveys are fast and fairly inexpensive, so this is how most salt domes are found. Seismic reflection can indicate the curved surface of the dome itself (box figure 3) or of the deformed rock layers above the dome. Seismic refraction can also locate a dome, for seismic waves speed up in salt. The arrival of waves at a recording station before they are expected indicates that a high-speed material, such as salt, exists below the surface.

Drilling, of course, is the final proof that a salt dome is there, and drilling is needed to extract the resources. Because drilling is so expensive, however, several types of geophysical surveys are usually run

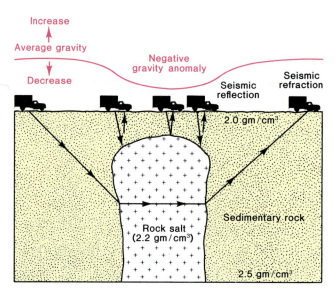

Box 21.1 Figure 3
A salt dome can be detected by a gravity survey, by seismic reflection, and by seismic refraction.

first. If the surveys indicate that a dome is probably there, then drilling is usually a good gamble (though many domes have no oil, gas, or sulfur).

the oil. Although this rise of oil is almost always carefully controlled today, spouts of oil, or *gushers,* were common in the past. Gas pressure gradually dies out, and oil is then pumped from the wells. Water or steam may be pumped down adjacent wells to help push the oil out. At a refinery the crude oil from underground is separated into natural gas, gasoline, kerosene, lubricating oil, fuel oil, grease, asphalt, and paraffin. *Petrochemicals* are manufactured from the components of the petroleum compounds. Petrochemicals include dyes, fertilizers, medicines, synthetic rubber, explosives, perfumes, paints, saccharin, solvents, synthetic fibers, and plastics used for such varied products as swimming pool liners, tube tents, cassettes, floor tile, and garbage bags.

As oil becomes increasingly difficult to find, the search for it is extended into more hostile environments. The development of the oil field on the North Slope of Alaska near the shore of the Arctic Ocean (figure 21.4) and the construction of the Alaska pipeline are examples of the great expense and difficulty involved in new oil discoveries. Offshore platforms extend the search for oil to the ocean's continental shelves and even beyond. More than one-quarter of the world's oil and almost one-fifth of the world's natural gas comes from offshore, even though offshore drilling is six to seven times as expensive as drilling on land.

Figure 21.3
Major oil fields in North America. The amount of oil in a field is not necessarily related to its areal extent on a map. It is also governed by the vertical "thickness" of the oil pools in a field. The fields with the most oil are in Alaska and east Texas.
From U.S. Geological Survey and other sources.

Figure 21.4
Drilling rig on Alaska's North Slope.
Photo by BP America, Inc.

Environmental effects Getting petroleum out of the ground and to the consumer can create environmental problems anywhere along the line. Pipelines carrying oil can be broken by faulting, mass-wasting processes, or acts of war, causing an *oil spill* (table 21.1). Tanker spillage from collisions or groundings (such as the *Argo Merchant* off Massachusetts in the winter of 1976–77, and the *Exxon Valdez* off Alaska in 1989) can create oil slicks at sea. Offshore platforms may also lose oil (such as the Santa Barbara blowout off southern California in 1969). Oil slicks can drift ashore, fouling the beaches. *Subsidence* of the ground can occur as oil is removed. The Wilmington field near Long Beach, California, has subsided 30 feet (9 meters) in fifty years; dikes have had to be built to prevent sea water from flooding the area. The refining and burning of petroleum and its products can cause *air pollution*. Advancing technology and strict laws are helping control some of these adverse environmental effects.

How Much Oil Do We Have Left?

Resources and reserves Two different terms are used to describe the amounts of geologic resources that have not yet been extracted from the earth. These terms apply not only to petroleum but to any geologic material. **Resources** is a broad term used for the total amount of a geologic material in all deposits, discovered and undiscovered. It includes both the deposits that can be economically extracted under present conditions and those that may be extracted economically in the future (figure 21.5). It is very difficult to estimate resources because educated guesses must be made about the existence and sizes of deposits yet undiscovered as well as about what type of deposit might someday be economical to extract. **Reserves** are a small part of resources. They are the *discovered* deposits that can be extracted economically and legally under present conditions. That is, they are the short-term supply of a geologic material.

TABLE 21.1

Notable Oil Spills

Date (Month/Year)	Source (Ship names in italic)	Barrels[1]	Oil Type	Location
3/67	*Torrey Canyon*	860,000	Crude	Southwest England
1/69	Well blowout	33,000	Crude	Santa Barbara, Calif.
9/69	*Florida*	4,500	Fuel oil	West Falmouth, Mass.
1/71	*Oregon Standard*	20,000	Fuel oil	San Francisco, Calif.
1/77	*Argo Merchant*	183,000	Crude	Nantucket Shoals, Mass.
4/77	Well blowout	210,000	Crude	North Sea
3/78	*Amoco Cadiz*	1,600,000	Crude	Normandy, France
6/79 to 3/80	Well blowout	2,700,000 (est.)	Crude	Bay of Campeche, Mexico
8/79	*Atlantic Express* and *Aegean Captain*	3,000,000	Crude	Off Barbados
1/87	*Stuyvesant*	14,000	Crude	Alaska
10/87	*Stuyvesant*	14,000	Crude	Alaska
3/89	*Exxon Valdez*	240,000	Crude	Alaska

[1]A *barrel* is 42 gallons.

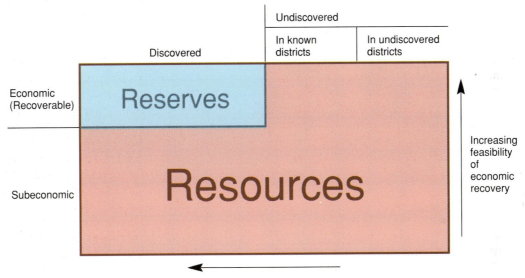

Figure 21.5
The difference between reserves and resources. Reserves (discovered deposits that are economically and legally extractable) form a small part of total resources (discovered and undiscovered deposits that are economical to extract now or may be economical in the future).

Once *resources* have been carefully estimated, the amount should not change from year to year, for an estimate of resources is basically an estimate of the total future supply. Estimates of *reserves,* however, change all the time. The extraction and use of a substance lowers reserves. New discoveries add to reserves, for part of the definition of reserves is that we have to know the deposit is there.

A deposit also has to be profitable to extract, and many factors determine whether a profit can be made. The costs of extraction, including workers' salaries and the fuel used to run equipment, are important. New inventions that make extraction cheaper can increase reserves. The final price a company can receive for its product is also important. When prices are kept artificially low by the government, as was the case recently for natural gas and oil,

reserves are naturally low because few deposits can be extracted profitably at that price. When prices rise, more deposits become economical, and reserves increase, even without new discoveries. Reducing the amount of taxes that extraction companies pay can also increase reserves.

Changes in laws can affect reserves. Large areas of government-owned land are off-limits to mining and drilling, so any geologic materials under these areas are not legally extractable and cannot be included in reserves. Opening more land to extraction can therefore increase reserves.

The world situation At this time the world's oil *reserves* are estimated to be 890 billion barrels (a barrel contains 42 gallons). Total *resources,* of course, are greater than reserves. One estimate is 3,000 billion barrels (including reserves).

Since the world's present consumption of oil is about 21 billion barrels per year, known reserves will yield a 42-year supply. New discoveries and recovery techniques, as well as price rises, will add to reserves in the future. How much they will add is difficult to say.

World *resources* should last well over 100 years. The world is not running out of oil.

Some petroleum cannot be recovered. The oil may be in a pool too small to warrant the expense of drilling. A pool may be too far from market or lie under regions in which drilling is forbidden, such as national parks or other public lands. During extraction, only about 30% to 40% of the known oil in a given pool can be brought to the surface. The rest remains underground because it is far too difficult and expensive to extract with present techniques and prices.

The outlook in the United States United States consumption of oil is 6.3 billion barrels per year (with 6% of the world population, the United States uses 30% of the world's oil production each year). Sixty-two percent of U.S. consumption is used for transportation. United States *reserves* are currently about 25 billion barrels. The United States imports about 42% of the oil it uses; domestic production is 3.6 billion barrels per year. At current production rates, therefore, we have about a 7-year supply. If the production rate goes up, the lifetime of the reserves goes down.

Since the reserve estimate changes each year, an estimate of *resources* is more important for planning purposes. Such estimates have fluctuated wildly in the past twenty-five years. Recent government and industry estimates of total resources are about 90 billion barrels (including reserves). This means that at our current production rate we have a 25-year supply of oil left, counting both reserves and resources. The United States is truly running out of oil and will need alternate fuels or more imported oil in the future.

The situation becomes more serious if the temporary ban on most new offshore drilling, enacted after the *Exxon Valdez* spill, becomes permanent. About 30 billion barrels of the 90-billion total are in undiscovered offshore oil. If this oil is placed off-limits, our oil resources will run out in 16 years at the current production rate.

Estimates of *natural gas* have been similarly gloomy, largely because estimators assumed that gas discoveries are linked to oil discoveries. In the past, gas has been found in a fixed ratio to oil, so some estimators assume that because we are running out of oil, we are also running out of gas.

The United States produces about 17 trillion cubic feet (TCF) of natural gas per year. Reserve estimates are 190 TCF, with total resources perhaps 600 TCF. By these estimates, natural gas will be gone in the United States in 35 years.

The recent decontrol of natural gas prices, however, has led to renewed interest in gas and revised estimates of resources. The new estimates are based on unconventional gas sources, not linked to oil. There may be 300 to 500 TCF of gas within coal beds, for example (the gas is a hazard to coal miners because it is toxic and tends to explode). Tightly compacted sandstones ("tight sands") in the western United States may hold 200 to 800 TCF. Devonian shales rich in organic matter in the eastern United States may hold 500 to 600 TCF. Gas can exist at greater depth than oil, so it is likely that deep drilling (beyond 5 kilometers) in sedimentary basins can produce substantial amounts of gas. Gas is also dissolved in hot salty water within highly fractured rock along the Gulf Coast ("geo-pressured zones") that may contain as much as 3,000 TCF of gas. Truly staggering amounts of gas may be tied up as gas hydrates, a peculiar solid that forms from natural gas and water in two unusual environments: (1) on the deep sea floor and (2) under permafrost. Soviet estimates of up to 35,000,000 TCF of gas in hydrate form (worldwide) have received some support in this country. All these estimates are for total gas in place, not for recoverable reserves, which would be far lower. Some geologists discount all these estimates as wildly optimistic and misleading. All these unconventional gas sources are getting serious attention today, however, and although some difficult extraction problems remain unsolved, there is at least cautious optimism that the United States may not be as short of gas as we used to think.

Heavy Crude and Oil Sands

Heavy crude is dense, viscous petroleum. It may flow into a well, but its rate of flow is too slow to be economical. As a result, heavy crude is left out of reserve and resource estimates of less viscous "light oil," or regular oil. Heavy crude can be made to flow faster by injecting steam or solvents down wells, and if it can be recovered, it can be refined into gasoline and many other products just as light oil is.

Oil sands (or *tar sands*) are asphalt-cemented sand or sandstone deposits. The asphalt is solid, so oil sands are often mined rather than drilled into, although the techniques for reducing the viscosity of heavy crude often work on oil sands as well.

The origin of heavy crude and oil sands is uncertain. They may form from regular oil if the lighter components are lost by evaporation or other processes. Oil sands and asphalt seeps at the earth's surface (such as the Rancho La Brea Tar Pits in Los Angeles) probably formed from evaporating oil. But some heavy crudes and oil sands are found as far as 4,000 meters underground. Most of them have much higher concentrations of sulfur and metals, such as nickel and vanadium, than does regular oil. These facts suggest that heavy crude and oil sands may have a somewhat different origin than light oil.

The best-known oil sand deposit is the Athabasca Oil Sand in northern Alberta, Canada (figure 21.6). The deposit, currently being strip-mined, contains 900 billion barrels of oil, of which at least 80 billion barrels are recoverable. The United States has more than 100 billion barrels of heavy crude and oil in oil sands, including about 30 billion in the form of Tertiary oil sands in Utah. Half of it may be ultimately recoverable. This means that heavy crude and oil sands may supply as much oil in the future as our light oil reserves, and the United States may be able to import both from Canada and Venezuela (which has more oil sand than Canada).

Oil Shale

Oil shale is a black or brown shale with a high content of solid organic matter from which oil may be extracted by distillation. The best-known oil shale in the United States is the Green River Formation, which covers more than 15,000 square miles (40,000 square kilometers) in Colorado, Wyoming, and Utah, with deposits up to 2,000 feet (650 meters) thick (figures 21.6 and 21.7). The oil shale, which includes numerous fossils of fish skeletons, formed from mud deposited on the bottom of large, shallow Eocene lakes. The organic matter came from algae and other organisms that lived in the lakes.

The Green River Formation includes more than 400 billion barrels of oil in rich beds that yield over 25 gallons of oil per ton of rock. Another 1,400 billion barrels of oil occur in lower-grade beds yielding 10 to 25 gallons per ton. An estimated 300 to 600 billion barrels may be recoverable.

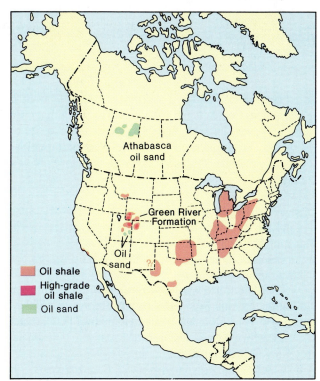

Figure 21.6
Distribution of major deposits of oil sand and oil shale in the United States and Canada.
From U.S. Geological Survey and other sources.

Figure 21.7
Cliffs of oil shale near Green River, Wyoming.
Photo by Frank M. Hanna.

Relatively low-grade oil shales in Montana contain another 180 billion barrels of recoverable oil in shale that should be economical to mine because of its high content of vanadium, nickel, and zinc. Therefore, oil shale can supply potentially vast amounts of oil in the future as our liquid petroleum runs out.

A few distillation plants extract shale oil, but the current low price for oil makes shale oil uneconomical. A price increase for oil may make large-scale production of shale oil feasible in the future.

The mining of oil shale has some environmental effects. There is a space problem, for during distillation the shale expands. Spent shale could be piled in valleys and compacted, but land reclamation would be troublesome.

Varieties (Ranks) of Coal

	Color	Water Content (%)	Other Volatiles (%)	Fixed Carbon[2] (%)	Approximate Heat Value (BTUs of heat per pound of dry coal)
Peat[1]	Brown	75	10	15	Varies
Lignite	Brown to brownish-black	45	25	30	7,000
Subbituminous coal	Black	25	35	40	10,000
Bituminous coal (soft coal)	Black	5 to 15	20 to 30	50 to 75	12,000 to 15,000
Anthracite (hard coal)	Black	5	5	90	14,000

1. Peat is not a coal.
2. "Fixed carbon" means solid combustible material left after water, volatiles, and ash (noncombustible solids) are removed.

A great amount of water is required, both for distillation and for reclamation, and water supply is always a problem in the arid West. New processing techniques that extract the oil in place without bringing the shale to the surface may eventually help solve some of the problems and lower the water requirements. It is possible to burn fractured oil shale in large underground excavations. The heat separates most of the oil from the rock, and it can be collected in liquid form. Another proposal involves heating the shale with radio waves or microwaves to separate the liquid oil from the rock.

Coal

Along with oil and natural gas, coal is our third major energy resource. It provided 90% of the energy used in the United States in 1900 but provides only 24% today. Coal use is now increasing as petroleum becomes scarcer and more expensive; it now provides more of our energy than natural gas.

More than 85% of the present use of coal in the United States is for generating electricity. Coal is also used to make *coke,* which is used in steel making. In the future, coal may be used instead of petroleum in the manufacture of some chemicals. *Coal gas* and *coal oil* made from coal also may replace petroleum in some other uses. These products are reasonable approximations of natural gas and oil and can be used for some of the same purposes, although they are still very expensive to produce. Coal can also be powdered and mixed with water to form a *slurry,* which can be transported through pipelines and burned as a liquid fuel.

Origin of Coal

Coal is a sedimentary rock that forms from the compaction of plant material that has not completely decayed. Rapid plant growth and deposition in water with a low oxygen content are needed, so shallow swamps or bogs in a temperate or tropical climate are likely environments of deposition. The plant fossils in coal beds include leaves, stems, tree trunks, and stumps with roots often extending into the underlying shales, so apparently most coal formed right at the place where the plants grew.

Partial decay of the abundant plant material uses up any oxygen in the swamp water, so the decay stops and the remaining organic matter is preserved. Burial by sediment compresses the plant material, gradually driving out any water or other volatile compounds. The coal changes from brown to black as the amount of carbon in it increases.

Table 21.2 shows the common varieties (ranks) of coal. *Peat,* a mat of unconsolidated plant material, is not coal but probably represents the initial stage of coal development. When dry, it can be burned as a fuel (peat fires used to dry malted barley give Scotch whisky its smoky flavor). With compaction, peat can become *lignite (brown coal),* which may still contain visible pieces of wood. Lignite is soft and often crumbles as it dries in air. It is subject to spontaneous combustion as it oxidizes in air, and this somewhat limits its use as a fuel. *Subbituminous coal* and *bituminous coal (soft coal)* are black and often banded with layers of different plant material. They are dusty to handle, ignite readily, and burn with a smoky flame. *Anthracite (hard coal)* is actually a metamorphic rock, generally formed only under the regional compression associated with folding. It is hard to ignite but is dust-free and smokeless.

Occurrence of Coal

Coal, a sedimentary rock, occurs in beds (figure 21.8) that range in thickness from less than 1 foot up to 100 feet (0.3 to 30 meters). If the beds are deeply buried, underground mines are dug to extract the coal (figure 21.9). If the coal beds are close to the land surface, the coal is mined in a **strip mine,** in which the overburden is removed to expose coal at the surface (figure 21.10). When a strip of coal has been uncovered and removed, the resulting trench is filled in with the overburden from the adjoining strip.

A

B

Figure 21.8
(*A*) A layer of peat being cut and dried for fuel, island of Mull, Scotland. Coal often forms from peat. (*B*) A bed of bituminous coal near Trinidad, Colorado.

Figure 21.9
Underground coal mine near Axial, Colorado.
Photo by E. F. Patterson, U.S. Geological Survey.

Figure 21.10
Coal strip mine, Pennsylvania. A dragline is removing a strip of overburden, which is then piled to the right. Smaller equipment removes the coal exposed by the dragline.
Photo by Frank M. Hanna.

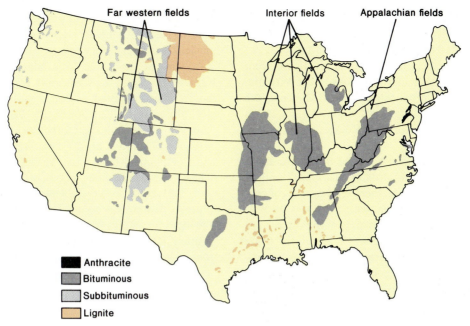

Figure 21.11
Coal fields of the United States. Alaska also has coal.
From U.S. Geological Survey.

Figure 21.11 shows the coal fields of the "lower 48" states. We will discuss three major regions, the Appalachian fields, the interior fields, and the far western fields.

The major coal-producing fields in the United States are the *Appalachian fields,* which stretch from Pennsylvania to Alabama and contain extensive beds of bituminous coal. The coals are mostly of Pennsylvanian age, with Mississippian and Permian rocks containing some coal as well. The coal beds, which thin westward, were included in the late Paleozoic folding and faulting of the Appalachian orogeny, so they are strongly deformed in the eastern part of the belt. Steeply dipping coal beds here are mostly mined underground. The folds are gentler to the west, where the coals can be extracted by both underground and strip mining. Northeastern Pennsylvania has some anthracite that resulted from intense folding.

There are twenty-five to fifty coal beds over most of this region, each bed generally 6 feet (2 meters) or less in thickness, although some are locally thicker. The coal lies within repeated sequences of sandstone, shale, and limestone, which indicate alternating continental and marine conditions. This implies a low-lying environment near the sea, such as lagoons, large deltas, and swampy coastal plains similar to those that exist in present-day Florida, Georgia, and South Carolina.

The *interior fields* extend from Michigan through Illinois to Texas and are extensions of the Appalachian rocks westward onto the continental interior. In Michigan and Illinois the rocks are preserved in large basins; the fields from Iowa to Texas are mostly horizontal. The coals are usually strip-mined, particularly near major industrial centers around the basin's edge in Illinois, Indiana, and Kentucky.

The *far western fields* extend from New Mexico northward through the Rocky Mountains to Montana and the Great Plains of North Dakota (and also into Canada). These coal beds are thicker and younger than eastern coals. They are generally of Cretaceous or Eocene age and are up to 100 feet (30 meters) thick. The coals occur in large basins and are generally of low rank, being either lignite or subbituminous coal, although some good-quality bituminous coals occur in Colorado and Utah. Many thick beds are very close to the surface and are strip-mined, although underground mining is common in some states. Western coals are attractive fuels, presently in very high demand, because they typically have less sulfur than eastern coals (sulfur compounds can pollute the air when coal burns).

Environmental Effects

Extracting and using coal creates environmental problems. The presence of a mine usually lowers the local water table as ground water is pumped out of the mine. The drainage out of the mines tends to be highly acid, polluting surface streams and water supplies. In the past, strip mines have been refilled as barren, unsightly piles, but new techniques of re-contouring the land and restoring removed overburden and topsoil help reclaim mined-out areas for other uses. When coal is burned, ash and sulfur gases can pollute air, but most of the harmful components can be removed with existing technology. Solving environmental problems associated with coal, of course, raises the cost of extraction and the price to the consumer.

Reserves and Resources

Coal production in the United States is about 1 billion tons per year (the United States uses 0.9 billion tons and exports the rest). More than 60% of the production is from surface strip mines. Although Kentucky, West Virginia, and Pennsylvania still produce almost half the coal, the rapid development of the huge beds of low-sulfur western coal may soon cause the Northern Rockies states to surpass the Appalachian states.

Recoverable *reserves* in the United States are about 300 billion tons (about two-thirds of this is in Montana and Wyoming). As you can see, there are centuries of coal left at the present rate of production.

Coal *resources* within 1,000 meters of the surface are 1,700 billion tons. A greater amount is estimated to lie deeper, so the total United States resources are an impressive 3,900 billion tons (much of which, of course, is not presently usable). Even as we step up extraction for the export market, it seems obvious that our coal supplies will last a very long time. The United States has been called the Saudi Arabia of coal.

Uranium

The metal *uranium,* which powers nuclear reactors, occurs as *pitchblende,* a black uranium oxide found in hydrothermal veins, or, much more commonly in the United States, as yellow *carnotite,* a complex hydrated oxide found as incrustations in sedimentary rocks. Ground water easily transports oxidized uranium, which is highly soluble. Organic matter reduces uranium, making it relatively insoluble, so uranium precipitates in association with organic matter.

Most of the easily recoverable uranium in the United States is found in sandstone in New Mexico and Wyoming, some of it in and near petrified wood. In the 1950s uranium boom, western prospectors looked for petrified logs and checked them with Geiger counters. Some individual logs contained tens of thousands of dollars worth of uranium. Most of the uranium is in sandstone channels that contain plant fragments.

Organic phosphorite deposits of marine origin in Idaho and Florida also contain uranium. The uranium is not very concentrated, but the deposits are so large that overall they contain a substantial amount of uranium. The black Devonian shales of the eastern United States also contain uranium. These shales are really low-grade oil shales (figure 21.6), and they contain large amounts of natural gas, as you have seen. Uranium may be recovered from phosphorites or shales as a by-product of another mining operation.

Uranium is used in nuclear reactors to produce electricity and in nuclear weapons and some naval craft. At present, nuclear reactors produce 7% of the energy needs of the United States (the United States produces one-third of the world's nuclear power). This figure may or may not rise appreciably in the future, depending on public acceptance of nuclear power. Nuclear plants produce long-lived waste products that remain dangerous for centuries, and

no suitable storage site yet exists, although Yucca Mountain in Nevada is being intensively studied as a possible nuclear waste repository (chapter titled "Ground Water"). Accidents at the Three Mile Island reactor in Pennsylvania in 1979 and the Chernobyl reactor near Kiev in the Soviet Union in 1986 caused a major rethinking of the desirability of nuclear power. In 1988 the United States had 108 operational reactors and 16 others under construction. This total of 124 compares with more than 230 built or planned before the Three Mile Island accident. There have been no new plants ordered since 1978.

Recoverable reserves of nearly 500,000 tons of uranium oxide in western sandstones seem adequate to power the existing reactors in the United States well into the next century. Phosphorites and black shales contain millions of tons of low-grade uranium-oxide deposits as resources.

Metals and Ores

The search for metals depends on finding rocks that contain metal-bearing minerals from which the metals can be extracted without too much difficulty and expense. A mining geologist looks for **ores,** which are naturally occurring materials that can be profitably mined. Whether or not a mineral (or rock) is considered an ore depends on its chemical composition, the percentage of extractable metal, and the market value of the metal. The mineral hematite (Fe_2O_3), for example, is a good *iron ore* because it contains 70% iron by weight. Limonite ($Fe_2O_3 \cdot nH_2O$) contains less iron than hematite and hence is not as extensively mined. Even a mineral containing a high percentage of metal is not described as an *ore* if the metal is too difficult to extract or the site is too far from a market; profit is part of what defines an ore. As the prices of metals and the energy used to extract them fluctuate, so do the potential profits from minerals. Some of the common ore minerals are listed in table 21.3.

Origin of Metallic Ore Deposits

Table 21.4 summarizes most of the ways ore deposits form. Several of the processes have been discussed in earlier chapters.

Ores Associated with Igneous Rocks

Crystal settling occurs as early-forming minerals crystallize and settle to the bottom of a cooling body of magma (figure 21.12). This process was described under differentiation in the chapter on igneous rocks. The metal chromium comes from chromite ore bodies near the base of sills and other intrusions. In South Africa a huge sill 8 kilometers thick and 500 kilometers long, called the Bushveldt Complex, contains up to twenty-five layers of chromite near its base, some of them up to 2 meters thick. Above the chromite are similar layers enriched in platinum. Most of the world's chromium and platinum come from this single intrusion. In Montana, another huge Precambrian sill called the Stillwater Complex contains similar but lower grade deposits of these two metals.

Alternative Sources of Energy

Several other sources may contribute enough energy in the future to help reduce the expected demand for oil, natural gas, coal, and uranium. *Hydroelectric power* contributes about 3% of the energy needs of the United States. Electricity is generated by turbines turned by water falling from dammed reservoirs. Hydroelectric power will probably not increase because most suitable rivers in the United States have already been dammed. Public pressure is growing to preserve most of the remaining undammed rivers in their wild state, despite the danger of floods.

Geothermal power (chapter on ground water) may contribute substantially to our power needs, particularly if successful techniques are developed for tapping the heat of areas not marked by surface hot springs. Currently, geothermal power contributes only 0.3% of our energy use. *Solar power* and *wind power* may contribute to our needs in the future, particularly if improved methods of storing the energy are devised. At the present, a great effort is being made to improve the technology for collecting solar energy. More exotic forms of generating energy include harnessing *tidal power, wave power, ocean current power,* and the energy represented by *vertical temperature differences in the sea. Nuclear fusion* reactors may be in operation before the end of the century. The burning of *hydrogen from the dissociation of water* also may be developed. However, major technological problems are slowing the adoption of these last two methods.

One great attraction of many of these sources of energy is that they are *renewable.* We do not use up sunlight, wind, or tides by harnessing their power. Widespread use of renewable resources in the next century may lessen the world's demand for fossil fuels.

Common Ore Minerals

Metal	Ore Mineral	Composition
Aluminum	Bauxite (a mineral mixture)	$Al_2O_3 \cdot nH_2O$
Chromium	Chromite	$FeCr_2O_4$
Copper	Native copper	Cu
	Chalcocite	Cu_2S
	Chalcopyrite	$CuFeS_2$
Gold	Native gold	Au
Iron	Hematite	Fe_2O_3
	Magnetite	Fe_3O_4
Lead	Galena	PbS
Manganese	Pyrolusite	MnO_2
Mercury	Cinnabar	HgS
Nickel	Pentlandite	$(Fe, Ni)S$
Silver	Native silver	Ag
	Argentite	Ag_2S
Tin	Cassiterite	SnO_2
Uranium	Pitchblende	U_3O_8
	Carnotite	$K(UO_2)_2(VO_4)_2 \cdot 3H_2O$
Zinc	Sphalerite	ZnS

Some Ways Ore Deposits Form

Type of Ore Deposit	Some Metals Found in This Type of Ore Deposit
Crystal settling within cooling magma	Chromium, platinum, iron
Hydrothermal deposits (contact metamorphism, hydrothermal veins, disseminated deposits, hot-spring deposits)	Copper, lead, zinc, gold, silver, iron, molybdenum, tungsten, tin, mercury, cobalt
Pegmatites	Lithium, rare metals
Chemical precipitation in layers	Iron, manganese, copper
Placer deposits	Gold, tin, platinum, titanium
Concentration by weathering and ground water	Aluminum, nickel, copper, silver, uranium, iron, manganese, lead, tin, mercury

Hydrothermal fluids are the most important source of metallic ore deposits. The hot water and other fluids might be part of the magma itself, injected into the surrounding country rock during the last stages of magma crystallization (figure 21.13*A*). Atoms of metals such as copper and gold, which do not fit into the growing crystals of feldspar and other minerals in the cooling pluton, would be concentrated residually in the remaining water-rich magma. Eventually a hot solution, rich in metals and silica

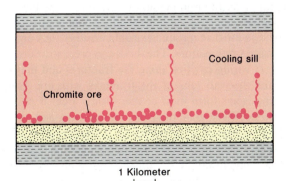

Figure 21.12
Early-forming minerals such as chromite may settle through magma to collect in layers near the bottom of a cooling sill.

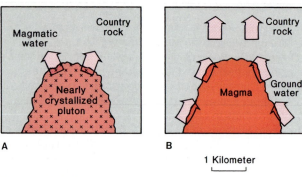

Figure 21.13
Two possible origins of hydrothermal fluids. (**A**) Residually concentrated magmatic water moves into country rock when magma is nearly all crystallized. (**B**) Ground water becomes heated by magma (or by a cooling solid pluton), and a convective circulation is set up.

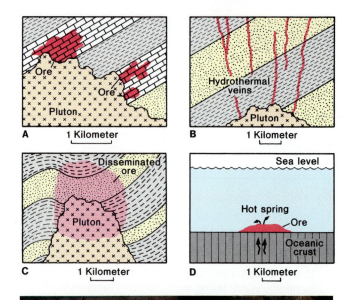

(quartz is the lowest-temperature mineral on Bowen's reaction series), could move into the country rock to create ore deposits.

Some hydrothermal fluids may be regular ground water, circulating in a convection pattern caused by the heat of the pluton. As water becomes heated by the pluton, it rises above it, drawing in new water from the sides to take its place. The new water then becomes heated and rises, and a continual circulation of water past the cooling pluton is set up (figure 21.13*B*). The hot water moving near and through the pluton could leach metals from the pluton and carry them upward to be deposited elsewhere as ore. However the hydrothermal solutions form, they tend to create four general types of hydrothermal ore deposits: (1) contact metamorphic deposits, (2) hydrothermal veins, (3) disseminated deposits, and (4) hot-spring deposits.

Contact metamorphism can create ores of copper, lead, zinc, silver, and other metals in country rock. The country rock may be completely or partially removed and replaced by ore (figure 21.14*A*). This is particularly true of limestone beds, which react readily with hydrothermal solutions. The ore bodies can be quite large and very rich. Most hydrothermal ores are metallic sulfides, often mixed with milky quartz. The origin of the sulfur is widely debated.

Hydrothermal veins are narrow ore bodies formed along joints and faults (figure 21.14*B*). They can extend great distances from their apparent plutonic sources. Some extend so far that it is questionable whether they are even associated with plutons. The fluids can precipitate ore (and quartz) within cavities along the fractures and may also replace the wall rock of the fractures with ore. Hydrothermal veins form most of the world's great deposits of lead, zinc, silver, gold, tungsten, tin, mercury, and, to some extent, copper.

Hot solutions can also form *disseminated deposits* in which metallic sulfide ore minerals are distributed in very low concentration through large volumes of rock, both above and within a pluton (figure 21.14*C*). Most of the world's copper comes from disseminated deposits (also

E

Figure 21.14
Hydrothermal ore deposits. (**A**) Contact metamorphism in which ore replaces limestone. (**B**) Ore emplaced in hydrothermal veins. (**C**) Disseminated ore within and above a pluton. (**D**) Ore precipitated around submarine hot spring (size of ore deposit exaggerated). (**E**) Hydrothermal veins of milky quartz in granite, northern California.

called *porphyry copper deposits* because the associated pluton is usually porphyritic). Along with the copper are deposited many other metals, such as lead, zinc, molybdenum, silver, and gold (and iron, though not in commercial quantities).

Where hot solutions rise to the earth's surface, *hot springs* form. Hot springs on land may contain large amounts of dissolved metals. Some California hot springs contain so much mercury that the water is unfit to drink.

Figure 21.15
Banded iron ores of Precambrian age probably accumulated in
shallow basins.

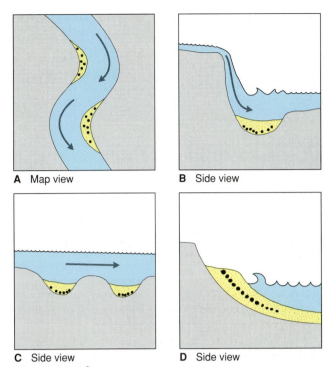

A Map view **B** Side view

C Side view **D** Side view

Figure 21.16
Types of placer deposits. (*A*) Stream bar. (*B*) Below waterfall.
(*C*) Depressions on stream bed. (*D*) Beach. Valuable grains
shown in black.

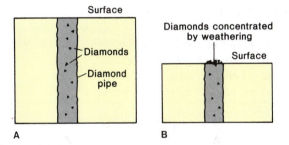

Figure 21.17
Residual concentration by weathering. (*A*) Cross-sectional view
of diamonds widely scattered within diamond pipe.
(*B*) Diamonds concentrated on surface by removal of rock by
weathering and erosion.

More impressive are hot springs on the sea floor (figure
21.14*D*), which can precipitate large mounds of metallic
sulfides, sometimes in commercial quantities. We will look
at submarine hot springs later in the chapter in connection
with plate tectonics.

Pegmatites (Box 4.1) are another type of ore deposit
associated with igneous rocks. They may contain important concentrations of minerals containing lithium, beryllium, and other rare metals, as well as gemstones such
as emeralds and sapphires.

Ores Formed by Surface Processes

Chemical precipitation in layers is the most common origin for ores of iron and manganese. A few copper ores form
in this way too. Banded iron ores, usually composed of alternating layers of iron minerals and chert, formed as sedimentary rocks in many parts of the world during the
Precambrian, apparently in shallow, water-filled basins
(figure 21.15). Later folding, faulting, metamorphism, and
solution have destroyed many of the original features of
the ore, so the origin of the ore is difficult to interpret. The
water may have been fresh or marine, and the iron may
have come from volcanic activity or deep weathering of
the surrounding continents. The alternating bands may
have been created by some rhythmic variation in volcanic
activity, river runoff, basin water circulation, growth of
organisms, or some other factor. Since banded iron ores
are all Precambrian, their origin might be connected to
an ancient atmosphere or ocean chemically different from
today's.

Placer deposits are found where running water or
waves have concentrated heavy sediment grains in a river
bar or on a beach (figure 21.16). Grains concentrated in
this manner include gold nuggets and dust, native platinum, diamonds and other gemstones, and worn pebbles
or sand grains composed of the heavy oxides of titanium
and tin.

Concentration by weathering can occur in several
ways. The concentration of diamonds by weathering is illustrated in figure 21.17. Diamonds are brought to the

surface of the earth in *diamond pipes,* columns of brec-
ciated ultramafic rock that have risen from the upper
mantle. Diamonds are widely scattered in diamond pipes
when they form. At the earth's surface the ultramafic rock
in the pipe may be weathered and eroded away. The diamonds, being more resistant to weathering, are left behind,
concentrated in rich deposits on top of the pipes. Rivers
may redistribute the diamonds into placer deposits, as in
South Africa.

Bauxite, the primary ore of aluminum, forms by lateritic weathering in tropical climates (chapter titled
"Weathering and Soil"), particularly on aluminum-rich
rocks (figure 21.18). Under tropical conditions of high
rainfall and high temperature, most weathering products
are soluble—even silica. The least soluble product is aluminum oxide, which remains on top of the weathering
rocks, forming bauxite in a soil very rich in aluminum.
Like the diamonds, the aluminum has been concentrated

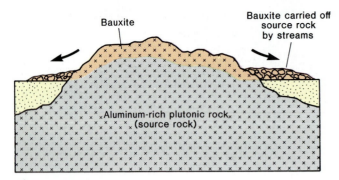

Figure 21.18
Bauxite forms by intense tropical weathering of a source rock.

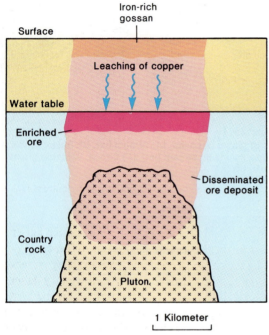

Figure 21.19
Supergene enrichment. Ground water leaches copper from upper part of disseminated deposit and precipitates it at or below the water table, forming rich ore.

residually by the removal of everything else. The aluminum ores may be redistributed slightly by running water (figure 21.18). Nickel ores can also form in laterites.

Another type of concentration by weathering is the *supergene enrichment* of disseminated ore deposits. The major ore mineral in a disseminated copper deposit is chalcopyrite, a copper-iron sulfide containing about 35% copper. Near the earth's surface, downward-moving ground water can leach copper and sulfur from the ore, leaving the iron behind (figure 21.19). At or below the water table the dissolved copper can react with chalcopyrite in the lower part of the disseminated deposit, forming a richer ore mineral such as chalcocite, which is about 80% copper:

$$3\ Cu^{++}\quad+\quad CuFeS_2\ \rightarrow\ 2\ Cu_2S\ +\ Fe^{++}$$

Copper dissolved   Chalcopyrite   Chalcocite   Iron in
in ground water   solution

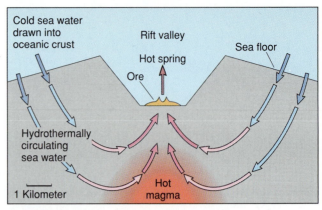

Figure 21.20
Hydrothermal circulation of sea water at ridge crest creates hot springs and metallic ores in rift valley. Cold sea water is drawn into fractured crust on ridge flanks. (Size of ore deposit exaggerated.)

In this way copper is removed from the top of the deposit and added to the lower part. The ore below the water table may be several times richer than in the rest of the deposit (silver can move with the copper). The iron left behind at the earth's surface forms a rusty cap called a *gossan*, which is a visible clue to the ore below.

Metal Ores and Plate Tectonics

There is currently much interest in the relationship between ore distribution and plate tectonics.

Diverging plate boundaries are often marked by lines of hot springs that carry and precipitate metals. Geologists in submersibles have observed hot springs in several localities along the rift valley of the mid-oceanic ridge. The hot springs, caused by the high heat flow and basaltic magma in the rift valley, range in temperature from about 20°C up to an estimated 350°C (660°F).

As the hot water rises in the rift valley, cold water is drawn in from the sides to take its place. This creates a circulation pattern in which cold sea water is actually drawn *downward* through cracks in the basaltic crust of the ridge flanks and then moves horizontally toward the rift valley, where it reemerges on the sea floor after being heated (figure 21.20). As the sea water moves through the crust, it dissolves metals and sulfur from the crustal rocks and magma. When the hot, metal-rich solutions contact cold sea water, metal sulfides are precipitated in a mound around the hot spring. This process has been filmed in the Pacific, where some springs spew clouds of fine-grained ore minerals that look like black smoke (figure 21.21).

The metals in rift-valley hot springs are predominantly iron, copper, and zinc, with smaller amounts of manganese, gold, and silver. Although the mounds are nearly solid metal sulfide, they are small and widely scattered on the sea floor, so commercial mining of them may not be practical. Occasionally, the ores may be concentrated in richer deposits. On the floor of the Red Sea metallic sediments worth billions of dollars have precipitated in basins filled with hot-spring solutions. Although the solutions are hot (up to 60°C or 140°F), they are very dense

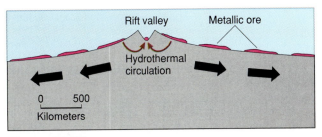

Figure 21.22
Sea-floor spreading carries metallic ores away from rift valley.
(Size of ore deposits exaggerated.)

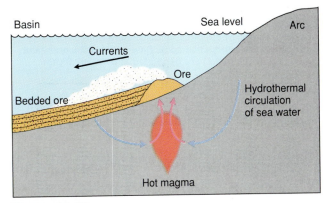

Figure 21.23
On island arcs metallic ores can form over
hot springs and be redistributed into layers by currents in
shallow basins.

Figure 21.21
"Black smoker" or submarine hot spring on the crest of the
mid-oceanic ridge in the Pacific Ocean near 21° north latitude.
The "smoke" is a hot plume of metallic sulfide minerals being
discharged into cold sea water from a chimney 0.5 meters
high. The large mounds around the chimney are metal
deposits. The instruments in the foreground are attached to the
small submarine from which the picture was taken.
Photo by U.S. Geological Survey.

because of their high salt content (they are seven times
saltier than sea water). The Red Sea brines are so dense
that they collect in sea-floor depressions instead of mixing
with the overlying sea water.

Hot metallic solutions are also found along some diverging continental boundaries. Near the Salton Sea in
southern California, which lies along the extension of the
mid-oceanic ridge inland, hot water very similar to the Red
Sea brines has been discovered underground. The hot
water is currently being used to run a geothermal power
plant. The high salt and metal content is corrosive to
equipment, but metals such as copper and silver may one
day be recovered as valuable by-products.

Sea-floor spreading carries the metallic ores away
from the ridge crest (figure 21.22), perhaps to be subducted beneath island arcs or continents at *converging
plate boundaries*. Slivers of *ophiolite* on land may contain
these rich ore deposits in relatively intact form. A notable
example of such ores occurs on the island of Cyprus in the
Mediterranean Sea. Banded chromite ores may also be
contained in the serpentinized ultramafic rock at the
bottom of ophiolites.

Volcanism at *island arcs* can also produce hot-spring
deposits on the flanks of the andesitic volcanoes. Pods of
very rich ore collect above local bodies of magma, and the
ore is sometimes distributed as sedimentary layers in
shallow basins (figure 21.23). The circulation pattern and
the ore-forming processes are quite similar to those of
spreading centers, but the island arc ores usually contain
more lead. Rich *massive sulfide deposits* overlying fractured volcanic rock in the Precambrian shield area of
Canada may have formed in this way on ancient island
arcs.

Subduction of the sea floor beneath a *continent* produces broad belts of metallic ore deposits on the edge of
the continent. Figure 21.24 shows how the distribution of
some metals in the western United States might be related
to depth along a subduction zone (the figure shows only
one of several competing models relating continental ore
deposits to plate tectonics). The pattern of ore belts has
probably been disturbed by changing subduction angles,
strike-slip faulting, and backarc spreading. (Each of these
processes may also have added new metals to the pattern
shown.) Similar patterns of ore belts occur in other subduction mountain ranges, notably the Andes.

The origin of the continental ores above a subduction
zone is not clear. The hot-spring deposits from the ridge
crest are subducted with oceanic crust and could become
remobilized to rise into the continent above. The ores may
also "distill" off other parts of the descending oceanic crust
or upper mantle. The metals may also derive from the

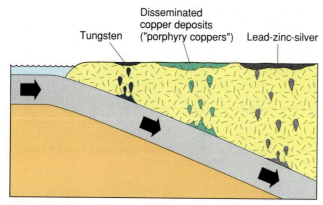

Figure 21.24
Possible relation of ore belts in the western United States to depth along the subduction zone. Different metallic ores (and different igneous rocks) are generated at different depths along a subducted plate.

continental crust itself or the asthenosphere below it. The metals may be concentrated somehow by the heat of a rising blob of magma or the hydrothermal circulation associated with it.

It is tempting to think that *mantle plumes* may cause ore concentration when they rise beneath continents, for plumes provide a magma source and hydrothermal circulation. The "Mississippi Valley-type" lead-zinc deposits of the continental interior are very puzzling features. Over broad areas metal ore has been emplaced in limestone and dolomite, both by cavity filling and replacement, and there is no obvious connection between the ores and any igneous rocks, which may be absent in the ore regions. One of many hypotheses of the origin of these ores is the movement of the continent over a mantle plume. There are at least two arguments against this hypothesis, however. One is that mantle plumes erupt large quantities of volcanic rock, and there are no volcanic rocks associated with the Mississippi Valley ores. The other is that where probable plumes exist today, as beneath Yellowstone, there are very few ore deposits.

Mining

Mining can be carried out on the earth's surface or underground (figure 21.25). Two forms of surface mines are *strip mines,* used for mining some beds of coal, and **open-pit mines,** in which ore is exposed in a large excavation. Some geologists use *strip mine* and *open-pit mine* interchangeably. However, strip mines generally expose coal or another resource in a long band that is later filled in, while open-pit mines are roughly circular and are excavated to extract huge amounts of material. They are not usually refilled. **Placer mines** are surface mines in which valuable sediment grains are extracted from stream bar or beach deposits.

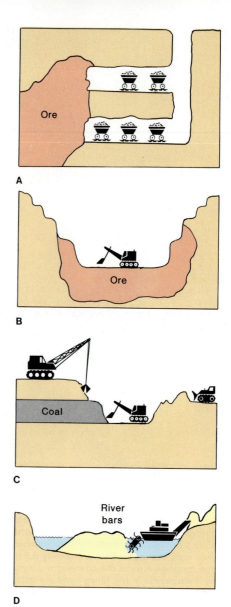

Figure 21.25
Types of mines. (*A*) Underground. (*B*) Open pit. (*C*) Strip. (*D*) Placer (being mined by a floating dredge).

Environmental Effects

Some of the environmental problems associated with mining can be partially solved if care is taken. For example, in the past *waste rock* was routinely left in unsightly heaps and dumps (figure 21.26*A*). The excavations for strip mines and placer mines can be filled in with waste rock, leveled or graded, and then covered over with topsoil to restore the land to usable condition. In some cases crops can be grown on reclaimed land within two or three years after mining operations are completed. Open pits, being larger, are rarely filled in, for the filling cannot be done gradually while mining is in progress. Underground mines are sometimes back-filled with waste rock to prevent land *subsidence* after ore is removed. Figure 21.26*B* shows extensive subsidence caused by mine collapse.

A

B

Figure 21.26
(*A*) Waste rock piles from coal mining in the 1950s, Sweetwater County, Wyoming. (*B*) Subsidence caused by the collapse of an underground coal mine, Acme, Wyoming. Mining continued from 1900 to 1943 (1976 photo).
Photo *A* by H. E. Malde, U.S. Geological Survey. Photo *B* by C. R. Dunrud, U.S. Geological Survey.

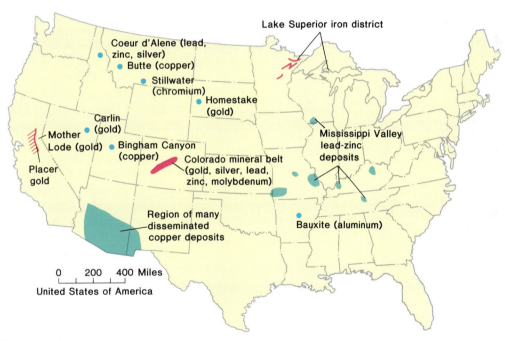

Figure 21.27
Some metallic ore localities mentioned in the text.

One of the more difficult problems to deal with is *acid drainage* from mines, caused by ground water running or being pumped out of a mine. Sulfide ore minerals are most often the source of the trouble (table 21.3). Ground water dissolves some of the sulfide, which oxidizes in air to form sulfuric acid. In some mines, expensive programs of holding and neutralizing drainage water prevent pollution of surface streams and harm to forests and wildlife. The worst problem is with long-abandoned mines that are still draining acid waters. Many of these may never be neutralized.

Some Important Metals

Iron

The basis for any modern, industrialized economy is *iron,* which is used to make steel. Iron and steel are used in a huge variety of products, from cast-iron frying pans to locomotives.

The major iron ore minerals are hematite and magnetite. Most of the iron in the United States comes from Minnesota and Michigan in the region around Lake Superior (figure 21.27). The ores are banded iron ores of Precambrian age, typical of most iron ores in the world. Mining is done mostly by open-pit methods.

Figure 21.28
Open-pit copper mine, Morenci, Arizona.

Copper

Less abundant is *copper,* another important metal for industry. More than half the copper used in the United States goes into electrical wire and equipment and one-third into the manufacture of brass, a copper-zinc alloy.

Most copper ores are sulfides. Chalcopyrite is the most important copper ore mineral. Some vein deposits of copper exist (as at Butte, Montana), but most major deposits are disseminated through large volumes of rock; so most copper mines are open pits (figure 21.28). Arizona, Utah, and a few other western states are the major producers in the United States (figure 21.27). The concentration of copper averages about 0.5% in most currently worked deposits; that is, 1 pound of copper is recovered for every 200 pounds of rock processed.

Aluminum

Widely used in the United States, *aluminum* is consumed in the manufacture of beer and soft drink cans, airplanes, electrical cable, and many other products. The use of aluminum is increasing rapidly.

The ore of aluminum is bauxite, which forms under tropical weathering conditions. The United States has very little bauxite, so it imports 90% of its aluminum ore from tropical countries. The largest mine in the United States is in Arkansas (figure 21.27). Open-pit mining is the usual technique for extracting bauxite.

Lead

The main use of *lead* (53%) is in batteries. Substantial amounts are recycled, largely from automobile batteries.

The most important ore of lead is galena. Major deposits occur in Missouri, Idaho, Utah, and Colorado (figure 21.27). The Missouri deposits are mostly found in limestone beds; their precise origin is a matter of some controversy. The ore is mined both underground and from open pits. Deposits in Idaho occur mostly as veins and are usually mined underground.

Zinc

Widely used in industry, *zinc* is necessary for galvanizing and the manufacture of brass and other alloys.

The major zinc ore is sphalerite. As sphalerite usually is found closely associated with galena, most lead mines also extract zinc. Zinc occurs without lead in some areas, however.

Silver

Coins, tableware, jewelry, photographic film, and many other products are made of *silver.*

Silver, found as a native metal and in sulfide ores, is a common by-product of lead and copper mining. The lead-zinc mines of Idaho (the Coeur d'Alene district) are the largest silver producers in the United States (figure 21.27).

Gold

The rare and valuable metal *gold* is used in coins, jewelry, decoration, dentistry, electronics, and the space program. Gold bars are stored to back national currency, although this use is rapidly disappearing.

Gold is most often found as a native element in the form of nuggets and grains. In some parts of the world these are concentrated in placer deposits (California's Gold Rush of 1849 was triggered by discoveries of placer gold). Gold nuggets, flakes, and dust can be separated from the other sediments by (1) *panning;* (2) *sluice boxes* (figure 21.29), which catch the heavy gold on the bottom of a box as gravel is washed through it; (3) *hydraulic mining* (figure 21.30), which washes gold-bearing gravel from a hillside into a sluice box; or (4) floating *dredges* (figure 21.31), which separate gold from gravel aboard a large barge, piling the spent gravel behind. When gold is found in hydrothermal veins associated with milky quartz, as it is in parts of Colorado and in California's Mother Lode (figure 21.27), it is mined underground. Disseminated gold is mined in open pits near Carlin, Nevada, and at many other localities in Nevada and California. The largest gold mine in the United States is the underground Homestake Mine in South Dakota, where finely disseminated gold is extracted from folded metamorphic rock.

The world has large reserves of iron and aluminum, moderate reserves of copper, lead, and zinc, and scanty reserves of gold and silver.

Other Metals

Many other metals are vital to our economy. *Chromium, nickel, cobalt, manganese, molybdenum, tungsten,* and *vanadium* are all important in the steel industry, particularly in the manufacture of specialty products such as stainless steel. Most of these metals have other uses as well. *Tin* is used in solder and for plating steel in "tin cans." *Mercury* is used in thermometers, silent electrical switches, medical compounds, and batteries. *Magnesium* is used in aircraft and flashbulbs. *Titanium,* as strong as steel but weighing half as much, is used in aircraft. *Platinum* is used in catalytic converters to clean automobile exhaust.

Figure 21.29
Sluice box used to separate gold from gravel, Alaska.
Photo by D. J. Miller, U.S. Geological Survey.

Figure 21.30
Hydraulic mining for gold, Alaska.
Photo by T. L. Pewe, U.S. Geological Survey.

Figure 21.31
A gold dredge separates gold from gravel.

Nonmetallic Resources

With the exception of the gemstones such as diamonds and rubies, nonmetallic resources do not have the glamour of many metals or energy resources. Nonmetallic resources are generally inexpensive and are needed in large quantities (again, except for gemstones). The large demand and low unit price means that these resources are best taken from local sources. Transportation over long distances would add significantly to the cost.

Construction Materials

Sand and *gravel* are both needed for the manufacture of concrete for building and highway construction. Sand is also used in mortar, which holds bricks and cement blocks together. The demand for sand and gravel in the United States has more than doubled in the last twenty-five years.

Sand dunes, river channel and bar deposits, glacial outwash, and beach deposits are common sources for sand and gravel. Cinder cones are mined for "gravel" in some areas. Sand and gravel are ordinarily mined in open pits (figure 21.32).

Stone refers to rock used in blocks to construct buildings or crushed to form roadbed. Most stone in buildings is limestone or granite, and most crushed stone is limestone. Huge quantities of stone are used each year in the United States. Stone is removed from open pits called *quarries* (figure 21.33).

Limestone has many uses other than building stone or crushed roadbed. Cement, used in concrete and mortar, is made from limestone. Pulverized limestone is in demand as a fertilizer and soil conditioner and is the principal ingredient of many chemical products.

Figure 21.32
Sand and gravel pit in a glacial esker, near Saranac Lake, New York.
Courtesy Ward's Natural Science Est., Inc., Rochester, NY.

Figure 21.34
Loading potassium minerals in an underground mine, Eddy County, New Mexico.
Photo by H. I. Smith, U.S. Geological Survey.

Fertilizers and Evaporites

Fertilizers (phosphate, nitrate, and potassium compounds) are extremely important to agriculture today, so much so that they are one of the few nonmetallic resources transported across the sea. *Phosphate* is produced from phosphorite, a sedimentary rock composed of certain remains of marine organisms. Major phosphate deposits in the United States are in Idaho, Wyoming, and Florida. *Nitrate* can form directly as an evaporite deposit but today is usually made from atmospheric nitrogen. *Potassium* compounds are often found as evaporites. Figure 21.34 shows an underground mine for potassium compounds.

Rock salt is coarsely crystalline halite formed as an evaporite. Salt beds are mined underground in New York, Ohio, and Michigan; underground salt domes are mined in Texas and Louisiana. (Some salt is also extracted from sea water by evaporation.) Rock salt is used in many ways—de-icing roads in winter, preserving food, as table salt, and in manufacturing hydrochloric acid and sodium compounds for baking soda, soap, and other products. Rock salt is heavily used by industry.

Gypsum forms as an evaporite. Beds of gypsum are mined in many states, notably California, Michigan, Iowa, and Texas. Gypsum, the essential ingredient of plaster and wallboard (sheet rock), is used mainly by the construction industry, although there are other uses.

Sulfur occurs in bright-yellow deposits of elemental sulfur. Most of its commercial production comes from the cap rock of salt domes (see Box 21.1). Sulfur is widely used in agriculture as a fungicide and fertilizer and by industry to manufacture sulfuric acid, matches, and many other products.

Figure 21.33
A limestone quarry in northern Illinois. The horizon marks the original land surface before the rock was removed.

Other Nonmetallics

Gemstones (called *gems* when cut and polished) include precious stones such as diamonds, rubies, emeralds, and sapphires and semiprecious stones such as beryl, garnet, jade, spinel, topaz, turquoise, and zircon. Gems are used for jewelry, phonograph needles, and bearings and abrasives (most are above 7 on Mohs' scale of hardness). Diamond drills and diamond saws are used to drill and cut rock. Watches often have hard gems at bearing points of friction ("17-jewel watches"). Gemstones are often found in pegmatites or in close association with other igneous intrusives. Some are recovered from placer deposits.

Asbestos is a fibrous variety of serpentine. The fibers can be separated and woven into fireproof fabric used for firefighters' clothes and theater curtains. It is also used in manufacturing ceiling and sound insulation, shingles, and brake linings, although the use of asbestos is being rapidly curtailed because of concern about its connection with lung cancer (Box 2.2). The United States produces little asbestos, mostly from belts of serpentinized ultramafic rocks in the Appalachians and the Pacific Coast states. Large amounts are mined in Canada, chiefly in Quebec. *Talc,* used in talcum powder and other products, is often found associated with asbestos (see Box 20.1).

Other nonmetallic resources are important. *Mica* is used in insulators. *Barite* ($BaSO_4$), because of its high specific gravity, is used to make heavy drilling mud to prevent oil gushers. *Borates* are boron-containing evaporites used in fiberglass, cleaning compounds, and ceramics. *Fluorite* (CaF_2) is used in toothpaste, Teflon finishes, and steel-smelting. *Clays* are used in ceramics and as filters and absorbents. *Diatomite* is used in swimming pool filters. *Glass sand,* which is over 95% quartz, is the main component of glass. *Graphite* is used in foundries, lubricants, steel-making, batteries, and pencil "lead."

Substitutes, Recycling, and Conservation

Substitutes for many geologic resources now exist, and others will be found. Aluminum is replacing the more expensive copper in many electrical uses, particularly in transmission lines. Glass fibers also are replacing copper for telephone lines. Aluminum and tin-coated steel are essentially interchangeable for beverage containers. Cotton and wool use could increase, replacing polyester and other petroleum-based synthetic fibers in clothes.

Suitable substitutes, however, seem unlikely for some resources. Nothing yet developed can take the place of steel in bridges or mercury in thermometers. Cobalt is vital for strong, permanent magnets. Although substitutes may help prolong the life of the supplies of some resources, they are not available for many others.

Recycling helps augment the supply of some resources such as aluminum, gold, silver, lead, and glass. No resource, however, receives even half its supply from recycling. Increased volunteer recycling on the part of the public and waste reclaiming of urban trash could increase these percentages. New ore will always be needed, however, because some uses of products prevent the material from being recyclable. A steel can that rusts away beside a road will never be recycled. The iron oxides are scattered in low percentage in the soil and can never be recovered. Many resources, such as petroleum and coal, are consumed during use and cannot be recycled.

Conservation of scarce resources is extremely important. The United States' use of oil from 1979 to 1985 declined as a result of conservation efforts such as the 55 mile-per-hour speed limit, more fuel-efficient automobiles, the upgrading of insulation in buildings, and the elimination of unneeded heating and lighting. But in 1986 oil use rose sharply as its price declined. Politically caused shortages and gluts of oil can obscure the fact that the supply of oil in the United States is severely limited. In the difficult times ahead as the United States faces up to this limit, the need for conserving fuel will be repeatedly stressed and the price of gasoline will rise substantially. Conservation of metals, particularly those imported in large quantities, will become increasingly important. Smaller automobiles and more durable appliances can help conserve metals.

Some Future Trends

Ocean mining, now rather uncommon, will increase in the near future. Mining of *manganese nodules* from the deep sea floor (chapter on "The Sea Floor") could provide the United States with substantial amounts of copper and nickel, together with far more manganese than United States industry can consume. The copper content of many nodules is four to five times higher than most deposits on land. *Metallic brines and deposits* of the Red Sea type may be a source of several metals in the future.

Several tools are of great help in mineral exploration on land. Highly sophisticated *geochemical tests* of soils and soil gases point to ore bodies underground. *Geophysical techniques* continue to be refined for resource exploration. Satellites and space-shuttle crews photograph the earth's surface in many different wavelengths of energy, and careful analysis of satellite imagery is proving to be of great help in prospecting. The economic returns from the satellite photography program will far outweigh its costs.

As the relation of plate tectonics to the distribution of resources such as metals and petroleum becomes clearer, selecting areas for exploring for these materials should become easier.

Summary

Geologic resources include energy resources, metals, and nonmetallic resources. All are nonrenewable, except ground water.

Petroleum (*oil* and *natural gas*) supplies almost two-thirds of the energy used by the United States.

The occurrence of oil and natural gas is limited to regions having these conditions together: *source rocks, reservoir rocks, traps,* such as anticlines, faults, stratigraphic traps, and salt domes, and *thermal maturity* from burial.

Reserves are known deposits that can be legally and economically recovered now—the short-term supply. *Resources* include reserves as well as other known and undiscovered deposits that may be economically extracted in the future.

At the present rate of consumption, the United States' oil supply will last 25 years.

Natural gas, *heavy crude, oil sand,* and *oil shale* may all help replace liquid petroleum in the future. Most of these resources are in the western states.

The United States has huge *coal* reserves, enough for centuries of use at the present rate. Coal, now used mostly for generating electricity, will probably be used more widely in the future as oil runs out. The United States now exports coal. Increased production will come largely from near-surface, low-sulfur western coals.

The United States has ample *uranium* for its reactor program, mostly in sandstones in western states.

Metallic ores, which can be profitably mined, are often associated with igneous rocks, particularly their hydrothermal fluids, which can form *contact metamorphic deposits, hydrothermal veins, disseminated deposits,* and submarine *hot-spring deposits.* Iron occurs in sedimentary layers, and aluminum ores form from weathering.

Metallic ores form from hot springs at diverging plate boundaries, on the flanks of island arcs, and in belts on the edges of continents above subduction zones. The association of ores with mantle plumes has been suggested.

Ores are mined *underground* and also at the earth's surface in *strip mines, open-pit mines,* and *placer mines.*

Metals are vital to an industrial economy, particularly *iron* for steel production and *copper* for electrical equipment.

Nonmetallic resources such as *sand and gravel* and *limestone* for crushed rock and cement are used in huge quantities. *Fertilizers, rock salt, gypsum, sulfur,* and *clays* are also widely used.

Substitutes, recycling, and conservation can help cut consumption of some resources but will not eliminate the need for finding new deposits.

Deep ocean mining and increasingly sophisticated exploration techniques will help supply some of our future resource needs.

Terms to Remember

crude oil
geologic resources
heavy crude
natural gas
nonrenewable resource
oil field
oil pool
oil sand
oil shale
oil trap

open-pit mine
ore
petroleum
placer mine
reserves
reservoir rock
resources
source rock
strip mine

Questions for Review

1. Name the three major classes of geologic resources. Give four examples of each class.
2. Discuss the United States' supplies and potential use of natural gas, heavy crude, oil sand, oil shale, and uranium.
3. List in decreasing order of use the energy resources used in the United States. Discuss possible future trends in this ranking.
4. What geologic conditions are necessary for the accumulation of oil and natural gas?
5. Differentiate between *reserves* and *resources*. Can reserves be increased? Can resources be increased?
6. Compare oil reserves with coal reserves. What might this indicate for the future use of each?
7. Describe several ways in which ore deposits form. Which are the most important?
8. Describe four ways in which resources are mined.
9. Discuss the environmental effects of oil extraction and coal mining.
10. Discuss common uses for iron, copper, lead, zinc, and aluminum.
11. Describe the potential of substitutes, conservation, recycling, and deep-ocean mining for meeting an increasing need for geologic resources.

Questions for Thought

1. Many underdeveloped countries would like to have the standard of living enjoyed by the United States, which uses 15% to 40% of the world production of many resources. As these countries become industrialized, what happens to the world demand for geologic resources? Where will these needed resources come from?

2. If driven 12,000 miles per year, how many more gallons of gasoline per year does a sports car rated at 16 miles per gallon use than a minicompact car rated at 54 mpg? Over 5 years how much more does it cost to buy gasoline at $1 per gallon for the low-mileage car? At $3 per gallon (the price in many European countries)?

Supplementary Readings

Bonatti, E. 1978. The origin of metal deposits in the oceanic lithosphere. *Scientific American* (February), pp. 54–61.

Brobst, D. A., and W. P. Pratt. 1973. *United States mineral resources.* U.S. Geological Survey Professional Paper 820.

Cameron, E. N. 1986. *At the crossroads—the mineral problems of the United States.* New York: John Wiley & Sons.

Dixon, C. J. 1979. *Atlas of economic mineral deposits.* Ithaca, N.Y.: Cornell University Press.

Guilbert, J. M., and C. F. Park. 1986. *The geology of ore deposits.* New York: W. H. Freeman.

Jensen, M. L., and A. M. Bateman. 1979. *Economic mineral deposits.* 3d ed. New York: John Wiley & Sons.

Levorsen, A. I. 1967. *Geology of petroleum.* 2d ed. New York: W. H. Freeman.

McKelvey, V. E. 1986. *Subsea mineral resources.* U.S. Geological Survey Professional Paper 1689-A.

Mero, J. L. 1965. *The mineral resources of the sea.* New York: American Elsevier.

North, F. K. 1985. *Petroleum geology.* Boston: Allen & Unwin.

Park, C. F., Jr. 1968. *Affluence in jeopardy: Minerals and the political economy.* New York: Freeman, Cooper.

Park, C. F., Jr. 1975. *Earthbound: Minerals, energy, and man's future.* New York: Freeman, Cooper.

Perry, H. 1983. *Coal in the United States: A status report. Science* 222:377–384.

Riley, C. M. 1959. *Our mineral resources.* Huntington, N.Y.: Robert E. Krieger.

Sawkins, F. J. 1984. *Metal deposits in relation to plate tectonics.* New York: Springer-Verlag.

Selley, R. C. 1985. *Elements of petroleum geology.* New York: W. H. Freeman.

Skinner, B. J. 1986. *Earth resources.* 3d ed. Englewood Cliffs, N.J.: Prentice-Hall.

U.S. Bureau of Mines. *Minerals yearbook* and *Mineral commodity summary* (published annually). Washington, D.C.: U.S. Government Printing Office.

Whitmore, E. C., Jr., and M. E. Williams, eds. 1982. *Resources for the twenty-first century.* U.S. Geological Survey Professional Paper 1193.

A

Identification of Minerals

Each mineral is identified by a unique set of physical or chemical properties. To determine some of these properties requires specialized equipment and techniques. Most common minerals, however, can be distinguished from one another by tests involving simple observations. Cleavage is an especially useful property. If cleavage is present, you should determine the number of cleavage directions, estimate the angles between cleavage directions, and note the quality of each direction of cleavage. Other easily performed tests and observations check for hardness (abbreviated H), luster, and color, and determining crystal form (if present). A simple chemical test can be made using dilute hydrochloric acid to see if the mineral effervesces.

The identification tables included here can be used to identify the most common minerals (the rock-forming minerals) and some of the most common ore minerals. For identifying less common minerals, refer to one of the books on mineralogy listed at the end of the chapter titled "Atoms, Elements, and Minerals." Mineral identification takes practice, and you will probably want to verify your mineral identifications with a geology instructor.

Because the common rock-forming minerals are the ones you are most likely to encounter, we have included a simple key for identifying them. The key is based on first determining whether or not the mineral is harder than glass and then checking other properties that should lead to identification of the mineral. You should verify your identification by seeing whether other properties of your sample correspond to those listed for the mineral in table A.1.

Ore minerals are usually distinctive enough that a key is unnecessary. To identify an ore mineral, read through table A.2 and determine which set of properties best fits the unknown mineral.

Key for Identifying Common Rock-Forming Minerals

Determine whether a fresh surface of the mineral is harder or softer than glass. If you can scratch the mineral with a knife blade, the mineral is softer than glass.

I. Harder than glass—knife will not scratch mineral. (If softer than glass, go to II.)
  A. Determine if cleavage is present or absent (this may require careful examination). If cleavage is absent proceed below; if cleavage is present, proceed to B.
    1. Vitreous luster
      a. Olive green or brown—*olivine*
      b. Reddish brown or in equidimensional crystals with twelve or more faces—*garnet*
      c. Usually light-colored or clear—*quartz*
    2. Metallic luster
      a. Bright yellow—*pyrite*
    3. Greasy or waxy luster
      a. Mottled green and black—*serpentine*
  B. Cleavage present. Determine the number of directions of cleavage in an individual grain or crystal.
    1. Two directions, good, at or near 90°—*feldspar*
      a. If striations are visible on cleavage surfaces—*plagioclase*
      b. If pink or salmon-colored—*potassium feldspar* (or *orthoclase*)
      c. If white or light gray without striations, it could be either type of feldspar
    2. Two directions, fair, at 90°
      a. Dark green to black—*pyroxene* (usually augite)
    3. Two directions, excellent, not near 90°
      a. Dark green to black—*amphibole* (usually hornblende)
II. Softer than glass—knife scratches mineral
  A. No cleavage detectable
    1. Earthy luster, in masses too fine to distinguish individual grains—*clay group* (for instance, *kaolinite*)
  B. Cleavage present
    1. One direction
      a. Perfect cleavage in flexible sheets—*mica:*
        Clear or white—*muscovite mica*
        Black or dark brown—*biotite mica*
    2. Three directions
      a. All three perfect and at 90° to each other (cubic cleavage)—*halite*
      b. All three perfect and not near 90° to each other:
        If effervesces in dilute acid—*calcite*
        If effervesces in dilute acid only after being pulverized—*dolomite*

Diagnostic Properties of the Common Rock-Forming Minerals

Name (mineral groups shown in capitals)	Chemical Composition	Chemical Group	Diagnostic Properties	Other Properties
AMPHIBOLE (A mineral group in which *hornblende* is the most common member.)	$XSi_8O_{22}(OH)_2$ (*X* is a combination of Ca, Na, Fe, Mg, Al)	Chain silicate	2 good cleavage directions at 60° (120°) to each other.	H = 5–6 (barely scratches glass). Hornblende is dark green to black; tends to form in needle-like or elongate crystals; vitreous luster.
Augite (see Pyroxene)				
Biotite (see Mica)				
Calcite	$CaCO_3$	Carbonate	3 excellent cleavage directions, *not* at right angles (they define a rhombohedron). H = 3. Effervesces vigorously in weak acid.	Usually white, gray, or colorless; vitreous luster. Clear crystals show double refraction.
CLAY MINERALS (*Kaolinite* is a common example of this large mineral group.)	Compositions include $XSi_4O_{10}(OH)_8$ (*X* is Al, Mg, Fe, Ca, Na, K)	Sheet silicate	Generally microscopic crystals. Masses of clay minerals are softer than fingernail. Earthy luster. Claylike smell when damp.	Seen as a chemical weathering product of feldspars and most other silicate minerals. A constituent of most soils.
Dolomite	$CaMg(CO_3)_2$	Carbonate	Identical to calcite (rhombohedral cleavage, H = 3) except effervesces in weak acid only when pulverized.	Usually white, gray, or colorless. Vitreous luster.
FELDSPAR (Most common group of minerals.) The group includes:	Framework	Framework silicates	H = 6 (scratch glass). 2 good cleavage directions at about 90° to each other.	Vitreous luster but surface may be weathered to clay, giving an earthy luster. Perfect crystal, shaped like an elongated box.
Potassium feldspar (orthoclase)	$KAlSi_3O_8$		White, pink, or salmon-colored.	Never has striations on cleavage surfaces.
Plagioclase (sodium and calcium feldspar)	Mixture of: $CaAl_2Si_2O_8$ and $NaAlSi_3O_8$		White, light to dark gray, rarely other colors. *May* have striations on cleavage surfaces.	Calcium-rich varieties generally a darker gray and may show an iridescent play of colors.
GARNET	$XSiO_4$ (*X* is a combination of Ca, Mg, Fe, Al, Mn)	Isolated silicate	No cleavage. Usually reddish brown. Tends to occur in perfect equidimensional crystals, usually 12 sided. H = 7.	Rarely yellow, green, or black. Usually found in metamorphic rocks. Vitreous luster.

Name (mineral groups shown in capitals)	Chemical Composition	Chemical Group	Diagnostic Properties	Other Properties
Gypsum	$CaSO_4 \cdot 2H_2O$	Sulfate	H = 2. 1 good and 2 perfect cleavage directions. Vitreous or silky luster.	Clear, white, or pastel colors. Flexible cleavage fragments.
Halite	NaCl	Halide	3 excellent cleavage directions at 90° to each other (cubic). H = 2½. Salty taste. Soluble in water.	Usually clear or white.
Hematite (*see* Ore mineral table)				
Hornblende (*see* Amphibole)				
Kaolinite (*see* Clay)				
MICA The group includes:	$K(X)(AlSi_3O_{10})(OH)_2$	Sheet silicate	1 perfect cleavage direction (splits easily into flexible sheets).	H = 2–3. Vitreous luster.
Biotite	(*X* is Mg, Fe, and Al)		Black or dark brown.	
Muscovite	(*X* is Al)		White or transparent.	
Olivine	X_2SiO_4 (*X* is Fe, Mg)	Isolated silicate	No cleavage. Generally olive green or brown. H = 6–7 (scratches glass). Vitreous luster.	Usually as small grains in mafic or ultramafic igneous rocks.
Orthoclase (*see* Feldspar)				
Plagioclase (*see* Feldspar)				
Pyrite ("fools gold")	FeS_2	Sulfide	H = 6 (scratches glass). Bright yellow, metallic luster. Black streak.	Commonly occurs as perfect crystals: cubes or crystals with five-sided faces. Weathers to brown.
PYROXENE (A mineral group; *Augite* is most common member.)	$XSiO_3$ (*X* is Fe, Mg, Al, Ca)	Chain silicate	2 fair cleavage directions at 90° to each other.	H = 6. Augite is dark green to black. Vitreous luster; usually stubby crystals.
Quartz	SiO_2	Framework silicate	H = 7. No cleavage. Vitreous luster. Does not weather to clay.	Almost any color but commonly white or clear. Good crystals have six-sided "column" with complex "pyramid" on top.
Serpentine	$Mg_6Si_4O_{10}(OH)_8$	Sheet silicate	Hardness variable but softer than glass. Mottled green and black. Greasy luster. Fractures along smooth curved surfaces.	Sometimes fibrous (asbestos).

Diagnostic Properties of the Most Common Ore Minerals

Name	Chemical Composition	Diagnostic Properties	Other Properties
Azurite	$Cu_3(CO_3)_2(OH)_2$	Azure blue; effervesces in weak acid.	$H = 3$–4.
Bauxite	$Al_2O_3 \cdot nH_2O$	Earthy luster. A variety of clay. Generally pea-sized spheres included in a fine-grained mass.	
Bornite	Cu_3FeS_4	Metallic luster, tarnishes to iridescent purple color.	Gray streak; $H = 3$ (softer than glass).
Chalcopyrite	$CuFeS_2$	Metallic luster, brass-yellow. Softer than glass.	Black streak.
Cinnabar	HgS	Scarlet red, bright red streak.	Softer than glass. Generally an earthy luster.
Galena	PbS	Metallic luster, gray; 3 directions of cleavage at 90° (cubic). High specific gravity.	Softer than glass; gray streak.
Gold	Au	Metallic luster, yellow. $H = 3$ (softer than glass, can be pounded into thin sheets, easily deformed).	Yellow streak; high specific gravity.
Halite	$NaCl$	Salty taste; 3 cleavage directions at 90° (cubic).	Clear or white; easily soluble in water.
Hematite	Fe_2O_3	Red-brown streak.	Either in earthy reddish masses or in metallic, silver-colored flakes or crystals.
Limonite	$Fe_2O_3 \cdot nH_2O$	Earthy luster; yellow-brown streak.	Yellow to brown color; softer than glass.
Magnetite	Fe_3O_4	Metallic luster, black; magnetic.	Harder than glass; black streak.
Malachite	$Cu_2(CO_3)(OH)_2$	Bright-green color and streak.	Softer than glass; effervesces in weak acid.
Sphalerite	ZnS	Brown to yellow color; 6 directions of cleavage.	Lusterlike resin; yellow or cream-colored streak; softer than glass.
Talc	$Mg_3Si_4O_{10}(OH)_2$	White, gray, or green; softer than fingernail ($H = 1$).	Greasy feel.

B

Identification of Rocks

Igneous Rocks

Igneous rocks are classified on the basis of texture and composition. For some rocks, texture alone suffices for naming the rock. For most igneous rocks, composition as well as texture must be taken into account. Ideally, the mineral content of the rock should be used to determine composition; but for fine-grained igneous rocks, accurate identification of minerals may require a polarizing microscope or other special equipment. In the absence of such equipment, we rely on the color of fine-grained rocks and assume the color is indicative of the minerals present.

To identify a common igneous rock, use either table 4.1 or follow the key given below.

Key for Identifying Common Igneous Rocks

I. What is the texture of the rock?
  A. Is it glassy (a very vitreous luster)? If so, it is *obsidian,* regardless of its chemical composition. Obsidian exhibits a pronounced conchoidal fracture.
  B. Does it have a frothy appearance? If so, it is *pumice,* regardless of its chemical composition. Pumice is light in weight and feels abrasive (it probably will float on water).
  C. Does it have angular fragments of rock embedded in a volcanic-derived matrix? If so, it is a *volcanic breccia.* If the precise nature of the rock fragments and matrix can be identified, modifiers may be used; for instance, the rock may be an *andesite* breccia or a *rhyolite* breccia.
  D. Is the rock composed of interlocking, very coarse-grained minerals? (The minerals should be more than 1 centimeter across.) If so, the rock is a *pegmatite.* Most pegmatites are mineralogically equivalent to granite, with feldspars and quartz being the predominant minerals.
  E. Is the rock entirely coarse-grained? (That is, does it have an interlocking crystalline texture in which nearly all grains are more than 1 mm across?) If so, go to part II of this key.
  F. Is the rock *entirely* fine-grained? (Are grains less than 1 mm across or too fine to distinguish with the naked eye?) If so, go to part III of this key.
  G. Is the matrix fine-grained with some coarse-grained minerals visible in the rock? If so, go to part III and add the adjective *porphyritic* to the name of the rock.

II. Igneous rocks composed of interlocking coarse-grained minerals.
  A. Is quartz present? If so, the rock is a *granite.* Confirmation: Granite should be composed predominantly of feldspar—generally white, light gray, or pink (indicating high amounts of potassium or sodium in the feldspar). Rarely are there more than 20% ferromagnesian minerals in a granite.
  B. Are quartz and feldspar absent? If so, the rock should be composed entirely of ferromagnesian minerals and is *ultramafic.* Confirmation: Identify the minerals as being olivine or pyroxene (or less commonly, amphibole or biotite).
  C. Does the rock have less than 50% feldspar and no quartz? If so, the rock should be a *gabbro.* Confirmation: Most of the rock should be ferromagnesian minerals. Plagioclase can be medium or dark gray. There would be no pink feldspars.
  D. Is the rock composed of 30% to 60% feldspar (and no quartz)? If so, the rock is a *diorite.* Confirmation: Feldspar (plagioclase) is usually white to medium gray but never pink.

III. Igneous rocks that are fine-grained.
  A. Can quartz be identified in the rock? If so, the rock is a *rhyolite.*
  B. If the rock is too fine-grained for you to determine whether quartz is present but is white, light gray, pink, or pale green, the rock is most likely a *rhyolite.*
  C. Is the rock composed predominantly of ferromagnesian minerals? If so, the rock is *basalt.*
  D. If the rock is too fine-grained to identify ferromagnesian minerals but is black or dark gray, the rock is probably a *basalt.*
    1. Does the rock have rounded holes in it? If so, it is a *vesicular basalt* or *scoria.*
  E. Is the rock composed of roughly equal amounts of white or gray feldspar and ferromagnesian minerals (but no quartz)? If so, the rock is an *andesite.* Confirmation: Most andesite is porphyritic, with numerous identifiable crystals of white or light-gray feldspar and lesser amounts of hornblende crystals within the darker, fine-grained matrix. Andesite is usually medium to dark gray or green.

Sedimentary Rocks

The following key shows how sedimentary rocks are classified on the basis of texture and composition. The descriptions of the rocks in the main body of the text provide additional information, such as common rock colors. *Equipment* needed for identification of sedimentary rocks includes a bottle of dilute hydrochloric acid, a hand lens or magnifying glass, a millimeter scale, a glass plate for hardness tests, and a pocketknife or rock hammer.

Begin by testing the rock for carbonate minerals by applying a small amount of dilute hydrochloric acid (0.1 molar HCl) to the surface of the rock (CAUTION–dilute HCl can burn eyes and clothing; use only with supervision).

1. The rock does not effervesce (fizz) in acid, or effervesces weakly, but when powdered by a knife or hammer, the powder effervesces strongly. If so, the rock is *dolomite*.
2. The rock does not effervesce at all, even when powdered, or effervesces only in some places, such as the cement between grains. Go to part I of this key.
3. The rock effervesces strongly. The rock is *limestone*. Go to part II of this key to determine limestone type.

I. With a hand lens or magnifying glass, determine if the rock has a clastic texture (grains cemented together) or a crystalline texture (visible, interlocking crystals).
   A. If clastic:
      1. Most grains are more than 2 mm in diameter.
         a. Angular grains—*sedimentary breccia*.
         b. Rounded grains—*conglomerate*.
      2. Most grains are between 1/16 and 2 mm in diameter. Rock feels gritty to the fingers. *Sandstone*.
         a. More than 90% of the grains are quartz—*quartz sandstone*.
         b. More than 25% of the grains are feldspar—*arkose*.
         c. More than 25% of the grains are fine-grained rock fragments, such as shale, slate, and basalt—*lithic sandstone*.
         d. More than 15% of the rock is fine-grained matrix—*graywacke*.
      3. Rock is fine-grained (grains less than 1/16 mm in diameter). Feels smooth to fingers.
         a. Grains visible with a hand lens—*siltstone*.
         b. Grains too small to see, even with a hand lens.
            1. Rock is laminated, fissile—*shale*.
            2. Rock is unlayered, blocky—*mudstone*.

   B. If crystalline:
      1. Crystals fine to coarse, hardness of 2—*rock gypsum*.
      2. Coarse crystals that dissolve in water—*rock salt*.
   C. Hard to determine if clastic or crystalline:
      1. Very fine-grained, smooth to touch, conchoidal fracture, hardness of 6 (scratches glass), nonporous—*chert* (*flint* if dark)
      2. Very fine-grained, smooth to touch, breaks into flat chips—*shale*.
      3. Black or dark brown, readily broken, soils fingers—*coal*.

II. *Limestone* may be clastic or crystalline, fine- or coarse-grained, and may or may not contain visible fossils. Usually gray, tan, buff, or white. Some distinctive varieties are:
   A. *Bioclastic limestone*—clastic texture, grains are whole or broken fossils. Two relatively rare varieties are:
      1. *Coquina*—very coarse, recognizable shells, much open pore space.
      2. *Chalk*—very fine-grained, white or tan, soft and powdery.
   B. *Oolitic limestone*—grains are small spheres (less than 2 mm in diameter), all about the same size.
   C. *Travertine*—coarsely crystalline, no pore space, often contains different-colored layers (bands).

Metamorphic Rocks

The characteristics of a metamorphic rock are largely governed by (1) the composition of the parent rock and (2) the particular combination of temperature, confining pressure, and directed pressure. These factors cause different textures in rocks formed under different sets of conditions. For this reason, texture is usually the main basis for naming a metamorphic rock. Determining the composition (e.g., mineral content) is necessary for naming some rocks (e.g., *quartzite*), but for others, the minerals present are used as adjectives to describe the rock completely (e.g., *biotite* schist).

Metamorphic rocks are identified by determining first whether the rock has a *foliated* or *nonfoliated* texture.

Key for Identifying Metamorphic Rocks

I. If the rock is *nonfoliated,* then it is identified on the basis of its mineral content.

   A. Does the rock consist of mostly quartz? If so, the rock is a *quartzite*. A quartzite has a mosaic texture of interlocking grains of quartz and will easily scratch glass.

   B. Is the rock composed of interlocking coarse grains of calcite or dolomite? If so, it is *marble*. (The individual grains should exhibit rhombohedral cleavage; the rock is softer than glass.)

   C. Is the rock a dense, dark mass of grains mostly too fine to identify with the naked eye? If so, it probably is a *hornfels*. A hornfels may have a few larger crystals of uncommon minerals enclosed in the fine-grained mass.

II. If the rock is *foliated,* determine the type of foliation and then, if possible, identify the minerals present.

   A. Is the rock very fine-grained and does it split into sheetlike slabs? If so, it is *slate*. Most slate is composed of extremely fine-grained clay minerals, and the rock has an earthy luster.

   B. Does the rock have a silky sheen but otherwise appear similar to slate? If so, it is a *phyllite*.

   C. Is the rock composed mostly of visible grains of platy or needlelike minerals that are approximately parallel to one another? If so, the rock is a *schist*. If the rock is composed mainly of mica, it is a *mica schist*. If it also contains garnet, it is called a *garnet mica schist*. If hornblende is the predominant mineral, the rock is a *hornblende schist*. If talc prevails, it is a *talc schist* (sometimes called soapstone). A schistose rock composed of serpentine is called a *serpentinite*.

   D. Are dark and light minerals found in separate lenses or layers? If so the rock is a *gneiss*. The light layers are composed of feldspars and perhaps quartz, whereas the darker layers commonly are formed of biotite, amphibole, or pyroxene. A gneiss may appear similar to granite or diorite but can be distinguished from these igneous rocks by the foliation.

C

The Elements Most Significant to Geology

Atomic Number	Name	Symbol	Atomic Weight	Some Usual Charges of Ions
1	Hydrogen	H	1.0	+1
2	Helium	He	4.0	0 inert
3	Lithium	Li	6.9	+1
4	Beryllium	Be	9.0	+2
5	Boron	B	10.8	+3
6	Carbon	C	12.0	+4
7	Nitrogen	N	14.0	+5
8	Oxygen	O	16.0	−2
9	Fluorine	F	19.0	−1
10	Neon	Ne	20.2	0 inert
11	Sodium	Na	23.0	+1
12	Magnesium	Mg	24.3	+2
13	Aluminum	Al	27.0	+3
14	Silicon	Si	28.1	+4
15	Phosphorus	P	31.0	+5
16	Sulfur	S	32.1	−2
17	Chlorine	Cl	35.5	−1
18	Argon	Ar	39.9	0 inert
19	Potassium	K	39.1	+1
20	Calcium	Ca	40.1	+2
22	Titanium	Ti	47.9	+4
23	Vanadium	V	50.9	
24	Chromium	Cr	52.0	
25	Manganese	Mn	54.9	+4, +3
26	Iron	Fe	55.8	+2, +3
27	Cobalt	Co	58.9	
28	Nickel	Ni	58.7	+2
29	Copper	Cu	63.5	+2
30	Zinc	Zn	65.4	+2
33	Arsenic	As	74.9	+3
35	Bromine	Br	79.9	−
37	Rubidium	Rb	85.5	+1
38	Strontium	Sr	87.3	+2
40	Zirconium	Zr	91.2	−
42	Molybdenum	Mo	95.9	+4
47	Silver	Ag	107.9	+1
48	Cadmium	Cd	112.4	−
50	Tin	Sn	118.7	+4
51	Antimony	Sb	121.8	+3
52	Tellurium	Te	127.6	−
55	Cesium	Cs	132.9	−
56	Barium	Ba	137.4	+2
60	Neodymium	Nd	144	
62	Samayium	Sm	150	
74	Tungsten	W	183.9	−
78	Platinum	Pt	195.2	−
79	Gold	Au	197.0	−
80	Mercury	Hg	200.6	+2
82	Lead	Pb	207.2	+2
83	Bismuth	Bi	209.0	−
86	Radon	Rn	222	0 inert
88	Radium	Ra	226.1	
90	Thorium	Th	232.1	−
92	Uranium	U	238.1	−
94	Plutonium	Pu	239.0	−

D

Periodic Table of Elements

1a	IIa	IIIb	IVb	Vb	VIb	VIIb	VIIIb			IB	IIIB	IIIa	IVa	Va	VIa	VIIa	0
1 H 1.008																	2 He 4.00
3 Li 6.94	4 Be 9.01											5 B 10.81	6 C 12.01	7 N 14.00	8 O 15.99	9 F 18.99	10 Ne 20.18
11 Na 22.99	12 Mg 24.31											13 Al 26.98	14 Si 28.09	15 P 30.97	16 S 32.06	17 Cl 35.45	18 Ar 39.95
19 K 39.10	20 Ca 40.08	21 Sc 44.96	22 Ti 47.90	23 V 50.94	24 Cr 51.99	25 Mn 54.94	26 Fe 55.85	27 Co 58.93	28 Ni 58.71	29 Cu 63.54	30 Zn 65.37	31 Ga 69.72	32 Ge 72.59	33 As 74.92	34 Se 78.96	35 Br 79.91	36 Kr 83.80
37 Rb 85.47	38 Sr 87.62	39 Y 88.91	40 Zr 91.22	41 Nb 92.91	42 Mo. 95.94	43 Tc (99)	44 Ru 101.97	45 Rh 102.91	46 Pd 106.4	47 Ag 107.87	48 Cd 112.40	49 In 114.82	50 Sn 118.69	51 Sb 121.75	52 Te 127.60	53 I 126.90	54 Xe 131.30
55 Cs 132.91	56 Ba 137.34	see below 57–71	72 Hf 178.49	73 Ta 180.95	74 W 183.85	75 Re 186.2	76 Os 190.2	77 Ir 192.2	78 Pt 195.09	79 Au 196.97	80 Hg 200.59	81 Tl 204.37	82 Pb 207.19	83 Bi 208.98	84 Po (210)	85 At (210)	86 Rn (222)
87 Fr (223)	88 Ra (226)	see below 89–103	104 Rf (261)	105 Ha (260)	106 263												

*newly produced

57 La 138.91	58 Ce 140.12	59 Pr 140.91	60 Nd 144.24	61 Pm (147)	62 Sm 150.35	63 Eu 151.96	64 Gd 157.25	65 Tb 158.92	66 Dy 162.50	67 Ho 164.93	68 Er 167.26	69 Tm 168.93	70 Yb 173.04	71 Lu 174.97
89 Ac (227)	90 Th 232.04	91 Pa (231)	92 U 238.03	93 Np (237)	94 Pu (242)	95 Am (243)	96 Cm (247)	97 Bk (247)	98 Cf (251)	99 Es (254)	100 Fm (253)	101 Md (256)	102 No (254)	103 Lw (257)

E

Selected Conversion Factors

	English Unit	Conversion Factor	Metric Unit	Conversion Factor	English Unit
Length and Distance	inch (in)	2.54	centimeters (cm)	0.4	inch (in)
	foot (ft)	0.3048	meter (m)	3.28	feet (ft)
	inch (in)	0.026	meter (m)	39.4	inches (in)
	mile, statute (mi)	1.61	kilometers (km)	0.62	mile (mi)
Area	square inch (in^2)	6.45	square centimeters (cm^2)	0.16	square inch (in^2)
	square foot (ft^2)	0.093	square meter (m^2)	10.8	square feet (ft^2)
	square mile (mi^2)	2.59	square kilometers (km^2)	0.39	square mile (mi^2)
	acre	0.4	hectare	2.47	acres
Volume	cubic inch (in^3)	16.4	cubic centimeters (cm^3)	0.06	cubic inch (in^3)
	cubic yard (yd^3)	0.76	cubic meter (m^3)	1.3	cubic yards (yd^3)
	cubic foot (ft^3)	0.0283	cubic meter (m^3)	35.3	cubic feet (ft^3)
	quart (qt)	0.95	liter	1.06	quarts (qt)
Weight	ounce (oz)	28.3	grams (g)	0.04	ounce (oz)
	pound (lb)	0.45	kilogram (kg)	2.2	pounds (lb)
	ton, short (2,000 lb)	907	kilograms (kg)	0.001	ton, short
	ton, short	0.91	ton, metric	1.1	tons, short
Temp.	degrees Fahrenheit (° F)	$-32° \times 5/9$	degrees Celsius (° C) (centigrade)	$\times\ 1.8 + 32°$	degrees Fahrenheit (° F)

F

Rock Symbols

Shown below are the rock symbols used in the text. In general, these symbols are used by all geologists, although they sometimes are modified slightly.

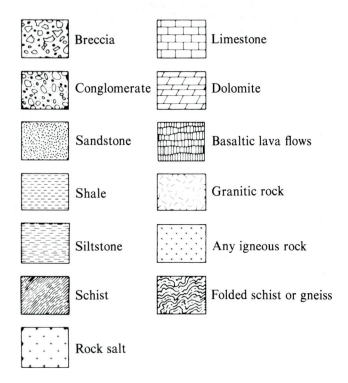

Breccia

Limestone

Conglomerate

Dolomite

Sandstone

Basaltic lava flows

Shale

Granitic rock

Siltstone

Any igneous rock

Schist

Folded schist or gneiss

Rock salt

Glossary

aa
A lava flow that solidifies with a spiny, rubbly surface.   56

abrasion
The grinding away of rock by friction and impact during transportation.   104, 207

absolute age
Age given in years or some other unit of time.   164

abyssal fan
Great fan-shaped deposit of sediment on the deep-sea floor at the base of many submarine canyons.   398

abyssal plain
Very flat region of the deep-sea floor, usually at the base of the continental rise.   400

accreted terrane
Terrane that appears to have formed essentially in place as a result of accretion and orogeny along a continent's margin.   466

accumulation stage
Stage in the evolution of major mountain belts characterized by the accumulation of great thicknesses (several kilometers) of sedimentary or volcanic rocks.   457

active continental margin
A margin consisting of a continental shelf, a continental slope, and an oceanic trench.   400

advancing glacier
Glacier with a positive budget, so that accumulation results in the lower edges being pushed outward and downward.   259

aftershock
Small earthquake that follows a main shock.   358

A horizon
The top layer of soil, characterized by the downward movement of water; also called *zone of leaching*.   110

alkali soil
Soil containing such a great quantity of sodium salts precipitated by evaporating ground water that it is generally unfit for plant growth.   113, 248

alluvial fan
Large fan-shaped pile of sediment that usually forms where a stream's velocity decreases as it emerges from a narrow canyon onto a flat plain at the foot of a mountain range.   218

alpine glaciation
Glaciation of a mountainous area.   256

amphibole group
Mineral group in which all members are double-chain silicates. 35

amphibolite
Amphibole (hornblende), plagioclase schist. 152

andesite
Fine-grained igneous rock of intermediate composition; composed of about equal amounts of ferromagnesian minerals and plagioclase feldspar. 65

angle of dip
A vertical angle measured downward from the horizontal plane to an inclined plane. 325

angular
Sharp-edged; lacking rounded edges or corners. 57

angular unconformity
An unconformity in which younger strata overlie an erosion surface on tilted or folded layered rock. 340

anorthosite
A crystalline rock composed almost entirely of calcium-rich plagioclase feldspar. 91

antecedent stream
A stream that maintains its original course despite later deformation of the land. 225

anticline
An arched fold in which the rock layers usually dip away from the axis of the fold. 327

aquifer
A body of saturated rock or sediment through which water can move readily. 238

arch (sea arch)
Bridge of rock left above an opening eroded in a headland by waves. 314

arête
A sharp ridge that separates adjacent glacially carved valleys. 266

arid region
An area with less than 25 cm of rain per year. 284

arkose
A sandstone in which more than 25% of the grains are feldspar. 123

artesian aquifer
An aquifer confined above, and sometimes below, by less permeable rocks. 240

artesian well
A well in which water rises above the aquifer. 240

artificial recharge
Ground-water recharge increased by engineering techniques. 245

aseismic ridge
Submarine ridge with which no earthquakes are associated. 404

ash (volcanic)
Fine pyroclasts (less than 4 mm). 57, 293

assimilation
The process in which very hot magma melts country rock and assimilates the newly molten material. 89

asthenosphere
A region of the earth's outer shell beneath the lithosphere. The asthenosphere is of indeterminate thickness and behaves plastically. 15, 378, 422

atoll
A circular reef surrounding a deeper lagoon. 405

atom
Smallest possible particle of an element that retains the properties of that element. 26

atomic mass number
The total number of neutrons and protons in an atom. 26

atomic number
The total number of protons in an atom. 27

atomic weight
The sum of the weight of the subatomic particles in an average atom of an element, given in atomic mass units. 27

augite
Mineral of the pyroxene group found in mafic igneous rocks. 35

aulacogen
An inactive, sediment-filled rift that forms above a rising mantle plume. The aulacogen becomes inactive (a "failed arm") as two other rifts widen to form an ocean. 440

aureole
Zone of contact metamorphism adjacent to a pluton. 148

axial plane
See hinge plane. 327

axis
See hinge line. 327

backarc spreading
A type of sea-floor spreading that occurs between an island arc and a continent, moving the island arc away from the continent. It may also occur inland of a magmatic arc on the edge of a continent, thinning and fracturing the continental crust. 433

backshore
Upper part of the beach, landward of the high-water line. 308

bajada
A broad, gently sloping, depositional surface formed at the base of a mountain range in a dry region by the coalescing of individual alluvial fans. 291

bar
A ridge of sediment, usually sand or gravel, that has been deposited in the middle or along the banks of a stream by a decrease in stream velocity. 209

barchan
A crescent-shaped dune with the horns of the crescent pointing downwind. 297

barrier island
Ridge of sand paralleling the shoreline and extending above sea level. 314

barrier reef
A reef separated from the shoreline by the deeper water of a lagoon. 404

basal sliding
Movement in which the entire glacier slides along as a single body on its base over the underlying rock. 261

basalt
A fine-grained, mafic, igneous rock made of ferromagnesian minerals and calcium-rich plagioclase feldspar. 65

base level
A theoretical limit for stream erosion of the earth's surface. 219

batholith
A large discordant pluton with an outcropping area greater than 100 square kilometers. 78

bauxite
The principal ore of aluminum; $Al_2O_3 \cdot nH_2O$. 114

baymouth bar
A ridge of sediment that cuts a bay off from the ocean. 309

beach
Strip of sediment, usually sand but sometimes pebbles, boulders, or mud, that extends from the low-water line inland to a cliff or zone of permanent vegetation.   308

beach face
The section of the beach exposed to wave action.   308

bedding
An arrangement of layers or beds of rock.   129

bedding plane
A nearly flat surface separating two beds of sedimentary rock.   129

bed load
Heavy or large sediment particles in a stream that travel near or on the stream bed.   207

bed rock
Solid rock that underlies soil.   110

Benioff zone
Distinct earthquake zone that begins at an oceanic trench and slopes landward and downward into the earth at an angle of about 30° to 60°.   360

bergschrund
The crevasse that develops where a glacier is pulling away from a cirque wall.   265

berm
Platform of wave-deposited sediment that is flat or slopes slightly landward.   308

B horizon
A soil layer characterized by the accumulation of material leached downward from the A horizon above; also called *zone of accumulation.*   112

bioclastic limestone
A limestone consisting of fragments of shells, corals, and algae.   125

biotite
Iron/magnesium-bearing mica.   35

block
Large angular pyroclast.   57

blowout
A depression on the land surface caused by wind erosion.   293

body wave
Seismic wave that travels through the earth's interior.   347

bomb
Large spindle- or lens-shaped pyroclast.   57

bonding
Attachment of an atom to one or more adjacent atoms.   28

bottomset bed
A delta deposit formed from the finest silt and clay, which are carried far out to sea by river flow or by sediments sliding downhill on the sea floor.   215

boulder
A sediment particle with a diameter greater than 256 mm.   118

Bowen's reaction series
The sequence in which minerals crystallize from a cooling basaltic magma.   87

braided stream
A stream that flows in a network of many interconnected rivulets around numerous bars.   210

breaker
A wave that has become so steep that the crest of the wave topples forward, moving faster than the main body of the wave.   305

breakwater
An offshore structure built to absorb the force of large breaking waves and provide quiet water near shore.   310

butte
A narrow pinnacle of resistant rock with a flat top and very steep sides.   288

calcareous
Containing calcium carbonate.

calcite
Mineral with the formula $CaCO_3$.   36

caldera
A volcanic depression much larger than the original crater.   55

capacity (of stream)
The total load that a stream can carry.   209

capillary action
The drawing of water upward into small openings as a result of surface tension.   236

capillary fringe
A thin zone near the water table in which capillary action causes water to rise above the zone of saturation.   236

carbonaceous chondrite
Stony meteorite containing chondrules and composed mostly of serpentine and large quantities of organic materials.   84

carbonic acid
H_2CO_3, a weak acid common in rain and surface waters.   108

cave (cavern)
Naturally formed underground chamber.   245

cement
The solid material that precipitates in the pore space of sediments, binding the grains together to form solid rock.   120

cementation
The chemical precipitation of material in the spaces between sediment grains, binding the grains together into a hard rock.   120

Cenozoic Era
The most recent of the eras; followed the Mesozoic Era.   172

chain silicate structure
Silicate structure in which two of each tetrahedron's oxygen ions are shared with adjacent tetrahedrons, resulting in a chain of tetrahedrons.   31

chalk
A very fine-grained bioclastic limestone.   125

channel (*Mars*)
Feature on the surface of the planet Mars that resembles very closely certain types of stream channels on the earth.   229

chaotic terrain (*Mars*)
Patch of jumbled and broken angular slabs and blocks on the surface of Mars.   191

chemical sedimentary rock
A rock composed of material precipitated directly from solution.   121

chemical weathering
The decomposition of rock resulting from exposure to water and atmospheric gases.   100

chert
A hard, compact, fine-grained sedimentary rock formed almost entirely of silica.   128

chill zone
In an intrusion, the finer-grained rock adjacent to a contact with country rock.   76

chondrule
Round silicate grain within some stony meteorites.   84

C horizon
A soil layer composed of incompletely weathered parent material.   112

cinder cone
A volcano constructed of loose rock fragments ejected from a central vent. 57

cinder (volcanic)
Pyroclast between 4 and 32 millimeters in diameter. 57

circum-Pacific belt
Major belt around the edge of the Pacific Ocean on which most composite volcanoes are located and where many earthquakes occur. 58, 360

cirque
A steep-sided, amphitheater-like hollow carved into a mountain at the head of a glacial valley. 265

clastic sedimentary rock
A sedimentary rock composed of fragments of preexisting rock. 121

clastic texture
An arrangement of rock fragments bound into a rigid network by cement. 121

clay
Sediment composed of particles with diameter less than 1/256 mm. 118

clay mineral
A hydrous aluminum-silicate that occurs as a platy grain of microscopic size with a sheet silicate structure. 108

clay mineral group
Collective term for several clay minerals. 35

cleavage
The ability of a mineral to break along preferred planes. 39

coal
A sedimentary rock formed from the consolidation of plant material. It is rich in carbon, usually black, and burns readily. 128

coarse-grained rock
Rock in which most of the grains are larger than 1 millimeter (igneous) or 2 millimeters (sedimentary). 80

coast
The land near the sea, including the beach and a strip of land inland from the beach. 311

coastal straightening
The gradual straightening of an irregular shoreline by wave erosion of headlands and wave deposition in bays. 311

cobble
A sediment particle with a diameter of 64 to 256 mm. 118

column
A dripstone feature formed when a stalactite growing downward and a stalagmite growing upward meet and join. 246

columnar structure
Volcanic rock in parallel, usually vertical columns, mostly six-sided; also called *columnar jointing*. 61

comet
Small object in space, no more than a few kilometers in diameter, composed of frozen methane, frozen ammonia, and water ice, with small solid particles and dust imbedded in the ices. 273

compaction
A loss in overall volume and pore space of a rock as the particles are packed closer together by the weight of overlying material. 120

competence
The largest particle that a stream can carry. 209

composite volcano (stratovolcano)
A volcano constructed of alternating layers of pyroclastics and rock solidified from lava flows. 58

compressive directed pressure
Directed pressure that tends to compress some portions of a body more than other portions. 145

compressive force
A force that tends to shorten a body. 323

conchoidal fracture
Curved fracture surfaces. 40

concordant
Parallel to layering or earlier developed planar structures. 78

concretion
Hard, rounded mass that develops when a considerable amount of cementing material precipitates locally in a rock, often around an organic nucleus. 248

cone of depression
A depression of the water table formed around a well when water is pumped out; it is shaped like an inverted cone. 239

confining pressure
Pressure applied equally on all surfaces of a body; also called *static pressure*. 145

conglomerate
A coarse-grained sedimentary rock formed by the cementation of rounded gravel. 122

consolidation
Any process that forms firm, coherent rock from sediment or from liquid.

contact
Boundary surface between two different rock types or ages of rocks. 134, 164

contact (thermal) metamorphism
Metamorphism under conditions in which high temperature is the dominant factor. 148

continental crust
The thick, granitic crust under continents. 13

continental drift
A concept suggesting that continents move over the earth's surface. 19, 414

continental glaciation
The covering of a large region of a continent by a sheet of glacial ice. 256

continental rise
A wedge of sediment that extends from the lower part of the continental slope to the deep sea floor. 399

continental shelf
A submarine platform at the edge of a continent, inclined very gently seaward generally at an angle of less than 1°. 396

continental slope
A relatively steep slope extending from a depth of 100 to 200 meters at the edge of the continental shelf down to oceanic depths. 397

contour current
A bottom current that flows parallel to the slopes of the continental margin (along the contour rather than down the slope). 399

contour line
A line on a topographic map connecting points of equal elevation.

convection (convection current)
A very slow circulation of a substance driven by differences in temperature and density within that substance. 13

converging plate boundary
Boundary between two plates that are moving toward each other. 17, 423

coquina
A limestone consisting of coarse shells. 124

core
The central zone of the earth. 13, 380

correlation
Determining age relationships between rock units or geologic events in separate areas. 168

country rock
Any rock that was older than and intruded by an igneous body. 76

covalent bonding
Bonding due to the sharing of electrons by adjacent atoms. 29

crater (of a volcano)
A basinlike depression over a vent at the summit of a volcanic cone. 55

craton
Portion of a continent that has been structurally stable for a prolonged period of time. 453

creep
Very slow, continuous downslope movement of soil or debris. 186

crest (of wave)
The high point of a wave. 304

crevasse
Open fissure in a glacier. 261

cross-bedding
An arrangement of relatively thin layers of rock inclined at an angle to the more nearly horizontal bedding planes of the larger rock unit. 130

cross-cutting relationship
A principal or law stating that a disrupted pattern is older than the cause of disruption. 165

cross section
See geologic cross section.

crude oil
A liquid mixture of naturally occurring hydrocarbons. 473

crust
The outer layer of rock, forming a thin skin over the earth's surface. 13, 375

crustal rebound
The rise of the earth's crust after the removal of glacial ice. 382

crystal
A homogeneous solid with an orderly internal atomic arrangement. 37

crystal form
Arrangement of various faces on a crystal in a definite geometric relationship to one another. 37

crystalline
Describing a substance in which the atoms are arranged in a regular, repeating, orderly pattern. 30

crystalline texture
An arrangement of interlocking crystals. 121

crystallization
Crystal development and growth. 121

crystal settling
The process whereby the minerals that crystallize at a high temperature in a cooling magma move downward in the magma chamber because they are denser than the magma. 89

cuesta
A ridge with a steep slope on one side and a gentle slope on the other side. 288

daughter product
The isotope produced by radioactive decay. 174

debris
Any unconsolidated material at the earth's surface. 184

debris avalanche
Very rapid and turbulent downslope movement of debris. 195

debris fall
A free-falling mass of debris. 195

debris slide
Rapid movement of debris as a coherent mass. 195

deflation
The removal of clay, silt, and sand particles from the land surface by wind. 293

delta
A body of sediment deposited at the mouth of a river when the river velocity decreases as it flows into a standing body of water. 214

dendritic pattern
Drainage pattern of a river and its tributaries, which resembles the branches of a tree or veins in a leaf. 224

deposition
The settling or coming to rest of transported material. 119

depth of focus
Distance between the focus and the epicenter of an earthquake. 350

desert
A region with low precipitation (usually defined as less than 25 cm per year). 284

desertification
The expansion of barren deserts into once-populated regions. 286

desert pavement
A thin layer of closely packed gravel that protects the underlying sediment from further deflation; also called *pebble armor*. 294

differential weathering
Varying rates of weathering resulting from some rocks in an area being more resistant to weathering than others. 100

differentiation
Separation of different ingredients from an originally homogeneous mixture. 87

dike
A tabular, discordant intrusive structure. 78

diorite
Coarse-grained igneous rock of intermediate composition. Composed of approximately equal amounts of plagioclase feldspar and ferromagnesian minerals. 85

dip
See angle of dip, direction of dip.

dip-slip fault
A fault in which movement is parallel to the dip of the fault surface. 333

directed pressure
Pressure applied unequally on the surfaces of a body; also called *dynamic pressure*. 145

direction of dip
The compass direction in which the angle of dip is measured. 326

discharge
In a stream, the volume of water that flows past a given point in a unit of time. 206

disconformity
A surface that represents missing rock strata but is parallel to beds above and below that surface. 340

discordant
Not parallel to any layering or parallel planes. 78

dissolved load
The portion of the total sediment load in a stream that is carried in solution. 208

distributary
Small shifting river channel that carries water away from the main river channel and distributes it over a delta's surface. 215

diverging plate boundary
Boundary separating two plates moving away from each other; a spreading center.   15, 423

dolomite
A sedimentary rock composed mostly of the mineral dolomite.   127

dolomitic marble
Marble in which dolomite, rather than calcite, is the prevalent mineral.   149

dome
See structural dome.

double refraction
The splitting of light into two components when it passes through certain crystalline substances.   42

downcutting
A valley-deepening process caused by erosion of a stream bed.   218

drainage basin
Total area drained by a stream and its tributaries.   202

drainage divide
Line dividing one drainage basin from another.   202

drainage pattern
The arrangement in map view of a river and its tributaries.   224

drawdown
The lowering of the water table near a pumped well.   239

dripstone
Deposits of calcite (and, rarely, other minerals) built up by dripping water in caves.   246

drumlin
A long, streamlined hill made of till.   270

dust (volcanic)
Finest-sized pyroclasts.   57

dynamic pressure
See directed pressure.

earthflow
Slow-to-rapid mass wasting in which debris moves downslope as a very viscous fluid.   188

earthquake
A trembling or shaking of the ground caused by the sudden release of energy stored in the rocks beneath the surface.   346

earthy luster
A luster giving a substance the appearance of unglazed pottery.   36

echo sounder
An instrument used to measure and record the depth to the sea floor.   395

elastic rebound theory
The sudden release of progressively stored strain in rocks results in movement along a fault.   346

elastic strain
Strain in which a deformed body recovers its original shape after the stress is released.   323

electron
A single, negative electric charge that contributes virtually no mass to an atom.   26

element
A substance that cannot be broken down to other substances by ordinary chemical methods. Each atom of an element possesses the same number of protons.   26

emergent coast
A coast in which land formerly under water has recently been placed above sea level, either by uplift of the land or by a drop in sea level.   318

end moraine
A ridge of till piled up along the front edge of a glacier.   270

environment of deposition
The location in which deposition occurs, usually marked by characteristic physical, chemical, or biological conditions.   120

epicenter
The point on the earth's surface directly above the focus of an earthquake.   347

epoch
Each period of the standard geologic time scale is divided into epochs (e.g., Pleistocene Epoch of the Quaternary Period).   172

equilibrium
Material is in equilibrium if it is adjusted to the physical and chemical conditions of its environment so that it does not change or alter with time.   18

era
Major subdivision of the standard geologic time scale (e.g., Mesozoic Era).   172

erosion
The physical removal of rock by an agent such as running water, glacial ice, or wind.   18, 100

erratic
An ice-transported boulder that does not derive from bed rock near its present site.   268

esker
A long, sinuous ridge of sediment deposited by glacial meltwater.   271

estuary
Drowned river mouth.   315

etch-pitted terrain (*Mars*)
A terrain on the surface of Mars characterized by small pits.   279

eugeosyncline
See volcanic portion of a geosyncline.   459

evaporite
Rock that forms from crystals precipitating during evaporation of water.   128

exfoliation
The stripping of concentric rock slabs from the outer surface of a rock mass.   105

exfoliation dome
A large, rounded landform developed in a massive rock, such as granite, by the process of exfoliation.   105

exotic terrane
Terrane that did not form at its present site on a continent.   466

expansive clay
Clay that increases in volume when water is added to it.   35

extrusive rock
Any igneous rock that forms at the earth's surface, whether it solidifies directly from a lava flow or is pyroclastic.   49

faceted
A rock fragment with one or more flat surfaces caused by erosive action.   264

fall
The situation in mass wasting that occurs when material free-falls or bounces down a cliff.   184

fault
A fracture in bed rock along which movement has taken place.   324

fault-block mountain range
A range created by uplift along normal or vertical faults.   462

faunal succession
A principle or law stating that fossil species succeed one another in a definite and recognizable order; in general, fossils in progressively older rock show increasingly greater differences from species living at present. 170

feldspars
Group of most common minerals of the earth's crust. All feldspars contain silicon, aluminum, and oxygen and may contain potassium, calcium, and sodium. 35

felsic rock
Silica-rich igneous rock with a relatively high content of potassium and sodium. 54

ferromagnesian mineral
Iron/magnesium-bearing mineral, such as augite, hornblende, olivine, or biotite. 36

fine-grained rock
A rock in which most of the mineral grains are less than one millimeter across. 65

fiord
A coastal inlet that is a glacially carved valley, the base of which is submerged. 278

firn
A compacted mass of granular snow, transitional between snow and glacier ice. 259

firn limit
See snow line.

fissility
The ability of a rock to split into thin layers. 123

flank eruption
An eruption in which lava erupts out of a vent on the side of a volcano. 55

flash flood
Flood of very high discharge and short duration; sudden and local in extent. 285

flood plain
A broad strip of land built up by sedimentation on either side of a stream channel. 212

flow
A type of movement that implies that a descending mass is moving downslope as a viscous fluid. 184

flowstone
Calcite precipitated by flowing water on cave walls and floors. 246

focus
The point within the earth from which seismic waves originate in an earthquake. 347

fold
Bend in layered bed rock. 326

fold axis
See axis.

foliation
Parallel alignment of textural and structural features of a rock. 146

footwall
The underlying surface of an inclined fault plane. 334

foreset bed
A sediment layer in the main part of a delta, deposited at an angle to the horizontal. 215

foreshock
Small earthquake that precedes a main shock. 367

foreshore
The zone that is regularly covered and uncovered by the rise and fall of tides. 308

formation
A body of rock of considerable thickness that has a recognizable unity or similarity making it distinguishable from adjacent rock units. Usually composed of one bed or several beds of sedimentary rock, although the term is also applied to units of metamorphic and igneous rock. A convenient unit for mapping, describing, or interpreting the geology of a region. 134

fossil
Traces of plants or animals preserved in rock. 132

fossil assemblage
Various different species of fossils in a rock. 170

fracture
The way a substance breaks where not controlled by cleavage. 40

fracture zone
Major line of weakness in the earth's crust that crosses the mid-oceanic ridge at approximately right angles. 403

fracturing
Cracking or rupturing of a body under stress. 323

framework silicate structure
Crystal structure in which all four oxygen ions of a silica tetrahedron are shared by adjacent ions. 32

fretted terrain (*Mars*)
Flat lowland with some scattered high plateaus on the surface of Mars. 191

fringing reef
A reef attached directly to shore. (*See* barrier reef.) 404

frost action
Mechanical weathering of rock by freezing water. 103

frost heaving
The lifting of rock or soil by the expansion of freezing water. 104

frost wedging
A type of frost action in which the expansion of freezing water pries a rock apart. 103

gabbro
A mafic, coarse-grained igneous rock composed predominantly of ferromagnesian minerals and calcium-rich plagioclase feldspar. 76

gaining stream
A stream that receives water from the zone of saturation. 236

geode
Partly hollow, globelike body found in limestone or other cavernous rock. 248

geologic cross section
A representation of a portion of the earth in a vertical plane. 326

geologic map
A map representing the geology of a given area. 324

geologic resources
Valuable materials of geologic origin that can be extracted from the earth. 472

geology
The scientific study of the earth. 6

geophysics
The application of physical laws and principles to a study of the earth. 374

geosyncline
A very long (thousands of kilometers) and relatively narrow (a few hundred kilometers) basin that acts as a collecting area for marine sedimentary and volcanic rocks. 459

geothermal energy
Energy produced by harnessing naturally occurring steam and hot water. 251

geothermal gradient
Rate of temperature increase associated with increasing depth beneath the surface of the earth (normally about 25° C/km). 86, 387

geyser
A type of hot spring that periodically erupts hot water and steam. 249

geyserite
A deposit of silica that forms around many geysers and hot springs. 250

glacier
A large, long-lasting mass of ice, formed on land by the compaction and recrystallization of snow, which moves because of its own weight. 256

glassy (or vitreous) luster
A luster that gives a substance a glazed, porcelainlike appearance. 36

glowing avalanche
Very hot flow of pyroclastics. 61

gneiss
A metamorphic rock composed of light and dark layers or lenses. 154

gneissic
The texture of a metamorphic rock in which minerals are separated into light and dark layers or lenses. 146

goethite
The commonest mineral in the limonite group; $Fe_2O_3 \cdot nH_2O$. 107

Gondwanaland
The southern part of *Pangaea* that formed South America, Africa, India, Australia, and Antarctica. 420

graben
A down-dropped block bounded by normal fault. 334

graded bed
A single bed with coarse grains at the bottom of the bed and progressively finer grains toward the top of the bed. 130

graded stream
A stream that exhibits a delicate balance between its transporting capacity and the sediment load available to it. 219

granite
A felsic, coarse-grained, intrusive igneous rock composed mostly of potassium with sodium-rich feldspars and quartz. 76

granitization
The process by which granite is created from other rock without a melt being involved. 158

gravel
Rounded particles coarser than sand. 118

gravity
The force of attraction that two bodies exert on each other that is proportional to the product of their masses and inversely proportional to the square of the distance from the centers of the two bodies. 6

gravity meter
An instrument that measures the gravitational attraction between the earth and a mass within the instrument. 382

graywacke
A sandstone with more than 15% fine-grained matrix between the sand grains. 123

groin
Short wall built perpendicular to shore to trap moving sand and widen a beach. 310

ground moraine
A blanket of till deposited by a glacier or released as glacier ice melted. 270

ground water
The water that lies beneath the ground surface, filling the cracks, crevices, and pore space of rocks. 234

guyot
Flat-topped seamount. 404

half-life
The time it takes for a given amount of a radioactive isotope to be reduced by one-half. 174

hanging valley
A smaller valley that terminates abruptly high above a main valley. 265

hanging wall
The overlying surface of an inclined fault plane. 334

hardness
The relative ease or difficulty with which a smooth surface of a mineral can be scratched; commonly measured by Mohs' scale. 36

headland
Point of land along a coast. 311

headward erosion
The lengthening of a valley in an uphill direction above its original source by gullying, mass wasting, and sheet erosion. 221

heat engine
A device that converts heat energy into mechanical energy. 6

heat flow
Gradual loss of heat (per unit of surface area) from the earth's interior out into space. 389

heavy crude
Dense, viscous petroleum that flows slowly or not at all. 480

hematite
A type of iron oxide that has a brick-red color when powdered; Fe_2O_3. 107

highland (*Moon*)
A rugged region of the lunar surface representing an early period in lunar history when intense meteorite bombardment formed craters. 150

hinge line
Line about which a fold appears to be hinged. Line of maximum curvature of a folded surface. 327

hinge plane
A plane containing all of the axes of a fold. 327

hogback
A sharp-topped ridge formed by the erosion of steeply dipping beds. 288

horn
A sharp peak attributed to cirques cut back into a mountain on several sides. 266

hornblende
Common amphibole frequently found in igneous and metamorphic rocks. 35

hornfels
A fine-grained, unfoliated metamorphic rock. 148

horst
An up-raised block bounded by normal faults.

hot spring
Spring with a water temperature warmer than human body temperature. 249

hydraulic action
The ability of water to pick up and move rock and sediment. 206

hydrologic cycle
The movement of water and water vapor from the sea to the atmosphere, to the land, and back to the sea and atmosphere again. 6, 234

hydrology
The study of water's properties, circulation, and distribution.

hydrothermal metamorphism
Alteration of a rock by hot water passing through it. 156

hydrothermal rock
Rock deposited by precipitation of ions from solution in hot water.   156

hypothesis
A tentative theory.   19

iceberg
Block of glacier-derived ice floating in water.   258

ice cap
A glacier covering a relatively small area of land but not restricted to a valley.   257

icefall
A chaotic jumble of crevasses that split glacier ice into pinnacles and blocks.   261

ice sheet
A glacier covering a large area (more than 50,000 square kilometers) of land.   257

igneous rock
A rock formed or apparently formed from solidification of magma.   17

incised meander
A meander that retains its sinuous curves as it cuts vertically downward below the level at which it originally formed.   228

inclusion
A fragment of rock that is distinct from the body of igneous rock in which it is enclosed.   76

index fossil
A fossil from a very short-lived species known to have existed during a specific period of geologic time.   170

intensity
A measure of an earthquake's size by its effect on people and buildings.   351

intermediate rock
Rock with a chemical content between felsic and mafic compositions.   55

intrusion (intrusive structure)
A body of intrusive rock classified on the basis of size, shape, and relationship to surrounding rocks.   77

intrusive rock
Rock that appears to have crystallized from magma emplaced in surrounding rock.   76

ion
An electrically charged atom or group of atoms.   28

ionic bonding
Bonding due to the attraction between positively charged ions and negatively charged ions.   29

iron meteorite
A meteorite composed principally of iron-nickel alloy.   84

island arc
A curved line of islands.   363, 431

isoclinal fold
A fold in which the limbs are parallel to one another.   330

isolated silicate structure
Silicate minerals that are structured so that none of the oxygen atoms are shared by silica tetrahedrons.   31

isostasy
The balance or equilibrium between adjacent blocks of crust resting on a plastic mantle.   381

isostatic adjustment
Concept of vertical movement of sections of the earth's crust to achieve balance or equilibrium.   381

isotherm
A line along which the temperature of rock (or other material) is the same.   155

isotopes
Atoms (of the same element) that have different numbers of neutrons but the same number of protons.   27, 172

jetty
Rock wall protruding above sea level, designed to protect the entrance of a harbor from sediment deposition and storm waves; usually built in pairs.   310

joint
A fracture or crack in bed rock along which essentially no displacement has occurred.   331

joint set
Joints oriented in one direction approximately parallel to one another.   331

karst topography
An area with many sinkholes and a cave system beneath the land surface and usually lacking a surface stream.   247

kettle
A depression caused by the melting of a stagnant block of ice that was surrounded by sediment.   271

KREEP (*Moon*)
A lunar basalt enriched in potassium (K), the rare earth elements (REE), and phosphorus (P).   91

laccolith
A concordant intrusive structure, similar to a sill, with the central portion thicker and domed upward.   78

laminar flow
Slow, smooth flow, with each drop of water traveling a smooth path parallel to its neighboring drops.

laminated terrain (*Mars*)
Area where series of alternating light and dark layers can be seen on the surface of Mars.   279

lamination
A thin layer in sedimentary rock (less than one centimeter thick).   123

landform
A characteristically shaped feature of the earth's surface, such as a hill or a valley.

landslide
The general term for a slowly to very rapidly descending rock or debris.   182

lateral erosion
Erosion and undercutting of stream banks caused by a stream swinging from side to side across its valley floor.   221

lateral moraine
A low ridgelike pile of till along the side of a glacier.   268

laterite
Highly leached soil that forms in regions of tropical climate with high temperatures and very abundant rainfall.   113

lava
Magma on the earth's surface.   48

left-lateral fault
A strike-slip fault in which the block seen across the fault appears displaced to the left.   334

limb
Portion of a fold shared by an anticline and a syncline.   327

limestone
A sedimentary rock composed mostly of calcite.   124

limonite
A type of iron oxide that is yellowish-brown when powdered; $Fe_2O_3 \cdot nH_2O$.   107

lithification
The consolidation of sediment into sedimentary rock.   18, 120

lithosphere
The rigid outer shell of the earth, approximately 100 kilometers thick.   15, 378, 422

loess
A fine-grained deposit of wind-blown dust.   271, 295

longitudinal dune (seif)
Large, symmetrical ridge of sand parallel to the wind direction.   298

longitudinal profile
A line showing a stream's slope, drawn along the length of the stream as if it were viewed from the side.   219

longshore current
A moving mass of water that develops parallel to a shoreline.   305

longshore drift
Movement of sediment parallel to shore when waves strike a shoreline at an angle.   309

losing stream
Stream that loses water to the zone of saturation.   236

low-velocity zone
Mantle zone at a depth of about 100 kilometers where seismic waves travel more slowly than in shallower layers of rock.   378

luster
The quality and intensity of light reflected from the surface of a mineral.   36

mafic rock
Silica-poor igneous rock with a relatively high content of magnesium, iron, and calcium.   54

magma
Molten rock, usually mostly silica. The liquid may contain dissolved gases as well as some solid minerals.   15, 48

magmatic arc
A line of batholiths or volcanoes. Generally the line, as seen from above, is curved.   457

magnetic anomaly
A deviation from the average strength of the earth's magnetic field. *See* negative magnetic anomaly, positive magnetic anomaly.   386

magnetic field
Region of magnetic force that surrounds the earth.   385

magnetic pole
An area where the strength of the magnetic field is greatest and where the magnetic lines of force appear to leave or enter the earth.   385

magnetic reversal
A change in the polarity of the earth's magnetic field.   385

magnetite
An iron oxide that is attracted to a magnet.   42

magnetometer
An instrument that measures the strength of the earth's magnetic field.   386

magnitude
A measure of the energy released during an earthquake.   351

major mountain belt
A long chain (thousands of kilometers) of mountain ranges.   450

mantle
A thick shell of rock that separates the earth's crust above from the core below.   13, 375

mantle plume
Narrow column of hot mantle rock that rises and spreads radially outward.   69, 440

marble
A coarse-grained rock composed of interlocking calcite crystals.   149

maria (*Moon*)
Lava plains on Moon's surface (singular, *mare*).   66

marine terrace
A broad, gently sloping platform that may be exposed at low tide.   308

mass wasting (or mass movement)
Movement, caused by gravity, in which bed rock, rock debris, or soil moves downslope in bulk.   182

matrix
Fine-grained material found in the pore space between larger sediment grains.   123

meander
A pronounced sinuous curve along a stream's course.   210

meander cutoff
A new, shorter channel across the narrow neck of a meander.   212

meander scar
An abandoned meander filled with sediment and vegetation.   210

mechanical weathering
The physical disintegration of rock into smaller pieces.   100

medial moraine
A single long ridge of till on a glacier, formed by adjacent lateral moraines joining and being carried downglacier.   270

Mediterranean-Himalayan belt (Mediterranean belt)
A major concentration of earthquakes and composite volcanoes that runs through the Mediterranean Sea, crosses the Mideast and the Himalayas, and passes through the East Indies.   59, 360

melt
Liquid rock resulting from melting in a laboratory.   76

mesa
A broad, flat-topped hill bounded by cliffs and capped with a resistant rock layer.   288

Mesozoic Era
The era that followed the Paleozoic Era and preceded the Cenozoic Era.   172

metallic bonding
Bonding, as in metals, whereby atoms are closely packed together and electrons move freely among atoms.   29

metallic luster
Luster giving a substance the appearance of being made of metal.   36

metamorphic facies
Pressure and temperature stability fields for metamorphic rocks as determined by mineral assemblages.   156

metamorphic rock
A rock produced by metamorphism.   17, 144

metamorphism
The transformation of preexisting rock into texturally or mineralogically distinct new rock as a result of high temperature, high pressure, or both, but without the rock melting in the process.   144

metasomatism
Metamorphism coupled with the introduction of ions from an external source.   157

meteor
Fragment that passes through the earth's atmosphere, heated to incandescence by friction; sometimes incorrectly called "shooting" or "falling" stars. 84

meteorite
Meteor that strikes the earth's surface. 84

meteoroid
Small solid particles of stone and/or metal orbiting the sun. 84

mica group
Group of minerals with a sheet silicate structure. 35

mid-oceanic ridge
A giant mountain range that lies under the ocean and extends around the world. 15, 401

migmatite
Mixed igneous and metamorphic rock. 154

mineral
A naturally occurring, inorganic, crystalline solid that has a definite chemical composition and possesses characteristic physical properties. 32

mineraloid
A substance that is not crystalline but otherwise would be considered a mineral. 32

miogeosyncline
See nonvolcanic portion of a geosyncline. 459

model
In science, a model is an image—graphic, mathematical, or verbal—that is consistent with the known data. 26

modified Mercalli scale
Scale expressing intensities of earthquakes (judged on amount of damage done) in Roman numerals ranging from I to XII. 351

Mohorovičić discontinuity
The boundary separating the crust from the mantle beneath it (also called *Moho*). 377

Mohs' hardness scale
Scale on which ten minerals are designated as standards of hardness. 36

monocline
A local steeping in a gentle regional dip; a steplike fold in rock. 288

moraine
A body of till either being carried on a glacier or left behind after a glacier has receded. 268

mountain range
A group of closely spaced mountains or parallel ridges. 450

mud
Term loosely used for silt and clay, usually wet. 118

mud crack
Polygonal crack formed in very fine-grained sediment as it dries. 130

mudflow
A flowing mixture of debris and water, usually moving down a channel. 192

mudstone
A fine-grained sedimentary rock that lacks shale's laminations and fissility. 123

multi-ringed basin (*Moon*)
Large lunar crater surrounded by a series of concentric rings with intervening lowlands. 151

muscovite
Transparent or white mica that lacks iron and magnesium. 35

natural gas
A gaseous mixture of naturally occurring hydrocarbons. 473

natural levee
Low ridges of flood-deposited sediment formed on either side of a stream channel, which thin away from the channel. 214

nebula
A large volume of interstellar gas and dust. 173

negative gravity anomaly
Less than normal gravitational attraction. 384

negative magnetic anomaly
Less than average strength of the earth's magnetic field. 387

neutron
A subatomic particle that contributes mass to an atom and is electrically neutral. 26

nonconformity
An unconformity in which an erosion surface on plutonic or metamorphic rock has been covered by younger sedimentary or volcanic rock. 340

nonmetallic luster
Luster that gives a substance the appearance of being made of something other than metal (e.g., glassy). 36

nonrenewable resource
A resource that forms at extremely slow rates compared to its rate of consumption. 472

nonvolcanic portion of a geosyncline
The part of a geosyncline formed of a thick sequence of limestone, sandstone, and shale, with little or no volcanic rock; called *miogeosyncline*. 459

normal fault
A fault in which the hanging-wall block moved down relative to the footwall block. 334

nucleus
Protons and neutrons form the nucleus of an atom. Although the nucleus occupies an extremely tiny fraction of the volume of the entire atom, practically all the mass of the atom is concentrated in the nucleus. 26

nuée ardente
Cloud of red-hot ash and dust caused by very explosive volcanic activity (French for "glowing cloud"). 60

oblique-slip fault
A fault with both strike-slip and dip-slip components. 334

obsidian
Volcanic glass. 65

oceanic crust
The thin, basaltic crust under oceans. 13

oceanic trench
A narrow, deep trough parallel to the edge of a continent or an island arc. 400

oil
See crude oil.

oil field
An area underlain by one or more oil pools. 475

oil pool
Underground accumulation of oil. 475

oil sand
Asphalt-cemented sand deposit. 480

oil shale
Shale with a high content of organic matter from which oil may be extracted by distillation. 480

oil trap
A set of conditions that hold petroleum in a reservoir rock and prevent its escape by migration. 473

olivine
A ferromagnesian mineral with the formula $(Fe, Mg)_2 SiO_4$. 31

oolite (ooid)
A small sphere of calcite precipitated from sea water. 125

oolitic limestone
A limestone formed from oolites. 125

open fold
A fold with gently dipping limbs. 329

open-pit mine
Mine in which ore is exposed at the surface in a large excavation. 490

ophiolite
A distinctive rock sequence found in many mountain ranges on continents. 408

ore
Naturally occurring material that can be profitably mined. 484

ore mineral
A mineral of commercial value. 36

organic sedimentary rock
Rock composed mostly of the remains of plants and animals. 121

original horizontality
The deposition of most water-laid sediment in horizontal or near-horizontal layers that are essentially parallel to the earth's surface. 129, 165

orogeny
An episode of intense deformation of the rocks in a region, generally accompanied by metamorphism and plutonic activity. 458

orthoclase (potassium) feldspar
A feldspar with the formula $KAlSi_3O_8$. 35

outcrop
A surface exposure of bare rock, not covered by soil or vegetation. 324

outwash
Material deposited by debris-laden meltwater from a glacier. 270

overburden
The upper part of a sedimentary deposit. Its weight causes compaction of the lower part. 120

overturned fold
A fold in which both limbs dip in the same direction. 330

oxbow lake
A crescent-shaped lake occupying the abandoned channel of a stream meander that is isolated from the present channel by a meander cutoff and sedimentation. 212

pahoehoe
A lava flow characterized by a ropy or billowy surface. 56

paired terraces
Stream terraces (see definition) that occur at the same elevation on each side of a river. 227

paleomagnetism
A study of ancient magnetic fields. 385

Paleozoic Era
The era that followed the Precambrian and began with the appearance of complex life, as indicated by fossils. 172

Pangaea
A supercontinent that broke apart 200 million years ago to form the present continents. 415

parabolic dune
A deeply curved dune in a region of abundant sand. The horns point upwind and are apt to be anchored by vegetation. 297

parallel retreat
The retreat of a slope or cliff that maintains a constant angle during retreat. 222

parent rock
Original rock before being metamorphosed. 144

partial melting
Melting of the components of a rock with the lowest melting temperatures. 69

passive continental margin
A margin that includes a continental shelf, continental slope, and continental rise that generally extends down to an abyssal plain at a depth of about 5 kilometers. 399

peat
A brown, lightweight, unconsolidated or semi-consolidated deposit of plant remains. 128

pebble
A sediment particle with a diameter of 2 to 64 mm. 118

pedalfer
A soil characterized by the downward movement of water through it, downward leaching, and abundant humus. Found in humid climates. 112

pediment
A gently sloping erosional surface cut into the solid rock of a mountain range in a dry region; usually covered with a thin veneer of gravel. 291

pedocal
A soil characterized by little leaching, scant humus, the upward movement of water through it, and the precipitation of salts. Found in dry climates. 112

pegmatite
Extremely coarse-grained igneous rock. 83

pelagic sediment
Sediment made up of fine-grained clay and the skeletons of microscopic organisms that settle slowly down through the ocean water. 407

peneplain
A nearly flat erosional surface presumably produced as mass wasting, sheet erosion, and stream erosion reduce a region almost to base level. 221

perched water table
A water table separated from the main water table beneath it by a zone that is not saturated. 236

period
Each era of the standard geologic time scale is subdivided into periods (e.g., the Cretaceous Period). 172

permafrost
Ground that remains permanently frozen for many years. 190

permeability
The capacity of a rock to transmit a fluid such as water or petroleum. 234

petrified wood
A material that forms as the organic matter of buried wood is replaced by inorganic silica carried in by ground water. 248

petroleum
Crude oil and natural gas. (Some geologists use petroleum as a synonym for oil.) 473

phenocryst
Any of the large crystals in porphyritic igneous rock. 65

phyllite
A metamorphic rock in which clay minerals have recrystallized into microscopic micas, giving the rock a silky sheen. 153

physical continuity
Being able to physically follow a rock unit between two places. 168

physical geology
A large division of geology concerned with earth materials, changes of the surface and interior of the earth, and the forces that cause those changes. 13

piezoelectricity
Electrical current generated when pressure is applied to certain minerals. 43

pillow structure
Rocks, generally basalt, formed in pillow-shaped masses fitting closely together; caused by underwater lava flows. 62

placer mine
Surface mines in which valuable mineral grains are extracted from stream bar or beach deposits. 490

plagioclase feldspar
A feldspar containing sodium and/or calcium in addition to aluminum, silicon, and oxygen. 35

plastic
Capable of being molded and bent under stress. 13, 144

plastic flow
Movement within a glacier in which the ice is not fractured. 261

plastic strain
Strain in which a body is molded or bent under stress and does not return to its original shape after the stress is released. 323

plate
A large, mobile slab of rock making up part of the earth's surface. 422

plateau
Broad, flat-topped area elevated above the surrounding land and bounded, at least in part, by cliffs. 288

plateau basalts
Layers of basalt flows that have built up to great thicknesses. 61

plate tectonics
A theory that the earth's surface is divided into a few large, thick plates that are slowly moving and changing in size. Intense geologic activity occurs at the plate boundaries. 15, 414

playa
A very flat, dry lake bed of hard, mud-cracked clay. 291

playa lake
A shallow temporary lake (following a rainstorm) on a flat valley floor in a dry region. 291

Pleistocene Epoch
An epoch of the Quaternary Period characterized by several glacial ages. 172

plunging fold
A fold in which the axis is not horizontal. 328

pluton
An igneous body that crystallized deep underground. 78

plutonic rock
Igneous rock formed at great depth. 78

pluvial lake
A lake formed during an earlier time of abundant rainfall. 277

point bar
A stream *bar* (see definition) deposited on the inside of a curve in the stream, where the water velocity is low. 210

polar wandering
An apparent movement of the earth's poles. 416

pore space
The total amount of space taken up by openings between sediment grains. 120

porosity
The percentage of a rock's volume that is taken up by openings. 234

porphyritic rock
An igneous rock in which large crystals are enclosed in a matrix (or ground mass) of much finer-grained minerals or obsidian. 65

positive gravity anomaly
Greater than normal gravitational attraction. 384

positive magnetic anomaly
Greater than average strength of the earth's magnetic field. 386

pothole
Depression eroded into the hard rock of a stream bed by the abrasive action of the stream's sediment load. 207

Precambrian
The vast amount of time that preceded the Paleozoic Era. 172

Precambrian shield
A complex of old Precambrian metamorphic and plutonic rocks exposed over a large area. 453

pressure release
A significant type of mechanical weathering that causes rocks to crack when overburden is removed. 104

principle of uniformitarianism
The hypothesis that geologic processes that are operating at present are the same processes that have operated in the geologic past; the present is the key to the past. 21

progressive metamorphism
Metamorphism in which progressively greater pressure and temperature act on a rock type with increasing depth in the earth's crust. 153

proton
A subatomic particle that contributes mass and a single positive electrical charge to an atom. 26

pumice
A frothy volcanic glass. 68

P wave
A compressional wave (seismic wave) in which rock vibrates parallel to the direction of wave propagation. 347

P-wave shadow zone
The region on the earth's surface, 103° to 142° away from an earthquake epicenter, in which P waves from the earthquake are absent. 379

pyroclast
Fragment of rock formed by volcanic explosion. 57

pyroclastic debris (tephra)
Rock fragments produced by volcanic explosion. 48

pyroxene group
Mineral group, all members of which are single-chain silicates. 35

quartz
Mineral with the formula SiO_2. 35

quartzite
A rock composed of sand-sized grains of quartz that have been welded together during metamorphism. 149

quartz sandstone
A sandstone in which more than 90% of the grains are quartz. 122

Quaternary Period
The youngest geologic period; includes the present time. 172

radial pattern
A drainage pattern in which streams diverge outward like spokes of a wheel. 224

radioactive decay
The spontaneous nuclear disintegration of certain isotopes. 174

radioactivity
The spontaneous nuclear disintegration of atoms of certain isotopes. 174

rain shadow
A region on the downwind side of mountains that has little or no rain because of the loss of moisture on the upwind side of the mountains. 284

rampart crater (*Mars*)
Meteorite crater that is surrounded by material that appears to have flowed from the point of impact. 191

rayed crater (*Moon*)
Crater with bright streaks radiating from it on the moon's surface. 150

receding glacier
A glacier with a negative budget, which causes the glacier to grow smaller as its edges melt back. 259

Recent (Holocene) Epoch
The present epoch of the Quarternary Period. 172

recessional moraine
An end moraine built during the retreat of a glacier. 270

recharge
The addition of new water to an aquifer or to the zone of saturation. 239

reclamation
Restoration of the land to usable condition after mining has ceased. 490

recrystallization
The development of new crystals in a rock, often of the same composition as the original grains. 125

rectangular pattern
A drainage pattern in which tributaries of a river change direction and join one another at right angles. 224

recumbent fold
A fold overturned to such an extent that the limbs are essentially horizontal. 330

reef
A resistant ridge of calcium carbonate formed on the sea floor by corals and coralline algae. 404

regional (dynamothermal) metamorphism
Metamorphism involving relatively high temperature and pressure (directed and confining). 152

regolith
Loose, unconsolidated rock material resting on bed rock. 151

relative time
The sequence in which events took place (not measured in time units). 164

relief
The vertical distance between points on the earth's surface. 185

reserves
The discovered deposits of a geologic material that are economically and legally feasible to recover under present circumstances. 477

reservoir rock
A rock that is sufficiently porous and permeable to store and transmit petroleum. 473

residual clay
Fine-grained particles left behind as insoluble residue when a limestone containing clay dissolves.

residual soil
Soil that develops directly from weathering of the rock below. 112

resources
The total amount of a geologic material in all its deposits, discovered and undiscovered (*see* reserves). 497

reverse fault
A fault in which the hanging-wall block moved up relative to the footwall block. 334

rhyolite
A fine-grained, felsic, igneous rock made up mostly of feldspar and quartz. 64

Richter scale
A numerical scale of earthquake magnitudes. 351

rift valley
A large crack, apparently of tensional origin, running down the crest of the mid-oceanic ridge. 401

right-lateral fault
A strike-slip fault in which the block seen across the fault appears displaced to the right. 334

rigid zone
Upper part of a glacier in which there is no plastic flow. 261

rille (*Moon*)
Elongate trenched or cracklike valley on the lunar surface. 66

rip current
Narrow currents that flow straight out to sea in the surf zone, returning water seaward that has been pushed ashore by breaking waves. 306

ripple mark
Any of the small ridges formed on sediment surfaces exposed to moving wind or water. The ridges form perpendicularly to the motion. 132

rock
Naturally formed, consolidated material composed of grains of one or more minerals. (There are a few exceptions to this definition.) 26

rock avalanche
A very rapidly moving, turbulent mass of broken-up bed rock. 193

rock-basin lake
A lake occupying a depression caused by glacial erosion of bed rock. 265

rock cycle
A theoretical concept relating tectonism, erosion, and various rock-forming processes to the common rock types. 44

rockfall
Rock falling freely or bouncing down a cliff. 193

rock flour
A powder of fine fragments of rock produced by glacial abrasion. 264

rock gypsum
An evaporite composed of gypsum. 128

rock salt
An evaporite composed of halite. 26, 128

rockslide
Rapid sliding of a mass of bed rock along an inclined surface of weakness. 193

rounding
The grinding away of sharp edges and corners of rock fragments during transportation. 118

rubble
Angular fragments coarser than sand. 118

saltation
A mode of transport that carries sediment downcurrent in a series of short leaps or bounces. 208

sand
Sediment composed of particles with a diameter of 1/16 mm. 118

sand dune
A mound of loose sand grains heaped up by the wind. 296

sandstone
A medium-grained sedimentary rock formed by the cementation of sand grains. 122

saturated zone
A subsurface zone in which all rock openings are filled with water. 235

scale
The relationship between distance on a map and the distance on the terrain being represented by that map.

schist
A metamorphic rock characterized by coarse-grained minerals oriented approximately parallel. 154

schistose
The texture of a rock in which visible platy or needle-shaped minerals have grown essentially parallel to each other under the influence of directed pressure. 146

scientific method
A means of gaining knowledge through objective procedures. 19

scoria
A basalt that is highly vesicular. 68

sea cave
A cavity eroded by wave action at the base of a sea cliff. 311

sea cliff
Steep slope that retreats inland by mass wasting as wave erosion undercuts it. 311

sea-floor spreading
The concept that the ocean floor is moving away from the mid-oceanic ridge and across the deep ocean basin, to disappear beneath continents and island arcs. 414

seamount
Conical mountain rising 1,000 meters or more above the sea floor. 404

seawall
A wall constructed along the base of retreating cliffs to prevent wave erosion. 312

sediment
Loose, solid particles that can originate by (1) weathering and erosion of pre-existing rocks, (2) chemical precipitation from solution, usually in water, and (3) secretion by organisms. 7, 118

sedimentary breccia
A coarse-grained sedimentary rock composed of lithified rubble. 121

sedimentary facies
Significantly different rock types occupying laterally distinct parts of the same layered rock unit.

sedimentary rock
Rock that has formed from (1) lithification of any type of sediment, (2) precipitation from solution, or (3) consolidation of the remains of plants or animals. 18, 121

sedimentary structure
A feature found within sedimentary rocks, usually formed during or shortly after deposition of the sediment and before lithification. 129

seismic profiler
An instrument that measures and records the subbottom structure of the sea floor. 396

seismic reflection
The return of part of the energy of seismic waves to the earth's surface after the waves bounce off a rock boundary. 374

seismic refraction
The bending of seismic waves as they pass from one material to another. 374

seismic wave
A wave of energy produced by an earthquake. 346

seismogram
Paper record of earth vibration. 349

seismograph
A seismometer with a recording device that produces a permanent record of earth motion. 348

seismometer
An instrument designed to detect seismic waves or earth motion. 348

serpentine
A magnesium silicate mineral. Most asbestos is a variety of serpentine. 456

shale
A fine-grained sedimentary rock with laminations and fissility. 123

shearing
Movement in which parts of a body slide relative to one another and parallel to the forces being exerted. 146

shear stress
Stress due to forces that tend to cause movement or strain parallel to the direction of the forces. 323

sheet erosion
The removal of a thin layer of surface material, usually topsoil, by a flowing sheet of water. 200

sheet-jointing
The development of cracks parallel to the outer surface of an expanding rock. 105

sheet silicate structure
Crystal structure in which each silica tetrahedron shares three oxygen ions. 32

sheetwash
Water flowing down a slope in a layer. 200

shield volcano
Broad, gently sloping cone constructed of solidified lava flows. 56

sial
Rock rich in silicon and aluminum; characteristic of continental crust. 377

silica
A term used for oxygen plus silicon. 28

silicate
A substance that contains silica as part of its chemical formula. 29

silica tetrahedron
Four-sided, pyramidal object that visually represents the four oxygen atoms surrounding a silicon atom; the basic building block of silicate minerals. 30

sill
A tabular intrusive structure concordant with the country rock. 78

silt
Sediment composed of particles with a diameter of $\frac{1}{256}$ to $\frac{1}{16}$ mm. 118

siltstone
A sedimentary rock consisting mostly of silt grains. 123

sima
Rock rich in silicon and magnesium; characteristic of oceanic crust. 377

sinkhole
A closed depression found on land surfaces underlain by limestone. 246

sinter
A deposit of silica that forms around some hot springs and geysers. 250

slate
A fine-grained rock that splits easily along flat, parallel planes. 153

slaty
Describing a rock that splits easily along nearly flat and parallel planes. 146

slaty cleavage
The ability of a rock to break along closely spaced parallel planes. 146

slide
In mass wasting, movement of a descending mass along a plane approximately parallel to the slope of the surface. 184

slip
In mass wasting, movement of a relatively coherent descending mass along one or more well-defined surfaces. 184

slip face
The steep, downwind slope of a dune; formed from loose, cascading sand that generally keeps the slope at the angle of repose (about 34°). 297

slump
In mass wasting, movement along a curved surface in which the upper part moves vertically downward while the lower part moves outward. 184

snow line
An irregular line marking the highest level to which the winter snow cover on a glacier is lost during a melt season. 260

soil
A layer of weathered, unconsolidated material on top of bed rock; often also defined as containing organic matter and being capable of supporting plant growth. 110

soil horizon
Any of the layers of soil that are distinguishable by characteristic physical or chemical properties. 110

solifluction
Flow of water-saturated debris over impermeable material. 189

solution
Usually slow but effective process of weathering and erosion in which rocks are dissolved by water. 207

sorting
Process of selection and separation of sediment grains according to their grain size (or grain shape or specific gravity). 118

source area
The locality that eroded to provide sediment to form a sedimentary rock. 135

source rock
A rock containing organic matter that is converted to petroleum by burial and other postdepositional changes. 473

spatter cone
A small, steep-sided cone built from lava spattering out of a vent. 56

specific gravity
The ratio of the mass of a substance to the mass of an equal volume of water, determined at a specified temperature. 40

spheroidally weathered boulder
Boulder that has been rounded by weathering from an initial blocky shape. 100

spit
A fingerlike ridge of sediment attached to land but extending out into open water. 309

spreading center
The crest of the mid-oceanic ridge, where sea floor is moving away in opposite directions on either side. 15

spring
A place where water flows naturally out of rock onto the land surface. 236

stable
Describing a mineral that will not react with or convert to a new mineral or substance, given enough time. 106

stack
Erosional remnant of a headland left behind as a wave-eroded coast retreats inland. 312

stalactite
Iciclelike pendant of dripstone formed on cave ceilings. 246

stalagmite
Cone-shaped mass of dripstone formed on cave floors, generally directly below a stalactite. 246

standard geologic time scale
A worldwide relative scale of geologic time divisions. 172

static pressure
See confining pressure.

stock
A small discordant pluton with an outcropping area of less than 100 square kilometers. 78

stony-iron meteorite
A meteorite composed of silicate minerals and nickel-iron alloy in approximately equal amounts. 84

stony meteorite
A meteorite made up mostly of plagioclase and iron-magnesium silicates. 84

stoping
Upward movement of a body of magma by fracturing of overlying country rock. Magma engulfs the blocks of fractured country rock as it moves upward. 90

storm surge
High sea level caused by the low pressure and high winds of hurricanes. 316

strain
Change in size (volume) or shape of a body (or rock unit) in response to stress. 322

stratovolcano
See composite volcano. 58

streak
Color of a pulverized substance; a useful property for mineral identification. 36

stream
A moving body of water, confined in a channel and running downhill under the influence of gravity. 200

stream capture
See stream piracy.

stream channel
A long, narrow depression, shaped and more or less filled by a stream. 200

stream discharge
Volume of water that flows past a given point in a unit of time. 206

stream gradient
Downhill slope of a stream's bed or the water surface, if the stream is very large. 204

stream headwaters
The upper part of a stream near the source. 200

stream mouth
The place where the stream enters the sea, a large lake, or a larger stream. 200

stream piracy
The natural diversion of the headwaters of one stream into the channel of another. 225

stream terrace
Steplike landform found above a stream and its flood plain. 227

stream velocity
The speed at which water in a stream travels. 204

stress
A force acting on a body, or rock unit, that tends to change the size or shape of that body, or rock unit. 322

striations
(1) On minerals, extremely straight, parallel lines; (2) Glacial—straight scratches in rock caused by abrasion by a moving glacier. 42

strike
The compass direction of a line formed by the intersection of an inclined plane (such as a bedding plane) with a horizontal plane. 325

strike-slip fault
A fault in which movement is parallel to the strike of the fault surface. 334

strip mine
A mine in which the valuable material is exposed at the surface by removing a strip of overburden. 481

structural basin
A structure in which the beds dip toward a central point. 328

structural dome
A structure in which beds dip away from a central point. 328

structural geology
The branch of geology concerned with the internal structure of bed rock and the shapes, arrangement, and interrelationships of rock units. 322

subduction
The sliding of the sea floor beneath a continent or island arc. 421

subduction zone
Elongate region in which subduction takes place. 17

submarine canyon
V-shaped valleys that run across the continental shelf and down the continental slope. 398

submergent coast
A coast in which formerly dry land has been recently drowned, either by land subsidence or a rise in sea level. 315

subsidence
Sinking or downwarping of a part of the earth's surface. 245

superimposed (superposed) stream
A river let down onto a buried geologic structure by erosion of overlying layers. 225

superposition
A principle or law stating that within a sequence of undisturbed sedimentary rocks, the oldest layers are on the bottom, the youngest on the top. 165

surf
Breaking waves. 305

surface wave
A seismic wave that travels on the earth's surface. 347

suspect terrane
A terrane that may not have formed at its present site. 466

suspended load
Sediment in a stream that is light enough in weight to remain lifted indefinitely above the bottom by water turbulence. 208

S wave
A seismic wave propagated by a shearing motion, which causes rock to vibrate perpendicular to the direction of wave propagation. 348

S-wave shadow zone
The region on the earth's surface (at any distance more than 103° from an earthquake epicenter) in which S waves from the earthquake are absent. 379

swelling clay
See expansive clay.

syncline
A fold in which the layered rock usually dips toward an axis. 327

talus
An accumulation of broken rock at the base of a cliff. 193

tectonic forces
Forces generated from within the earth that result in uplift, movement, or deformation of part of the earth's crust. 15

tectonostratigraphic terrane
See terrane.

tensional force
A force that tends to elongate or pull apart a body. 323

tephra
Fragments of rock produced by volcanic explosion. 48

terminal moraine
An end moraine marking the farthest advance of a glacier. 270

terminus
The lower edge of a glacier. 261

terrane (tectonostratigraphic terrane)
A region within which there is geologic continuity. 465

terrigenous sediment
Land-derived sediment that has found its way to the sea floor. 407

theory
An explanation for observed phenomena that has a high possibility of being true. 19

theory of glacial ages
At times in the past, colder climates prevailed during which significantly more of the land surface of the earth was glaciated than at present. 256

thermal metamorphism
See contact (thermal) metamorphism. 148

thrust fault
A reverse fault in which the dip of the fault plane is at a low angle to horizontal. 334

tidal delta
A submerged body of sediment formed by tidal currents passing through gaps in barrier islands. 314

till
Unsorted and unlayered rock debris carried by a glacier. 268

tillite
Lithified till. 279

time-transgressive rock unit
An apparently continuous rock layer in which different portions formed at different times.

tombolo
A bar of marine sediment connecting a former island or stack to the mainland. 309

topographic map
A map on which elevations are shown by means of contour lines.

topset bed
In a delta, a nearly horizontal sediment bed of varying grain size formed by distributaries shifting across the delta surface. 215

traction
Movement by rolling, sliding, or dragging of sediment fragments along a stream bottom. 208

transform plate boundary
Boundary between two plates that are sliding past each other. 15, 423

transform fault
The portion of a fracture zone between two offset segments of a mid-oceanic ridge crest. 427

transportation
The movement of eroded particles by agents such as rivers, waves, glaciers, or wind. 100

transported soil
Soil not formed from the local rock but from parent material brought in from some other region and deposited, usually by running water, wind, or glacial ice. 112

transverse dune
A relatively straight, elongate dune oriented perpendicular to the wind. 297

travel-time curve
A plot of seismic-wave arrival times against distance. 350

travertine
A porous deposit of calcite that often forms around hot springs. 250

trellis pattern
A drainage pattern consisting of parallel main streams with short tributaries meeting them at right angles. 224

tributary
Small stream flowing into a large stream, adding water to the large stream. 202

trough (of wave)
The low point of a wave. 304

truncated spur
Triangular facet where the lower end of a ridge has been eroded by glacial ice. 264

tsunami
Huge ocean wave produced by displacement of the sea floor; also called seismic sea wave. 358

tuff
A rock formed from fine-grained pyroclastic particles (ash and dust). 68

turbidity current
A flowing mass of sediment-laden water that is heavier than clear water and therefore flows downslope along the bottom of the sea or a lake. 123, 398

turbulent flow
Eddying, swirling flow in which water drops travel along erratically curved paths that cross the paths of neighboring drops. 208

ultramafic rock
Rock composed entirely or almost entirely of ferromagnesian minerals. 84

unconformity
A surface that represents a break in the geologic record, with the rock unit immediately above it being considerably younger than the rock beneath. 167, 340

underplating
The pooling of magmas at the base of the continental crust. 95

unconsolidated
In referring to sediment grains, loose, separate, or unattached to one another. 118

uniformitarianism
See principle of uniformitarianism. 164

unloading
The removal of a great weight of rock. 104

unpaired terraces
Stream terraces (see definition) that do not have the same elevation on opposite sides of a river. 227

unsaturated zone
A subsurface zone in which rock openings are filled partly with air and partly with water; above the saturated zone. 235

U-shaped valley
Characteristic cross-profile of a valley carved by glacial erosion. 264

valley glacier
A glacier confined to a valley. The ice flows from a higher to a lower elevation. 257

Van der Waal's bonds or forces
Weak bonds in crystals, such as those that hold adjacent sheets of mica together. 29

varve
Two thin layers of sediment, one dark and the other light in color, representing one year's deposition in a lake. 271

vent
The opening in the earth's surface through which a volcanic eruption takes place. 55

ventifact
Boulder, cobble, or pebble with flat surfaces caused by the abrasion of wind-blown sand. 293

vertical exaggeration
An artificial steepening of slope angles on a topographic profile caused by using a vertical scale that differs from the horizontal scale. 396

vesicle
A cavity in volcanic rock caused by gas in a lava. 68

viscosity
Resistance to flow. 52

vitreous luster
See glassy luster.

volcanic breccia
Rock formed from large pieces of volcanic rock (cinders, blocks, bombs). 68

volcanic dome
A steep-sided, dome- or spine-shaped mass of volcanic rock formed from viscous lava that solidifies in or immediately above a volcanic vent. 60

volcanic neck
An intrusive structure that apparently represents magma that solidified within the throat of a volcano. 77

volcanic portion of a geosyncline
A thick sequence of submarine lava flows and marine sedimentary rock, much of which derives from volcanic ash and from sediment eroded from volcanic rocks; called *eugeosyncline*. 459

volcanism
Volcanic activity, including the eruption of lava and rock fragments and gas explosions. 48

volcano
A hill or mountain constructed by the extrusion of lava or rock fragments from a vent. 49

wastage
Glacial ice or snow that is lost by melting, evaporation, or breaking off into icebergs. 258

water table
The upper surface of the zone of saturation. 235

wave crest
See crest.

wave-cut platform
A horizontal bench of rock formed beneath the surf zone as a coast retreats because of wave erosion. 312

wave height
The vertical distance between the crest (the high point of a wave) and the trough (the low point). 304

wavelength
The horizontal distance between two wave crests (or two troughs). 304

wave refraction
Change in direction of waves due to slowing as they enter shallow water. 305

wave trough
See trough.

weathering
The group of processes that change rock at or near the earth's surface. 18, 100

welded tuff
A rock composed of pyroclasts welded together. 61

well
A hole, generally cylindrical and usually walled or lined with pipe, that is dug or drilled into the ground to penetrate an aquifer below the zone of saturation. 238

wind ripple
Small, low ridge of sand produced by the saltation of windblown sand. 297

wrinkle ridge (*Moon*)
Wrinkle on lunar maria surface. 66

zone of accumulation
(1) That portion of a glacier with a perennial snow cover; (2) *See* B horizon (a soil layer). 260

zone of aeration
See unsaturated zone.

zone of leaching
See A horizon (a soil layer).

zone of saturation
See saturated zone.

zone of wastage
That portion of a glacier in which ice is lost or wasted. 260

Credits

Illustrator Credits

Box Figures:

Chapter Opening Illustrations:

Index